Marine Faunal
Diversity in India

Taxonomy, Ecology and Conservation

Marine Faunal Diversity in India

Taxonomy, Ecology and Conservation

Edited by

Krishnamoorthy Venkataraman
Zoological Survey of India
Kolkata, West Bengal, India

Chandrakasan Sivaperuman
Zoological Survey of India
Andaman and Nicobar Regional Centre
Port Blair, Andaman and Nicobar Islands, India

AMSTERDAM • BOSTON • HEIDELBERG • LONDON
NEW YORK • OXFORD • PARIS • SAN DIEGO
SAN FRANCISCO • SINGAPORE • SYDNEY • TOKYO

Academic Press is an Imprint of Elsevier

Academic Press is an imprint of Elsevier
32 Jamestown Road, London NW1 7BY, UK
525 B Street, Suite 1800, San Diego, CA 92101-4495, USA
225 Wyman Street, Waltham, MA 02451, USA
The Boulevard, Langford Lane, Kidlington, Oxford OX5 1GB, UK

Notice
No responsibility is assumed by the publisher for any injury and/or damage to persons
or property as a matter of products liability, negligence or otherwise, or from any use or
operation of any methods, products, instructions or ideas contained in the material herein.
Because of rapid advances in the medical sciences, in particular, independent verification of
diagnoses and drug dosages should be made

British Library Cataloguing-in-Publication Data
A catalogue record for this book is available from the British Library

Library of Congress Cataloging-in-Publication Data
A catalog record for this book is available from the Library of Congress

ISBN: 978-0-12-801948-1

For information on all Academic Press publications
visit our website at http://store.elsevier.com/

Typeset by Thomson Digital

Printed and bound in United States of America

Working together
to grow libraries in
developing countries

www.elsevier.com • www.bookaid.org

Contents

9. Coral Reef Associated Macrofaunal Communities of Rutland Island, Andaman and Nicobar Archipelago

C. Raghunathan and K. Venkataraman

10. Diversity and Distribution of Sea Grass Associated Macrofauna in Gulf of Mannar Biosphere Reserve, Southern India

K. Paramasivam, K. Venkataraman, C. Venkatraman, R. Rajkumar and S. Shrinivaasu

11. Diversity and Ecology of Sedentary Ascidians of the Gulf of Mannar, Southeast Coast of India

G. Ananthan, S. Mohamed Hussain, A. Selva Prabhu and T. Balasubramanian

12. Diversity of Marine Fish of India

K.C. Gopi and S.S. Mishra

18. **Diversity of Marine Mammals of India—Status, Threats, Conservation Strategies and Future Scope of Research**

    S. Venu and B. Malakar

19. **Coastal and Marine Biodiversity of India**

    K. Venkataraman and C. Raghunathan

Part II
Ecology and Conservation

20. **DNA Barcoding of Marine Venomous and Poisonous Fish of Families Scorpaenidae and Tetraodontidae from Andaman Waters**

    V. Sachithanandam, P.M. Mohan and N. Muruganandam

21. Molecular Taxonomy of Serranidae, Subfamily
Epinephelinae, Genus *Plectropomus* (Oken, 1817)
of Andaman Waters by DNA Barcoding Using COI
Gene Sequence

*V. Sachithanandam, P.M. Mohan, N. Muruganandam,
I.K. Chaaithanya and R. Baskaran*

22. Diversity of Antagonistic *Streptomyces* Species
in Mangrove Sediments of Andaman Island, India

R. Baskaran, P.M. Mohan, R. Vijayakumar and V. Sachithanandam

23. Impact of Anthropogenic Activity and Natural Calamities
on Fringing Reef of North Bay, South Andaman

R. Raghuraman and C. Raghunathan

24. Lucrative Business Opportunities with Shrimp
Brood Stocks

V.S. Gowri and P. Nammalwar

Color plates

Foreword

India is one of the recognized megadiverse countries of the world. Dense and handsome mangrove forests of Sundarbans, the world's largest congregations of nesting turtles in Orissa, delicate and beautiful sea grass beds in Palk Bay, enigmatic sea cows in the Gulf of Mannar, majestic whale sharks in the Gulf of Kutchchh, and some of the world's most beautiful and striking coral reefs in the Andaman and Nicobar Islands and Lakshadweep are just a few examples of the rare treasures of India's coastal and marine biodiversity. Besides being repositories of biological diversity, coastal regions in India are home to a large proportion of the human population.

We owe much to the coastal and marine environment for our sustenance. Vast, seemingly limitless and indestructible, the oceans have been feared, fathomed and revered with awe and gratitude. Marine and coastal biodiversity benefits all of humanity: fisheries provide over 15 percent of animal protein in the global diet; resilient coastal ecosystems protect coastal communities from natural disasters occurring at sea; substances derived from the sea are key components in many commercial products, such as creams, paints, paper and medicines; and marine phytoplankton release half of all the oxygen into the atmosphere.

In celebration of these benefits, the Zoological Survey of India, a 98-year-old organization, continues to identify, describe and catalogue the coastal and marine species of the country. The present book, *Marine Faunal Diversity in India: Taxonomy, Ecology and Conservation*, contains 29 research papers of a high standard grouped into two thematic areas: "Marine Faunal Diversity" and "Ecology and Conservation." Marine taxonomy and ecology are still in their infancy in India, and there is a pressing need for all-round nurturing of these fields of study. The present book is a welcome step in this direction. I congratulate the editors Dr K. Venkataraman and Dr C. Sivaperuman for their earnest effort to bring this volume, with its treasure of knowledge, to the public domain.

I am happy to present this book during India's Presidency of Conference of the Parties (CoP) to the Convention on Biological Diversity (CBD) and the United Nations Decade on Biodiversity 2011–2020.

At the CBD-COP 10 held at Nagoya in 2010, 20 global Aichi Biodiversity targets were adopted. Out of these, target 6 is the avoidance of overfishing, target 10 calls for minimization of ocean acidification, and target 11 is the conservation of 10 percent of coastal and marine areas that are of particular importance for biodiversity and ecosystem services. The present publication, I hope, will contribute to our efforts to achieve these targets.

Hem Pande
Additional Secretary
Government of India
Ministry of Environment, Forests and Climate Change

Preface

The world coastline extends for about 356,000 km, and coastal areas account for more than 10 percent of the land surface in 123 countries around the world. India has a coastline of 8118 km, with an Exclusive Economic Zone (EEZ) of 2.02 million km^2 and a continental shelf area of 468,000 km^2, spread across ten coastal States and seven Union Territories, including the islands of Andaman and Nicobar and of Lakshadweep. The coastal and marine ecosystem supports about 35 percent of the nation's human population, who depend on the rich exploitable coastal and marine resources.

Marine and coastal ecosystems of India provide supporting services in the form of a wide range of habitats. Marine ecosystems such as estuaries, coral reefs, marshes, lagoons, sandy and rocky beaches, mangrove forests and sea grass beds are all known for their high biological productivity, providing a wide range of habitats for many aquatic flora and fauna. The Indian coastal and marine ecosystem supports 17,795 species of fauna and flora, which is about 6.75 percent of global distribution. To protect these resources, the Government of India has declared 683 areas as National Parks, Wildlife Sanctuaries, Community Reserves and Conservation Reserves. In addition, 17 Biosphere Reserves have also been declared to protect entire ecosystems. A total of 885 species of marine fauna belonging to eight phyla—Poriferea, Coelentrata, Arthropoda, Mollusca, Echinodermata, Pisces, Reptilia and Mammalia—were listed under different schedules of the Indian Wildlife (Protection) Act, 1972. Globally threatened marine and coastal bird species such as the spot-billed pelican, *Pelecanus phillipensis*, and the lesser adjutant, *Leptoptilos javanicus*, are found in India. Important Birds Areas (IBAs) along coasts include the Gulf of Kachchh, Gulf of Mannar, Pulicat, Chilika Lake, etc. About 25 species of marine mammal are known to occur in Indian waters.

This volume contains 29 chapters detailing studies by reputed researchers working in the field of marine faunal communities of India. It is divided into two parts: "Marine Faunal Communities" and "Ecology and Conservation." The first covers a wide range of marine faunal communities—sponges, octocorals, molluscs, crabs, bryozoa, sea urchins, ascidians, fish, sea snakes, coastal and marine birds, mammals—and coastal and marine biodiversity in general. The second part deals with subjects such as DNA barcoding and molecular taxonomy of fish, potential exploitation of mangrove sediment bacteria to produce antibiotics, impact of anthropogenic activity and natural calamities on coral reefs, faunal diversity and its conservation on shipwrecks, human–crocodile conflict, and conservation of marine and mangrove ecosystem fauna.

This book presents unique information on marine faunal diversity in India. We sincerely hope that it will be of great help to researchers and field scientists in the area of marine biodiversity in India and in other countries.

K. Venkataraman, Kolkata
C. Sivaperuman, Port Blair

Acknowledgments

We express our heartfelt gratitude to all those who helped in different ways to complete this work. Our sincere thanks go to the officials of the Ministry of Environment, Forests and Climate Change, Government of India for their constant support and encouragement. We would like to thank Professor P.M. Mohan, Department of Ocean Studies and Marine Biology, Pondicherry University, Port Blair campus for his valuable comments during the preparation of the draft. We also thank all the authors who have contributed the various articles for this book.

Contributors

D. Adhavan, Department of Ocean Studies and Marine Biology, Pondicherry University, Port Blair, Andaman and Nicobar Islands, India

S. Ajmal Khan, Centre of Advanced Study in Marine Biology, Faculty of Marine Sciences, Annamalai University, Parangipettai, Tamil Nadu, India

G. Ananthan, Centre of Advanced Study in Marine Biology, Faculty of Marine Sciences, Annamalai University, Parangipettai, Tamil Nadu, India

D. Apte, Bombay Natural History Society, Mumbai, Maharashtra, India

T. Balasubramanian, Centre of Advanced Study in Marine Biology, Faculty of Marine Sciences, Annamalai University, Parangipettai, Tamil Nadu, India

R. Balasubramanian, Centre of Advanced Study in Marine Biology, Faculty of Marine Sciences, Annamalai University, Parangipettai, Tamil Nadu, India

R. Baskaran, Department of Ocean Studies and Marine Biology, Pondicherry University, Port Blair, Andaman and Nicobar Islands, India

I.K. Chaaithanya, Regional Medical Research Centre, Indian Council of Medical Research, Port Blair, Andaman and Nicobar Islands, India

K. Devi, Andaman and Nicobar Regional Centre, Zoological Survey of India, Port Blair, Andaman and Nicobar Islands, India

S. Geetha, Department of Zoology, Kamaraj College, Thoothukudi, Tamil Nadu, India

K.C. Gopi, Fish Division, Zoological Survey of India, Kolkata, West Bengal, India

V.S. Gowri, Institute for Ocean Management, Anna University, Chennai, Tamil Nadu, India

T. Immanuel, Central Island Agricultural Research Institute, Port Blair, Andaman and Nicobar Islands, India

A. Joseph, Department of Marine Biology, Microbiology & Biochemistry, Cochin University of Science & Technology, Kochi, Kerala, India

K. Kathiresan, Centre of Advanced Study in Marine Biology, Faculty of Marine Sciences, Annamalai University, Parangipettai, Tamil Nadu, India

P. Krishnan, Central Island Agricultural Research Institute, Port Blair, Andaman and Nicobar Islands, India

A. Kulkarni, Department of Zoology, Gogate Jogalekar College, University of Mumbai, Ratnagiri, Maharashtra, India

N.P. Kumar, Department of Ocean Studies and Marine Biology, Pondicherry University, Port Blair, Andaman and Nicobar Islands, India

P. Lakshmi Devi, Department of Marine Biology, Microbiology & Biochemistry, Cochin University of Science & Technology, Kochi, Kerala, India

B. Malakar, Department of Ocean Studies and Marine Biology, Pondicherry University, Port Blair, Andaman and Nicobar Islands, India

M. Mankeshwar, Department of Zoology, Gogate Jogalekar College, University of Mumbai, Ratnagiri, Maharashtra, India

J.K. Mishra, Department of Ocean Studies and Marine Biology, Pondicherry University, Port Blair, Andaman and Nicobar Islands, India

A. Mishra, Department of Ocean Studies and Marine Biology, Pondicherry University, Port Blair, Andaman and Nicobar Islands, India

S.S. Mishra, Fish Division, Zoological Survey of India, Kolkata, West Bengal, India

P.M. Mohan, Department of Ocean Studies and Marine Biology, Pondicherry University, Port Blair, Andaman and Nicobar Islands, India

S. Mohamed Hussain, Centre of Advanced Study in Marine Biology, Faculty of Marine Sciences, Annamalai University, Parangipettai, Tamil Nadu, India

R. Mohanraju, Department of Ocean Studies and Marine Biology, Pondicherry University, Port Blair, Andaman and Nicobar Islands, India

S.K. Mohanty, Chilika Development Authority, Bhubaneswar, Odisha, India

A. Mohapatra, Marine Aquarium and Regional Centre, Zoological Survey of India, Digha, West Bengal, India

A.K. Mukhopadhyay, Zoological Survey of India, Kolkata, West Bengal, India

N. Muruganandam, Regional Medical Research Centre (ICMR), Indian Council of Medical Research, Port Blair, Andaman Islands, India

P. Nammalwar, Institute for Ocean Management, Anna University, Chennai, Tamil Nadu, India

P. Padmanaban, Marine Biology Regional Centre, Zoological Survey of India, Chennai, Tamil Nadu, India

K. Paramasivam, Marine Biology Regional Centre, Zoological Survey of India, Chennai, Tamil Nadu, India

C. Raghunathan, Andaman and Nicobar Regional Centre, Zoological Survey of India, Port Blair, Andaman and Nicobar Islands, India

R. Rajkumar, Marine Biology Regional Centre, Zoological Survey of India, Chennai, Tamil Nadu, India

P.T. Rajan, Andaman and Nicobar Regional Centre, Zoological Survey of India, Port Blair, Andaman and Nicobar Islands, India

D. Ray, Marine Aquarium & Regional Centre, Zoological Survey of India, Digha, West Bengal, India

V. Sachithanandam, Department of Ocean Studies and Marine Biology, Pondicherry University, Port Blair, Andaman and Nicobar Islands, India

K. Sadhukhan, Andaman and Nicobar Regional Centre, Zoological Survey of India, Port Blair, Andaman and Nicobar Islands, India

A. Selva Prabhu, Centre of Advanced Study in Marine Biology, Faculty of Marine Sciences, Annamalai University, Parangipettai, Tamil Nadu, India

S. Kumar Shah, Andaman and Nicobar Regional Centre, Zoological Survey of India, Port Blair, Andaman and Nicobar Islands, India

S. Shrinivaasu, Marine Biology Regional Centre, Zoological Survey of India, Chennai, Tamil Nadu, India

J. Sinduja, Department of Ocean Studies and Marine Biology, Pondicherry University, Port Blair, Andaman and Nicobar Islands, India

C. Sivaperuman, Andaman and Nicobar Regional Centre, Zoological Survey of India, Port Blair, Andaman and Nicobar Islands, India

B. Tripathy, Zoological Survey of India, Kolkata, West Bengal, India

P. Tudu, Marine Aquarium & Regional Centre, Zoological Survey of India, Digha, West Bengal, India

N. Veerappan, Centre of Advanced Study in Marine Biology, Faculty of Marine Sciences, Annamalai University, Parangipettai, Tamil Nadu, India

K. Venkataraman, Zoological Survey of India, Kolkata, West Bengal, India

C. Venkatraman, Marine Biology Regional Centre, Zoological Survey of India, Chennai, Tamil Nadu, India

S. Venu, Department of Ocean Studies and Marine Biology, Pondicherry University, Port Blair, Andaman and Nicobar Islands, India

R. Vijayakumar, Department of Microbiology, Bharathidasan University Constituent College for Women, Orathanadu, Tamil Nadu, India

Yasmin, Department of Ocean Studies and Marine Biology, Pondicherry University, Port Blair, Andaman and Nicobar Islands, India

P. Yennawar, Freshwater Biology Regional Centre, Zoological Survey of India, Hyderabad, Andhra Pradesh, India

J.S. Yogesh Kumar, Andaman and Nicobar Regional Centre, Zoological Survey of India, Port Blair, Andaman and Nicobar Islands, India

Part I

Marine Faunal Diversity

Chapter 1

An Updated Report on the Diversity of Marine Sponges of the Andaman and Nicobar Islands

T. Immanuel*, P. Krishnan* and C. Raghunathan†

*Central Island Agricultural Research Institute, Port Blair, Andaman and Nicobar Islands, India;
†Andaman and Nicobar Regional Centre, Zoological Survey of India, Port Blair, Andaman and Nicobar Islands, India

INTRODUCTION

The phylum Porifera, commonly known as sponges, designates the most primitive of the multicellular animals (more than 500 million years old) (Müller, 1995), with a most ancient geological history. The sponges are a unique group of organisms: although multicellular they lack tissue grade of construction (Bergquist, 1978). Most of them are sedentary or immobile as adults but possess mobile larval forms. These invertebrates do not have a nervous, digestive, or circulating system but rely on constant water flow through their bodies to obtain food, oxygen and to remove waste. Sponges form an important biotic component of the coral reef ecosystem (Reswig, 1973; Wulff, 2006) and constitute one of the most abundant and diverse groups of marine benthic communities around the world (Hartman, 1977). In fact they are more diverse than corals in many coral reef ecosystems around the world (Diaz and Rützler, 2001; Wulff, 2006). They play several ecologically important roles, such as binding live corals to the reef frame, facilitating regeneration of broken reefs, and harboring nitrifying and photosynthesizing microbial symbionts, or intervening in erosion processes (Diaz and Rützler, 2001; Wulff, 2001, 2006). Sponges have been the focus of much recent interest, as they are a rich source of active secondary metabolites (Bergmann and Feeney, 1950). Sponges account as a source for around 37 percent of the biomedical compounds obtained from the marine environment worldwide (Jha and Zi-rong, 2004).

The coral reefs of the Andaman and Nicobar Islands (ANI) are among the most diverse, but also most threatened reefs in the world. Accurate baseline studies on the constituent taxa and their environmental conditions will aid in conservation and management of the reefs (Mora et al., 2003). Study of spatial

Marine Faunal Diversity in India. DOI: 10.1016/B978-0-12-801948-1.00001-X and seasonal variations in environmental parameters is essential for an understanding of how the environmental processes interact to construct the marine assemblages. Sponges are one of the least studied of the major phyla of the ANI. The continental shelf of the islands plays host to large areas of coral reefs which harbor rich Poriferan diversity. Studies describe 88 species of sponges from the ANI (Box 1.1 and Figure 1.1); however, details of these sedentary

---

Box 1.1 Checklist of Sponges Described as from the Andaman and Nicobar Islands

1. *Cinachyrella arabica* (Carter, 1869)
2. *Cinachyrella tarentina* (Pulitzer-Finali, 1983)
3. *Paratetilla bacca* (Selenka, 1867)
4. *Craniella cranium* (Muller, 1776)
5. *Tetilla dactyloidea* (Carter, 1869)
6. *Ecionemia acervus* (Bowerbank, 1864)
7. *Stelletta clavosa* (Ridley, 1884)
8. *Stelletta purpurea* (Ridley, 1884)
9. *Stelletta cavernosa* (Dendy, 1910)
10. *Stelletta orientalis* (Thiele, 1898)
11. *Stelletta validissima* (Thiele, 1898)
12. *Rhabdastrella globostellata* (Carter, 1883)
13. *Erylus lendenfeldi* (Sollas, 1888)
14. *Dercitus (Stoeba) simplex* (Carter, 1880)
15. *Poecillastra eccentrica* (Dendy & Burton, 1926)
16. *Poecillastra tenuilaminaris* (Sollas, 1886)
17. *Thenea andamanensis* (Dendy & Burton, 1926)
18. *Cliona ensifera* (Sollas, 1878)
19. *Cliona kempi* (Annandale, 1915)
20. *Cliona lobata* (Hancock, 1849)
21. *Cliona mucronata* (Sollas, 1878)
22. *Cliothosa quadrata* (Hancock, 1849)
23. *Cliothosa hancocki* (Topsent, 1888)
24. *Pione vastifica* (Hancock, 1849)
25. *Pione carpenteri* (Hancock, 1867)
26. *Spirastrella andamanensis* (Pattanayak, 2006)
27. *Spheciospongia inconstans* (Dendy, 1887)
28. *Tethya andamanensis* (Dendy & Burton, 1926)
29. *Tethya diploderma* (Schmidt, 1870)
30. *Tethya repens* (Schmidt, 1870)
31. *Tethya robusta* (Bowerbank, 1873)
32. *Discodermia gorgonoides* (Burton, 1928)
33. *Discodermia papillata* (Carter, 1880)
34. *Theonella swinhoei* (Gray, 1868)
35. *Leiodermatium pfeifferae* (Carter, 1876)
36. *Damiria toxifera* (van Soest, Zea & Kielman, 1994)
37. *Clathria (Microciona) atrasanguinea* (Bowerbank, 1862)

Box 1.1 Checklist of Sponges Described as from the Andaman and Nicobar Islands *(cont.)*

38. *Clathria (Thalysias) vulpina* (Lamarck, 1814)
39. *Echinochalina (Echinochalina) barba* (Lamarck, 1813)
40. *Echinodictyum asperum* (Ridley & Dendy, 1886)
41. *Raspailia (Raspailia) viminalis* (Schmidt, 1862)
42. *Rhabderemia prolifera* (Annandale, 1915)
43. *Monanchora enigmatica* (Burton & Rao, 1932)
44. *Kirkpatrickia spiculophila* (Burton & Rao, 1932)
45. *Psammochela elegans* (Dendy, 1916)
46. *Damiriopsis brondstedi* (Burton, 1928)
47. *Iotrochota baculifera* (Ridley, 1884)
48. *Tedania (Tedania) anhelans* (Lieberkühn, 1859)
49. *Biemna liposigma* (Burton, 1928)
50. *Biemna tubulata* (Dendy, 1905)
51. *Mycale (Rhaphidotheca) coronata* (Dendy, 1926)
52. *Mycale (Aegogropila) crassissima* (Dendy, 1905)
53. *Mycale (Mycale) indica* (Carter, 1887)
54. *Auletta andamanensis* (Pattanayak, 2006)
55. *Axinella acanthelloides* (Pattanayak, 2006)
56. *Axinella tenuidigitata* (Dendy, 1905)
57. *Phakellia columnata* (Burton, 1928)
58. *Amorphinopsis foetida* (Dendy, 1889)
59. *Petromica (Petromica) massalis* (Dendy, 1905)
60. *Spongosorites andamanensis* Pattanayak, 2006)
61. *Topsentia halichondrioides* (Dendy, 1905)
62. *Haliclona (Gellius) flagellifera* (Ridley and Dendy, 1886)
63. *Haliclona (Gellius) megastoma* (Burton, 1928)
64. *Gelliodes fibulata* (Carter, 1881)
65. *Calyx clavata* (Burton, 1928)
66. *Xestospongia testudinaria* (Lamarck, 1815)
67. *Carteriospongia foliascens* (Pallas, 1766)
68. *Clathrina coriacea* (Montagu, 1818)
69. *Pericharax heteroraphis* (Poléjaeff, 1883)
70. *Hyalonema (Ijimaonema) aculeatum* (Schulze, 1895)
71. *Hyalonema (Hyalonema) sieboldii* (Gray, 1835)
72. *Hyalonema (Coscinonema) indicum* (Schulze, 1895)
73. *Hyalonema (Coscinonema) lamella* (Schulze, 1900)
74. *Hyalonema (Cyliconema) martabanense* (Schulze, 1900)
75. *Hyalonema (Cyliconema) masoni* (Schulze, 1895)
76. *Hyalonema (Cyliconema) nicobaricum* (Schulze, 1904)
77. *Hyalonema (Cyliconema) rapa* (Schulze, 1900)
78. *Hyalonema (Cyliconema) apertum apertum* (Schulze, 1886)
79. *Lophophysema inflatum* (Schulze, 1900)
80. *Pheronema raphanus* (Schulze, 1895)
81. *Semperella cucumis* (Schulze, 1895)
82. *Aphrocallistes beatrix* (Gray, 1858)
83. *Farrea occa* (Bowerbank, 1862)

(Continued)

> **Box 1.1 Checklist of Sponges Described as from the Andaman and Nicobar Islands** *(cont.)*
>
> **84.** *Tretodictyum minor* (Dendy & Burton, 1926)
> **85.** *Euplectella aspergillum* (Owen, 1841)
> **86.** *Euplectella aspergillum regalis* (Schulze, 1900)
> **87.** *Euplectella simplex* (Schulze, 1896)
> **88.** *Lophocalyx spinosa* (Schulze, 1900)

Acanthella cavernosa (Dendy, 1922)   *Amphimedon chloros* (Ilan, Gugel & van-Soest, 2004)   *Clathria (Thalysias) vulpina* (Lamark, 1814)

Cliona varians (Duchassaing & Michelotti, 1864)   *Liosina paradoxa* (Thiele, 1899)   *Monanchora unguiculata* (Dendy, 1922)

Neopetrosia exigua (Kirkpatrick, 1900)   *Oceanapia sagittaria* (Sollas, 1902)   *Plakortis simplex* (Schulze, 1880)

Rhabderemia prolifera (Annandale, 1915)   *Stylissa massa* (Carter, 1887)   *Xestospongia testudinaria* (Lamarck, 1815)

FIGURE 1.1 Sponges of the Andaman and Nicobar Islands. (Please see color plate at the back of the book.)

invertebrates are very scanty. Though works on the diversity of sponges in Andaman date back to deep water scientific voyages in 1902, studies since then have been very scanty, and shallow water sponges from the ANI have not been thoroughly studied. This study is intended to bridge this scientific gap by documenting the rich diversity of marine sponges in these islands, especially in the coral reefs.

METHODS

Study Sites

The collections were made using SCUBA gear from four sites: North Bay, Chidiyatappu, Pongibalu, and Ritchie's Archipelago. The field photograph was taken using a Canon digital camera with an Ikelite waterproof housing.

North Bay (11° 42′ 09.11″ N; 92° 45′ 12.80″ E): North Bay beach is located 5 km from Phoenix Bay, and its coral reefs are the closest to the capital city of Port Blair. Water enters the bay, area approximately 1 km^2, through the open sea on the southeastern side. Massive corals (*Porites* sp.) dominate the benthic biota here. The bay is home to a considerable amount of mangroves which, along with the runoff, cause considerable sedimentation load on corals.

Chidiyatappu (11° 29′ 11.50″ N; 92° 42′ 33.46″ E): A popular tourist destination located 25 km from Port Blair, it is the southernmost tip of South Andaman, also known for its diversity of corals and other reef-associated fauna. It is located at the mouth of a creek and is thus subjected to much freshwater load during the rainy season.

Pongibalu (11° 30′ N; 92° 39′ E): It is situated 25 km from Port Blair and is endowed with richly diversified scleractinia and other reef-associated organisms. It is also characterized by very high currents as it is located in a channel. It falls under the Mahatma Gandhi Marine National Park (MGMNP).

Ritchie's Archipelago (12° 02′ 31″ N; 92° 58′ 56″ E): Ritchie's Archipelago is a cluster of smaller islands which lie about 30 km east of Port Blair. The archipelago comprises 11 islands and a few smaller islets. It is known to possess some of the most popular dive locations in the Andaman and Nicobar Islands. Rani Jhansi Marine National Park has been established with the aim of preserving the huge natural biodiversity of these islands.

Specimen Preservation and Identification

The specimens were preserved in absolute alcohol as soon as possible after collection. Forty-eight hours later the alcohol previously used was discarded and fresh alcohol was added for final preservation. The marine sponges were identified on the basis of taxonomic characters, using available literature (Rao, 1941; Pattanayak and Manna, 2001; Pattanayak, 2006). Sponge sections were made using a razor blade and then kept in xylene until the sections were translucent. Spicule preparations were made using nitric acid to dissolve the organic material in the sponge tissue, leaving only the inorganic skeleton. The sections and the spicules were observed under a compound light microscope and photographed using a Canon Digital Camera. Spicule measurements were taken using an ocular micrometer which was calibrated using a stage micrometer.

RESULTS

A total of 49 species of sponge were identified from South Andaman, belonging to the orders Homosclerophorida, Spirophorida, Agelasida, Dendroceratida and Verongida (1 species each), Astrophorida and Chondrosida (2 each), Dictyoceratida (4), Hadromerida (6), Halichondrida (7), Poecilosclerida (10) and Haplosclerida (13) (Table 1.1). Poecilosclerida and Haplosclerida are the most diverse orders recorded from the four sites. North Bay accounted for 31 species, Chidiyatappu 28, Pongibalu 35 and Ritchie's Archipelago 46. Ritchie's Archipelago has the highest diversity among the sites surveyed, probably because of the larger area it covers. Of the 49 species of sponge identified during this study, 34 were new records to the Andaman and Nicobar Islands. Twenty-two sponges have been added to the Indian fauna through this study.

DISCUSSION

There are only a few reports on the taxonomy and diversity of the marine sponge in India in general (Pattanayak, 2006) and the ANI in particular (Pattanayak, 2006). This could be primarily attributed to the difficulties in their taxonomical characterization (Pattanayak and Manna, 2001). Though the Zoological Survey of India has collections spanning decades, most of them have yet to be described taxonomically.

Pattanayak (2006) described 75 species from the ANI, which included 18 new distributional records and 4 newly described species. The sponges described were from collections made over a period of 100 years, with major contributions from RIMS "Investigator," and some by Scientists of the Zoological Survey of India. A total of 56 species from the order Demospongiae, 2 from Calcarea and 17 from Hexactinellida were described. Other major contributions to the sponge diversity of these Islands were made by Schulze (1902), Dendy and Burton (1926), Burton (1928) and Burton and Rao (1932).

The Andaman and Nicobar Islands are known for their remarkably high biodiversity and affinities to the Indo-Malayan and Indo-Chinese regions (Smith, 1930). Studies on the Zoogeographical patterns of distribution of sponges in the Andaman and Nicobar region with reference to adjacent regions (Burma, Sumatra, Indonesia and Gulf of Mannar) reveal that these islands, acting as a barrier between the Bay of Bengal and the Andaman Sea, harbor sponge species from both seas. The need for further investigation of the sponges of these islands is emphasized by the fact that they bear similarity to sponge species from by far the richest region in the Indonesian Archipelago, harboring 786 species (Hooper *et al.*, 2000). The inventory of the sponges of these islands is by no means over and will continue to expand with growing taxonomic expertise.

TABLE 1.1 Sponges Reported in this Study with Distributional Records of the Four Sites in South Andaman

Sl. No.	Species Name	North Bay	Chidiya-tappu	Pon-gibalu	Ritchie's Archipelago
	Order: Homosclerophorida (Dendy, 1905)				
	Family: Plakinidae (Schulze, 1880)				
1.	*Plakortis simplex* (Schulze, 1880)	√	√	√	√
	Order: Spirophorida (Bergquist & Hogg, 1969)				
	Family: Tetillidae (Sollas, 1886)				
2.	*Paratetilla bacca* (Selenka, 1867)	√	√	√	√
	Order: Astrophorida (Sollas, 1888)				
	Family: Ancorinidae (Schmidt, 1870)				
3.	*Rhabdastrella globostellata* (Carter, 1883)	√	√		√
4.	*Ecionemia acervus* (Bowerbank, 1864)	√	√	√	√
	Order: Hadromerida (Topsent, 1894)				
	Family: Clionaidae (D'Orbigny, 1851)				
5.	*Cliona varians* (Duchassaing & Michelotti, 1864)	√		√	√
6.	*Cliona ensifera* (Sollas, 1878)	√	√	√	√
7.	*Spheciospongia vagabunda* (Ridley, 1884)				√
	Family: Spirastrellidae (Ridley & Dendy, 1886)				
8.	*Spirastrella cunctatrix* (Schmidt, 1868)	√	√	√	√
	Family: Suberitidae (Schmidt, 1870)				
9.	*Terpios gelatinosa* (Bowerbank, 1866)				√
	Family: Tethyidae (Gray, 1848)				
10.	*Tethya diploderma* (Schmidt, 1870)	√	√		√
	Order: Chondrisida (Boury-Esnault & Lopes, 1985)				
	Family: Chondrillidae (Gray, 1872)				
11.	*Chondrilla australiensis* (Carter, 1873)	√	√	√	√

(Continued)

TABLE 1.1 Sponges Reported in this Study with Distributional Records of the Four Sites in South Andaman *(cont.)*

Sl. No.	Species Name	North Bay	Chidiya-tappu	Pon-gibalu	Ritchie's Archipelago
12.	*Chondrilla grandistellata* (Thiele, 1900)				√
	Order: Poecilosclerida (Topsent, 1928)				
	Family: Microcionidae (Carter, 1875)				
13.	*Clathria (Thalysias) vulpine* (Lamark, 1814)			√	√
14.	*Clathria (Thalysias) cervicornis* (Thiele, 1903)	√		√	√
	Family: Rhabderemiidae (Topsent, 1928)				
15.	*Rhabderemia prolifera* (Annandale, 1915)	√	√	√	√
	Family: Crambeidae (Levi, 1963)				
16.	*Monanchora unguiculata* (Dendy, 1922)			√	√
	Family: Crellidae (Dendy, 1922)				
17.	*Crella (Grayella) cyathophora* (Carter, 1869)			√	
	Family: Iotrochotidae				
18.	*Iotrochota baculifera* (Ridley, 1884)	√		√	√
	Family: Tedanidae (Ridley & Dendy, 1886)				
19.	*Tedania (Tedania) anhelans* (Lieberkühn, 1859)			√	
	Family: Mycalidae (Lundbeck, 1905)				
20.	*Mycale (Aegogropila) crassissima* (Dendy, 1905)			√	√
	Family: Isodictyidae (Dendy, 1924)				
21.	*Coelocarteria singaporensis* (Carter, 1883)	√			
	Family: Podospongiidae (de Laubenfels, 1936)				
22.	*Diacarnus megaspinorhabdosa* (Kelly-Borges & Vacelet, 1995)	√		√	√
	Order: Halichondrida (Gray, 1867)				

TABLE 1.1 Sponges Reported in this Study with Distributional Records of the Four Sites in South Andaman *(cont.)*

Sl. No.	Species Name	North Bay	Chidiya-tappu	Pon-gibalu	Ritchie's Archipelago
	Family: Axinellidae (Carter, 1867)				
23.	*Axinella acanthelloides* (Pattanayak, 2006)	√		√	√
24.	*Axinella cannabina* (Esper, 1794)			√	√
	Family: Dictyonellidae (Van Soest, Diaz & Pomponi, 1990)				
25.	*Acanthella cavernosa* (Dendy, 1922)	√	√	√	√
26.	*Liosina paradoxa* (Thiele, 1899)	√	√	√	√
27.	*Stylissa carteri* (Dendy, 1889)	√	√	√	√
28.	*Stylissa massa* (Carter, 1887)	√	√	√	√
	Family: Heteroxyidae (Dendy, 1905)				
29.	*Myrmekioderma granulatum* (Esper, 1794)				√
	Order: Agelasida (Hartman, 1980)				
	Family: Agelasidae (Verrill, 1907)				
30.	*Agelas axifera* (Hentschel, 1911)		√		√
	Order: Haplosclerida (Topsent, 1928)				
	Family: Callyspongidae (de Laubenfels, 1936)				
31	*Callyspongia (Toxochalina) multiformis* (Pulitzer-Finali, 1986)			√	√
32.	*Callyspongia (Euplacella) australis* (Lendenfeld, 1887)		√		√
33.	*Callyspongia (Cladochalina) subarmigera* (Ridley, 1884)				√
	Family: Chalinidae (Gray, 1867)				
34	*Chalinula nematifera* (de Laubenfels, 1954)	√	√	√	√
35.	*Haliclona (Gellius) cymaeformis* (Esper, 1794)		√		√
36.	*Haliclona (Reniera) fascigera* (Hentschel, 1912)				√
	Family: Niphatidae (Van Soest, 1980)				
37	*Gelliodes fibulata* (Carter, 1881)	√	√	√	√

(Continued)

TABLE 1.1 Sponges Reported in this Study with Distributional Records of the Four Sites in South Andaman *(cont.)*

Sl. No.	Species Name	North Bay	Chidiya-tappu	Pon-gibalu	Ritchie's Archipelago
38.	*Amphimedon chloros* (Ilan, Gugel & van-Soest, 2004)	√		√	√
	Family: Phloeodictyidae (Carter, 1882)				
39.	*Oceanapia sagittaria* (Sollas, 1902)	√	√	√	√
	Family: Petrosiidae (Van Soest, 1980)				
40.	*Xestospongia testudinaria* (Lamarck, 1815)	√	√	√	√
41.	*Petrosia (Strongylophora) strongylata* (Thiele, 1903)				√
42.	*Neopetrosia exigua* (Kirkpatrick, 1900)	√	√	√	√
43.	*Neopetrosia carbonaria* (Lamarck, 1814)	√	√	√	√
	Order: Dictyoceratida (Minchin, 1900)				
	Family: Ircinidae (Gray, 1867)				
44.	*Ircinia strobilina* (Lamarck, 1816)	√	√	√	√
	Family: Thorectidae (Bergquist, 1978)				
45.	*Hyrtios erectus* (Keller, 1889)	√	√	√	√
46.	*Carteriospongia foliascens* (Pallas, 1766)	√	√	√	√
	Family: Dysideidae (Gray, 1867)				
47.	*Lamellodysidea herbacea* (Keller, 1889)	√	√	√	√
	Order: Dendroceratida (Minchin, 1900)				
	Family: Darwinellidae (Merejkowsky, 1879)				
48.	*Aplysilla rosea* (Barrois, 1876)		√		√
	Order: Verongida (Bergquist, 1978)				
	Family: Pseudoceratinidae (Carter, 1885)				
49.	*Pseudoceratina purpurea* (Carter, 1880)	√	√	√	√

REFERENCES

Bergmann, W., Feeney, R.J., 1950. The isolation of a new thymine pentoside from sponges. J. American Chem. Soc. 72, 2809–2810.

Bergquist, P.R., 1978. Sponges. London & University of California Press, Hutchinson, Berkeley & Los Angeles. p. 268.

Burton, M., 1928. Report on some deep sea sponges from the Indian Museum collected by the R.I.M.S "Investigator" part II. Tetraxonida (concluded) and Euceratosa. Rec. Indian Mus. 30, 109–138.

Burton, M., Rao, H.S., 1932. Report on the shallow-water marine sponges in the collection of the Indian Museum. Rec. Indian Mus. 34, 299–356.

Dendy, A., Burton, M., 1926. Report on some deep sea sponges from the Indian Museum collected by the R.I.M.S. "Investigator" Part-I Hexactinellida & Tetraoxinida (Pars.). Rec. Indian Mus. 28, 225–248.

Diaz, M.C., Rützler, K., 2001. Sponges: an essential component of Caribbean coral reefs. Bull. Mar. Sci. 69, 535–546.

Hartman, W.D. 1977. Sponges as reef builders and shapers. PP. 127–134. In: Frost, S.H., Weiss, M.P., Saunders, J.B. (Eds) Reefs and related carbonates - ecology and sedimentology. Tulsa, OK: Am Assoc of Pet Geol Bull.

Hooper, J.N.A., Kennedy, J.A., van Soest, R.W.M., 2000. Annotated checklist of sponges (Porifera) of the South China Sea region. Raffles Bull. Zool. Suppl 8, 125–207.

Jha, R.K., Zi-rong, X., 2004. Biomedical compounds from marine organisms. Mar. Drugs. 2, 123–146.

Mora, C., Chittaro, P.M., Sale, P.F., Kritzer, J., Ludsin, S.A., 2003. Patterns and processes in reef fish diversity. Nature 421, 933–936.

Müller, W.E.G., 1995. Molecular phylogeny of metazoa (animals): monophyletic origin. Naturwissenschaften 82, 321–329.

Pattanayak, J.G., 2006. Marine sponges of Andaman and Nicobar Islands, India. Rec. Zool. Surv. India, Occasional Paper No. 255, 1–152.

Pattanayak, J.G., Manna, B., 2001. Distribution of Marine sponges (Porifera) in India. Proc. Zool. Soc. Calcutta 54 (1), 73–101.

Rao, H.S., 1941. Indian and Ceylon sponges in the naturalistoriska Riksmuseet, Stockholm, collected by K. Fistedt. Rec. Indian Mus. 43, 417–496.

Reswig, H.M., 1973. Population dynamics of the three Jamaican Demospongiae. Bull. Mar. Sci. 23, 191–226.

Schulze, F.E., 1902. An account of the Indian Triaxonia collected by the Royal Indian Marine Survey Ship "Investigator" translated into English by R. Von Lendenfeld. Trustees of the Indian Museum, Calcutta. p. 113.

Smith, M.A., 1930. The Reptilia and the Amphibia of the Malay Peninsula from the Isthmus of Kra to Singapur, including adjacent islands. Bull. Raffles. Mus. 3, 1–149.

Wulff, J.L., 2001. Assessing and monitoring coral reef sponges: why and how? Bull. Mar. Sci. 69, 831–864.

Wulff, J.L., 2006. Rapid diversity and abundance decline in a Caribbean coral reef sponges community. Biol. Conserv. 127, 167–176.

Chapter 2

Abundance of Shallow Water Octocorals in the Andaman and Nicobar Archipelago, India

J.S. Yogesh Kumar,* C. Raghunathan* and K. Venkataraman†
*Andaman and Nicobar Regional Centre, Zoological Survey of India, Port Blair, Andaman and Nicobar Islands, India; †Zoological Survey of India, Kolkata, West Bengal, India

INTRODUCTION

Oceanic islands have fascinated explorers and scientists since the earliest times because of their spectacular geological settings and the extravagant and exotic life forms found there. Andaman and Nicobar are a group of islands in the Indian Ocean. The marine fauna of these islands have been the subject of scientific interest especially for their biogeographic relevance and biodiversity (for example, dispersal patterns and endemism). Islands are bound ecosystems and serve as excellent laboratories for assessing changes in community structure and biodiversity (Sadler, 1999). The Andaman and Nicobar Islands (ANI) have been the target of numerous expeditions to explore their species richness (Soundararajan, 2002); nevertheless the octocorals were virtually unknown until recent years. This paper summarizes the present knowledge of the octocoral fauna of the ANI, describing both achievements in this field and the needs to be considered in future comprehensive biodiversity studies.

Octocorals, known as soft corals, sea fans and sea pens, are sedentary, mostly colonial marine animals. Octocorals belong to the anthozoan subclass Octocorallia, which comprises colonies of polyps that bear eight tentacles, almost always pinnate, and eight mesenteries in the gastrovascular cavity. The colonies present skeletal elements of calcium carbonate, called sclerites, embedded in their tissue. Many taxa also have proteinaceous and calcified axial skeletons. Octocorallia is composed of four orders: Helioporacea (Blue Coral), Pennatulacea (Sea Pens), Gorgonacea (Sea Fans) and Alcyonacea (Soft Corals and Telestids) (Fabricius and Alderslade, 2000; Grasshoff, 2000). Pennatulaceans and alcyonaceans are found in the Andaman and Nicobar coast. Their diversity and abundance is at present being studied. The pennatulaceans are colonies that show bilateral

Marine Faunal Diversity in India. DOI: 10.1016/B978-0-12-801948-1.00002-1

15

symmetry and polyp dimorphism. They are formed by a very large polyp called the oozooid, on the wall of which the coenenchyme spreads with numerous small (secondary) polyps; the large primary polyp may be additionally supported by a horny or calcium carbonate axis. Part of the oozooid forms the peduncle that anchors the colony in sand or soft substrates. The other part of the oozooid forms the rachis, which bears other kinds of polyps: autozooids and siphonozooids. In some species the emergent part looks like a feather (Williams, 1990).

The alcyonaceans include the soft corals in the group Alcyoniina, stoloniferous octocorals in the group Stolonifera, and the gorgonians in the suborder Holaxonia. Alcyonians (soft corals) form fleshy colonies characterized by having polyps aggregated or concentrated into polyparies. An internal medulla or axis is absent and the polyps are embedded into a soft coenenchymal tissue, which may or may not contain sclerites (Williams, 1992). Stolonifera include a group of octocorals that consist of individual tubular polyps that arise separately from ribbon-like stolons. They present a series of transitional forms, from those where the polyps are not united to those where they are joined at their bases in a common coenenchyme (Williams, 1993). Gorgonians include sea rods, sea whips, sea candelabra, sea feather plumes, and sea fans. They present very diverse growth forms: incrusting colonies, upright fans, bushes with slender branches, and simple whips. Gorgonian colonies have a central axial skeleton composed of a collagenous matrix, called gorgonin, and calcifications within the collagen interstitial spaces (Jeyasuria and Lewis, 1987). A layer of coenenchyme with sclerites and polyps surrounds it. Species of octocorals are identified according to colony and sclerite morphology. A combination of characteristics of the colony—branching pattern, color and shape—and of the sclerites—sizes, colors, forms and abundance of the different types of sclerites in the samples—determine the species. Even though these characteristics can be modified by the environment, they are sufficiently consistent to diagnose a species (Breedy and Guzman, 2003). Bayer (1981) estimated that there are 3000 valid species of octocorals worldwide, although new species and even genera continue to be described at a rapid rate.

In India very few comprehensive works are available on the octocorals. This study on the octocoral fauna in the Andaman and Nicobar Archipelago (Figure 2.1) was conducted from 2009 through 2012. More than 500 specimens were collected from 23 islands (Table 2.1).

METHODS

The line intercept transect method was employed by SCUBA diving from 5 m to 40 m depth (English *et al.*, 1997). The collected specimens were identified on the basis of the morphological characteristics of the colonies and the sclerites structure. Sclerites were extracted using 5 percent sodium hypochlorite (Bayer, 1961). Diversity indices were analyzed using statistical software (Ludwig and Reynolds, 1988).

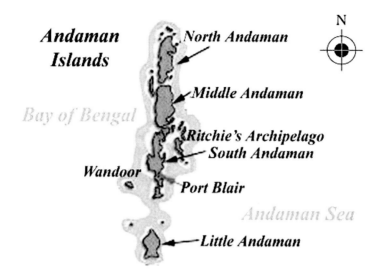

Map of Andaman and Nicobar Islands

FIGURE 2.1 Map showing the study area.

RESULTS

The shallow water (<40 m) octocoral species have been reported in literature from the Andaman and Nicobar Islands (Yogesh Kumar *et al.*, 2014a,b,c). During the survey, more than 500 samples were collected for identification and 65 species under 38 genera belonging to 16 families were identified. The

TABLE 2.1 Survey Locations in Andaman and Nicobar Islands

Island/location	Coordinates	No. of islands
South Andaman		11
Pongibalu	Lat. 11° 31.030′ N; Long. 92° 39.159′ E	
Rutland Island	Lat. 11° 28.541′ N; Long. 92° 40.371′E	
Jolly Buoy Island	Lat. 11° 30.251′ N; Long. 92° 32.591′ E	
Twins Island	Lat. 11° 23.773′ N; Long. 92° 33.097′ E	
Havelock Island	Lat. 12° 00.005′ N; Long. 92° 56.808′ E	
Henry Lawrence Island	Lat. 12° 05.000′ N; Long. 93° 06.312′ E	
Inglis Island	Lat. 12° 08.639′ N; Long. 93° 06.786′ E	
John Lawrence Island	Lat. 12° 04.075′ N; Long. 93° 00.398′ E	
Outram Island	Lat. 12° 00.574′ N; Long. 92° 56.808′ E	
South Button Island	Lat. 12° 13.467′ N; Long. 92° 01.334′ E	
Sister Island	Lat. 10° 55.830′ N; Long. 92° 07.023′ E	
Middle Andaman		3
Quaiter Island	Lat. 12° 20.323′ N; Long. 92° 54.529′ E	
Long Island	Lat. 12° 21.749′ N; Long. 92° 55.410′ E	
North Passage	Lat. 12° 18.121′ N; Long. 92° 55.718′ E	
North Andaman		4
Plastic Telkri	Lat. 13° 24.427′ N; Long. 93° 04.265′ E	
Nadiyel Telkri	Lat. 13° 26.076′ N; Long. 93° 05.708′ E	
East Island	Lat. 13° 37.575′ N; Long. 93° 02.469′ E	
Smith Island	Lat. 13° 18.532′ N; Long. 93° 04.314′ E	
Nicobar Islands		5
Malacca	Lat. 09° 10.490′ N; Long. 92° 49.714′ E	
Trinket Island	Lat. 08° 02.806′ N; Long. 93° 34.556′ E	
Kamorta Island - Kardip	Lat. 08° 02.151′ N; Long. 93° 33.182′ E	
Munak Gate	Lat. 07° 59.806′ N; Long. 93° 29.852′ E	
Katchal island	Lat. 08° 00.002′ N; Long. 93° 24.362′E	

identified specimens were deposited at the Natural Zoological Collection at Zoological Survey of India, Port Blair. The geographical distribution of the samples was: 45 species under 29 genera, 13 families from South Andaman; 26 species under 23 genera, 12 families from Nicobar region; 26 species under 18 genera, 7 families from Middle Andaman Islands; 12 species under 12 genera, 7 families from North Andaman Islands (Table 2.2 and Figures 2.2 to 2.9).

TABLE 2.2 Abundance of Octocorals from Andaman and Nicobar Archipelago

Sl. No.	Family	Species	NA	MA	SA	NI
1	Acanthogorgiidae	Acanthogorgia breviflora	–	–	+	+
2		Acanthogorgia spinosa	–	–	+	+
3		Anthogorgia ochracea	–	–	–	+
4		Muricella paraplectana	–	–	+	–
5		Muricella ramose	–	–	+	–
6	Clavulariidae	Carijoa riisei	–	–	+	–
7	Ellisellidae	Dichotella gemmacea	–	+	+	–
8		Ellisella azilia	–	+	–	–
9		Ellisella cercidia	+	+	+	–
10		Ellisella marisrubri	–	–	+	–
11		Ellisella nuctenea	–	–	+	–
12		Jenceella juncea	+	–	+	+
13		Junceella delicate	–	–	+	–
14		Junceella eunicelloides	–	–	–	+
15		Nicella flabellata	–	–	+	–
16		Nicella laxa	–	–	+	–
17		Verrucella cerasina	–	+	+	–
18		Verrucella corona	–	+	+	–
19		Verrucella diadema	–	–	+	–
20		Verrucella gubalensis	–	+	–	–
21		Verrucella klunzingeri	–	+	+	–
22		Verucella cerasina	–	+	–	–
23		Viminella crassa	–	+	+	+
24		Viminella junceelloides	–	+	–	–
25	Gorgoniidae	Hicksonella princeps	–	+	–	–
26		Rumphella aggregate	–	+	+	–
27		Rumphella torta	–	+	+	–
28	Helioporidae	Heliopora coerulea	–	–		+
29	Isididae	Isis hippuris	+	–	+	+
30	Melithaeidae	Acabaria cinquemiglia	–	–	+	–
31		Acabaria ouvea	–	+	–	–
32		Melithaea caledonica	–	–	+	+
33		Melithaea ochracea	–	–	+	–

(Continued)

TABLE 2.2 Abundance of Octocorals from Andaman and Nicobar Archipelago *(cont.)*

Sl. No.	Family	Species	NA	MA	SA	NI
34		*Mopsella rubeola*	–	–	+	–
35		*Wrightella braueri*	–	–	–	+
36	Nidaliidae	*Siphonogorgia media*	–	–	+	–
37	Pennatulidae	*Pteroeides esperi*	–	–	+	–
38	Plexauridae	*Bebryce sirene*	–	–	–	+
39		*Bebryce studeri*	–	–	+	–
40		*Echinogorgia flora*	–	–	+	–
41		*Echinogorgia toombo*	–	–	+	–
42		*Echinomuricea indica*	–	–	+	–
43		*Echinomuricea indomalaccensis*	–	–	+	–
44		*Euplexaura amerea*	–	–	–	+
45		*Euplexaura rhipidalis*	–	+	–	–
46		*Menella indica*	–	–	+	–
47		*Menella kanisa*	–	–	+	+
48		*Menella kouare*	+	+	+	–
49		*Menella woodin*	–	+	+	–
50		*Trimuricea caledonica*	–	–	–	+
51		*Villogorgia tenuis*	–	+	–	–
52	Subergorgiidae	*Annella mollis*	–	–	+	+
53		*Annella reticulate*	–	–	+	+
54		*Subergorgia rubra*	–	+	–	–
55		*Subergorgia suberosa*	+	+	+	+
56	Tubiporidae	*Tubipora musica*	+	–	–	+
57	Veretillidae	*Cavernularia pusilla*	–	–	–	+
58	Virgulariidae	*Virgularia gustaviana*	–	–	+	–
59		*Virgularia mirabilis*	–	–	–	+
60	Nephtheidae	*Nephthea* sp.	+	+	+	+
61		*Dendronephthya* sp.	+	+	+	+
62	Alcyoniidae	*Sinularia* sp.	+	+	+	+
63		*Cladiella* sp.	+	+	+	+
64		*Sarcophyton* sp.	+	+	+	+
65		*Lobophytum* sp.	+	+	+	+

NA, North Andaman; MA, Middle Andaman; SA, South Andaman; NI, Nicobar Islands; + present; – absent

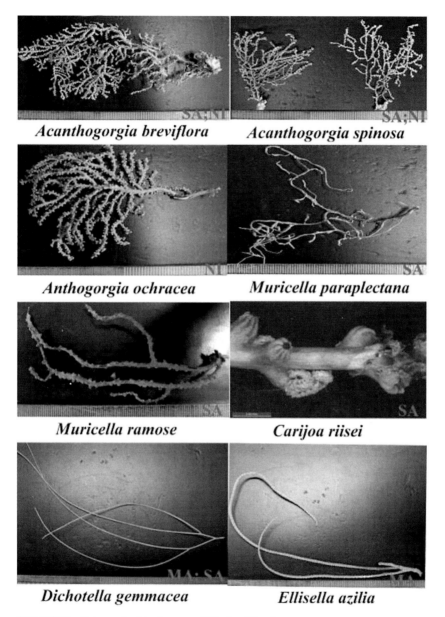

Acanthogorgia breviflora   *Acanthogorgia spinosa*

Anthogorgia ochracea   *Muricella paraplectana*

Muricella ramose   *Carijoa riisei*

Dichotella gemmacea   *Ellisella azilia*

FIGURE 2.2   Octocorals of Andaman and Nicobar Islands.

The survey of this region resulted in the observation of ubiquitous distribution of invasive octocoral species, mostly at 5 m depth or more. Alcyonacea and Helioporacea were predominantly found at 5–20 m on reef areas, and Pennatulacea and Gorgonacea were habitually noted at 5–40 m on reef and sand substratum.

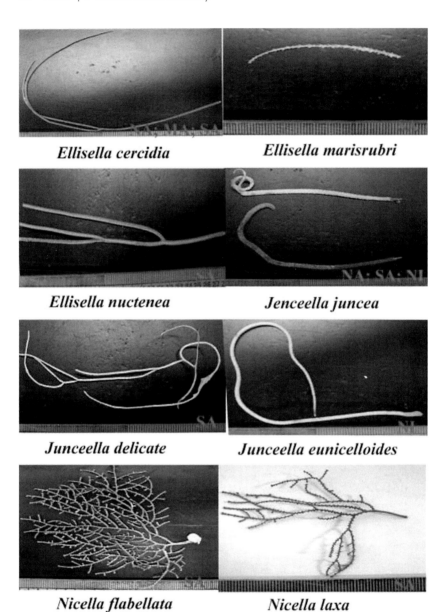

FIGURE 2.3   Octocorals of Andaman and Nicobar Islands.

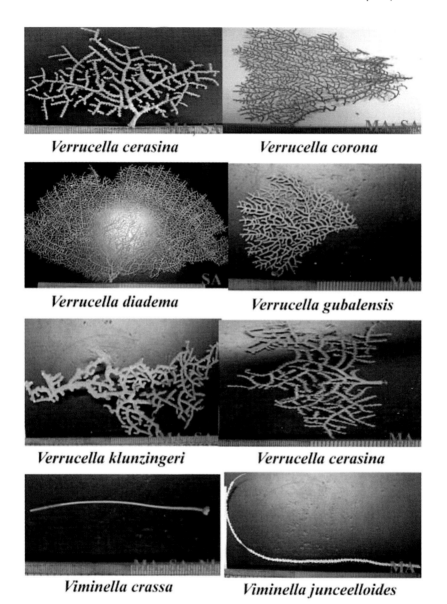

FIGURE 2.4 Octocorals of Andaman and Nicobar Islands.

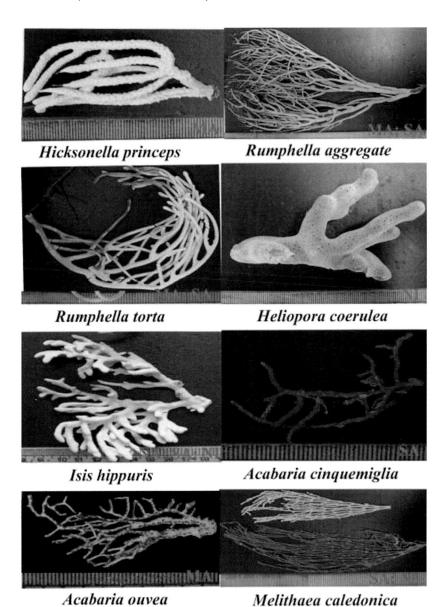

Hicksonella princeps   *Rumphella aggregate*

Rumphella torta   *Heliopora coerulea*

Isis hippuris   *Acabaria cinquemiglia*

Acabaria ouvea   *Melithaea caledonica*

FIGURE 2.5   Octocorals of Andaman and Nicobar Islands.

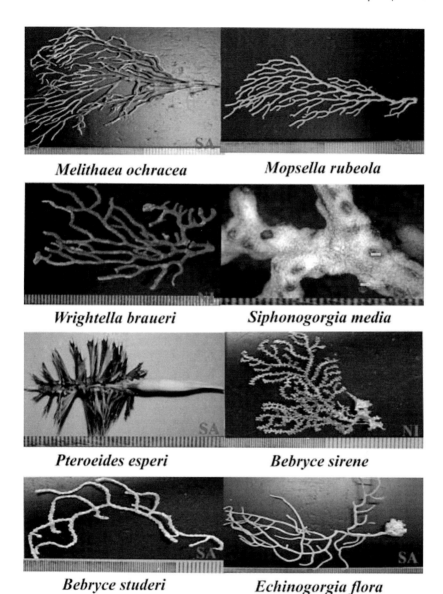

Melithaea ochracea   *Mopsella rubeola*

Wrightella braueri   *Siphonogorgia media*

Pteroeides esperi   *Bebryce sirene*

Bebryce studeri   *Echinogorgia flora*

FIGURE 2.6   Octocorals of Andaman and Nicobar Islands.

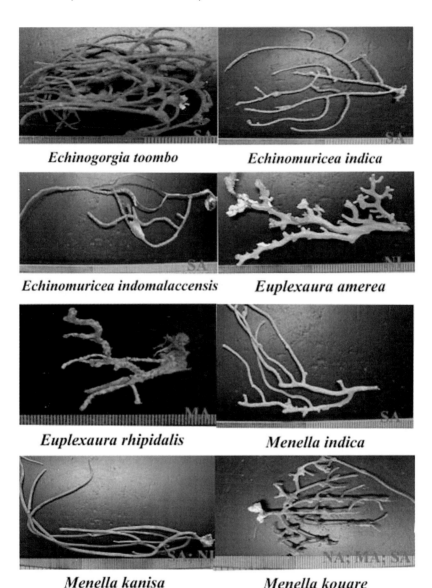

Echinogorgia toombo   *Echinomuricea indica*

Echinomuricea indomalaccensis   *Euplexaura amerea*

Euplexaura rhipidalis   *Menella indica*

Menella kanisa   *Menella kouare*

FIGURE 2.7   Octocorals of Andaman and Nicobar Islands.

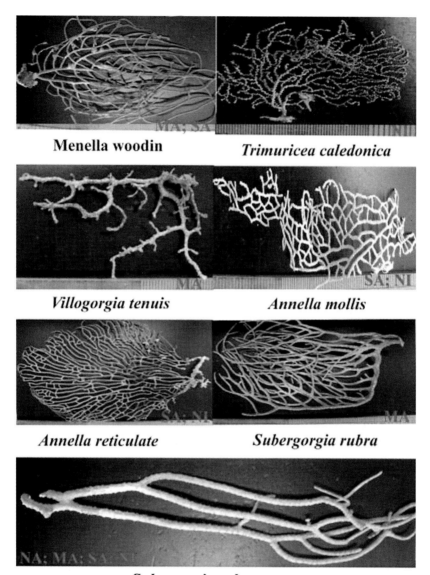

FIGURE 2.8   Octocorals of Andaman and Nicobar Islands.

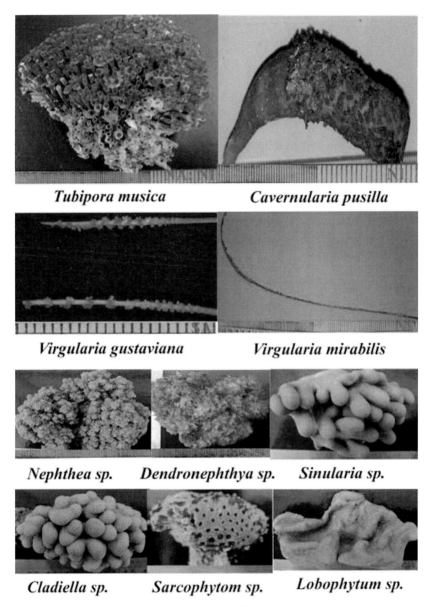

FIGURE 2.9   Octocorals of Andaman and Nicobar Islands.

The relative frequency of occurrence of each genus of octocorals is shown in Figure 2.10 for four regions of the Andaman and Nicobar Islands. The most frequently occurring genera in North Andaman were *Lobophytum* (20.9%), *Sinularia* and *Isis* (17.9%), and *Subergorgia* (8.9%). In Middle Andaman,

Sinularia (23.5%) occurred most frequently, followed by *Sarcophyton* (17.5%), *Lobophytum* (17.5%) and *Nephthea* (11.1%). In South Andaman, *Sinularia* (16.5%) was also most frequently found, followed by *Lobophytum* (14.5%), *Nephthea* (10.3%), *Sarcophyton* (9.2%) and *Verrucella* (6.5%). In the Nicobar Islands, *Lobophytum* (22.8%) occurred most frequently, followed by *Sinularia* (20.5%) and *Sarcophyton* (16.9%). The diversity indices of gorgonians at ANI are given in Figure 2.11. The highest species diversity, evenness and richness were recorded in the South Andaman Islands (4.11, 0.86 and 0.92, respectively) and lowest values were reported in the North Andaman Islands (3.06, 0.88 and 0.86).

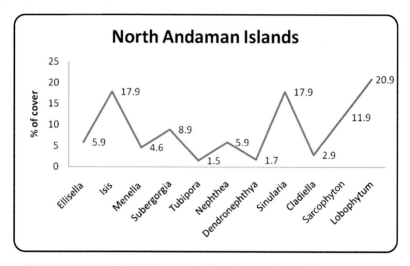

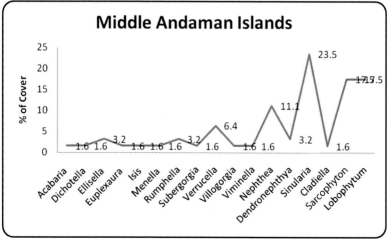

FIGURE 2.10   Relative frequency (%) of occurrence of octocorals in Andaman and Nicobar Archipelago.

(Continued)

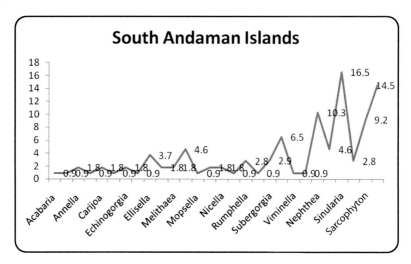

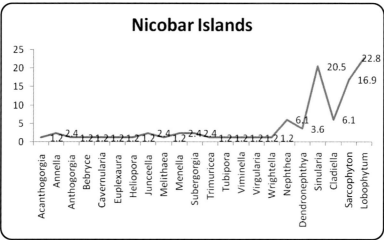

FIGURE 2.10 *(cont.)*

DISCUSSION

The occurrence and distribution of octocoral species are presented in Table 2.2 and Figure 2.10, and examples are shown in Figures 2.2 to 2.9, but more exploration is required, thus this list constitutes a preliminary report. Currently, Alcyoniidae, Nephtheidae and Subergorgiidae are the families with the widest distribution around the island. Studies of octocoral diversity in Indian waters are scarce; most taxonomic research has been done in the Indo-Pacific shallow waters, and gorgonians in this region have been listed by several authors (Stiasny, 1941; Mai-Bao-Thu and Domantay, 1970, 1971; Muzik and Wainwright, 1977;

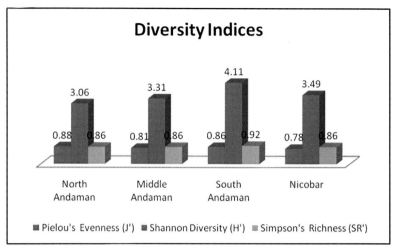

FIGURE 2.11   Diversity Indices of gorgonian octocorals in Andaman and Nicobar Archipelago.

Zou and Scott, 1980; Lasker and Coffroth, 1983; Zou and Chen, 1984; Alderslade, 1986; Van Ofwegan, 1987, 1994; Goh and Chou, 1995, 1996; Williams, 1992; Williams and Lindo, 1997; Grasshoff and Alderslade, 1997; Grasshoff, 1999; Dautova, 2007).

India and its adjacent waters were studied by Pratt (1903), Thompson and Crane (1909), Thomson and Simpson (1909) and Thomas *et al.* (1995). In Andaman and Nicobar, 26 species were reported by Jayasree *et al.* (1994); later, 85 species belonging to 13 families were recorded from ANI (Venkataraman *et al.*, 2004). However, a recent survey conducted during 2009–11 in ANI revealed 51 species of gorgonians belonging to 25 genera, 8 families, and 3 suborders (Yogesh Kumar *et al.*, 2012). More studies are needed to enhance the assessment of the diversity of octocorals in India as well as the Andaman and Nicobar islands, including deep water exploration, and also oceanographic and ecological research to determine reasons for diversity and abundance in shallow waters.

ACKNOWLEDGEMENTS

The authors are thankful to the Ministry of Environment and Forests, Government of India for providing financial support.

REFERENCES

Alderslade, P., 1986. An unusual leaf like gorgonian (Coelenterata: Octocorallia) from the Great Barrier Reef, Australia. The Beagle (Occasional papers of the Northern Territory Museum of Art and Sciences) 3 (1), 81–93.

Bayer, F.M., 1961. The shallow water octocorallia of the West Indian region. Stud. Fauna Curacao 12, 1–373.

Bayer, F.M., 1981. Status of Knowledge of Octocorals of World Seas. Seminários de Biologia Marinha, Academia Brasileira de Ciências, Rio de Janeiro, 1-102.

Breedy, O., Guzman, H.M., 2003. The genus *Pacifigorgia* (Octocorallia: Gorgonacea) in Costa Rica (Coelenterata: Octocorallia: Gorgoniidae). Zootaxa 281, 1–60.

Dautova, T.N., 2007. Gorgonians (Anthozoa: Octocorallia) of the Northwestern Sea of Japan. Russian J. Mar. Biol. 33, 297–304.

English, S., Wilkinson, C., Baker, V., 1997. Survey manual for Tropical Marine Resources. Australian Institute of Marine Science, Townsville, Australia, 390.

Fabricius, K., Alderslade, P., 2000. Soft Corals and Sea Fans: A Comprehensive Guide to the Tropical Shallow Water Genera of the Central-West Pacific, the Indian Ocean and the Red Sea. Australian Institute of Marine Sciences, Townsville, Australia, p. 264.

Goh, N.K.C., Chou, L.M., 1995. Growth of five species of gorgonians (Sub-class Octocorallia) in the sedimented waters of Singapore. Mar. Ecol. 16, 337–346.

Goh, N.K.C., Chou, L.M., 1996. An annotated checklist of the gorgonians of Singapore, with a discussion of gorgonian diversity in the Indo-West Pacific. Raff. Bull. Zool. 44, 435–459.

Grasshoff, M., 1999. The shallow water gorgonians of New Caledonia and adjacent islands. Senckenbergiana Biol. 78 (1/2), 1–121.

Grasshoff, M., 2000. The gorgonians of the Sinai coast and the strait of Gubal, Red Sea (Coelenterata, Octocorallia). Cour. Forsch - Inst. Senckenbergiana, 1–224.

Grasshoff, M., Alderslade, P., 1997. Gorgoniidae of Indo-Pacific reefs with description of two new genera (Coelenterata: Octocorallia). Senckenbergiana Biol. 77 (1), 23–35.

Jayasree, V., Bhat, K.L., Parulekar, A.H., 1994. *Sarcophyton andamanensis* a new species of soft coral from Andaman Islands. J. Andaman Sci. 10 (1&2), 107–111.

Jeyasuria, P., Lewis, J.C., 1987. Mechanical properties of the axial skeleton in gorgonians. Coral Reefs 5, 213–219.

Lasker, H.R., Coffroth, M.A., 1983. Octocoral distributions at Carrie Bow Cay. Belize. Mar. Ecol. Prog. Ser. 13, 21–28.

Ludwig, J.A., Reynolds, J.F., 1988. Statistical Ecology. Wiley, New York, p. 337.

Mai-Bao-Thu, F., Domantay, J.S., 1970. Taxonomic studies of the Philippine gorgonaceans in the collections of the University of Santo Tomas, Manila. Acta Manilana 6, 25–78.

Mai-Bao-Thu, F., Domantay, J.S., 1971. Taxonomic studies of the Philippine gorgonaceans in the collections of the University of Santo Tomas, Manila (cont'd). Acta Manilana 7, 3–77.

Muzik, K., Wainwright, S., 1977. Morphology and habitat of five Fijian sea fans. Bull. Mari. Sci. 27 (2), 308–337.

Pratt, M., 1903. The Alcyonaria of the Maldives. In: Gardiner, J.S., (ed.) The fauna and geography of the Maldive and Laccadive Archipelagoes, 2: 503–509.

Sadler, J.P., 1999. Biodiversity on oceanic Islands: a palaeoecological assessment. J. Biogeogr 26, 75–87.

Soundararajan, R., 2002. Mass bleaching of coral reefs in the A & N Islands in 1998. SANE Newsl.

Stiasny, G., 1941. Octocorallia from Philippine waters. Phil. J. Sci. 76, 67–73.

Thomas, P.A., George, R.M., Lazarus, S., 1995. Distribution of Gorgonids in the northeast coast of India with particular reference to *Heterogorgia flabellum*. J. Mar. Biol. Assoc. India 37, 134–142.

Thompson, A., Crane, G., 1909. The alcyonarians of Okhalmandal. In: Hornell, J., (Ed.) Marine Zoology of Okamandal in Kattiawar, 1: 125-135.

Thomson, J.A., Simpson, J.J., 1909. An account of the Alcyonarians collected by the Royal Indian Marine Survey Ship Investigator in the Indian Ocean. II. The Alcynonarians of littoral area, The Indian Museum, Calcutta, 1-319.

Van Ofwegan, L.P., 1987. Melithaeidae from the Indian Ocean and the Malay Archipelago. Zool Verh Leiden 239, 1–57.

Van Ofwegan, L.P., 1994. *Pseudothelogorgia*, a new genus of Gorgonacean Octocorals from the Indian Ocean. Precious Coral and Octocoral Research 3, 19–22.

Venkataraman, K., Jeyabaskaran, R., Raghuram, K.P., Alfred, J.R.B., 2004. Bibliography and Checklist of Coral and Associated Organisms of India. Zoological Survey of India, Occasional Paper 226, 1–468.

Williams, G.C., 1990. The Pennatulacea of southern Africa (Coelenterata, Anthozoa). Ann. South. African. Mus. 99, 31–119.

Williams, G.C., 1992. Biogeography of the octocorallian coelenterate fauna of southern Africa. Biol J Linnean Soc 46, 351–401.

Williams, G.C., 1993. Coral Reef Octocorals - An Illustrated Guide to the Soft Corals, Sea Fans and Sea Pens inhabiting the Coral Reefs of Northern Natal. Durban Natural Science Museum, Durban, p. 64.

Williams, G.C., Lindo, K.G., 1997. A review of the octocorallian genus *Leptogorgia* (Anthozoa: Gorgoniidae) in the Indian Ocean and Subantarctic, with description of a new species and comparisons with related taxa. Proceedings of the California Academy of Sciences 49, 499–521.

Yogesh Kumar, J.S., Raghunathan, C., Venkataraman, K., 2012. Studies on new findings of Gorgoniidae from Ritchie's Archipelago Andaman and Nicobar Islands. Internat. J. Sci. Nat. 3 (2), 395–405.

Yogesh Kumar, J.S., Raghunathan, R. Raghuraman, C.R. Sreeraj and Venkataraman, K. 2014a. Handbook on Gorgonians (Octocorallia) of Andaman and Nicobar Islands. Published by the Director, Zoological Survey of India, Kolkata, 1-119.

Yogesh Kumar, J.S., C. Raghunathan, and Venkataraman K. 2014b. New records of Octocorallia (Order: Pennatulacea) from Indian waters. International Journal of Applied Biology and Pharmaceutical Technology, (5) 2, 52-56.

Yogesh Kumar, J.S., C. Raghunathan, S. Geetha and Venkataraman K. 2014c. New species of soft corals (Octocorallia: Alcyonacea) on the coral reef of Andaman and Nicobar Islands. International Journal of Integrative Sciences, Innovation and Technology, (3) 1, 8-11.

Zou, R.L., Chen, Y., 1984. Study of the shallow water Gorgonacea from the coast of Guangdong. Nanhai Studi Marina Sinica 5, 67–75.

Zou, R.L., Scott, P.J.B., 1980. The gorgonacea of Hong Kong. In: Morton, B.S., Tseng, C.K. (Eds.), In: Proceedings of the first international Marine Biology Workshop; the Marine Flora and Fauna of Hong Kong and Southern China. Univ Press, Hong Kong, pp. 135–159.

Chapter 3

Occurrence of Brown Paper Nautilus *Argonauta hians* (Lightfoot, 1786) at Inglis Island, South Andaman

K. Devi

Andaman and Nicobar Regional Centre, Zoological Survey of India, Port Blair, Andaman and Nicobar Islands, India

INTRODUCTION

Cephalopods are exclusively marine molluscs, and there are about 660 species, diverse in form, size and nature, in the world's oceans (Voss and Williamson, 1971; Worms, 1983). Of these, about 80 species, which are commercially and scientifically important, are distributed in the Indian seas (Silas, 1968; Sarvesan, 1974). Silas (1968) provided a comprehensive list of species reported from the Indian Ocean until 1968.

Argonauta, known commonly as the paper nautilus or the paper shelled nautilus, are distributed throughout the warm and temperate waters of the world, and the systematics of the species of genus *Argonauta* are still unsettled. Nesis (1982) recognized only four valid species: *Argonauta argo, Argonauta nodosa, Argonauta boettgeri* and *Argonauta hians.* There are two species *Argonauta argo* Linnaeus and *Argonauta hians* Lightfoot known from the Indian Sea (Robson, 1929; Silas, 1968). Hitherto, 29 species of cephalopods within 18 genera and 11 families, including a single species of *Argonauta boettgeri*, have been known to occur in the Andaman Islands (Subba Rao and Dey, 2000). The systematic study of the specimen of *Argonauta hians* found reveals that this species is a new record to the Andaman and Nicobar Islands.

METHODS

A single specimen of *Argonauta hians* was collected at Inglis Island on 2 January 2009 from 3 m depth on coral substratum between (Lat. 12° 08′–12° 11′ N; Long. 93° 07′– 93° 10′ E). Identification was based on the morphological

Marine Faunal Diversity in India. DOI: 10.1016/B978-0-12-801948-1.00003-3

35

features of the specimen (Olive, 1975). The identified species has been deposited in the National Zoological Collection of Zoological Survey of India, Port Blair (Reg. No. 4360).

RESULTS AND DISCUSSION

Systematic Account

See Box 3.1 and Figure 3.1.

ACKNOWLEDGEMENTS

The author wishes to thank the Director, Zoological Survey of India, Kolkata and Officer-in-Charge, Zoological Survey of India, Andaman and Nicobar Regional Centre, Port Blair for providing necessary facilities and encouragement. Thanks also to the Chief Wildlife Warden, Department of Environment and Forests, the Divisional Forest Officer and the Range Officer, Havelock Forest Division for their permission and logistic support. The valuable help and excellent co-operation extended in the field by Shri G. Ponnuswamy, Photographer and A. Polycap, Field Collector are also gratefully acknowledged.

Box 3.1 Systematic Account of *Argonauta hians*

Class Cephalopoda (Cuvier, 1797)
Order Octopoda (Leach, 1818)
Family **ARGONAUTIDAE** (Tryon, 1879)
Genus *Argonauta* (Linnaeus, 1758)
Species *Argonauta hians* (Lightfoot, 1786)
Common name: Brown Paper Nautilus

Description: The shell is highly variable in sculpture. It is rather small, the aperture is wide and strongly inflated, about 50 percent of shell length. The keel is wide and bears the characteristic 15 to 20 prominent large and blunt nodules placed in pairs over the keel. Great variations exist in size and form of the nodules. The ribs on the sides of the shell extending from the cranial knobs are less numerous than in the other species, with less bifurcation. There are two types of ribs present: 11–13 long and full sized ribs extending from the nodules to the nucleus which are in regular alternating series with the secondary ones; the latter 10–12 are short and extend up to halfway from the nodules. The color of the shells varies from white with brownish black tint on the nodules and adjacent ribs to light brown with sooty brown pigmentation over most of the surface of the shell (Voss and Williamson, 1971).

Habitat: Coral reefs
Distribution: Taiwan, Hong Kong, Japan, Philippines, Australia
Ecology: Continental shelf, Epipelagic, Nektonic

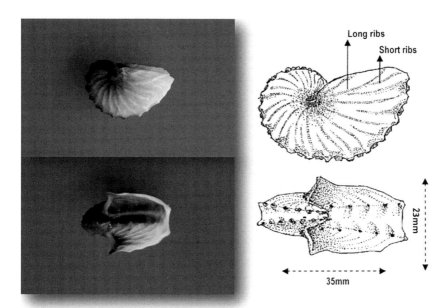

FIGURE 3.1   Schematic diagram of *Argonauta hians* (Lightfoot, 1786).

REFERENCES

Nesis, K.N., 1982. Cephalopods of the World. English Translation from Russian. Levitov, B.S. (Transl.) Burgess, L.A. (Eds.) (1987) Neptune City, T.F.H. Publications Inc. p. 351.

Olive, A.P.H., 1975. The Hamlyn Guide to Shells of the World. pp. 1–320.

Robson, G.V., 1929. A Monograph of the Recent Cephalopoda. Part I Oeteposme. British Museum (Nat. Hist.) London, p. 236.

Sarvesan R. 1974. Cephalopods in the Commerical Molluscs of India. CMFRI Bulletin No.25: pp. 63–83.

Silas, E.G., 1968. Cephalopoda of the west coast of India, collected during the cruises of the *R.V. Varuna* with a catalogue of the species known from the Indian Ocean. Proc Symp. on Mollusca. Mar. Biol. Assoc. India Part-I, 277–359.

Subba Rao, N.V., Dey, A., 2000. Catalogue of Marine Molluscus of Andaman and Nicobar Islands. Rec. Zool. Surv. India, Occ. Paper 187, 1–323.

Voss, O.L., Williamson, Q.R., 1971. Cephalopoda of Hong Kong. Govecoott Dt Press, Hongkong, p. 138.

Worms, J., 1983. World fisheries for cephalopods: A synoptic overview. In: Caddy, J.E. (eds.) Advances in Assessment of world cephalopod Resources, FAO Fish Technical Paper 231: pp. 1–20.

Chapter 4

Marine Molluscan Diversity in India

B. Tripathy and A.K. Mukhopadhyay
Zoological Survey of India, Kolkata, West Bengal, India

INTRODUCTION

Mollusca constitute an important component of marine biodiversity of the mainland and island coasts of India. Marine molluscs occur in diverse habitats: rocky coasts, sandy beaches, sea grass beds, coral reef ecosystems, mangroves and also at abyssal depths in the sea. However, in terms of molluscan species diversity and abundance, rocky intertidal zones and coral reef areas of India are rich. Mollusca are classified into seven classes, of which five are represented in India. Of the 586 global families, 279 are represented in the Indian region, including about 3600 species, of which approximately 2300 (65%) are marine. As there are no proper estimates of the number of molluscan species occurring in the marine and coastal ecosystems of India, this is an attempt towards assessment of marine molluscan diversity in India. This work was based primarily on the earlier literature records of Indian marine molluscs, materials present in the National Collection of the Zoological Survey of India (ZSI), the materials received by the Mollusca Section of the ZSI from different parts of the country for identification, and publications on marine Mollusca of India other than those of ZSI. This chapter attempts to present an overview of the diversity of Indian marine molluscs and their current status, to identify gaps in knowledge with a view to future research and conservation of the malacofauna of India.

HISTORICAL REVIEW

The important contributions to marine molluscan taxonomical research in India are those of Adam (1939), Ray (1948a,b, 1952, 1954), Hornell and Tomlin (1951), Subrahmanyam *et al.* (1952), Patil (1954), Satyamurti (1956), Patil and Gopalkrishnan (1960), Tikader (1964), Cheriyan (1968), Narayanan (1968, 1969, 1972), Prabhakara Rao (1968), Subba Rao (1968, 1970, 1980, 2003), Joshi (1969), Radhakrishna and Ganapati (1969), Nagabhushanam and Rao (1972), Satyanarayana Rao and Sundaram (1972), Virabhadra Rao and Krishna

Kumary (1973, 1974), Rajagopal and Subba Rao (1974), Starmuehlner (1974), Virabhadra Rao *et al.* (1974), Subba Rao and Dey (1975, 1984, 1986, 2000), Subba Rao and Mookherjee (1975), Das *et al.* (1977, 1981), Kohn (1978), Rajagopal and Mookherjee (1978, 1982), Namboodiri and Sivadas (1979), Gopinadha Pillai and Appukuttan (1980), Subba Rao and Surya Rao (1980, 1991, 1993), Subba Rao *et al.* (1983, 1987, 1991, 1992, 1993, 1995), Kasinathan and Shanmugam (1985), Mookherjee (1985), Tikader and Das (1985), Tikader *et al.* (1986), Jothinayagam (1987), Mookherjee and Barua (1989), Apte (1992, 1993, 1997, 1998, 2009), Babu Philip and Appukuttan (1995), Surya Rao and Maitra (1998), Mahapatra (2001, 2008, 2010), Surya Rao *et al.* (2004), Venkataraman *et al.* (2004), Subba Rao and Sastry (2005), Dey *et al.* (2005), Dey (2006, 2008), Arularasan and Kasinathan (2007), Ramakrishna *et al.* (2007, 2010), Rao and Sastry (2007), Venkitesan (2007), Roy *et al.* (2008), Apte *et al.* (2010), Raghunathan *et al.* (2010), Rao (2010), Venkitesan and Mukherjee (2011). In addition, there are several anecdotal accounts on the ecology, taxonomy, diversity and distribution of the marine mollusca of India in studies carried out by universities and other academic and research institutions.

TAXONOMIC DIVERSITY OF MARINE MOLLUSCA OF INDIA

Classification of mollusca has been and remains problematic. For example, a phylogenetic study suggests that the Polyplacophora form a clade with a monophyletic Aplacophora. Additionally, it suggests a sister taxon relationship exists between the Bivalvia and the Gastropoda. Therefore, the authors have taken the liberty of using classification as suggested by different authors for different classes of mollusca. The classifications for Polyplacophora (Okusu *et al.*, 2003), the Scaphopoda as suggested by Steiner and Dreyer (2003), Gastropoda and Bivalvia by Vaught (1989) and Cephalopoda by Nixon and Young (2003) were followed.

CLASS POLYPLACOPHORA

Globally, the class Polyplacophora is represented by 1 order, 3 suborders, 9 families and 49 genera; of these, 21 species belonging to 10 genera under 5 families within 2 suborders and 1 order are found in India, as summarized below.

Ischnochitonidae. This family is further divided into 6 subfamilies, which include 19 genera and 15 subgenera; of these, 3 subfamilies, 4 genera, 2 subgenera and 5 species are represented from the Indian subcontinent.

Mopaliidae. This family includes 6 genera and 2 subgenera, of which 1 genus and 1 species is represented from the Indian subcontinent.

Chitonidae. This family is represented by 3 subfamilies, 7 genera and 5 subgenera, of which 3 subfamilies, 4 genera and 9 species are represented from Indian waters.

Cryptoplacidae. This family has a single genus, *Cryptoplax*, and 9 species, including 1 from India.

Acanthochitonidae. This family is subdivided into 2 subfamilies which include 6 genera and 2 subgenera, of which 1 subfamily, 2 genera and 3 species are represented from the Indian subcontinent.

CLASS GASTROPODA

Class Gastropoda or snails are torted or detorted, unsegmented, asymmetrical molluscs, typically provided with a spirally coiled univalve shell, with a distinct head, which has unspecialized forms, has eyes and tentacles and is more or less fused with the foot. The cerebral and pleural ganglion is distinct. Organs of pallial complex are re-oriented in relation to their positions in conjectural primitive molluscs due to torsion, which in some forms is a definite episode to observe in early ontogeny and in others is inferred to have taken on ancestor forms. The shell if present is calcareous, univalve, closed apically, endogastric when spiral and not divided regularly in chambers. Gastropods, except for a few, are more or less symmetrical, but asymmetry is obvious in most forms in which the shell is coiled. In *Patella* and other similar genera, the conical shell is symmetrical, but there is asymmetry in the digestive, nervous, excretory and reproductive organs. In *Fissurella* the shell is symmetric in the adult, but it is coiled and asymmetrical in the early developmental stages. In Pteropods, there are several genera with bilaterally symmetrical shells, but their digestive, circulatory and reproductive organs lack symmetry. The foot is sole like and adapted for creeping, but this is not strictly applicable even to all benthonic gastropods. In Pteropods and Heteropods, the foot is much modified for purposes of propulsion. In *Stilifer* (a parasitic genus) the foot is much reduced. The foot has completely atrophied in the endoparasitic family Entoconchidae. A radula is present in the great majority of gastropods but is obsolete in parasitic families such as Eulimidae and Pyramidellidae. The tentacles are absent from the head in families like Gadiniidae and Siphonariidae, while eyes are absent in most of Cocculinacea (deep sea forms).

Order Archaeogastropoda

Archaeogatropods are lower prosobranchs, with two auricles, two bipectinate internal gills, two osphradia and two nephridia, but there are some families in which one auricle, gill, osphradium and nephridium is reduced or wanting. The members of this order lack proboscis, siphon, operculum (with some exceptions), penis and prostatic gland. The nervous system is not highly concentrated and the pedal ganglia are drawn out into the pedal cords and numerous cross connections. The gonad discharges through the right nephridia directly into the sea. Archaeogatropods are herbivorous molluscs. They are familiar as intertidal species and abundant in intertidal regions. Some have limpet-like shells and lack an operculum, while others have top-like or spindle shaped shells with an operculum. Often the shell has a nacreous lining.

Superfamily Pleurotomarioidea Many of the members of this group are primitive among all living gastropods, have a slit, a hole or a series of holes in the shell. They are recognized by three families: Pleruotomariidae, Scissurellidae and Haliotidae, of which the Haliotidae are reported from Indian waters.

Haliotidae. This group is represented by a single genus, *Haliotis*, and 13 sub-genera, of which 3 subgenera—*Haliotis, Ovinotis, Sanhaliotis*—and 8 species are found in Indian waters and the occurrence of two species is doubtful.

Superfamily Fissurelloidea The superfamily is represented by only one family, Fissurellidae, which is also represented in Indian waters.

Fissurellidae. This family is subdivided into 4 subfamilies, which include 30 genera and 19 subgenera, of which 4 subfamilies, 8 genera, 2 subgenera and 30 species are represented from the Indian subcontinent.

Superfamily Patelloidea Patelloidea are the true limpets, bilaterally symmetrical, lacking the orifice or marginal slit on the shell possessed by other limpet-like groups, and having a horseshoe shaped columellar muscle scar. The radula is docoglossate type; the rectum lacks a pericardial cavity; the epipodium is lacking; there are two nephridia, the left smaller than the right. The animals browse on algae. This superfamily contains four families: Acmaeidae, Lottiidae, Patellidae and Lepetidae, which are common on rocky shores.

Acmaeidae. This family is subdivided into 2 subfamilies which include 2 genera, of which 1 subfamily, 1 genus and 4 species are represented from the Indian subcontinent. Subfamily: Acmaeinae is represented from Indian waters.

Lottiidae. This family includes 3 subfamilies and 9 genera, of which 3 species under 2 genera and 1 subfamily are from Indian waters.

Patellidae. This family has 2 subfamilies, 4 genera and 13 subgenera, of which 2 subfamilies, 2 genera, 1 subgenus and 3 species are represented from India.

Superfamily Lepetelloidea This is a small limpet type with the subcentral apex directed backwards. Eyeless snails dwell primarily in deep waters; single gills present in the nuchal region, radula rhipidoglossate type or degraded forms by the loss of teeth.

Cocculinellidae. This family has 6 genera, of which 2 genera and 2 species are represented from India.

Superfamily Trochoidea This superfamily is recognized as having eight families: Trochidae, Tegulidae, Pelycidiidae, Stomatellidae, Skeneidae, Cyclostrematidae, Turbinidae, Phasianellidae and Tricoliidae, of which all but Pelycidiidae are represented from India.

Trochidae. This family has 10 subfamilies, 66 genera and 107 subgenera, of which 7 subfamilies, 14 genera, 10 subgenera and 58 species are represented from India.

Cyclostrematidae. This family has 9 genera and 4 subgenera, of which 1 genus and 3 species are represented from India.

Turbinidae. This family contains 6 subfamilies, 23 genera and 33 subgenera, of which 5 subfamilies, 6 genera, 3 subgenera and 27 species are represented from India.

Phasianellidae. This small family has a single genus which is also represented from India.

Tricolidae. This small family has a single genus and 4 subgenera, represented by a single species from India.

Superfamily Neritoidea

Neritidae. This family contains 3 subfamilies, 11 genera and 27 subgenera, of which 1 subfamily with 4 genera, 6 subgenera and 26 species is represented from India.

Septariidae. This family includes a single genus and 3 subgenera, represented by 1 genus and 2 species from India.

Phenaciolepadidae. This small family contains 2 genera and 3 subgenera, of which only 1 genus and 2 species are represented from India.

Order Mesogastropoda
Superfamily Littorinoidea

Littorinidae. This family has 5 subfamilies, 23 genera and 14 subgenera, of which 3 subfamilies with 5 genera, 5 subgenera and 14 species are represented from India.

Superfamily Rissoidea

Stenothyridae. This family has 2 genera and 1 subgenus, of which 2 genera and 7 species are represented from India.

Assimineidae. This family contains 5 subfamilies, 23 genera and 14 subgenera, of which 3 subfamilies with 5 genera, 5 subgenera and 15 species are represented from India.

Superfamily Cerithioidea

Planaxidae. This family has 6 genera, of which 2 genera, 1 subgenus and 6 species are represented from India.

Modulidae. This small family has a single genus which is also represented from India.

Cerithiidae. This family includes 2 subfamilies, 21 genera and 9 subgenera, of which 1 subfamily with 5 genera, 3 subgenera and 31 species is represented from India.

Diastomatidae. This small family contains 7 genera and 3 subgenera, of which 1 genus and 1 species is represented from India.

Potamididae. This family has 2 subfamilies, 8 genera and 5 subgenera, of which 2 subfamilies with 4 genera, 2 subgenera and 8 species are represented from India.

Fossaridae. This family contains 6 genera and 5 subgenera, of which 2 genera, 1 subgenus and 2 species are represented from India.

Turritellidae. This family includes 5 subfamilies, 18 genera and 11 subgenera, of which 3 subfamilies with 3 genera and 13 species are represented from India.

Siliquariidae. This monogeneric family has a single genus and 2 subgenera, of which 5 species are represented from India.

Superfamily Stromboidea

Strombidae. This family contains 6 genera and 15 subgenera, of which 5 genera, 12 subgenera and 29 species are represented from India.

Superfamily Hipponicoidea

Hipponicidae. This family contains 6 genera and 1 subgenus, of which 2 genera and 3 species are represented from India.

Vanikoridae. This small family has a single genus of which 5 species are represented from India.

Superfamily Crepiduloidea
Like the superfamily Hipponicoidea the members of this family have limpet-like shells, but they have a spirally coiled apex and internally some of them have a ledge-like or cup-shaped process as the septum. Operculum is absent.

Crepidulidae. This family contains 6 genera and 5 subgenera, of which 3 genera and 5 species are represented from India.

Capulidae. This family has 3 genera and 3 subgenera, of which 1 genus and 4 species are represented from India.

Superfamily Xenophoroidea

Xenophoridae. This family has a single genus and 2 subgenera, of which both subgenera and 7 species are represented from India.

Superfamily Cypraeoidea

Cypraeidae. This family has a single genus and 53 subgenera, of which one subgenus and 51 species are represented from India.

Ovulidae. This family has a single subfamily, 39 genera and 10 subgenera, of which 6 genera, 2 subgenera and 13 species are represented from India.

Triviidae. This family comprises 280 species approximately, under 2 subfamilies, 39 genera and 10 subgenera, of which 6 species, under 2 subfamilies, 4 genera and 4 subgenera are recorded from India.

Lamellariidae. This family includes 2 subfamilies, 9 genera and 6 subgenera, of which a single subfamily, genus, subgenus and species is represented from Indian waters.

Superfamily Naticoidea
Naticidae. This family comprises 5 subfamilies, 20 genera and 24 subgenera, of which 3 subfamilies, 5 genera, 8 subgenera and 52 species are represented from India.

Superfamily Tonnoidea
Tonnidae. This family has 3 genera and 1 subgenus, of which 2 genera and 9 species are represented from India.

Ficidae. This family contains 2 genera, of which 1 genus and 4 species are represented from India.

Cassidae. This family includes 3 subfamilies, 12 genera and 5 subgenera and about 60 species, of which 2 subfamilies, 6 genera, 1 subgenus and 11 species are represented from India.

Ranellidae. This family contains 3 subfamilies, 12 genera and 5 subgenera, of which 3 subfamilies, 5 genera, 7 subgenera and 37 species are represented from India.

Bursidae. This family has 4 genera and 3 subgenera, of which 3 genera, 2 subgenera and 11species are represented from India.

Superfamily Epitonioidea
Epitoniidae. This family contains 3 subfamilies, 65 genera and 24 subgenera, of which 1 subfamily, 4 genera, 2 subgenera and 19 species are represented from India.

Janthinidae. This family has 2 genera and 5 subgenera, of which 2 genera and 3 species are represented from India.

Superfamily Eulimoidea
Eulimidae. This family comprises 66 genera and 4 subgenera, of which 4 genera and 11 species are represented from India.

Stiliferidae. This family contains 12 genera, of which a single genus with a single species is represented from India.

Aclididae. This small family comprises a single genus *Acilis* represented by 2 species from India.

Order Neogastropoda
Superfamily Muricoidea
Muricidae. The subfamily Thaidinae comprises the common rock shells. The family contains 8 subfamilies, 97 genera and 48 subgenera, of which 7 subfamilies with 25 genera, 8 subgenera and 89 species are represented from India.

Buccinidae. This family has 4 subfamilies, 93 genera and 53 subgenera, of which 2 subfamilies with 8 genera, 1 subgenus and 28 species are represented from India.

Columbellidae. This family includes 2 subfamilies, 37 genera and 36 subgenera, of which 2 subfamilies with 4 genera and 24 species are represented from India.

Nassariidae. This family has 3 subfamilies, 12 genera and 27 subgenera, of which 3 subfamilies with 4 genera, 10 subgenera and 72 species are represented from India.

Melongenidae. This family contains 6 genera and 7 subgenera, of which 2 genera, 1 subgenus and 4 species are represented from India.

Fasciolariidae. This family includes 4 subfamilies, 32 genera and 18 subgenera, of which 4 subfamilies with 6 genera, 1 subgenus and 26 species are represented from India.

Volutidae. This family has 9 subfamilies, 467 genera and 23 subgenera, of which 3 subfamilies with 3 genera and 10 species are represented from India.

Harpidae. This family has 2 subfamilies, 3 genera and 2 subgenera, of which 1 subfamily with 1 genus and 5 species are represented from India.

Vasidae. This family contains 4 subfamilies, 14 genera and 9 subgenera, of which 2 subfamilies with 3 genera, 1 subgenus and 4 species are represented from India.

Olividae. This family includes 4 subfamilies, 16 genera and 50 subgenera, of which 4 subfamilies with 5 genera and 33 species are represented from India.

Marginellidae. This family comprises 3 subfamilies, 32 genera and 30 subgenera, of which 1 subfamily with 2 genera and 10 species are represented from India.

Mitridae. This family has 3 subfamilies, 9 genera and 8 subgenera, of which 3 subfamilies with 7 genera, 6 subgenera and 56 species are represented from India.

Costellariidae. This family contains 7 genera and 2 subgenera, of which 2 genera, 1 subgenus and 43 species are represented from India.

Superfamily Cancellarioidea

Cancellariidae. This family has 2 subfamilies, 39 genera and 13 subgenera, of which 1 subfamily, 3 genera and 10 species are represented from India.

Superfamily Conoidea

Conidae. This family has 1 genus and 44 subgenera globally, of which 82 species are represented from India.

Turridae. This family comprises 15 subfamilies, 229 genera and 70 subgenera, of which 7 subfamilies with 26 genera, 2 subgenera and 68 species are represented from India.

Terebridae. This family has 8 genera and 20 subgenera, of which 5 genera and 44 species are represented from India.

Subclass Heterobranchia

Order Allogastropoda

Superfamily Architectonicoidea

Architectonicidae. This family comprises 11 genera and 9 subgenera, of which 3 genera and 9 species are represented from India.

Superfamily Pyramidelloidea

Pyramidellidae. The family includes 4 subfamilies, 47 genera and 95 subgenera, of which 3 subfamilies with 7 genera, 3 subgenera and 29 species are represented from India.

Amathinidae. This family contains 7 genera, of which 1 genus and 5 species are represented from India.

Subclass Opisthobranchia

Shell is absent, or external or internal if present; ovate to cylindrical in shape with a broad aperture and shortened whorls. In the family Julidae the shell has two valves. Operculum presents in the juvenile but is absent in the adult. The head has a cephalic shield; mantle cavity is reduced. A bipectinate ctenidium is present in some cases, but in many it is replaced by secondary respiratory structures in the form of dorsal cerata, anal tufts or anal leaflets, etc. Nervous system is primitive streptoneurous, but secondarily euthyneurous in advanced forms.

Opisthobranchs feed on a variety of organisms. Most of them are highly specialized herbivores or carnivores, and many restrict their diet to a single genus or species of plant or animals. Some species such as *Philine aperta* graze on algae or microscopic animals. Most opisthobranchs feed using the jaws and radula. The configuration and structure of the radula varies greatly in the opisthobranchs. The radula is broad in the case of grazing herbivores like *Aplysia* and *Chromodoris*, which rasp the tissue from sponges. In *Elysia* there is a single tooth per row of radular teeth. To supplement their diet, sacoglossan retain the chloroplast in intracellular organelles which are responsible for photosynthesis. These chloroplasts remain active in the translucent skin of the digestive branches and continue to produce sugars which are consumed by the animal. Sexes are united with internal fertilization. Opisthobranchs require copulation between two individuals, and the sperm is stored in a seminal receptacle for several months. When the eggs mature they pass the seminal receptacle and are fertilized. The fertilized eggs are encapsulated by a membrane, with generally a single egg per capsule but in some cases more: up to 45 eggs per capsule have been recorded. This egg mass is covered by a thick layer of mucus when it passes through the female reproductive opening, linking them into a single matrix. The egg mass is then deposited on the substratum. Within the egg mass the embryos develop and hatch into larvae. Opisthobranchs are found exclusively in the marine environment. Included are 9 orders, 15 superfamilies and 115 families.

Cephalaspidea

This includes 2 superfamilies: Philinoidea, with 13 families, and Runcinoidea with 2 families.

Acteonidae. This comprises 16 genera and 2 subgenera, of which 2 genera and 4 species are recorded from India.

Ringiculidae. This small neglected monogeneric family suck in tiny creatures through their siphon while gliding through muddy sand.

Hydatinidae. This is a small family

Bullinidae. This small monogeneric family contains 3 species recorded in India, whose status is undetermined.

Scaphandridae. This includes 12 genera and 10 subgenera, of which 4 genera and 13 species are recorded from India.

Philinidae. This is a monogeneric family with a cosmopolitan distribution. It is represented by 2 species from India and one species whose status remains undetermined.

Aglajidae. This comprises 6 genera and one subgenus, of which 4 genera and 9 species are recorded from India. The occurrence of one species, *Aglaja tricolorata* Renier, from India is doubtful.

Gastropteridae. This small family contains 2 genera, represented by single species from India.

Bullidae. This is a monogeneric family represented by one species from India.

Hamineidae. This family comprises 2 subfamilies, 18 genera and 1 subgenus, of which 1 subfamily, 3 genera and 18 species are recorded from India. The occurrence of one species, *Haminoea ovalis* Pease, from India is doubtful.

Smaragdinellidae. Both genera, with 1 species each, are recorded from India.

Retusidae. This includes 4 genera and 5 subgenera, of which 2 genera, one subgenus and 4 species are recorded from India.

Order Saccoglossa

Herbivorous slugs possess special feeding structures to pierce and suck out the contents of algal cells. Many retain the chloroplast in their bodies, using it to synthesize food by photosynthesis. These slugs vary in size from 2 to 50 mm. Shell when present is usually univalve and rarely bivalve. The univalve shell has a large aperture. The animal is slug-like and small: up to 10 mm in length. The head bears paired rhinophores or lobes. The radula is single blade-like with a seriated central tooth. Sexes are united. Usually they are associated with algae.

Cylindrobullidae. This is a small monogeneric family with a subgenus, represented by a subgenus and 1 species from India.

Volvatellidae. This is a small monogeneric family represented by single species from India.

Julidae. This family contains 2 subfamilies and 5 genera, of which 2 subfamilies, 3 genera and 3 species are recorded from India.

Caliphyliidae. This comprises 7 genera, of which one genus and one species are recorded from India.

Stiligeridae. This has 2 subfamilies, 11 genera and 1 subgenus, of which 1 genus and 1 species is recorded from India.

Elysiidae. This includes 4 genera and 2 subgenera, of which 2 genera and 13 species are recorded from India. Among these, one species, *Elysia abei* Baba, is an erroneous identification.

Order Aplysiomorpha

The animal is medium to large in size with a small corneous internal shell covered by mantle. Cephalic shield is usually absent. The head has two pairs of tentacles, one anterior on the marginal ridges and the other posterior pair known as rhinophores behind the eyes. The foot is well developed lateral parapodia directed upwards. Radula is multiseriate with a central tooth and numerous lateral teeth. Sexes are united. They are commonly known as sea hares and herbivorous in habit. Represented by 1 superfamily Aplysioidea and 2 families Aplysidae and Notarchidae; both are represented from India.

Suborder Aplysiodiea

Aplysiidae. This comprises 3 subfamilies, 6 genera and 5 subgenera, of which 3 subfamilies, 5 genera, one subgenus and 12 species are recorded from India. One species, *Aplysia rudmani* Bebbington, remains undetermined.

Order Notaspidea

The animal is medium sized to moderately large with an ovately elongated body and with external, internal, or without shell. Operculum is present in the juvenile and absent in the adult. Radula is with or without a central tooth, but with numerous lateral teeth. Gills are situated on the right side of the body. Most of them are carnivorous and use their strong jaw and broad radula to scrape tissue from the sponges or tunicates. Sexes are united. Epifaunal and exclusively marine.

Superfamily Umbraculoidea

Umbraculidae. This is a monogeneric family widely distributed in warm tropical seas.

Superfamily Pleurobranchoidea

Pleurobranchaeidae. The animals have very soft bodies with a well developed oral veil and moderately developed cylindrical oral tentacles; rhinophores are usually large and deeply cleft longitudinally. The back is covered with thick or thin mantle; projected freely over the head anteriorly and over the tail posteriorly. Most of the species have a more or less large internal shell, present within mantle. Gill is tripectinate, feather like (size 50 to 150 mm) on the right side of the body between the mantle and foot, and below it lies the urinary aperture. Foot is moderately large and overhung by the mantle; the sole of the tail region of the foot often bears a gland. A single genital aperture lies at the front end of the right side of the body; above or below of this, but still anterior to the gill, a probranchial opening or papilla leads into a sac. Many of them secrete chemicals to protect themselves from predators.

Order Thecosomata

The shell is thin, fragile and transparent, needle shaped, cigar shaped, or tubular, bilaterally symmetrical without an operculum or spirally coiled with an operculum. Foot is modified as fins or epipodia, which are separate and dorsal to the mouth. Mantle cavity and osphridium are present. The radula when present has a central and a pair of lateral teeth. Hermaphrodites have separate testis and ovary. Pteropods are commonly known as sea-butterflies. Pelagic, they occur on open seas and are cosmopolitan in distribution. They feed on minute organisms. A test, using aragonite, for their presence in sea sediment is helpful in palaeoclimatic interpretation.

Suborder Euthecosmata

Pteropods are divided into two groups: one shelled and the other naked. The shelled forms are further subdivided into Euthecosomata and Pesudothecosomata. The Euthecosomata comprise 2 families: Limacinidae and Cavoliniidae, both of which are represented from India.

Limacinidae. This has a single genus and 2 subgenera, of which 3 species are represented from India.

Cavoliniidae. This includes 3 subfamilies and 8 genera, of which 2 subfamilies, 6 genera and 15 species are recorded from India.

Suborder Pseudothecosomata

The suborder Pseudothecosomata is recognized by the snout or proboscis formed by the fusion of median parts of the foot around the mouth and concomitantly by the fusion of two parapodia below the mouth. Mantle cavity is generally dorsal except for the genus *Peraclis*, where it is dorsal. The suborder comprises 3 families: Peraclidae, Cymbuliidae and Desmopteridae, of which only Peraclidae is represented from India.

Peraclidae. This family has a calcareous spiral shell, an operculum and a gill. It includes 2 genera, of which 1 genus and 1 species is recorded from India.

Order Nudibranchia

Nudibranchs are marine molluscs, devoid of shell, mantle cavity, osphradium and internal gill; visceral mass, mantle and foot are incorporated into a single body, which is bilaterally symmetrical externally and bears cerata or other outgrowths for respiratory purpose. They exhibit an incredible range of external form. The order includes 4 suborders, 7 superfamilies and 66 families, 33 of which are recorded from India.

Suborder Doridoidea

The doridoidean nudibranchs are dorsoventrally flattened with the edge of the back (notum) projecting beyond the foot, with prominent rhinophores, anus middorsal partly or wholly encircled with feathery external gills, midgut gland unbranched and compact and mesenchymal spicules. Represented by four superfamilies: Gmathodoridoidea, Anadoridoidea, Eudoridoidea and Phyllidioidea.

Superfamily Anadoridoidea

Goniodorididae. This family contains 3 genera, of which 1 genus and 3 species are recorded from India.

Okeniidae. This family comprises 2 genera, of which 1 genus and 1 species is recorded from India.

Triophidae. This family has 2 subfamilies, 7 genera and 1 subgenus, of which 1 subfamily, 1 genus and 1 species is recorded from India.

Notodorididae. This family contains 4 genera, of which 1 genus and 2 species are recorded from India.

Polyceridae. This family includes 11 genera, of which 1 genus and 2 species are recorded from India.

Gymnodorididae. This family has 8 genera and 1 subgenus, of which 3 genera and 8 species are recorded from India, but the Indian occurrence of 3 species is doubtful.

Superfamily Eudoridoidea The animals have rhinophoral sheaths into which the rhinophores can retract. The notal edge is reduced, indicated by a row of cerata, ranging from simple to greatly branched sometimes accompanied with gills. Cnidosac is present or absent. Midgut gland is compact and unbranched or highly branched.

Hexaabranchidae. This monogeneric family is represented by 1 species from India.

Cadlinidae. Contains 2 genera, of which 1 genus and 1 species is recorded from India.

Chromodorididae. Comprises 18 genera, of which 10 genera and 39 species are recorded from India. Out of these, the occurrence of 8 species from India is doubtful.

Actinocyclidae. Includes 2 genera, of which 1 genus and 1 species is recorded from India.

Aldisidae. Includes 2 genera, of which 1 genus and 1 species is recorded from India.

Dorididae. Comprises 2 subfamilies and 10 genera, of which 1 subfamily, 2 genera and 4 species are recorded from India.

Archidorididae. Includes 7 genera, of which 3 genera and 4 species are recorded from India, out of which the occurrence of 2 species from India is doubtful.

Discodorididae. Has 3 subfamilies and 17 genera, of which 2 subfamilies, 5 genera and 9 species are recorded from India. Occurrence of 1 species from India is doubtful.

Kentrodorididae. Includes 3 genera, of which 2 genera and 2 species are recorded from India.

Halgerdidae. Comprises 4 genera, of which 3 genera and 7 species are recorded from India.

Platydorididae. Has 2 subfamilies, 5 genera, of which 1 subfamily, 1 genus and 4 species are recorded from India.

Superfamily Phyllidioidea

Phyllidiidae. Includes 4 genera and a subgenus, of which 2 genera, 1 subgenus and 16 species are recorded from India, of which the occurrence of 6 species from India is doubtful.

Dendrodorididae. Consists of 2 genera, of which both genera and 11 species are recorded from India.

Suborder Dendronotoidea

Contains the superfamily Dendronotoidea, of which 5 families are represented from India.

Superfamily Dendronotoidea Comprises the families Tritoniidae, Marianinidae, Lohanotidae, Scyllaeidae, Hancockiidae, Dendronotidae, Boronellidae, Tethyidae, Dotidae and Phylliroidae.

Tritoniidae. Includes 5 genera and 4 subgenera, of which 2 genera and 2 species are recorded from India.

Scyllaeidae. Contains 3 genera, of which 2 genera and 2 species are recorded from India.

Hancockiidae. Monogeneric family of which 1 species is recorded from India.

Bornellidae. Has 3 genera, of which 1 genus and 3 species are recorded from India.

Tethyidae. Includes 2 genera and a subgenus, of which 1 genus and 3 species are recorded from India, out of which the occurrence of 1 species from India is doubtful. Besides these 3 recorded species the status of 1 species remains undetermined.

Suborder Arminoidea

This contains 2 superfamilies: Arminoidea and Metarminoidea, the former with 3 families and the latter with 6 families, of which 1 family from each superfamily is recorded from India.

Superfamily Arminoidea This is a heterogeneous group containing forms that resemble doridoideans and other eolidadeans. Rhinophores are retractile or non-retractile; sheaths are lacking, frontal sail is present, but cephalic tentacles are absent. Comprises three families: Heterodoridae, Arminidae and Leminoidae.

Arminidae. Includes 5 genera and 2 subgenera, of which 4 genera, 1 subgenus and 8 species are recorded from India; out of these the status of 2 species remains undetermined.

Madrellidae. Has 2 genera, of which 1 genus and 1 species is recorded from India

Suborder Aeolidoidea

This is the second largest group of nudibranchs, with well developed cerata. Most of them store pneumatocysts in the cerebral tips. The stinging cells are collected from their cnidarian prey and remain in their bodies, used for defensive purposes. Most aeolids have a single radular tooth per row. In the case of Eubranchus and Flabellina the lateral teeth are a pair on either side of the central or rachidian tooth. They have a pair of strong chitinous jaws. Most aeolids feed on the hydroids, while some of them feed on other opisthobranchs or their eggs, bryozoans, sea anemones, corals or gorgonians. The group contains 1 superfamily and 17 families, of which 7 families are recorded from India.

Superfamily Euaeolidioidea This comprises the families Notaeolididae, Coryphellidae, Paracoryphellidae, Nossidae, Flabellinidae, Protaeolidieludae, Pleurolidiidae, Eubranchidae, Pseudovermidae, Tergipedidae, Fionidae, Calamidae, Glaucidae, Facelinidae, Aeolidiidae, Hybrininidae and Spurilidae, of which Flabellinidae, Eubranchidae, Pseudovermidae, Tergipedidae, Glaucidae, Facelinidae and Aeolidiidae are recorded from India.

Flabellinidae. Includes 5 genera, of which 1 genus and 4 species are recorded from India.

Eubranchidae. Contains 2 subfamilies and 3 genera, of which 1 subfamily, 1 genus and 5 species are recorded from India.

Pseudovermidae. This monogeneric family has 2 species recorded from India.

Tergipedidae. Includes 7 genera, of which 2 genera and 5 species are recorded from India.

Glaucidae. Contains 7 genera, of which 2 genera and 5 species are recorded from India.

Facelinidae. Comprises 5 subfamilies, 27 genera and 1 subgenus, of which 4 subfamilies, 5 genera and 8 species are recorded from India.

Aeolidiidae. Has 8 genera and 1 subgenus, of which 2 genera and 4 species are recorded from India, out of which the occurrence of 2 species from India is doubtful.

Subclass Gynomorpha

Order Systellommatophora

Superfamily Onchidioidea

Onchidiidae. Contains 12 genera, of which 1 genus and 5 species are recorded from India, the status of 1 species of which remains undetermined.

Subclass Pulmonata

Order Archaeopulmonata

Superfamily Ellobioidea

Ellobiidae. Comprises 3 subfamilies, 21 genera and 9 subgenera, of which 3 subfamilies, 5 genera and 18 species are recorded from India.

Order Basommatophora

Superfamily Siphonarioidea

Siphonariidae. This family has 4 genera and 11 subgenera, of which 1 genus and 8 species are recorded from India, of which the occurrence of 1 species, *Siphonaria lineolata* Sowerby, is doubtful.

Superfamily Amphiboloidea

Amphibolidae. Includes 2 genera, of which 1 genus and 1 species is recorded from India.

CLASS BIVALVIA

The Bivalvia are aquatic molluscs with bilateral symmetry and lateral compression. They are elongated anterio-posteriorly and invariably provided with shell, two wholly or partly calcified valves lying on the left and right side of the body (Ramakrishna and Dey, 2010). Diversity of bivalves is not enumerated comprehensively as per the available literature; 652 species of marine bivalves are recorded from the Indian subcontinent, of which 173 genera, 69 families, 11 orders, 4 subclasses and 88 species are endemic to India.

Superfamily Solemyoidea This superfamily is represented by a single family, 1 genus and 2 species also present in Indian waters. Most of the shells are oblong and distinct, with a broad fringe of shiny periostracum. Beaks are closer to the posterior margin. Hinge teeth are absent with prominent ligament.

Superfamily Nuculoidea Shell is trigonal, subovate and small; teeth are taxodont, typically V shaped in section, long, pointed and sharp. The internal margin may be serrated or smooth. Represented by 2 families of which 1 family and 8 species are recorded from India.

Superfamily Nuculanoidea Includes 3 families, 3 genera and 10 species: family Naculanidae with 7 species and 1 genus; family Yoldiidae with 1 genus and 2 species; and family Malletiidae with 1 genus and 1 species recorded from India.

Superfamily Arcoidea

This superfamily contains 4 families, of which Arcidae, Neotiidae and Cuculacidae are present in India. Arcidae is characterized by a distinctive chevron pattern of obliquely aligned ligament bands. Arcidae is subdivided into 2 subfamilies on the basis of 2 distinct ecological groups. This superfamily includes 37 species and 5 genera. The Neotiidae is a less diverse group with vertically aligned bands; it comprises 5 species and 3 genera. The Cuculacidae have the chevron ligament, but the hinge teeth are sub-parallel and the aductor muscles are attached to elevated buttresses; the family comprises 2 species and 1 genus.

Superfamily Limopsoidea Limopsoidea has two distinct families: Limopsidae containing 1 genus and 3 species; and Glycymerididae containing 1 genus and 5 species from India. These groups are poor burrowers, usually living in mobile, often shallow water within fairly coarse sand and gravel.

Superfamily Mytiloidea Commonly called mussels, these have a thin shell, smooth, elongate and anteriorly beaked; ribs are weak in some species. Included are a single family, 10 genera and 32 species, of which 5 species are endemic to India: 3 from the Andaman and Nicobar Islands, 1 from Lakshadweep and 1 from the west coast.

Superfamily Pterioidea This superfamily includes the pearl, wing and hammer oysters found attached with byssal thread. The shell has a deep anterior notch, with the beaks close to anterior. Sculpture shows scaly growths, lined internally with mother of pearl. This superfamily contains three families: Pteriidae, Malleidae and Isognomonidae. Pteridae, the wing oysters, comprises 12 species and 3 genera. Family Malleidae includes 5 species and 2 genera, and Isognomonidae contains 5 species, 3 varieties and one genus.

Superfamily Pinnoidea The animals in this group are fan shaped. The fan shells are large, thin and brittle bivalves, triangular in shape, with a hinge along the straight dorsal margin. Hinge teeth are absent. Pinnoidea are semi-infaunal in habit, living embedded in sand and muddy substratum, attached with byssus threads to the sediment particles. The superfamily comprises a single family, 2 genera and 7 species.

Superfamily Limoidea This group is known as file shells, and they resemble the scallops in having shells with small ears at the side of the hinge. This superfamily is represented by a single family, Limidae, containing 3 genera and 8 species.

Superfamily Ostreoidea Representative animals are known as true oysters, epifaunal in habit, living cemented with the substratum by the left valve. Shell shape varies, often influenced by and conforming to the substratum. Two families, Ostreidae and Gryphaeidae, are recorded from Indian waters, Ostreidae having 4 genera and 11 species and Gryphaeida having a single genus and single species.

Superfamily Plicatuloidea The superfamily Plicatuloidea is represented by a single family, Picatulidae, containing a single genus and 2 species from Indian waters. The animals are cemented and possess an isodont hinge. They have much smaller shells, sculptured with divaricating, smooth or imbricate radial ribs. The attachment area ranges widely from being restricted to the umbo to involving the whole of the cemented valve.

Superfamily Pectinoidea Comprises 4 families, of which 3, Pectinidae, Propeamusiidae and Spondylidae (thorny oysters), are recorded from Indian waters. Family Pectinidae are free living or attached; 31 species within 13 genera are reported from Indian waters. Family Propeamussidae includes 11 species within 2 genera, and family Spondylidae has 6 species from a single genus.

Superfamily Anomioidea This superfamily include 2 families, Anomiidae and Placunidae, and both are present in Indian waters. Family Anomiidae has 2 genera and 4 species, and family Placunidae has a single genus and a single species. The famous windowpane oyster is included in this family.

Subclass Heterodonta

This subclass comprises 2 orders, Veneroidea and Myoidea, containing 18 superfamilies, 33 families, 97 genera and 411 species. Veneroida is the largest order of bivalvia, comprising 15 superfamilies, 26 families, 87 genera and 250 species.

Superfamily Lucinoidea The superfamily Lucinoidea contains 3 families, 9 genera and 22 species, of which10 species are endemic to India: 3 from Andaman and 7 from the east coast of India. Among the other 12 species, 4 are represented from the east coast and Andaman and Nicobar islands (ANI), 2 from the east coast, west coast and ANI, 1 from Lakshadweep and ANI, and 1 from ANI, Lakshadweep and both coasts.

Superfamily Galeommatoidea Contains 3 families, 4 genera and 15 species, among which 8 species are endemic to India. Out of the remaining 7 species, 4 are represented from the east coast, 2 from ANI, and 1 from both coasts. Among the 8 endemic species, 6 are from ANI and 2 from the east coast of India.

Superfamily Carditoidea This superfamily comprises 1 family, 3 genera and 7 species, of which one species is endemic to ANI. Of the other 6 species, 2 are represented from ANI, 1 from the east coast, 1 common to both coasts, 1 from both coasts and ANI, and 1 from both coasts, Lakshadweep and ANI.

Superfamily *Chamoidea* This superfamily has 1 family, 2 genera and 6 species. Among these 6 species all are found in ANI, 2 at Lakshadweep, 3 on the east coast and 2 on the west coast of India.

Superfamily *Crassatelloidea* Contains a single family, 3 genera and 3 species. All 3 species are endemic to India: 1 from the east coast, 1 from the west coast, and 1 from both coasts, Lakshadweep and ANI.

Superfamily *Cardioidea* This superfamily has a single family, 8 genera and 25 species, among which 1 species is endemic to ANI. Among the other 24 species, 10 are reported from ANI, 5 each from the east coast and west coast, 3 from Lakshadweep, 2 are common to both coasts; 2 from ANI as well as the east coast, 1 from ANI and Lakshadweep, 1 from east and west coasts, and 1 from east coast, west coast, ANI and Lakshadweep.

Superfamily *Tridacnoidea* Contains a single family, 2 genera and 4 species, of which 3 are recorded in ANI and 1 from the east and west coasts as well as ANI.

Superfamily *Mactroidea* Contains 2 families, 8 genera and 32 species, among which 3 species are endemic to India: 2 from ANI and 1 from the east coast. Among the other 32 species, 6 are recorded from the west coast, 6 from ANI, 5 from the east and west coasts, 5 from ANI as well as the east and west coasts, 4 from the east coast, 4 from east coast and ANI, 1 from west coast and ANI, and 1 from east, west, ANI and Lakshadweep islands.

Superfamily *Solenoidea* The superfamily Solenoidea has 2 families, 7 genera and 19 species, of which 2 species are endemic to the east coast of India. Among the other 17 species, 7 are reported from the east coast of India, 5 from both the coast and ANI, 4 from both coasts, 1 from ANI, and 1 from the east coast and ANI.

Superfamily *Tellinoidea* This is the largest superfamily of the bivalves, having 4 families, 19 genera and 112 species. Eleven species are endemic to India: 6 from ANI, 4 from the east coast of India, and 1 from the west coast. Among the other 101 species, 28 are reported from ANI, 13 from the west coast, 15 from the east coast, 10 common to both coasts, and 15 from ANI and both coasts. They are more frequently found on the east coast of India than on the west.

Superfamily *Dreissenoidea* This smallest superfamily has a single family and species from Andhra Pradesh, east coast of India.

Superfamily Arcticoidea The superfamily Arcticoidea has one family, 2 genera and 3 species, of which 2 species are reported from ANI and one from the east and west coasts as well as ANI.

Superfamily Glossoidea This superfamily has 2 families, 2 genera and 5 species, of which 2 species are endemic to India: one from the east coast and one from the west coast. Of the remaining 3 species, 1 is reported from the east coast, 1 from ANI, and 1 from both the coast and ANI.

Superfamily Corbiculoidea This superfamily has a single family, 2 genera and 2 species: 1 from ANI and one from the east coast as well as ANI.

Superfamily Veneroidea This superfamily is the second largest after the Tellinids, having 3 families 27 genera and 90 species reported from India. Among these, 15 species are endemic to India: 4 from ANI, 7 from the east coast, 4 from the west coast. Among the remaining 75 species, 17 are reported from the east coast, 6 from the west coast, 20 from ANI, 14 from both coasts, 17 from both coasts and ANI, 3 from ANI and the east coast, and 5 from both coasts, Lakshadweep and ANI. The greatest concentration is reported from the east coast of India.

Superfamily Myoidea This superfamily has 2 families, 3 genera and 13 species, of which 6 species are endemic to India: 3 from ANI and 2 from the east coast, 1 from the west coast. Among the other 7 species, the east coast of India is represented by 5 species, 1 species from the east coast, west coast and ANI, 1 species from the east coast and the west coast.

Superfamily Gastrochaenoidea These are known as flask shells. They are found within the cavity of corals, rocks and crevices; others build a case or tube round themselves. The end of the flask is very narrow and forms a figure of eight. The shell elongates, with a large ventral gape and without hinge teeth. The animals burrow by mechanical as well as chemical process. This superfamily has a single family, 1 genus and 5 species: 3 from the east coast of India, 1 from the east coast, west coast, ANI and Lakshadweep, and 1 from both coasts.

Superfamily Pholadoidea Comprises 3 families, 17 genera and 43 species, of which 2 are endemic to the east coast and 1 to ANI. Of the remaining 40 species, 1 is from Lakshadweep, 2 from ANI, 5 from the east coast, 1 from the west coast, 5 from both coasts, 6 from the east coast and ANI, 12 from both coasts and ANI, one from both coasts, ANI and Lakshadweep, 1 from the east coast and Lakshadweep, and 7 from both coasts and Lakshadweep.

Subclass Anomalodesmata

Superfamily Pandoroidea Pandoroidea comprises 5 families, 5 genera and 14 species, of which 2 species are endemic to India: one from ANI and one from the west coast.

Superfamily Clavagelloidea Members of this superfamily are unusual among the bivalves, looking like watering-pots. The superfamily is represented by 3 species from the east coast of India.

Superfamily Poromyoidea This superfamily comprises 3 families, 6 genera and 12 species, of which 9 are endemic to India: 5 from ANI, 1 from Lakshadweep, and 3 from the east and west coasts of India. Of the remaining 4 species, 2 are distributed in the east coast and 2 in ANI.

CLASS SCAPHOPODA

This is a small but distinctive class of the phylum Mollusca and poorly represented in Indian waters. Eighteen species are reported from India under 2 orders, 2 families and 2 genera. The order Dentalida is represented by 1 family, 1 genus and 17 species, and there is 1 species under the order Cadilida. The order Dentalida comprises family Dentalidae with 1 genus and 17 species, of which 4 species are endemic to India: 3 from ANI and 1 from the east coast. Of the remaining 13 species, 2 are reported from the east coast, 2 from the east coast and ANI, 2 from both coasts and ANI, 2 from the west coast, 2 from both coasts, 1 from the west coast and ANI, 1 from west coast and Lakshadweep, and 1 from ANI. The order Cadilida consists of the family Cadilidae having only 1 species under genus *Cadulus*, from the east coast and ANI only.

CLASS CEPHALOPODA

The class cephalopoda contains 650 species of nautilus, cuttlefish, squid and octopus. A total of 62 species belonging to 17 genera and 10 families of the class Cephalopoda are reported by various research team workers from the Zoological Survey of India, Kolkata. The documentation of cephalopods of the west coast of India by R.V. Varuna (Silas, 1968) with a catalogue of species known from the Indian Ocean is the only study on cephalopod resources of the west coast of India. It highlights the taxonomic confusion regarding many species recorded from India in the past, and there are still taxonomic confusions regarding many other species: for example, *Abralia lineata, Stigmatoteuthis japonica, (Doratosepion) andreanoides, Sepia esculenta* and *Aurosepina arabica.* All the living cephalopods belong to two subclasses—Nautiloidea and the Coleoidea—which occur in the Indian coastal waters.

Subclass Nautiloidea

Nautiloids are a large and diverse group of marine cephalopods belonging to the subclass Nautiloidea. They have an external shell (coiled and chambered), circumoral appendages without suckers, two pairs of gills and a bilobed funnel. The only living representatives of this subclass belong to a single family, Nautilidae and genus *Nautilus*.

Subclass Coleoidea

Subclass Coleoidea includes all living cephalopods other than chambered nautilus. Their shell is internal (except in the family Argonautidae), embedded in tissue, calcareous, chitinous or cartilaginous and has 8 or 10 circumoral appendages with suckers, one pair of gills, and a tube-like funnel. Coleoidea comprises 2 superorders, 6 orders, 46 families, 140 genera and about 800 species.

Order Sepiida

The families Sepiidae and Sepiadariidae come under the order Sepiida. This order contains the cuttlefishes, characterized by an oval, dorso-ventrally flattened body. They have calcareous internal shell (cuttlebone/sepion), either straight and laminated, or coiled and chambered, or vestigial and chitinous, or even absent. Fins are narrow and extend along the entire length of the mantle, fin lobes are free posteriorly. Other major features are eight sessile and two tentacular arms which are contractile and retractile into pockets on ventro-lateral sides of the head. The eyes of the animals are covered with a corneal membrane. These species are present in variable color patterns due to the great complexity of chromatophores.

Order Sepiolida

Order Sepiolida contains small to tiny squids, ranging in shape from the spherical bobtail and bottle squids to the elongate pygmy squids. They have a short mantle rounded posteriorly. The internal shell (gladius) is chitinous and may be rudimentary. Anteriorly the dorsal mantle may be free or fused with the head. The head is wide and the large eyes are covered with corneal membranes. Fins are rounded or kidney shaped and do not extend to the full length of the mantle. The anterior and posterior lobes of the fins are free, the posterior lobes being broadly separated. Arms are short and bear suckers. The tentacles present are retractile. The order Sepiolida contains two families, one of which is the Sepiolidae (bobtail squid) and other is Idiosepiidae (pigmy squid) with 14 genera and over 50 species occurring in tropical, temperate and subpolar waters of all oceans (Norman and Lu, 2000).

Order Teuthida

The order Teuthida includes two suborders: the Myopsida (covered eye squids) and the Oegopsida (open eyed squids). The general characters of this group are the following: shell (gladius or open) internal, chitinous, feather or rod-shaped; eight sessile arms; two tentacular arms contractile but not retractile, pockets absent, tentacles lost secondarily in some; suckers stalked and with or without hooks and fin lobes fused posteriorly. Eyes are either covered or open and without supplementary eyelid. The suborder Myopsida contains a single family, Loliginidae, comprising seven genera and around 50 species, occurring in all temperate and tropical oceans.

Order Octopoda

Octopuses are the members of the order Octopoda. Internal shell is vestigial and cartilaginous except in females of Argonauta which has an external, calcified shell. They possess eight arms, suckers without stalks and chitinous rings, tentacles and fins absent except in a few deep water species and have no light organs. 12 species belong to this order.

STATUS OF MARINE MOLLUSCA IN INDIA

MARINE MOLLUSCA ALONG THE EAST COAST OF INDIA

Species Diversity

As per available literature, 2199 species of molluscs under 588 genera and 185 families have been recorded from the east coast of India:

- 17 species of Polyplacophora under 12 genera and 6 families;
- 1487 species of Gastropoda under 344 genera and 104 families;
- 54 species of Cephalopoda under 23 genera and 11 families;
- 632 species of Bivalvia under 207 genera and 63 families;
- 9 species of Scaphopoda under 2 genera and one family (Venkataraman *et al.*, 2012).

Of the 1487 listed species of gastropods, 222 are repeated and 7, under 3 genera and 3 families, are freshwater molluscs listed along with the marine fauna. However, the authenticity of the data is debatable since there are several typographic errors, including distribution records from both the coasts of India and islands.

Molluscan Hotspots along East Coast

Here molluscan hotspot refers to areas with diversity within a special and limited geographical territory. In terms of molluscan diversity, some of the coastal stretches have a unique distribution of marine mollusca. On the east coast of India, some of the mollusc-rich areas in terms of species diversity and abundance are: the coastal stretches along the Sudarban mangroves including Bakkhali, Digha and Talsari Beaches along the upper east coast, the estuarine mouth of Subarnarekha, the mudflats of Chandipur, the lagoon area of Chilika, the wide beaches of southern Odisha coast including Rushikulya, Gopalpur, the rocky coast of Bimunipatnam and Visakhapatnam up to Konada, the Kakinada Bay, the riverine mouth of Godavari and Krishna, the Pennaru estuary, the Pulicate lake, Gulf of Mannar and Palk Bay, Nagapattinum, Tuticorin and Kannyakumari coast.

Endemism along East Coast

Endemism among marine mollusca is not known adequately due to incomplete surveys and difficulty in collecting from deeper waters. Nevertheless, as per the available records, 3 species of cephalopods, 36 species of bivalves, 1 each of scaphopoda and polyplacophora, and 25 species of gastropods are considered to be endemic to the east coast of India. One species of gastropod, spiral tudicla *Tudicla spirillus,* which is also included as scheduled mollusca in the WLPA (1972) of the Government of India is endemic to the Bay of Bengal and east coast of India.

Threatened Mollusca along East Coast

Several species of marine mollusca in India are well known to be widely used as food, ornaments and decorative items including use in clothing, pharmaceuticals and in the lime industry. Overexploitation of mollusca, due to their wide social application, has led to many species becoming threatened in their natural habitat and being designated as endangered and vulnerable. For example, the sacred chank *Turbinella pyrum*, which occurs along the coast of Tinnevelly, Sivaganga, Ramnathapuram, Tanjavur, South Arcot and Chingleput, Nagapattinum and Kanyakumari area, is threatened by the age-old practice of fishery for its market value. Similarly, the windowpane oyster *Placuna placenta* is a source of lime, its meat is edible, and the right valve is exported in good quantities to be used for glazing windows (Menon and Pillai, 1996). Pearl can be produced by this oyster also. This species occurs along the east coast of India; however, heavy exploitation takes place from the coastal areas of Tuticorin, Mandapam, Nagapattanam, Chennai and Kakinada Bay.

The other species of conservation importance which occur along the east coast of India are gastropods such as gold-banded volute *Harpulina arusiaca*, trapezium conch *Fasciolaria trapezium*, cowrie *Cypraea limacina*, spiral tudicla *Tudicla spirillus*, Glory of India *Conus milnedewardsi* and also the king shell *Cassis cornuta* besides one cephalopod, the chambered nautilus *Nautilus pompilus*. All these are included in the Schedules of Widlife (Protection) Act, 1972 of the Government of India, which prohibits trade, collection or hunting from the wild, keeping them in custody in any form, or marketing of the products and byproducts of these shells and imposes punishment or fine or both as per law.

MARINE MOLLUSCA ALONG THE WEST COAST OF INDIA

Species Diversity

In spite of a diversified ecosystem—coral reef, mangrove, estuaries and rocky patches—molluscan diversity along the west coast of India is less than that of the east coast. Available literature indicates that 707 species of gastropods, 248 species of bivalves, 2 species of scaphopoda, 80 species of cephalopods, and 9 species of polyplacophora occur along the west coast. The reason for such apparently poor diversity could be meagre inventory by faunal survey organizations.

Molluscan Hotspots along West Coast

The west coast of India has a good coral reef ecosystem, the Gulf of Kutchch and Malvan, estuarine ecosystems of Mandvi, Juari, Narmada, Tapi and several backwaters in Kerala, and sandy coastal stretches of Karnataka (Gangoli, Kundapur), Kerala (Alapuzha, Kovalam) which are rich in molluscan fauna. There are new records of bivalves and gastropods from these areas.

Endemism along West Coast

Among the gastropoda, out of the 707 species, 35 are endemic to the west coast of India. Similarly, 8 species of marine bivalves are endemic. However, there is no endemic species of scaphopoda, cephalopoda or opithobranchia known adequately from the west coast.

Threatened Mollusca along West Coast

The present state of knowledge does not provide any specific information on threatened marine mollusca from the west coast of India. Many of the species were not recollected after their original discovery or their first collection. Authentic data on the status of various species of molluscs are not available, and there are no documented data on the exploitation of marine molluscs from the west coast. Nevertheless, species like *Placuna placenta* are heavily exploited along the Maharashtra and Goa coast (Murud-Jinjira, Nauxim Bay), and also from Gujarat (Pouchitra, Raida, Goomara of Gulf of Kutch). These areas were once abundant with *Placuna placenta* beds but, because of over-exploitation by shell industries, the population has declined heavily. A similar species, *Meretrix casta*, is now heavily exploited by the local fisheries along all the west coast of India for its nutritive food value. This species has been found to sell abundantly in local markets of Karwar, Murud, Margao, Daman, Udipi, Malvan, Beypore Cochin, and Vizhingam along the west coast of India, and over-exploitation of undersized clams has been witnessed which may lead to stock depletion and habitat damage and ultimately to extinction of the species.

The species *Paphia malabarica* and *Villorita* spp. are also at stake because of commercial over-exploitation from the sub-fossil sources of lakes such as Kalanadi, Vembanad and Ashtamudi, and stock depletion is an inevitable result. The situation is similar for *Gafrarium divericatum*, a common venerid used for food along the coast of Maharashtra and popular among those dwelling near to the coast. Additionally, species such as the windowpane oyster *Placuna placenta*, *Nautilus pompilius*, Sibbald conch *Strombus plicatus sibbaldi*, which are included in the schedules of Wildlife (Protection) Act, 1972, are found along the west coast of India.

MARINE MOLLUSCA ALONG ISLAND COASTS OF INDIA

Andaman and Nicobar Islands

Species Diversity

A total of 1147 species belonging to 384 genera and 143 families representing five classes of molluscs are reported from the Andaman and Nicobar Islands (ANI). Of these, 1057 species are present in the National Zoological Collections (NZC) of ZSI and the rest are reported by authors outside ZSI. These species belong to 372 genera and 141 families. The class Polyplacophora is represented by only 12 species belonging to 7 genera and 4 families. The class Cephalopoda includes 33 species belonging to 18 genera and 11 families, and these islands are the type locality for at least 7 species of cephalopods. About 350 species of bivalves have been recorded from the islands, under 150 genera and 54 families. The class Scaphopoda is represented by 7 species belonging to a single family and genus; all the species are restricted in distribution to ANI.

Endemism in the Andaman and Nicobar Islands

The ANI are the type locality for at least 67 species of mollusca (49 gastropods and 18 bivalves). Of the 12 species of Polyplacophora, 3 are endemic to ANI. As many as 177 species are endemic to the islands (~19%). About 45 species of mollusca of Indo-Pacific distribution occur in ANI, and the species trumpet shell *Charonia tritonis* has restricted distribution in Nicobar only, while *Turbinella pyrum fuscus* occurs only in Andaman.

Threatened Mollusca from Andaman and Nicobar Islands

Turban shell *Turbo marmoratus*, top shell *Trochus niloticus*, trumpet shell *Charonia tritonis* and giant clam *Tridacna maxima* are among the threatened species of mollusca unique to the ANI, and the turban shell is on the verge of extinction because of over-exploitation (Raghunathan *et al.*, 2010). Out of the 24 species of protected mollusca as per schedule I and IV of the Wildlife (Protection) Act, 1972 of the Government of India, at least 19 species are found in the ANI, of which 7 are in Schedule I of the Act.

Lakshadweep

Species Diversity

Nagabhushanam and Rao (1972) made an ecological survey of marine fauna of Minicoy atoll (Lakshadweep Archipelago) and reported 191 marine mollusca belonging to 94 genera. Of these, 3 species belong to 3 genera of Polyplacophora; 130 species belong to 67 genera of Gastropoda, 7 species belong to 5 genera of Cephalopoda, and 51 species belong to 19 genera of Bivalvia. Subsequently, Surya Rao and Subba Rao (1991) provided a consolidated list of molluscs from Lakshadweep islands in which 424 species under 201 genera and 105 families were reported. Among these were 4 species of polyplacophora under 3 genera and 1 family, 303 species of gastropods under 138 genera and 70 families, 11 species of cephalopods under 7 genera and 5 families, and 107 species of bivalves under 53 genera and 29 families.

MARINE MOLLUSCA OF PROTECTED AREAS OF INDIA

There are about 31 Marine Protected Areas (MPAs) in India, covering an approximate area of 627 km^2 with an average size each of 20 km^2. These areas are ecologically important, and the Government of India initiated action through the state governments to create a network of MPAs under the Wild Life Protection Act, 1972. Marine mollusca play an important role in MPAs as indicator species, keystone species as well as targeted species of conservation importance.

Marine Mollusca of Gulf of Mannar Marine National Park and Biosphere Reserve

Satyamurti (1952) published a comprehensive and descriptive report on the molluscs of Krusadai Island and Gulf of Mannar, in which he has reported 258 species of polyplacophora and gastropoda of which 8 species of polyplacophora are under 3 families and 250 species of gastropods belong to 63 families. Among the gastropoda, 219 species belong to the subclass Prosobranchia, 29 species belong to the subclass Opisthobranchia, and 2 species belong to the subclass Pulmonata.

Marine Mollusca of Sundarbans Mangroves and Biosphere Reserves

Dey (2006) published a handbook on mangrove associated molluscs of Sundarbans, in which 56 species of molluscs, under 42 genera and 30 families, of which 31 species are gastropods belonging to 20 genera and 15 families, have been reported from the mangrove areas including the coastal areas of the Sundarbans Biosphere Reserve.

Marine Mollusca of Bhitarkanika National Park and Gahirmatha Sanctuary

A total of 238 species of marine molluscs have been reported from the Bhitarkanika coast of Odisha, including 1 species of polyplacophores, 108 species of gastropods under 65 genera and 53 families, 9 species of cephalopods under 6 genera and 4 families, 121 species of bivalves under 41 genera and 36 families, and 1 species of scaphopods.

Marine Mollusca of Chilka lagoon

Molluscan faunal diversity in the Chilka lagoon is very rich. Studies on the molluscan diversity of Chilka Lake date back to 1924. The first expedition of faunal inventory of Chilka by Annandale (1924) recorded only 75 species of marine mollusca. Subsequently, Subba Rao *et al.* (1995) recorded 136 species of marine molluscs under 96 genera and 66 families from the lagoon. A total of 59 species of gastropods under 41 genera and 30 families, 69 species of bivalves under 47 genera and 28 families, and a single family, genus and species of cephalopods have been reported. Among the bivalves *Meretrix meretrix, Meretrix casta, Neosolen aquaedulcioris, Macoma birmanica, Theora opalina, Clementias vatheleti, Modiolus undulates* and *Modiolus striatulus* are the most abundant in the lake. Nevertheless, endemism is very high in the lagoon at about 39 percent, which is unique in terms of molluscan diversity. As many as 33 species of mollusca have their type localities in the Chilka lagoon and so far are not reported from anywhere else.

Marine Mollusca of Gulf of Kachchh

Subba Rao and Sastry (2005) provided the basic information on the biodiversity of the Gulf of Kachchh Marine National Park and listed 178 species of molluscs under 123 genera and 72 families, including 1 species of polyplacophores and 124 species of gastropods under 85 genera and 52 families, 3 species of cephalopods under a single genus and family, and 50 species of bivalves under 36 genera and 18 families.

NEED FOR CONSERVATION OF MARINE MOLLUSCA

Exploitation of Marine Mollusca in India

Insufficient information on population (wild stocks) and level of exploitation of marine mollusca in India is known to allow determination of the status of a particular species in order to provide protection and conservation. Data on the life history, abundance, productivity and rates of exploitation from specific localities are required for virtually every species involved in the shell trade. Nevertheless, anecdotal evidence suggests that conservation problems are on the increase and makes it possible to predict which areas and species are most vulnerable. The available literature reveals that depletion of the molluscan population appears to be occurring on a local scale in India.

There are several reports of illegal collection of mollusca in West Bengal in areas where collectors concentrate their efforts in order to meet tourist demand. For example, *Ancilla ampla*, a lustrous Olividae which was once very common on the sandy beach at Digha on the upper east coast of India has now become very scarce due to rampant collection from the beach for use as curios. Poushitra, an area in the Gulf of Kachchh, was once famous for a *Placuna placenta* bed, but now the population of this species is declining heavily as a result of over-explotation for calcium and cement. Similarly, the Operculum of *Pugilina (Hemifusus) cochlidium*, a crown conch, is used for its medicinal value "Sanka Bhasma" by the coastal dwellers of West Bengal, especially fisherfolk in Sunderbans. The bivalve *Meretrix meretrix* is extensively used in the manufacture of poultry feed and lime and is collected by the tonne from the Subarnarekha River in Odisha, East Medinipur and 24 Parganas (South) of West Bengal. *Crassostrea gryphoides* and *Anadara granosa* are also used dead or alive as a source of calcium in the poultry industry.

The population of the larger species of giant clam (Tridacnidae: *Tridacna maxima, Tridacna squamosa*) declined in many parts of the Indo-Pacific, including the ANI and Lakshaweep islands, as a result of over-exploitation both for shell and for flesh. Species like giant clam attain late reproductive maturity, have a comparatively short larval life span and have poor recruitment to adult population, and this makes them vulnerable to depopulation. The giant triton *Charonia tritonis*, which occurs naturally at low densities, has become rare through over-collection. The gastropod sacred chank *Turbinella pyrum* has a very restricted distribution, occurring at a depth of 30–45 m in the Gulf of Mannar and in the Gulf of Kachchh region. The rights for fishing are leased out by the respective state governments. In the life cycle of the sacred chank, a free swimming veliger stage of development is absent and therefore adults are unable to swim to new territories. Over-exploitation of species such as this will automatically lead to decline in their stocks because of their life history pattern.

Ornamental mollusca *Trochus niloticus* and *Turbo marmoratus* are the two commercially important gastropods and have distribution only in the ANI but, because of overfishing, there is substantial depletion in the natural stock. The two rare cones Glory of the Sea *Conus gloriamaris* and Glory of India *Conus milne-edwardsi* (Jousseaume, 1889) are rare among the marine mollusca because of over-exploitation in the past. At present, relatively few of the widely distributed ornamental species are known to be similarly affected but the future for all these species is not so bright and may verge towards depletion and even to extinction.

Anthropogenic Pressure on Molluscan Diversity in India

Various activities connected with shell collection can alter or degrade the marine environment and corals and molluscan habitats. A common type of disturbance includes trampling and rock removal. Corals are also deliberately or inadvertently broken in order to remove shells. Many fishermen use a bottom trawl net to catch more fish; it drags across sand and rubble areas of the sea where molluscs live, taking huge amounts of molluscs along with fish. The fishermen dry this on the coast to make poultry feed, but the practice has evidently declined to some extent, partly because the habitat was altered so drastically that even fish were found to have been affected. The collectors have also been forced to collect from more remote areas because of decline in yield in the vicinity of India, which is leading to destruction of specialized habitats for mollusca, including other species of corals and associated fauna.

Little attention has been paid to the consequences of selective removal of shells on the ecosystem as a whole. It has been suggested, for example, that over-collection of the Giant Triton *Charonia tritonis,* which preys on large starfish, has contributed to population explosions of the Crown of Thorns starfish *Acanthaster planci* (Endean and Cameron, 1985). Plagues of the starfish have caused extensive damage to coral reefs in many parts of the Indo-Pacific. A comparable sequence of events may be responsible for upsurges in numbers of sea urchins, which glaze the surface of coral and, in doing so, cause structural damage to reefs.

Conservation Measures for Marine Mollusca of India

In-situ Protection of Mollusca

Protection of molluscs in the long run may pave the way for the restoration of species balance in nature, especially in the ecologically sensitive, biotically rich, fragile areas such as coral reef ecosystems. There are several rules and regulations formulated by the Government of India to protect marine environments, including specialized animals and habitat of coastal stretches of India, in the islands and the coastal and marine waters of India. However, protection of these molluscs by the rules and regulations is difficult for the policy makers, planners and protection agencies because of social, economic and political constraints. There are several courses of action that can be taken to control trade in shells, and thus avoid over-exploitation and habitat damage. Conservation problems should not exist if the fisheries are properly managed on an ecologically sound, sustainable yield basis. The problems would also be lessened if demand for ornamental shells declined. A greater public awareness of the conservation issues on marine mollusca could help in this respect. The considerable knowledge of many collectors themselves about the distribution and biology of exploited species such as mollusca could be put to use, and they could be encouraged to build on existing traditional techniques of management. Measures taken to ensure conservation of species or habitats are much more likely to succeed if people who are affected by management decisions are involved in their formulation.

Legal Protection of Marine Mollusca

The Ministry of Environment and Forest of the Government of India brought 24 species of marine mollusca under Schedule I (9 species) and IV (15 species) of the Wildlife (Protection) Act, 1972 (Table 4.1). The Convention on International Trade in Endangered Species of Wild Fauna and Flora (CITES) provides a means of controlling international trade in species considered to be seriously threatened. At present, the only marine molluscs listed are the giant clams (family Tridacnidae). These are listed in Appendix II, which means that a valid export licence is required from the country of origin before the shells, meat or live animals can be traded between the parties to the convention. This provides a

TABLE 4.1 Marine Mollusca Listed in the Wildlife (Protection) Act, 1972

Schedule I, Part IV (B)	Schedule IV, Part 19
Cassis cornuta (King Shell)	*Cypraea limacina* (Limacina Cowrie)
Charonia tritonis (Trumpet Shell)	*Cypraea mappa* (Map Cowrie)
Conus milne-edwardsi (Glory of India)	*Cypraea talpa* (Mole Cowrie)
Cypraecassis rufa (Queen Shell)	*Fasciolaria trapezium* (Trapezium Conch)
Hippopus hippopus (Horse's Hoof Clam)	*Harpulina arausiaca* (Vexillate volute)
Nautilus pompilius (Chambered Nautilus)	*Lambis (Harpago) chiragra chiragra* (Chiragra Spider Conch)
Tridacna maxima (Elongated Giant Clam)	*Lambis (Harpago) chiragra arthritica* (Arthritic Spider Conch)
Tridacna squamosa (Fluted Giant Clam)	*Lambis (Lambis) crocata crocata* (Orange Spider Conch)
Tudicla spirillus (Spiral Tudicla)	*Lambis (Millepes) millepeda* (Millipede Spider Conch)
	Lambis (Millepes) scorpio scorpio (Scorpio Conch)
	Lambis (Lambis) truncata truncata (Truncate Spider Conch)
	Placuna placenta (Windowpane Oyster)
	Strombus (Dolomena) plicatus sibbaldi (Pigeon Conch)
	Trochus niloticus (Top Shell)
	Turbo marmoratus (Great Green Turban)

useful means of monitoring trade, particularly as custom and fishery statistics tend to be so poor for molluscs.

Status Survey of Marine Mollusca of India

It is essential to carry out research/fieldwork on status survey and inventory of mollusca from the coastal stretches of the mainland and islands of India. Based on information from surveys and inventory, it will be possible to reassess stock abundance estimates and thus amend the total allowable catch (TAC) quota for future harvests of any species of mollusc deemed to be threatened. For example, based on the status survey carried out by ZSI and as per the present stock of *Trochus niloticus* in Andaman and Nicobar, it was suggested that a ban on fishing of these economically important gastropods may be lifted initially for 3 years on a temporary basis for the rational utilization of these shellfish resources.

Ex-situ and Mariculture of Marine Mollusca

Considerable success has been achieved with several marine molluscan species through mariculture: larvae and juveniles are reared in hatcheries, and the adults are kept in tanks for production of spawn and ultimately for harvesting. Mariculture clearly has potential for positive benefits on marine molluscan diversity and conservation through sea ranching.

ACKNOWLEDGEMENTS

The authors are thankful to the Director, Zoological Survey of India for encouragement and support. Thanks are also due to the anonymous reviewers in bringing this publication to its final stage. We are grateful to the Staff of Mollusca Division, Zoological Survey of India, Kolkata for their help in literature searching and verification of the voucher specimens.

REFERENCES

Adam, W., 1939. Cephalopoda, II, Revision des especes Indo-Malaises du genre Sepia Linnaeus 1758. III – Revision du genre Sepiella (Gary) Steenstrup, 1880. Siboga-Expeditie. Resultatats des expedition Zoologiques, Botaniques, Oceanographiques et enterprises aux Indes Neerfandaises Orientales en 1899-1900, 55b: 35–122.

Annandale, N., 1924. Fauna of Chilika Lake. Mollusca: Gastropoda (Revision). Mem. Indian Mus. 5, 853–873.

Apte, D., 1992. A unusual species of *Turbinella pyrum* (L). J. Bombay nat. Hist. Soc. 89, 267.

Apte, D., 1993. Marine Gastropoda of Bombay - A recent survey. J. Bombay nat. Hist. Soc. 90, 537–539.

Apte, D., 1997. Record of *Homalocanthus secunda* (Lamarck, 1822) from Okha in Gulf of Kutch. J. Bombay Nat. Hist. Soc. 95 (3), 526–527.

Apte, D., 1998. The book of Indian shells. Bombay Natural History Society, p.114.

Apte, D., 2009. Opisthobranch fauna of Lakshadweep Islands, India with 52 new records to Lakshadweep and 40 new records from India: Part 1. J. Bombay Nat. Hist. Soc. 106 (2), 162–175.

Apte, D., Bhave, V., Parasharya, D., 2010. An annotated and illustrated checklist of the Opisthobranch fauna of Gulf of Kutch, Gujarat, India with 21 new records from Gujarat and 13 new records from India: Part 1. J. Bombay Nat. Hist. Soc. 107 (1), 14–23.

Arularasan, S., Kasinathan, R., 2007. Molluscan composition at Vellar estuary, Porto Novo coast. Zoo's Print J. 22 (1), 2546.

Babu Philip, M., Appukuttan, K.K., 1995. A check list of gastropods landed at Santhikulangara-Neendakara area. Mar. Fish. Inf. Ser. T & E., No. 138, 9–10.

Cheriyan, P.V., 1968. A collection of Molluscs from the Cochin Harbour area. Proc. Symp. on Mollusca, 121–136, Marine Biological Association of India, Part 1.

Das, A.K., Mitra, S.C., Mukhopadhyaya, S., 1977. Notes on the occurrence of *Telescopium mauritsi* Butot (Mesogastyropoda: Mollusca) from India. Newsl. Zool. Surv. India 3 (3), 129–130.

Das, A.K., Mitra, S.C., Mukhopadhyaya, S., 1981. Studies on some molluscan collection by the "Golden Crown" from the Bay of Bengal with a note on camouflage habit of a gastropod, *Xenophora pallidula* (Reeve). Proc. Zool. Soc. Calcutta 32, 79–87.

Dey, A., 2006. Contribution to the knowledge of Indian Marine molluscs (Part – IV) Family Tellindiae. Rec. Zool. Surv. India., Occ. Paper No 249, 1–124, pls 38.

Dey, A., 2008. Commercial and medicinal important molluscs of Sunderbans. Rec. Zool. Surv. India Occ. Paper No. 286, 1–54.

Dey, M., Jamadar, Y.A., Mitra, A., 2005. Distribution of intertidal malacofauna at Sagar Island. Rec. Zool. Surv. India 105 (1-2), 25–35.

Endean, R., Cameron, A.M., 1985. Ecocatastrophe on the Great Barrier Reef. Proc. 5th International Coral Reef Congress 5, 309–314.

Gopinadha Pillai, C.S., Appukuttan, K.K., 1980. Distribution of molluscs in and around the coral reefs of the southeastern coast of India. J. Bombay Nat. Hist. Soc. 77 (1), 26–48.

Hornell, J., Tomlin, J.R.K.B., 1951. Checklist of marine and fluvialite mollusca of Bombay and neighbourhood. Appendix In: Indian molluscs. Bombay Natural History Society, Bombay, pp. 83-97.

Joshi, M.C., 1969. The marine Mollusca of the Konkan Coast. J. Shivaji Univ. (Sci.) 2 (4), 47–54.

Jothinayagam, J.T., 1987. Cephalopoda of the Madras Coast. Technical Monograph Zoological Survey of India, Kolkata, 15: 1–85.

Kasinathan, R., Shanmugam, A. (1985). Molluscan fauna of Pitchavaram Mangroves, Tamil Nadu. Proc. Nat. Symp. Biol. Util. Cons. Mangroves (L.J. Bhosle (Ed.) pp. 438–443.

Kohn, A.J., 1978. The Conidae (Mollusca: Conidae) of India. J. Nat. Hist. 12, 295–335.

Mahapatra, A., 2001. Molluscan fauna of Godavari estuary: Estuarine Ecosystem Series 4 55–82, Zoological Survey of India, Kolkata.

Mahapatra, A., 2008. Molluscan fauna of Krishna estuary: Estuarine Ecosystem Series 5 105–173, Zoological Survey of India, Kolkata.

Mahapatra, A., 2010. Mollusca of Vamsadhara and Nzagavali estuaries, Andhra Pradesh. Zoological Survey of India, Kolkata. Estuarine Ecosystem Series 6, 47–72.

Menon, M.G., Pillai, C.S.G., 1996. Marine biodiversity conservation and management. Central Marine Fisheries Research Institute, Spl. Publ., Cochin, p. 204.

Mookherjee, H.P., 1985. Contributions to the molluscan fauna of India, Part III. Marine molluscs of the Coromandel coast, Palk strait and Gulf of Mannar-Gastropoda: Mesogastropoda (Pt.2). Rec. Zool. Surv. India, Occ. Paper No. 75, 1–93.

Mookherjee, H.P., Barua, S., 1989. Molluscan fauna of Manauli Island in relation to environmental niche. Rec. Zool. Surv. India 85 (4), 527–531.

Nagabhushanam, A.K., Rao, G.C., 1972. An ecological survey of the marine Fauna of Minicoy Atoll (Laccadive Archipelago, Arabian Sea). Mitt. Zool. Mus., Berlin 48 (2), 266–324.

Namboodiri, P.N., Sivadas, P., 1979. Monation of molluscan assemblage at Kavaratti Atoll (Laccadive). Mahsagar Bull. Nat. Inst. Oceanogr. 12 (4), 1239–1246.

Narayanan, K.R., 1968. On the opisthobranchiate Fauna of the Gulf of Kutch. Proc. Symp. Molluscs, Mar. Biol. Assoc. India Part 1, 188–213.

Narayanan, K.R., 1969. On two Doridacean Nudibranchs (Mollusca: Gastropoda) from the Gulf of Kutch, new to the Indian Coast. J. Bombay Nat. Hist. Soc. 68 (1), 280–281.

Narayanan, K.R., 1972. The sacred chank of India. Seafood J. 4 (3), 25–27.

Nixon, M., Young, J.Z., 2003. The Brains and Lives of Cephalopods. Oxford University Press, New York, pp. 392.

Norman, M.D., Lu, C.C., 2000. Preliminary checklist of the cephalopods of the South China Sea. Raffles B. Zool. 8, 539–567.

Okusu, A., Schwabe, E., Eernisse, D.J., Giribet, G., 2003. Towards a phylogeny of chitons (Mollusca, Polyplacophora) based on combined analysis of five molecular loci. Organisms, Diversity & Evolution, 3, 281–302.

Patil, A.M., 1954. Study of marine fauna of the Karwar Coast and neighbouring Islands. J. Bombay Nat. Hist. Soc. 50, 549–558.

Patil, A.M., Gopalkrishnan, M., 1960. Molluscan shells washed on the sandy beach at Suratkal, South Kanara. Proc. Indian Science Congress 47 (3), 468.

Prabhakara Rao, K., 1968. On a new genus and some new species of opisthobranchiate gastropods of the family Eubranchidae from the Gulf of Kutch. Proc. Symp. Molluscs, Mar. Biol. Assoc. India, 51–60, Part 1.

Radhakrishna, Y., Ganapati, P.N., 1969. Fauna of Kakinada Bay. Bull. Nat. Inst. Sci. India 38, 689–999.

Raghunathan, C., Sivaperuman, C., Ramakrishna, 2010. Diversity of littoral corals and their associated Molluscs and Echinoderms in Andaman Sea, South Andaman. In: Ramakrishna, Raghunathan, C., Sivaperuman, C. (Eds.), Recent trends in Biodiversity of Andaman and Nicobar Islands. Zoological Survey of India, Kolkata, pp. 249–273.

Raghunathan, C., Sivaperuman, C., Ramakrishna., 2010. An account of newly recorded five species of nudibranchs (Opisthobranchia, Gastropoda) in Andaman and Nicobar Islands. In: Ramakrishna, Raghunathan, C., Sivaperuman, C. (Eds.), Recent trends in Biodiversity of Andaman and Nicobar Islands. Zoological Survey of India, Kolkata, pp. 283–288.

Rajagopal, A.S., Subba Rao, N.V., 1974. On chitons from the Andaman and Nicobar Islands. J. Mar. Biol. Assoc. India 16 (2), 398–411.

Rajagopal, A.S., Mookherjee, H.P., 1978. Contribution to the molluscan fauna of India Part I. Marine molluscs of Coromandel Coast, Palk Strait and Gulf of Mannar - Gastropoda: Archaeogastropoda. Rec. Zool. Surv. India, Occ. Paper 12, 1–48.

Rajagopal, A.S., Mookherjee, H.P., 1982. Contribution to the molluscan fauna of India Part II. Marine molluscs of Coromandel Coast, Plak strait and Gulf of Mannar - Gastropoda: Mesogastropoda (Partim). Rec. Zool. Surv. India, Occ. Paper 28, 1–53.

Ramakrishna, Dey, A., 2010. Manual on identification of Schedule mollusks from India. Zoological Survey of India, p. 40.

Ramakrishna, Dey, A., Barua, S., Mukhopadhya, A., 2007. Fauna of Andhra Pradesh, Marine molluscs: Polyplacophora and Gastropoda. State Fauna Series, Fauna of Andhra Pradesh, 1–148, Part 7.

Ramakrishna, Sreeraj, C.R., Raghunathan, C., Sivaperuman, C., Yogesh Kumar, J.S., Raghuraman, R., Immanuel, Titus, Rajan, P.T., 2010. Opisthobranchs of Andaman and Nicobar Islands. Zoological Survey of India, p. 196.

Rao, D.V., 2010. Field Guide to Coral and Coral Associates of Andaman and Nicobar Islands. Zoological Survey of India, p. 83.

Rao, D.V., Sastry, D.R.K., 2007. Fauna of Button Island National Parks, South Andamans, Bay of Bengal. Rec. Zool. Surv. India, Occ. Paper 270, 1–54.

Ray, H.C., 1948a. On a collection of Mollusca from Coromandel coast of India. Rec. Indian Mus. 46, 87–122.

Ray, H.C., 1948b. Revision of Cypracea in the collection of Zoological Survey of India. Part 1. The families Triviidae, Eratoidae and Pediculariidae. Rec. Indian Mus. 46, 183–213.

Ray, H.C., 1952. Cowries (Mollusca: Gastropoda: Family Cypraeidae). J. Bombay Nat. Hist. Soc. 49 (2), 663–669.

Ray, H.C., 1954. Mitres of Indian waters. Mem. Indian Mus. 14, 1–72.

Roy, M., Dey, A., Banerjee, S., Nandi, N.C., 2008. Molluscan Macrobenthic diversity of Backwater Wetlands in West Bengal. Zoological Research in Human Welfare, 25–34, Paper-I.

Satyamurti, S.T., 1952. The mollusca of the Krusadai Island (In the Gulf of Manaar). Bull. Madras Govt. Mus. New Ser. (Nat. Hist) 1 (2), 1–267, 6.

Satyamurti, S.T., 1956. The Mollusca of Krusadai Island (in the Gulf of Mannar) II. Scaphoda, Pelecypoda and Cephalopoda. Bull. Madra Govt. Mus. N.S. Nat. Hist. Sec. 7, 1–202.

Satyanarayana Rao, K., Sundaram, K.S., 1972. Ecology of intertidal molluscs of Gulf of Mannar and Palk Bay. Proc. Indian Nat. Sci. Acad. 38B (5&6), 462–474.

Silas, E.G., 1968. Cephalopoda of Indian Ocean. Proc. Symp. Moll. Mar. Biol. Assoc. India, 277–359, Part 1.

Starmuehlner, F., 1974. Beitrage zur Kenntnis der Mollusken-Fauna im Littoral von sudindien und Ceylon. J. Mar. Biol. Assoc. India 16 (1), 49–82.

Steiner, G., Dreyer, H., 2003. Molecular phylogeny of Scaphopoda (Mollusca) inferred from 18S rDNA sequences: support for a Scaphopoda–Cephalopoda clade. Zoologica Scripta 32, 343–356.

Subba Rao, N.V., 1968. Report on a collection of wood boring molluscs from Mahanadi estuary, Orissa, India. Proc. Symp. Moll. Mar. Biol. Assoc. India, 85–93, Part 1.

Subba Rao, N.V., 1970. On the collection of Strombidae (Mollusca: Gastropoda) from Bay of Bengal, Arabian sea and Western Indian ocean, with some new records, 1. Genus *Strombus*. J. Mar. Biol. Assoc. India 12 (1&2), 109–124.

Subba Rao, N.V., 1980. On the Conidae of Andaman and Nicobar Islands. Rec. Zool. Surv. India 77, 39–50.

Subba Rao, N.V., 2003. Indian Seashells (Part I), Polyplacophora and Gastropoda. Rec. Zool. Surv. India, Occ. Paper 192, 1–416.

Subba Rao, N.V., Dey, A., 1975. Studies on Indian Mitridae (Mollusca: Gastropoda: Stenoglossa). Newsl. Zool. Surv. India 1 (4), 79–80.

Subba Rao, N.V., Dey, A., 1984. Contribution to the knowledge of Indian marine molluscs. 1. Family Mitridae. Zool. Surv. India, Occ. Paper 61, 1–48.

Subba Rao, N.V., Dey, A., 1986. Contribution to the knowledge of Indian marine molluscs. 2. Family Donacidae. Zool. Surv. India, Occ. Paper 91, 1–30.

Subba Rao, N.V., Dey, A., 2000. Catalogue of marine molluscs of Andaman and Nicobar Islands. Rec. Zool. Surv. India, Occ. Paper 187, 1–323.

Subba Rao, N.V., Mookherjee, H.P., 1975. On a collection of Mollusca from the Mahanadi Estuary, Orissa. In: Natarajan, R. (Ed.), Recent Research in Estuarine Biology. Hindustan Publishing Corp, Delhi, pp. 165–176.

Subba Rao, N.V., Sastry, D.R.K., 2005. Fauna of Marine National Park, Gulf of Kachchh, (Gujarat) an overview. Conservation Area Series 23, 1–79, Zoological Survey of India, Kolkata.

Subba Rao, N.V., Surya Rao, K.V., 1980. On a rare Nudibranch, *Thordisa crosslandi* Eliot (Mollusca: Dorididae) from the west coast of India. Bull. Zool. Surv. India 2 (2&3), 219.

Subba Rao, N.V., Surya Rao, K.V., 1991. Mollusca of Lakshadweep. State Fauna Series, 2: Fauna of Lakshadweep. Zoological Survey of India, Kolkata, pp. 273–362.

Subba Rao, N.V., Surya Rao, K.V., 1993. Contribution to the knowledge of Indian marine molluscs. 3. Family Muricidae. Zool. Surv. India, Occ. Paper 153, 1–133.

Subba Rao, N.V., Dey, A., Barua, S., 1983. Studies on the malacofauna of Muriganga estuary, Sunderbans, West Bengal. Bull. Zool. Surv. India 5 (1), 47–56.

Subba Rao, N.V., Dey, A., Barua, S., 1992. Estuarine and marine molluscs of West Bengal: State Fauna Series 3, Fauna of West Bengal (Part-9). Zoological Survey of India, Kolkata, 129–268.

Subba Rao, N.V., Dey, A., Barua, S., 1993. Mollusca. Estuarine Ecosystem Series Part 2, Hugli-Matla Estuary. Zoological Survey of India, Calcutta, pp. 41–91.

Subba Rao, N.V., Surya Rao, K.V., Mitra, S.C., 1987. Malacological notes on Sagar Island. Bull. Zool. Surv. India 8 (1-3), 149–158.

Subba Rao, N.V., Surya Rao, K.V., Maitra, S., 1991. Marine molluscs of Orissa. Zool. Surv. India: State Fauna series, 1: Fauna of Orissa (Part 3): 1–175.

Subba Rao, N.V., Surya Rao, K.V., Manna, R.N., 1995. Molluscs of Chilka Lake. Wetland Ecosystem Series, 1, Fauna of Chilka Lake, Zoological Survey of India, Kolkata, pp. 391–468.

Subrahmanyam, T.V., Karandikar, K.R., Murti, N.N., 1952. Marine gastropoda of Bombay, Part II. J. Univ. Bombay 21 (3), 26–73.

Subrahmanyam, T.V., Karandikar, K.R., Murti, N.N., 1952. Marine gastropoda of Bombay. J. Univ. Bombay 20, 21–34.

Surya Rao, K.V., Maitra, S., 1998. Fauna of Mahanadi Estuary: Mollusca. Estuarine Ecosystem Series, 3, Fauna of Mahanadi Estuary. Zoological Survey of India, Kolkata, 161–197.

Surya Rao, K.V., Subba Rao, N.V., 1991. Molluscan fauna of Lakshadweep. Zool. Surv. India: State Fauna series, 2, Fauna of Lakshadweep: 273–362.

Surya Rao, K.V., Maitra, S., Barua, S., Ramakrishna, 2004. Marine Molluscs of Gujarat (Part I : Polyplacophora, Gastropoda and Scaphopoda). State Fauna Series 8, Fauna of Gujarat. Zoological Survey of India, Kolkata, pp. 263–331.

Tikader, B.K., 1964. Marine fauna of Deogad Coast (Ratnagiri District), Maharashtra, 1. Mollusca. J. Univ. Poona (Sci. Technol.) 28, 43–48.

Tikader, B.K., Das, A.K., 1985. Glimpses of animal life of Andaman and Nicobar Islands. Zoological Survey of India, Kolkata, p. 170.

Tikader, B.K., Daniel, A., Subba Rao, N.V., 1986. Sea shore animals of Andaman and Nicobar Islands. Zoological Survey of India, Kolkata, p. 188.

Venkataraman, K., Jeyabaskaran, R., Raghuram, K.P., Alfred, J.R.B., 2004. Bibliography and Checklist of Corals and Coral Reef associated organisms of India. Rec. Zool. Surv. India, 1–468, Occ. Paper 226.

Venkataraman, K., Raghunathan, C., Sreeraj, C.R., Raghuraman, R., 2012. Guide to the dangerous and venomous marine animals of India. Zoological Survey of India, Kolkata, p. 104.

Venkitesan, R., Mukherjee, A.K., 2011. New records of *Comitas albicincta* (Adams and Reeve, 1830) and *Turritella bicingulata* (Lamarck, 1822) (Mollusca: Gastropoda: Turritellidae) from India. Rec. Zool. Surv. India 111 (2), 95–97.

Virabhadra Rao, K., Krishna Kumary, L., 1973. On a new species of *Dendrodoris* Ehrenberg from Goa: (Molluscs: Nudibranchiata). J. Mar. Biol. Assoc. India 15 (1), 242–250.

Virabhadra Rao, K., Krishna Kumary, L., 1974. On some aspects of taxonomy, structure and early development of the nudibranchiate gastropod, *Discodoris fragilis* (Alder and Hancock). J. Mar. Biol. Assoc. India 16 (3), 689–699.

Virabhadra Rao, K., Sivadas, P., Krishna Kumary, L., 1974. On three rare Nudibranch Molluscs from Kavaratti Lagoon, Laccadive Islands. J. Mar. Biol. Assoc. India 16 (1), 113–125.

Chapter 5

Diversity of Brachyuran Crabs of Cochin Backwaters, Kerala, India

P. Lakshmi Devi,* A. Joseph* and S. Ajmal Khan†
*Department of Marine Biology, Microbiology & Biochemistry, Cochin University
of Science & Technology, Kochi, Kerala, India; †Centre of Advanced Study in Marine Biology,
Annamalai University, Parangipettai, Tamil Nadu, India

INTRODUCTION

Among the decapods, crustaceans, the brachyuran crabs belonging to the infraorder Brachyura, are important in view of their rich diversity besides ecological and economic consequences. These crabs generally have a hard exoskeleton and are armed with a single pair of chelae. They typically have a short projecting tail (the reduced abdomen) which is entirely hidden under the thorax. They range in size from the pea crab of few millimetres wide to the Japanese spider crab, with a leg span of up to 4 metres (13 feet). They are adapted to various biotopes, such as marine, estuarine, fresh water, intertidal and even terrestrial habitats. Brachyurans are quite diverse, comprising about 700 genera and 5000 species worldwide. About 640 species of crabs are listed from Indian waters, among which 12 species are regarded as commercially important, inhabiting the coastal waters and adjoining brackish water environment. They occupy an important place in the crustacean fishery on account of their export potential and high nutritive value. The crab meat is not inferior to any sea food item, as it is rich in vitamins, glycogen, protein, fats and minerals. Crab meat is also considered to possess some medicinal value. It is believed to cure asthma and chronic fevers (Chopra, 1939; Chidambaram and Raman, 1944). Among crustaceans, crabs occupy the third position in terms of external market demand. Even though the volume of export is not comparable to that of shrimps, crabs support a sustenance fishery of appreciable importance (Rao *et al.*, 1973). In the mangrove environment, crabs represent predominant taxa. They play a significant role in detritus formation, nutrient recycling and dynamics of the ecosystem. Their burrowing behavior enhances aeration and facilitates free circulation of

Marine Faunal Diversity in India. DOI: 10.1016/B978-0-12-801948-1.00005-7

water (Ajmal Khan and Ravichandran, 2009). They are also found to have commensalism with algae, seaweeds, coelenterates, barnacles, bivalves and holothurians.

It is quite surprising that, despite so much of importance, this group is not given priority in investigations. In particular the diversity of brachyuran crabs in mangrove areas is poorly known. In view of this fact, the present study was undertaken to document the brachyuran crab diversity in the mangrove patches of the Cochin backwaters.

METHODS

Study Area

The study area, Cochin backwaters situated at the tip of northern Vembanad Lake, is a tropical positive estuarine system extending between 9° 40′ and 10° 12′ N and 76° 10′ and 76° 30′ E, with its northern boundary at Azheekode and southern boundary at Thannirmukham bund (Figure 5.1). Fifteen stations were selected in the study area on the basis of salinity. Crab specimens were collected from the above stations every month. They were collected from local markets, local fishermen, stake net operators, Chinese dip net operators and the indigenous 'Njandara' operators. Various methods were employed to collect the crabs from mangrove regions, including handpicking, scoop nets, knots made of coconut leaflets, etc. Sampling was also done every month using a boat at selected stations of the study area to collect the bottom crawlers using a Van Veen grab.

The specimens collected were then brought alive to the laboratory and details regarding their color, carapace width, carapace length, and other morphological features like shape of the carapace, markings on the body, number of anterolateral teeth, features of frontal lobe, pubescence, spination on the chelae and other legs etc. were noted. The specimens were preserved in 10 percent formaldehyde for further identification. Identification was done up to species level following Chhapgar (1957), Crane (1975), Sakai (1976), Lucas (1980), Sethuramalingam and Ajmal Khan (1991), Keenan *et al.* (1998), Jayabaskaran, (1999); Ng *et al.* (2008), Ajmal Khan and Ravichandran, (2009).

RESULTS

A total of 24 species of brachyuran crabs belonging to 16 genera and 8 families was recorded from Cochin backwaters. Among the families, the highest number of species was recorded from the family Portunidae (9 species) followed by Grapsidae (7 species).

Systematics

Phylum: Arthropoda
Superclass: Crustacea Pennant, 1777

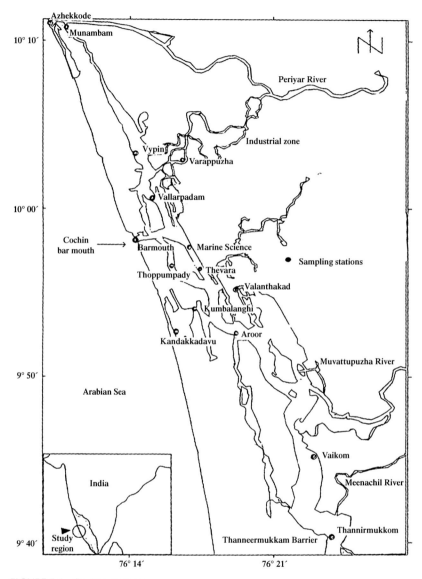

FIGURE 5.1   Study area map.

Class: Malacostraca Latreille, 1806
Subclass: Eumalacostraca Calman, 1904
Order: Decapoda Latreille, 1803
Suborder: Pleocyemata Burken Road, 1963
Infraorder: Brachyura Latreille, 1802

Description of the Species

Scylla serrata (Forskal)

Family: Portunidae
Subfamily: Portuninae

It is an edible crab, commonly known as the mangrove or mud crab and found to be the largest known species from the near shore and brackish water habitats of India. Carapace is pale brown to greenish brown, hands red colored at the outer surface and the dactylus of chela greenish. Frontal margin is usually with sharp teeth. The anterolateral border is cut into nine narrow spines. The chelipeds are very strong and the carpus of the chelipeds has two spines on the distal half of the outer margin. The fifth pair of legs is paddle shaped and adapted for swimming. It occurs fairly abundantly throughout the year in the backwaters and the associated mangrove mud flats and has a great tolerance to fluctuations in salinity. It constitutes an important fishery stock and is caught in crab traps, gill nets, Chinese dip nets, and stake nets.

Scylla tranquebarica (Fabricius)

Family: Portunidae
Subfamily: Portuninae

It is an edible crab and also known as mud crab. Carapace is greyish green in color. Frontal lobe is sharp and acuminate. Anterolateral border is cut into nine broad teeth. Carpus of the chelipeds has two obvious spines on the distal half of the outer margin. The frontal margin is usually with rounded teeth. The chelipeds and the limbs exhibit conspicuous polygonal patterns in both sexes. The fifth pair of legs is paddle shaped and adapted for swimming. Like *S. serrata*, this crab is also present abundantly throughout the year in the backwaters and the mangrove mud flats. It is adapted to a wide range of salinity and it contributes to the crab fishery of the region. It is caught using crab traps, gill nets, Chinese dip nets and stake nets.

Scylla sp.

Family: Portunidae
Subfamily: Portuninae

It is comparatively small in size with morphological characters distinct from the other two *Scylla* species. Research works are in progress to determine whether it is a separate species or a subspecies. It usually occurs in the mangrove mud flats. It is caught in crab traps and Gill nets.

Portunus (Portunus) pelagicus (Linnaeus)

Family: Portunidae
Subfamily: Portuninae

It is an edible crab, commonly known as the blue swimmer crab. Carapace is blue colored in males and sand colored in females, with an extensive spread of irregular white spots. The distal segments of the legs are pinkish in color. The anterolateral border of the carapace has nine teeth of which the most posterior one is much larger and laterally prolonged. The last pair of legs is paddle shaped and adapted for swimming. It is a marine migrant species and found to occur in the backwaters during late post-monsoon and pre-monsoon period, when the salinity is comparatively high. It is caught in stake nets, gill nets and Chinese dip nets.

Portunus (Portunus) sanguinolentus (Herbst)

Family: Portunidae
Subfamily: Portuninae

It is an edible crab, commonly known as blood-spotted swimmer crab. It is comparatively small in size. Carapace is sand colored and characterized by three large reddish round spots on the posterior half of the carapace, of which one is median and the other two lateral, each spot encircled by a white ring. The carapace is much broader than long. Its anterolateral margin is cut into nine spines, of which the most posterior is the longest. The fifth pair of legs is paddle shaped and adapted for swimming. Like *P. pelagicus*, it is also marine, and occurs in the back waters during post-monsoon and pre-monsoon period, when salinity is high. It is caught in Chinese dip nets, stake nets and gill nets.

Charybdis (Charybdis) feriata (Linnaeus)

Family: Portunidae
Subfamily: Portuninae

It is an edible crab, commonly known as crucifix crab, and is fairly large in size. Carapace is brown in color with a purple tinge and conspicuous yellow markings, the central one resembling a cross; chelipeds and limbs are brownish with yellow and white spots and the tips brownish pink. Carapace is broad, slightly convex and smooth, length as much as its width. Anterolateral borders have six spines, the first truncated and notched anteriorly and the rest acuminate. The last pair of legs is paddle shaped and adapted for swimming. It inhabits offshore waters, but is found to occur in the backwater during the pre-monsoon period, when the salinity is relatively high. Though it is an edible species, it is not consumed in some parts of the country, while it contributes to the crab fishery of the Cochin region. It is caught in Chinese dip nets from the backwaters.

Charybdis (Charybdis) lucifera (Fabricius)

Family: Portunidae
Subfamily: Portuninae

Carapace is yellowish brown with two large white spots on either branchial region. Chelipeds are scarlet pink, the fingers brownish and the extreme tips whitish. The right cheliped is markedly larger than the left cheliped. The last pair of legs is paddle shaped, for swimming, and the posterior border of its propodite is found to be serrated. They are well distinguished with the presence of a sharp median lobule on the lower border of the orbit. There is no spine on the posterior margin of carpus of natatory leg. It is found to occur during the post-monsoon and pre-monsoon period and is caught in stake nets and Chinese dip nets from the backwaters.

Charybdis (Goniohellenus) hoplites (Wood Mason)

Family: Portunidae
Subfamily: Portuninae

Carapace of the species is light brown in color. The most distinguishing character of this species is that the posterior border of the carapace forms an angular junction with the posterolateral borders and the last tooth of the antero-lateral borders is a long spine, about twice as long as those in front of it, which is usually observed in the Genus *Portunus*. It is also found to possess a spine on the posterior border of the arms of the cheliped. The last pair of legs is paddle shaped and adapted for swimming. The specimen was caught in a Chinese dip net during the pre-monsoon season.

Thalamita crenata (Latreille)

Family: Portunidae
Subfamily: Portuninae

Color of the carapace is brownish or greenish grey, claws pinkish, tips brown and the extreme tips white. The front is cut into six lobes, excluding the inner supra-orbital tooth. The anterolateral borders are cut into five equal teeth. The transverse ridge on the carapace appears faint. The outer surface of the propodus of the cheliped is smooth. The last pair of legs is paddle shaped and adapted for swimming. This estuarine crab inhabits mud flats, sandy beaches and mangroves. It occurs throughout the year and is caught in stake nets, Chinese dip nets and rarely in a Van Veen grab.

Uca (Celuca) lactea annulipes (H. Milne Edwards)

Family: Ocypodidae
Subfamily: Ocypodinae

It is commonly known as the 'Porcelain fiddler crab'. Color of the carapace is black with white or yellow stripes and the chelipeds pinkish or white. Carapace is subquadrilateral with moderately convergent lateral borders. Front is broad, supraorbital border oblique and the external orbital angle pointed obliquely outward. The most striking feature of this species is the enlarged pincers of the males, which it uses for courtship. The pincers or the major cheliped is smooth on the outer surface, while the lower border has a faint ridge, extending up to the base of the immovable lower finger. There is a wide gape between the fingers, and the movable upper finger is curved inward at the tip and extends past the immovable lower finger. It inhabits damp ground, found on the shores and near the mangroves. It is caught by handpicking and by using knots made of coconut leaflets from the study area.

Macrophthalmus depressus (Ruppell)

Family: Ocypodidae
Subfamily: Macrophthalminae

Carapace is grey in color; broader than long; rectangular in shape; surface studded with minute pearly granules. The anterolateral borders are parallel and the anterolateral angles have a square cut lobe, rather than a tooth. Eyestalks are remarkably long and slender and almost reach the end of the orbital angles. The inner surface of the palm is smooth, and ornamented with thick hairs. Hair is also present on the ambulatory legs. It is found to be a mud dweller and occurs in mangrove regions. It is caught in stake nets. Handpicking is also used more often.

Nanosesarma (Beanium) batavicum (Moreira)

Family: Grapsidae
Subfamily: Sesarminae

Color of the carapace is mottled grey, chelipeds cherry red. Front is broad and deflexed. Upper surface of the palm of cheliped has two oblique pectinated crests. The distinguishing feature of the species is the presence of three acute spines on the posterodistal border of pereopod four. It is found to occur in the mangrove regions and collected by hand picking and with knots made of coconut leaflets.

Nanosesarma (Beanium) andersonii (De Man)

Family: Grapsidae
Subfamily: Sesarminae

Color of the carapace is dark brown, chelipeds reddish. Front is broad and deflexed. Upper surface of the palm of cheliped has numerous striae, one of which forms a pectinated crest. It is distinguished by the presence of four strong spines on the posterodistal border of merus of pereopod four. It is abundantly found in the mangrove regions. Handpicking and knots made of coconut leaflets are the methods employed to collect this species.

Neoepisesarma (Neoepisesarma) mederi (H. Milne Edwards)

Family: Grapsidae
Subfamily: Sesarminae

Color of the carapace is black or dark brown, Chelipeds reddish. Front is broad and deflexed. Carapace is quadrangular, with an anterolateral tooth, slightly convergent posteriorly. The surface of the carapace has tufts of hair on the anterior portion. The distinguishing feature of this species is that, above the transverse dactylar tubercles, a sulcus runs about one third of the total length of the tubercles and the vertical granular crest of the inner palm is salient. It is found in the muddy substratum of mangrove regions and is caught by handpicking and knots made of coconut leaflets.

Neoepisesarma (Selatium) brockii (De Haan)
Family: Grapsidae
Subfamily: Sesarminae

Color of the carapace is dark brown, chelipeds orange colored with white tips. The distinguishable feature of the species is that the pectinated crests on the upper palm reach the margin and the dactylar tubercles are well separated from one another without any transverse sulcus above. It is a semi-terrestrial form, found in mangrove mud, fissures of rocks, stone embankments, backwater shores, etc. It is very vigilant and intelligent and it trusts its speed and craft to escape its enemies, thus it is hard to pursue. It is captured by handpicking.

Metapograpsus messor (Forskal)
Family: Grapsidae
Subfamily: Grapsinae

Color of the carapace is bottle green, chelipeds appears light violet. The legs are striped with alternate dark and light bands. Carapace is four-fifths as long as broad, and the lateral margins markedly converge backwards. Front is more than half the width of the carapace. The post-frontal region appears to have some transverse markings. The propodus of the chelae is inflated. Walking legs are short, and the dactylus of the legs are nearly as long as the propodus. The last segment of the abdomen appears triangular in males. It is a mud dweller and inhabits the mangroves, subtidal region, etc. It is caught by handpicking, knots made of coconut leaflets, etc.

Metapograpsus maculatus (H. Milne Edwards)
Family: Grapsidae
Subfamily: Grapsinae

Carapace is black in color and chelipeds bright violet. Carapace is comparatively elongated, being seven-eighths as long as broad, with less convergent sides. The post-frontal region is devoid of any transverse markings or ridges. The fingers of the chelipeds are much longer than the upper border of the palm. Walking legs are larger, and the dactylus of the legs distinctly shorter than the propodus. The striking feature of this species is that the last segment of the male abdomen is trilobed. It inhabits the mangroves and the intertidal region. The specimens were caught by handpicking.

Varuna litterata (Fabricius)

Family: Grapsidae
Subfamily: Varuninae

It is commonly known as the 'herring bow crab'. Color of the carapace is light brown to brownish grey. Carapace is squarish, with a smooth surface. Anterolateral borders are cut into three broad, but less sharp teeth. Dactylus, propodus and carpus of legs are laterally flattened like paddles and are fringed with long, densely packed hair. With its legs shaped as paddles used for swimming, it is sometimes called the 'Paddler crab'. It is an estuarine species and is found to occur in the mangrove regions and intertidal areas. It is caught in large numbers in stake nets and Chinese dip nets throughout the year.

Heteropanope indica (De Man)

Family: Xanthidae
Subfamily: Pilumnidae

It is a small crab, with transversely oval shaped carapace, markedly oval and glabrous. Frontal lobe is straight and truncate. Anterolateral borders are cut into four teeth. The dorsal surface of the carapace has two parallel transverse ridges, which are beaded with fine granules. Chelipeds are extremely unequal, with the carpus and the propodus of the chelipeds being studded with pearly granules. The ambulatory legs are slender and sparingly haired. It is estuarine and found to inhabit the mangrove regions. The specimen was caught in a Van Veen grab.

Xenophthalmus pinnotheroides (White)

Family: Pinnotheridae
Subfamily: Xenophthalminae

Color of the carapace is dirty white. Carapace is subtrapezoid in shape, slightly broader than long, with the anterolateral angle bluntly rounded. Frontal region is narrow and strongly deflexed. Orbits are situated longitudinally and parallel to each other. Chelipeds are found to be symmetrical, with palm longer than the fingers. Propodus of the first walking leg is as long as broad. Carpus and propodus of second leg are armed with a tuft of dense pubescence. The third and the last legs are longer and more slender and covered with short hair. The third leg is the longest of all. The crab is caught in large numbers from the backwater, where the substratum was found to be muddy, throughout the year. It was caught in a Van Veen grab as well as in stake nets.

Doclea gracilipes (Stimpson)

Family: Majidae
Subfamily: Pisinae

Carapace is discoid and has a series of blunt spine-like structures arranged longitudinally along the middle line. The body and limbs are covered with thick hair, and the chelipeds are shorter than the other legs. The second pair of legs is the longest and is about three to four times the length of the carapace. Color of the carapace and the segments of the legs covered with hair is greenish brown, while the segments devoid of hair are pink in color. It is a marine species, occurring in 30–50 m depth in sandy or muddy substrata. The specimen was caught in a Chinese dip net from the Cochin backwaters during the pre-monsoon season, when the salinity was high in the sampling location.

Ebalia malefactrix (Kemp)

Family: Leucosiidae
Subfamily: Ebaliinae

The carapace is subcircular in shape, with a separate facet on the side wall of the carapace at the hepatic region. Eyes and orbits are inconspicuous. Chelipeds are stouter and longer than the legs. This tiny crab inhabits the sediments of the backwaters and is collected by operating a Van Veen grab. It occurs in large numbers during the post-monsoon and the pre-monsoon period, when the salinity is comparatively high.

Halicarcinus messor (Stimpson)

Family: Hymenosomatidae

It is a small sized, benthic crab. Carapace is suboval and as long as broad. Lateral walls of the carapace have two teeth, one at the anterolateral angle and the second at the posterolateral angle. The front region is extended to form a rostrum. The rostrum is trilobed, the lateral lobes are small and acute and the median one is long and spatuliform, broadly rounded apically. The median lobe is broadest at halfway along its length and devoid of any setae. Chelipeds are heavy, slightly longer than the legs. Palm of the chelipeds is markedly swollen. It is a benthic species caught in Van Veen grabs.

Elamenopsis alcocki (Kemp)

Family: Hymenosomatidae

It is a small benthic crab with a circular carapace. Frontal region is extended to form a rostrum. Excluding the rostrum, the carapace width to length ratio is 1:1. The anterolateral border is cut into a single tooth. Chelipeds are heavy and the fingers of the chelae have a wide gape at the base. The ambulatory legs are long and slender. It is caught from the muddy sediments of the backwaters using a Van Veen grab.

DISCUSSION

Although much work has been carried out on the benthic fauna, mangrove diversity and fish diversity in the Cochin backwaters, a comprehensive study on the brachyuran crab fauna has not been attempted. Some of the previous works reported that the Cochin backwater supports a low diversity of crabs (Pillai, 1951; Rao and Kathirvel, 1972; Devasia and Balakrishnan, 1985). During the present study 24 species of brachyuran crabs were recorded. Among the 24 species, six of them— *S. serrata, S. tranquebarica, Scylla* sp. (?), *P. pelagicus, P. sanguinolentus* and *C. feriata*—are edible and commercially important. Sheeba (2000) recognized four species—*S. serrata, P. pelagicus, P. sanguinolentus* and *C. feriata*—as the commercially important crab species of Cochin backwaters, and Menon *et al.* (2000) reported that *S. serrata, P. pelagicus* and *P. sanguinolentus* constitute 4 percent of the commercial fishery of Vembanad Lake. Kurup (1990) reported the occurrence of *S. serrata* in the backwaters on a year round basis where a mesohaline condition is prevailing, but in the present study *S. serrata* was found to be available throughout the year, even during monsoon season, where the backwater turns into a freshwater basin. Rao and Kathirvel (1972) reported the occurrence of *P. pelagicus* in the backwaters during the post-monsoon, while Kurup (1990) reported its appearance during the high-saline pre-monsoon period, which is in agreement with the findings of the present study. This species disappears when the salinity falls during the monsoon season. Menon (1953) reported the landing of *P. sanguinolentus* in the Malabar Coast in large numbers during the period from January to April. In the present study too, *P. sanguinolentus* was found to be caught in large numbers during the pre-monsoon as well as post-monsoon period from the Cochin backwaters. Raj (2006) claims that *C. feriata* comes up occasionally in the shore seines and are rarely caught in the Pulicut lake, and after the 2-year sampling in Cochin backwaters, only a few samples were caught, during the pre-monsoon period, when the salinity had almost reached the marine condition. Kurup (1982) recorded the occurrence of the marine crab *C. lucifera* in the Vemabanad Lake during the warmer months. In the present study too, *C. lucifera* is found to occur in the Cochin backwaters during the pre-monsoon period. Manickaraja and Balasubramaniam (2009) reported *T. crenata* to be a deep sea crab, occurring in deep sea gill net operation, never formed a fishery, and lacked any economic importance. In the present study, *T. crenata* was found to occur in the backwaters in significant numbers and were largely caught in gill nets and stake nets from the backwaters, but without any commercial importance. They are not being consumed in the Cochin area.

Pillai (1977) reported six benthic species from the sediments of the Cochin backwaters—*Litocheira* sp., *Viaderiana* sp., *Rynchoplax* sp., *Macrophthalmus* sp., *E. malefactrix, Eriphia smithii*—while Batcha (1984) could find only two benthic species: *Rynchoplax* sp. and *Viaderiana* sp. In the present study, six benthic species were observed—*E. malefactrix, E. alcocki, H. messor,*

H. indica, X. pinnotheroides and *M. depressus*—from the sediments of the Cochin backwaters, mainly where the substratum is muddy.

Roy and Nandi (2008) reported 18 species from the Vembanad Lake and pointed out the requirement of a season-wise survey for about 2 years to assess the exact brachyuran crab diversity status of Vembanad Lake. Radhakrishnan and Samuel (1983) reported the occurrence a subspecies of *Scylla* from the Cochin backwaters and designated as *S. serrata serrata*. However, they could not prove their findings with strong scientific evidence. Joel and Raj (1983) besides Kathirvel and Srinivasagam (1992) proved the third variety to be *S. tranquebarica* itself. However, in the present study, a third variety of Genus *Scylla* has been observed. The specimens are being studied in detail to confirm whether it is a new species or subspecies. The present study was carried out for a period of 2 years, and survey was conducted every month, rather than season wise, and could record 24 species from the Cochin backwaters. However, destruction of mangroves, filling of lands, waste disposal and other human interventions are leading to the impairment of the backwaters. This would affect the diversity of brachyuran crabs in the near future itself, since their habitat is being destroyed.

ACKNOWLEDGEMENTS

The first author is thankful to Kerala State Council for Science, Technology and Environments, Thiruvananthapuram for KSCSTE Research Fellowship. The authors are thankful to the Department of Marine Biology, Microbiology & Biochemistry, School of Marine Sciences, Cochin University of Science & Technology for providing the necessary facilities to carry out this work.

REFERENCES

Ajmal Khan, S., Ravichandran, 2009. Brachyuran crabs. ENVIS publication. CAS in Marine Biology, Annamalai University, Parangipettai. pp. 322–336.

Batcha, A.S.M., 1984. Studies on the bottom fauna of Northern Vembanad Lake Ph.D Thesis, Cochin University of Science and Technology, Cochin India. 157.

Chhapgar. B.F., 1957. Marine crabs of Bombay State. Contribution No.1, Taraporevala Marine Biological Station, Bombay. p. 89.

Chidambaram, K., Raman, R.S.V., 1944. Prawn and crab fishery in Madras. Indian Farming 5, 454–455.

Chopra, B.N., 1939. Some food prawns and crabs of India and their fisheries. J. Bombay Nat. Hist. Soc. 41 (2), 221–234.

Crane, J., 1975. Fiddler crabs of the world: Ocypodidae: Genus *Uca*. Princeton University Press, New York, NY. USA, p. 736.

Devasia, K.V., Balakrishnan, K.P., 1985. Fishery of the edible crab *Scylla serrata* (Forskal) (Decapoda, Brachyura) in the Cochin backwater. In: Ravindran, K., Unnikrishnan Nair N., Perigreen, P.A., Madhavan P., Gopalakrishnan Pillai AG, Panikkar PA and Mary Thomas (eds) Proc Symp On Harvest and Post Harvest Technology of fish. Society of Fisheries Technologists, Cochin, pp 52–56.

Jayabaskaran, R., 1999. Brachyuran crabs of Gulf of Mannar. CAS in Marine Biology Publication, Annamalai University, India, p. 99.

Joel, D.R., Raj, P.J.S., 1983. Taxonomic remarks on two species of the genus *Scylla* de Haan (Portunidae : Brachyura) from Pulicat lake. Indian J. Fisheries 30, 13–26.

Kathirvel, M., Srinivasagam, S., 1992. Taxonomical status of Mud crab *Scylla serrata* (Forskal) from India. In: The Mud crab, Report of the seminar on the Mud Crab Culture and Trade. Surat Thani, Thailand, November 5–8, 1991. Bay of Bengal Program, BOBP/REP/51, Madras, India.

Keenan, C.P., Davie, J.F., Mann, D.L., 1998. A revision of the genus *Scylla* de Haan, 1833 (Crustacea: Decapoda: Brachyua: Portunidae). The Raffles Bull. Zoology 46 (1), 217–245.

Kurup, B.M., 1982. Studies on the systematics biology of fishes of Vembanad lake. Ph.D. Thesis. Cochin University of Science and Technology.

Kurup, B.M., May 27–31, 1990. Fishery and biology of edible crabs of Vembanad Lake. Proc. Sec. Indian Fish Forum, 169–173.

Lucas, J.S., 1980. Spider crabs of the family Hymenosomatidae (Crustacaea; Brachyura) with particular reference to Australian species: systematic and biology. Rec. Australian Mus. 33 (4), 148–247.

Manickaraja, M., Balasubramaniam, T.S., 2009. Occurrence of the deep sea crab *Thalamita crenata* in the shallow water gill net (Mural valai) operation at Tharuvaikulam, Off Tuticorin. Marine fisheries Information service, Technical and extension series (201): p. 28.

Menon, M.K., 1953. A note on the bionomics and fishery of the swimming crab *Neptunus sanguinolentus* (Herbst) on Malabar coast. J. Zool. Soc. India 4 (2), 177–184.

Menon, N.N, Balchand, A.N., Menon, N.R., 2000. Hydrobiology of the Cochin backwater System - A Review. Hydrobiologia 430, 149–183.

Ng, P.K.L., Guinot, D., Davie, P.J.F., 2008. Systema Brachyrorum : Part 1. An annotated list of extant brachyuran crabs of the world. The Raffles Bull. Zool. (17), 286.

Pillai, N.G.K., 1951. Decapaoda (Brachyura) from Travancore. Bull. Cent. Res. Inst., University Travancore, Series C2(1): 1–46.

Pillai, N.G.K., 1977. Distribution and abundance of macrobenthos of the Cochin backwaters. Indian J. Fisheries 6, 1–5.

Radhakrishnan, C.K., Samuel, C.T., 1982. Report on the occurrence of one subspecies of *Scylla serrata* (Forskal) in Cochin backwaters. Fish Technol. 19, 5–7.

Raj, P.J.S., 2006. Macrofauna of Pulicat Lake. NBA Bulletin. 6. National Bioodiversity Authority, Chennai, Tamil Nadu, India. pp. 23–30.

Rao, P.V., Thomas, M.M., Rao, G.S., 1973. Crab fishery resources of India. Proc Symp On living resources of the seas around India, pp. 581–591.

Rao, P.V., Kathirvel, M., 1972. On the seasonal occurrence of *Palinurus polyphagus* (Herbst) and *Portunus pelagicus* (Linnaeus) in the Cochin backwater. Indian J. Fisheries 14, 112–134.

Roy, M.K.D., Nandi, N.C., 2008. Brachyuran Biodiversity of Some Selected Brackishwater Lakes of India. In: Proceedings of Taal 2007: The 12th World Lake Conference, India. (Eds.) Sengupta, M. and Dalwani R. Ministry of Environment and Forests, Government of India, New Delhi, pp. 496–499.

Sakai, T., 1976. Crabs of Japan and adjacent seas. Kodansha Ltd., Tokyo, p. 773.

Sethuramalingam, S., Ajmal Khan, S., 1991. Brachyuran Crabs of Parangipettai Coast. CAS in Marine Biology Publication, Annamalai University, India, p. 92.

Sheeba, P., 2000. Distribution of benthic infauna in the Cochin backwaters in relation to the environmental parameters. Ph.D. Thesis. Cochin University of Science and Technology. p. 453.

Chapter 6

Status of Horseshoe Crabs at Digha, Northern East Coast of India

P. Yennawar

Freshwater Biology Regional Centre, Zoological Survey of India, Hyderabad, Andhra Pradesh, India

INTRODUCTION

Horseshoe crabs are threatened in many areas as a result of worldwide exploitation for limulus amoebocyte lysate (LAL), an important diagnostic reagent prepared from the blue blood of the animal (Rudloe, 1979). This reagent is capable of detecting extremely minute quantities of pyrogenic endotoxin produced by Gram-negative bacteria, permitting its use to test for endotoxin on medical devices, implants and vaccines. Collection of this animal, extraction of blood and transportation can be stressful to the estimated 250,000 animals annually. A 2008 report in *Marine News* indicated that two species—*Carcinoscorpius rotundicauda* and *Tachypleus gigas* (Müller, 1785)—are being poached in their thousands by children engaged by local fishermen to collect the crabs, which are then sold to pharmaceutical companies (Pati, 2008). In addition, harvesting for commercial purposes is considered a serious threat to the survival of these species in Malaysia (Christianus and Saad, 2007). With the increasing interest in *C. rotundicauda* for medical research (Ding *et al.,* 2005; Ng *et al.*, 2007) and the additional threats of habitat loss due to coastal development in Southeast Asia, such information is urgently needed to assess the ecological requirements and conservation status of this species and to manage and conserve it. Loss of habitat is listed as one of the two main threats to horseshoe crabs in Singapore (Davidson *et al.*, 2008); the other is pollution. According to the IUCN Red Data Book, one species is near threatened (*Limulus polyphemus*), and data on the other three are deficient (IUCN, 2010). The Singapore Red Data book released in November 2008 classifies the mangrove horseshoe crab *C. rotundicauda* as vulnerable (Davidson *et al.*, 2008).

Out of four horseshoe crab species in the world, two are reported in India: *Tachypleus gigas* (Müller, 1785) and *Carcinoscorpius rotundicauda* (Latreille,

Marine Faunal Diversity in India. DOI: 10.1016/B978-0-12-801948-1.00006-9
89

1802). *T. gigas* is reported in the Bay of Bengal, particularly along the coast of Odisha to Indo-China, North Vietnam, Borneo and Celebes. *C. rotundicauda* is reported on the northern shores of the Bay of Bengal in Odisha and Sunderban of West Bengal to the southern coast of the Philippines (Chatterji, 1994; Chaterji *et al.*, 1992, 1996; Mishra, 2009). The Digha coast, which lies on the border of Odisha and West Bengal in the northern part of the Bay of Bengal, hosts both of these species. However, there is no information available on the population, abundance, etc. of these species in the area, which was largely ignored in previous studies. There is little recent information on abundance, sex ratios, population structure or densities of *C. rotundicauda*. Unavailability of population trends of horseshoe crabs in the area also hinders the conservation initiatives of their natural habitats as well as awareness for their over-exploitation. The previous observations in the area show that there was a fair population of these species along this coast. But the present status is largely poor, and the cause of reduction of the natural population may be habitat destruction by natural or anthropogenic factors.

METHODS

Digha beach is situated close to the Gangetic mouths on the east coast of India facing the Bay of Bengal (21° 36' N and 87° 30' E). Here, the sea is quite shallow with very little wave action on the beach, and an extensive area of about 250 m of the intertidal zone is exposed during low tides. The beach slope is very low up to the low water mark. The shore was subjected to considerable erosion in the recent past. Digha and the surrounding coastal area are variable habitats. The coastline here is straight and the beach, composed of sand grains mixed with variable proportions of silt, is flat and compact. Digha has potential coastline of about 15 km which offers scope for more effective exploitation of marine fishery resources. The present study was conducted at five different locations: Talsari, Udaipore, Digha, Mohana and Shankarpur (Figure 6.1). The observations of the abundance of horseshoe crabs were made in an intertidal stretch of a kilometre at each study location at monthly intervals during the collection of live ornamental fauna for display in an aquarium. The number of horseshoe crabs was counted in the sampling area. The average number each year was used for comparison of population in the study area.

RESULTS AND DISCUSSION

During the study period, it was observed that two species of horseshoe crab—*Tachypleus gigas* (Müller, 1785) and *Carcinoscorpius rotundicauda* (Latreille, 1802)—regularly occur in the area. The average number of sightings of *T. gigas* and *C. rotundicauda* on Digha coast is shown in Figure 6.2.

Of the five study locations, Udaipur and Talsari showed the highest number of horseshoe crab sightings. This may be due to less disturbance in the

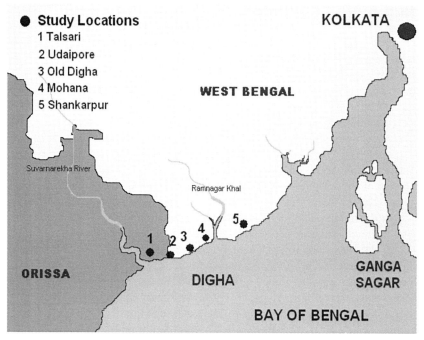

FIGURE 6.1   Study locations on Digha coast.

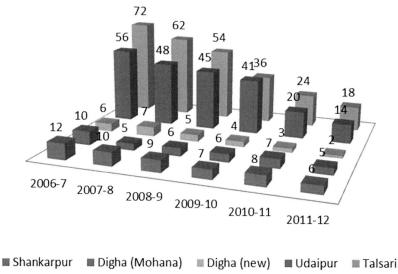

■ Shankarpur   ■ Digha (Mohana)   ■ Digha (new)   ■ Udaipur   ■ Talsari

FIGURE 6.2   Average number of sightings of horseshoe crabs.

habitat as well as to the topography of the area. During the study period of 6 years, the population decreased considerably in all locations. It was observed that the mature pairs of both species migrate towards the shores during their breeding period. *Carcinoscorpius rotundicauda* prefers nesting in muddy mangrove swamps whereas *Tachypleus gigas* breeds on clean sandy beaches. As the number of sightings of both species varied little throughout the year, we infer that horseshoe crabs are engaged in breeding activity all year round. Digha coast's muddy and sandy beaches provide a suitable habitat for breeding, and it hosts both species in the area. Taylor *et al.* (2009) also observed breeding of *C. rotundicauda* throughout the year during their study in tropical waters. The American horseshoe crab *Limulus polyphemus* shows marked seasonal breeding patterns in the temperate latitudes where tidal amplitudes are high, but this species shows no such seasonal breeding pattern farther south off Florida, in estuaries and microtidal lagoons with low tidal amplitudes and warmer waters (Ehlinger and Tankersley, 2009). In these southern waters, as at Kranji, which is also an estuary with low tidal amplitudes and warm waters, no large spawning aggregations were seen, and spawning is not triggered by environmental cues (Ehlinger and Tankersley, 2007); instead spawning in Florida was protracted and occurred year round but was either aperiodic (Ehlinger and Tankersley, 2009) or showed episodes of increased mating activity in early spring (Ehlinger *et al.*, 2003). *C. rotundicauda* in the tropics seems to display a similar pattern (Taylor *et al.,* 2009).

As per the Singapore Red Data book (2008), *C. rotundicauda* is classified as vulnerable, and both species classify for the Data Deficient category as per the IUCN Red List. The present study on the population status of such threatened species will be useful for management and conservation of these species on the Digha coast. The study may also be useful for artificial propagation of these species in nature, since the Digha coast is a suitable habitat for the horseshoe crab.

ACKNOWLEDGEMENTS

I am thankful to Dr K. Venkataraman, Director, Zoological Survey of India, Kolkata for providing facility and guidance during the study.

REFERENCES

Chatterji, A., 1994. The Horseshoe Crab - A Living Fossil, A Project. Swariya Publication, 1994, p. 157.

Chatterji, A., Vijaykumar, R., Parulekar, A.H., 1992. Spawning migration of the horseshoe crab, Tachypleus gigas (Muller), in relation to lunar cycle. Asian Fisheries Science 5, 123–128.

Chatterji, A., Parulekar, A.H., Qasim, S.Z., 1996. Nesting behavior of the Indian horseshoe crab, *Tachypleus gigas* (Muller) (Xiphosura). In: Qasim, S.Z., Roonwal, G.S. (Eds.), India's Exclusive Economic Zone. Omega Scientific Publishers, New Delhi, 1996, pp. 142–148.

Christianus, A., Saad, C.R., 2007. Horseshoe crabs in Malaysia and the world. Fish Mail 16, 8–9.

Davidson, G.W.H., Ng, P.K.L., Ho, H.C., 2008. The Singapore Red Data Book: Threatened Plants and Animals of Singapore. The Nature Society (Singapore), Singapore, p. 285.

Ding, J.L., Tan, K.C., Thangamani, S., Kusuma, N., 2005. Spatial and temporal co-ordination of expression of immune response genes during *Pseudomonas* infection of horseshoe crab. Genes Immun. 6, 557–574.

Ehlinger, G.S., Tankersley, R.A., 2007. Reproductive ecology of the American horseshoe crab *Limulus polyphemus* in the Indian River Lagoon: an overview. Fla. Sci. 70, 449–463.

Ehlinger, G.S., Tankersley, R.A., 2009. Ecology of horseshoe crabs in microtidal lagoons. In: Tanacredi, J.T., Botton, M.L., Smith, D.R. (Eds.), Biology and conservation of horseshoe crabs. Springer, New York, pp. 149–162.

Ehlinger, G.S., Tankersley, R.A., Bush, M.B., 2003. Spatial and temporal patterns of spawning and larval hatching by the horseshoe crab, *Limulus polyphemus*, in a microtidal coastal lagoon. Estuaries 26, 631–640.

IUCN 2010. Red List of Threatened Species. Version 2010:2. Downloaded in August 2010.

Mishra, J.K., 2009. Horseshoe crabs, their eco-biological status along the north-east coast of India and the necessity for ecological conservation. In: Tanacredi, J.T., Botton, M.L., Smith, D.R. (Eds.), Biology and conservation of horseshoe crabs. Springer, New York, pp. 89–96.

Ng, P.M.L., Saux, A., Le, C.M., Lee, N.S., Tan, J., Lu, S., Thiel, B., Ho, J.K., Ding, 2007. C-reactive protein collaborates with plasma lectins to boost immune response against bacteria. EMBO J 26, 3431–3440.

Pati, B., 2008. Horseshoe crabs galloping towards extinction. *Marine News*, 24 June 2008. www.merinews.com/article/horseshoe-crabs-galloping-towards-extinction/136265.html

Rudloe, A., 1979. *Limulus polyphemus.* A review of the ecological significant literature. In: Cohen, E. (Ed.), Biomedial application of the horseshoe crab (Limulidae). Liss, New York, pp. 27–35.

Taylor, L.C., Julian, L., Chia, C.H., 2009. Population structure and breeding pattern of the mangrove horseshoe crab *Carcinoscorpius rotundicauda* in Singapore. Aquatic Biology 8, 61–69.

Chapter 7

Diversity of Bryozoans of India with New Records from Maharashtra

M. Mankeshwar,* A. Kulkarni* and D. Apte[†]
*Department of Zoology, Gogate Jogalekar College, University of Mumbai, Ratnagiri, Maharashtra, India; [†]Bombay Natural History Society, Mumbai, Maharashtra, India

INTRODUCTION

Bryozoans are the dominant colonial invertebrates of the intertidal zone. Bryozoan studies from the coastal waters of Maharashtra have been undertaken by many workers. In an important work, Pillai (1978) described *Hippoporina indica* as a new species and a subspecies *Parasmittina crosslandi serrata* from Mumbai waters. Studies on bryozoans as components of the fouling community have been carried out by Swami and Udhayakumar (2010) and Swami and Karande (1987). Life-histories of cheilostome bryozoans have also been documented (Karande and Udhayakumar, 1992). A study by Gaonkar *et al.* (2010) to investigate the species composition of benthic sessile fauna at Mumbai harbor listed 49 species of bryozoans. Study of bryozoans through artificial recruitment has thus been prevailing, and taxonomic works, although present, are limited (Chapgar and Sane, 1996; Pillai, 1978, 1981; Raveendran *et al.,* 1990).

In the context of Maharashtra, studies on this phylum of invertebrates have been concentrated on the waters around Mumbai, hence leaving most of the state's vast coastline still unexplored. Apart from a reference work by Alan *et al.* (1988) on marine biofouling at Ratnagiri, Maharashtra, the present study is the first of its kind undertaken on the Konkan coast and gives an account of four new bryozoan records to Maharashtra along with their range and key. This chapter further includes a literature review of the recent marine bryozoans from Indian waters.

LITERATURE REVIEW

On account of their resilient nature and complex structure, bryozoans have been of interest to a number of classical taxonomists. Accounts of bryozoan studies from Indian waters can be found from as early as 1887 when Hincks described seven species from the Mergui archipelago. Thornely (1905, 1906, 1907, 1912,

Marine Faunal Diversity in India. DOI: 10.1016/B978-0-12-801948-1.00007-0
95

and 1916) documented bryozoans from the offshore waters of Orissa and Andhra Pradesh to the Gulf of Mannar, Sri Lanka, and Okhamandal (Gujarat) and described sixteen new species and one new genus. Intertidal and offshore species of bryozoans from the Bay of Bengal were described by Robertson (1921) along with new species of abyssal bryozoans, *Kinetoskias* and *Farciminaria*. Studies by Annandale (1906, 1907a,b, 1908, 1911a,b, 1912) also added greatly to knowledge of the bryozoan fauna of India. Much of his work has been on brackish and fresh water bryozoans such as *Bowerbankia* and *Victorella*, which have been described in much detail.

A few of the earlier studies on bryozoans as part of the fouling community were carried out by Kurien (1950), Danial (1954) and Antony Raja (1959). Their succession and seasonal distribution were studied, along with those of other sedentary organisms.

Regarding the more recent taxonomical works, contributions by Menon (1967, 1972b,c,d, 1973, 1974a,b) and Menon and Nair (1967a,b, 1969a,b, 1970a,b, 1971, 1972b, 1973, 1974a) have been tremendously good in documenting the bryozoan fauna of the south, southeast and southwest coast of India. Long term studies were undertaken to understand bryozoan diversity through artificial recruitment, dredging, trawling and random sampling. Four species of the genus *Rhynchozoon* (1974a) and nine species of the genus *Scrupocellaria* were described by Menon (1972a), six of them being new species. A new species of *Schizoporella*, three species of *Tremogasterina* and six species of *Bugula* were described by Menon and Nair (1970a,b, 1972b). They also described 22 species of Malacostegan bryozoans of which *Electra crustulenta* subspecies *borgii* was new to science (1975). Effects of salinity on two species of bryozoans were also studied (Menon and Nair, 1974b) along with the growth rate of four species of intertidal bryozoans from the backwaters of Cochin (Menon and Nair, 1972a). Nair (1989, 1991) has done considerable work, among which he has listed 15 species of Bryozoans from the Vellar estuary in Tamil Nadu. Menon *et al.* (1977) also compared the settling of oysters, hydroids and polychaetes with bryozoans at the Mangalore harbor. Geetha (1994) made toxicological observations on the effect of copper, cadmium and mercury on three species of bryozoans—*Victorella pavida, Electra crustulenta* and *Electra bengalensis*—along with taxonomic reports on Indian (southwest) and Antarctic bryozoans. Soja (2006) studied taxonomy, bionomics and biofouling of bryozoans onboard FORV Sampada along the west coast of India, recording 102 species of which three were new to science. A monograph on the taxonomy of bryozoans from the Indian EEZ by Menon and Menon (2006) is the most concise work on bryozoan taxonomy from India, with illustrated descriptions of 128 species. *Biflustra perambulata* was the new alien species from Cochin described by Soja and Menon (2009).

Katti and Rao (1976) noted *Bugula neritina* as an important fouler on marine trawlers. From Vishakhapatnam on the southeast coast of India, works by Ganapati *et al.* (1969) and Rao and Ganapati (1972b, 1975, 1978) on bicellariellid bryozoa and littoral bryozoans are prevalent. In an earlier account, Ganapati and Rao (1968) reported nine species of bryozoans from various sites including the Visakhapatnam naval base. In their works they have listed four species of *Electra* for the first time from Visakhapatnam (1972b) and also have provided an account on the epizoic fauna on *Thalamoporella var. indica* and *Pherusclla tubulosa* (1980). Rao and Viswanadham (1984) gave the first ancestrular study of *Thalamoporella stapifera* and notes on its astogeny. Rao (1975) has also documented fauna from the northeast coast of India.

Joseph (1978) in his work on seaweed-associated fauna, listed three species— *Electra indica, Thalamoporella rozierii* and *T. hamata*—which commonly encrusted algal fronds from Palk Bay. Other than the aforementioned studies on the west coast of India, research on bryozoans has been carried out on and around the Mumbai coastline. As stated in the introduction, taxonomic studies have been undertaken by Pillai and Santhakumaran (1972) and Pillai (1978, 1981). New species *Hippoporina indica* and a subspecies *Parasmittina crosslandi serrata* were described by Pillai (1978). Chapgar and Sane (1996) listed eleven previously unrecorded species of bryozoan from Mumbai. A study by Swami and Karande (1987) on the polluted coastal waters of Mumbai documents seven species of cheilostome bryozoans; *Electra bengalensis* was found to be the most prevalent, occurring throughout the year. They also recorded seven species of encrusting bryozoans from the waters of Karwar, on the central west coast of India (1994). They have carried out extensive research on biofouling along the shores of Mumbai (Karande and Swami 1988; Swami and Gaokar 1998). Bryozoans from the offshore waters of Mumbai were studied by Raveendran *et al.* (1990) where they recorded seven cheilostome and one cyclostome bryozoa. From the Zuari estuary at Goa a study on the benthic fouling community was undertaken by Anil and Wagh (1988). As per the Zoological Survey of India (2013) an estimated 200 species of bryozoans have currently been documented, forming 6 percent of the total global bryozoan diversity.

METHODS

Direct searches were made at the rocky shores of Ratnagiri (Mandavi and Mirya). Bryozoans were collected from the undersides of rocks from the tidal pools by scraping the colony using a surgical knife. The collected colonies were cleaned in sodium hypochlorite and then stored in 70 percent ethanol. For identification, colonies were observed under a stereomicroscope (Zeiss) and a compound microscope (Zeiss). The specimens were deposited in the collections of The Bombay Natural History Society.

TAXONOMY

Bowerbankia gracilis

    Order: Ctenostomata (Busk, 1852)
    Family: Vesiculariidae (Hincks, 1880)
    Genus: *Bowerbankia* (Farre, 1837)
    Species: *gracilis* (Leidy, 1855)

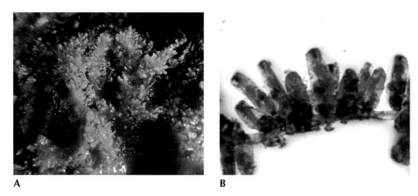

A   B

FIGURE 7.1   (a) *Bowerbankia gracilis* colony seen growing along *Watersipora subovoidea.* (b) *Bowerbankia gracilis* showing arrangement of zooids around the internode (Magn: 10×).

Description (Figure 7.1a&b)   One of the most abundant species in Ratnagiri. Found in shallow rock pools in association with other ctenostomes. Colonies straw colored. Zooids arising from creeping stolons with irregular branching. Zooids seen either in pairs or clumped around the stolon. Stolons more tufted near the start of a colony. Young colonies transparent but become brown with age and more so because of detritus. Colorless lophophores with 10 tentacles. Zooids tubular with squarish aperture. Zooids with polypides longer than the empty ones. Caudate processes seen in few zooids. Gizzard present. Embryo pink in color.

Remarks   The present form resembles the one described by Menon (1967) with respect to the presence of zooids with caudate appendages, although few and the diagonally positioned bundles of parietal muscles could be distinctly observed in a few clear zooids.

Previous Records from India   Kerela (Menon, 1972b; Menon and Nair, 1967a, 1971; Soja, 2006)

Distribution   Pacific coast of North America (Osburn, 1953), Japan (Mawatari, 1953), Atlantic coast of Florida (Winston, 1982), New Zealand (Gordon and Mawatari, 1992).

Catalogue number   BNHS BRY. 070

Bugulella clavata

> Order Cheilostomata (Busk, 1852)
> Family Bugulidae (Gray, 1848)
> Genus *Bugulella* (Verrill, 1879)
> Species *clavata* (Hincks, 1887)

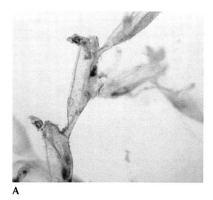

A   B

FIGURE 7.2   (a) *Bugulella clavata*: Autozooids with the beak-shaped avicularia and posteriorly placed globose points of lateral branching (Magn: 10×). (b) *Bugulella clavata*: Colony growing in the wild.

Description (Figure 7.2a&b)   Colony white. Chains of zooids interconnected and forming a well-spaced meshwork on the substrate. Growth of zooids more bent towards the substratum than erect. Zooids elongate with a narrow tubular proximal end and the aperture covering most of the distal end. The tubular extension of the zooid joins the previous one at the back of the aperture. Two blunt projections arise disto-laterally from the aperture. A pair of beak-shaped avicularia is articulated disto-laterally. The rostrum curving steeply and the mandible with a bend at its tip. Points of branching can be made out from swollen masses appearing laterally. Branching zooids arising at right angles. No ovicells noticed.

Remarks   The opesia in the present material and the aviculariaare more like *Beania* (Menon and Nair, 1967a) although spines are not observed except for the distal orificial processes. Also the present form resembles greatly the description of *Beania klugie* (Tilbrook *et al.*, 2001), and so the possibility of *B. klugie* being a junior synonym of *B. clavata* remains (Cook, 1985).

Previous records from India   Gulf of Mannar (Menon, 1967)

Distribution   Mergui archipelago, Bay of Bengal (Hincks, 1887)

Catalogue number   BNHS BRY. 054

Electra indica

Order Cheilostomata (Busk, 1852)
Family Electridae (d'Orbigny, 1851)
Genus *Electra* (Lamoroux, 1816)
Species *indica* (Menon & Nair, 1967)

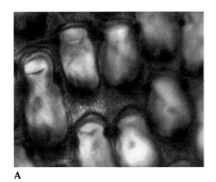 

A   B

FIGURE 7.3 (a) *Electra indica*: Zooids showing gymnocystal pores and three marginal tubercles (Magn: 10×). (b) *Electra indica*: Colony in the wild encrusting on *Sargassum* sp.

Description: (Figure 7.3a&b) Colony glistening white. Encrusting mainly on *Sargassam* sp. and *Gracilaria* sp. Zooids growing in well demarcated rows centrally with a tendency to bifurcate towards the edges. Oval opesia. The gymnocyst is more developed proximally and marked with defined pores. The margin of the zooids is occupied by three tubercles: two placed disto-laterally and one median on the proximal side, the median tubercle being the longest and sometimes being twice the zooid size. A membrane stretches between the two tubercles giving the zooid a hooded appearance. Cryptocyst is present.

Previous Records from India First described from Kovalam by Menon and Nair (1967a), Southwest coast (Menon and Nair, 1975), Kochi (Soja, 2006), Palk Bay (Joseph, 1978).

Distribution To date the species has only been recorded from Indian waters.

Catalogue number BNHS BRY.035

Thalamoporella rozierii
    Order Cheilostomata (Busk, 1852)
    Family Thalamoporellidae (Levinson, 1909)
    Genus *Thalamoporella*
    Species *rozierii* (Audouin, 1826)

Description: (Figure 7.4) Colony white, encrusting and unilaminar. The black polypides give a spotted appearance. Zooids elongated hexagonal. Thick margins separate the zooids. Orifice is broad and circular. Cryptocyst present around the orificial rim and raised proximally. Lateral tuberosities are continuous with the orificial rim. These are not seen in ovicellate zooids. Ovicells bilobate. Cryptocyst granulated finely with perforating opesiules present suborally laterally. Some being more elongate than others. Vicarious avicularia

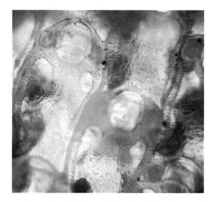

FIGURE 7.4 *Thalamoporella rozierii*: Autozooids showing the proximally raised cryptocystal margin (Magn: 10×).

unseen. Three types of spicules are noticed: calipers, compasses and a third 'U' shaped one.

Remarks The presence of a third 'U' shaped spicule is of importance in identifying this species and in agreement with earlier described specimen from India (Menon and Nair, 1967a).

Previous Records from India Arabian Sea (Menon, 1967; Soja, 2006), Indian Ocean (Thornely, 1907), Bay of Bengal (Robertson, 1921), Gulf of Kachchh (Rao and Sastry, 2005), Palk Bay (Mohan, 1978).

Distribution Red Sea (Audoin, 1826); Atlantic Ocean (Hincks, 1880).

Catalogue number BNHS BRY.072

DISCUSSION

All the species reported above are new bryozoan records to the Konkan coast. A taxonomic key to species is presented in Box 7.1. *B. clavata* requires more detailed examination to confirm its systematics. The described species favor distinct niches of the rocky intertidal area: *B. gracilis* occurring in open shallow rock pools, *E. indica* encrusts floating algae mostly at low tide mark, whereas *T. rozeirii* forms patches in shady and comparatively cooler rock pools. Hence a more comprehensive study to determine the ecology and associations of individual species is required, along with study on reproductive behavior. As understood from the review of the literature, considerable work has been done on the taxonomical as well as from the biofouling aspect; but, apart from a few, most of the work is limited to certain pockets on the coastline, and still a vast stretch remains unexplored. Studies, such as the present one, from other parts will certainly add to the knowledge on bryozoan diversity in India.

Box 7.1 Key Identifying Features

1. **Ctenostome:** Zooids membranous or gelatinous. Orifice terminal. No avicularia or ovicells.
2. **Cheilostome:** Zooids calcified. Orifice frontal with variously developed hinged operculum
3. **Vesiculariidae:** Creeping. Cylindrical zooids uniform anteriorly and posteriorly. Membrane very thin
4. ***Bowerbankia gracilis:*** Tentacles 10. Clumped masses of zooids. Gizzard present. Pink embryo
5. ***Electridae:*** Encrusting. Opesia wide with spines. Extensive gymnocyst. Ovicells and avicularia absent
6. ***Electra indica:*** Gymnocyst with pores. Three tubercles
7. **Bugulidae:** Colony formed by lateral branching of zooids. Aperture covering the front. Articulated beak-shaped avicularia
8. ***Bugulella clavata:*** Distal lateral avicularia in pairs. Swelling of membrane at point of branching
9. **Thalamoporellidae:** Body cavity containing spicules. Well-developed median process. Opesiules present. Ovicellsbilobate. Various avicularia
10. ***Thalamoporella rozierii:*** Avicularia absent. Assymetrical opesiules

ACKNOWLEDGEMENTS

This study would not have been possible without the constant support of members of the Bombay Natural History Society. We are grateful to the Principal, Dr S. Deo, Gogate Jogalekar College for providing us with their laboratory facilities and for his support throughout. We would also like to thank Ms Reshma Pitale for her help with collections of bryozoan colonies on field.

REFERENCES

Alan, S.M., Khan, A.K., Nagabhushanam, R., 1988. Marine biofouling at Ratnagiri Coast, India. In: Thompson, M.P., Sarojini, R., Nagabhushanam, R. (Eds.), Marine Biodeterioration. Oxford Publishing Company, pp. 539–550.

Anil, A.C., Wagh, A.B., 1988. Aspects of biofouling community development in the Zuari Estuary, Goa, India. In: Thompson, M.P., Sarojini, R., Nagabhushanam, R. (Eds.), Marine Biodeterioration. Oxford Publishing Company, pp. 529–538.

Annandale, N., 1906. Notes on the fresh water fauna of India, No. 11. Affinities of I Hislopia. Proc. Asiatic Soc. Bengal 2, 59–63.

Annandale, N., 1907a. The fauna of brackish ponds at Port Canning, Lower Bengal, Pt. 1. Int. and preliminary account of the fauna. Rec. Ind. Mus. 1, 35–43.

Annandale, N., 1907b. Fauna of Chilka Lake. Polyzoa of the lake and of brackish water from the Gangetic delta. Rec. Indian Mus. 1, 179–196.

Annandale, N., 1908. The fauna of brackish ponds at Port Canning, Lower Bengal. Pt. 7. Further observations on the Polyzoa with the description of a new genus of Entoprocta. Rec. Ind. Mus. 2, 11–19.

Annandale, N., 1911a. Fresh water sponges, hydroids and Polyzoa. Fauna of British India 3, 161–251.

Annandale, N., 1911b. Systematic notes on the Ctenostomatous Polyzoa of freshwater. Rec. Ind. Mus. 6, 193–201.

Annandale, N., 1912. Fauna Symbiotica Indica. Polyzoa attached to Indo-Pacific Stomatopods. Rec. Ind. Mus. 7, 147–150.

Antony Raja, B.T., 1959. Studies on the distribution and succession of sedentary organisms on the Madras Harbour. J. Mar. Biol. Assoc. India 1, 180–197.

Audouin, J.V., 1826. Explication sommaire des Planches de Polypes de l'Egypte. Hist. Nat. 1.

Chapgar, B.F., Sane, S.R., 1996. Intertidal entoprocta and ectoprocta (bryozoa) of Bombay. J. Bombay Nat. Hist. Soc. 63, 449–454.

Cook, P.L., 1985. Bryozoa from Ghana. Annales du Musée Royal de l'Afrique Centrale. Sciences Zoologique 283, 1–135.

Danial, A., 1954. Seasonal variation and distribution of the fouling communities in the Madras harbor waters. J. Madras Uni. 24, 189–212.

Ganapati, P.N., Rao, K.S., 1968. Fouling Bryozoans in Visakhapatnam Harbour. Curr. Sci. 37, 81–83.

Ganapati, P.N., Rao, K.S., Rao, M., 1969. Record of Kinetoskias sp. (Bicellarielids, Polyzoa) from Visakhapatnam coast, Bay of Bengal. Current Science 38 (16), 387.

Gaonkar, C.A., Sawant, S.S., Anil, A.C., Krishnamurthy, V., Harkantra, S.N., 2010. Changes in the occurrence of hard substratum fauna: A case study from Mumbai harbor India. Indian J. Mar. Sci. 39 (1), 74–84.

Geetha, P., 1994. Indian and Antarctic bryozoans: taxonomy and observations on toxicology. Ph.D. Thesis. p. 234.

Gordon, D.P., Mawatari, S.F., 1992. Atlas of marine fouling Bryozoa of New Zealand ports and harbours. Misc. Publ. NZOI. 107, 1–52.

Hincks, T., 1880. A History of British Marine Polyzoa. Van Voorst, London, p. 601.

Hincks, T., 1887. On the Polyzoa and Hydroids of the Mergui archipelago collected for the trustees of the Indian museum. J. Linn. Soc. London (Zool.) 21, 121–136.

Joseph, M.M., 1978. Ecological studies on the Fauna Associated with Economic Seaweeds of South India—I. Species composition, feeding habits and interrelationships. Seaweed Research Utilization 3 (1), 9–25.

Karande, A.A., Swami, B.S., 1988. Importance of test coupons in the assessment of marine biofouling community development in coastal waters of Bombay. Indian J. Mar. Sci. 17, 317–321.

Karande, A.A., Udhayakumar, M., 1992. Consequences of crowding on life-histories of bryozoans in Bombay waters. Indian J. Mar. Sci. 21, 133–136.

Katti, R.J., Rao, D.K., 1976. Notes on *Bugula neritina* a fouling Bryozoan from Indian waters. Current Research 5, 104–105.

Kurien, G.K., 1950. The fouling organisms of Pearl Oyster Cages. J. Bombay Hist. Soc. 49, 90–92.

Mawatari, S., 1953. On *Electra angulata* Levinson, one of the fouling bryozoans in Japan. Miscellaneous Reports of the Research Institute of Natural Resources 32, 17–27.

Menon, N.R., 1967. Studies on the Polyzoa of the south-west coast of India. Ph.D Thesis, University of Kerala, p. 548.

Menon, N.R., 1972a. Species of the Genus *Scrupocellaria* Van Beneden (Bryozoa, Anasca) from Indian waters. Int. Revue. Ges. Hydrobiol. 57 (5), 801–819.

Menon, N.R., 1972b. Species of the suborder Ctenostomata Busk (Bryozoa) from Indian waters. Int. Revue. Ges. Hydrobiol. 57 (4), 599–629.

Menon, N.R., 1972c. Species of the genus *Parasmittina* (Bryozoa: Ascophora) from Indian waters. Mar. Biol. 14 (1), 72–84.

Menon, N.R., 1972d. Vertical and horizontal distribution of fouling bryozoans in Cochin backwaters, west Coast of India. Recent Researches in Bryozoa, Academic Press. London 14, 153–164.

Menon, N.R., 1973. Species of the sub-order Ctenostomata Busk (Bryozoa) from Indian waters - distributional aspects. In: Zeitschel, B. (Ed.), The biology of the Indian Ocean - Ecological Studies III. Springer Verlag, New York, pp. 407–408.

Menon, N.R., 1974a. Four species of Bryozoa belonging to the genus *Rhynchozoon* from the Indian waters. J. Bombay Nat. Hist. Soc., 115–119.

Menon, N.R., 1974b. Notes on two species of *Cleidochasma* (Bryozoa) from the Indian waters. J. Bombay Nat. Hist. Soc., 109–111.

Menon, N.R., Menon, N.N., 2006. Taxonomy of bryozoans from the Indian EEZ. Monograph. Ocean Science and Technology Cell. CUSAT. 1–263

Menon, N.R., Nair, N.B., 1967a. The ectoproctous bryozoans of the Indian waters. J. Mar. Biol. Assoc. India 9 (2), 12–17.

Menon, N.R., Nair, N.B., 1967b. Observations on the structure and ecology of *Victorella pavida*, Kent from the southwest coast of India. Int. Revue Ges. Hydrobiol. 52 (2), 237–256.

Menon, N.R., Nair, N.B., 1969a. Rediscovery of *Bugulella elavata* Hincks, 1887 (Ectoprocta) from the Indian Ocean. Current Science 5, 116–117.

Menon, N.R., Nair, N.B., 1969b. Notes on *Alcyonidium erectum* Silen, from the Indian Ocean. Current Science 38, 439–440.

Menon, N.R., Nair, N.B., 1970a. On a new species of the genus *Schizoporella* Hincks from the Indian Ocean. Current Science 39, 238–239.

Menon, N.R., Nair, N.B., 1970b. Three species of the genus *Tremogasterina* Canu (Bryozoa) from the Indian Ocean. Current Science 39, 258–261.

Menon, N.R., Nair, N.B., 1971. Ecology of fouling bryozoans in Cochin waters. Mar. Biol. 8, 280–307.

Menon, N.R., Nair, N.B., 1972a. The growth rates of four species of intertidal bryozoans in Cochin backwaters. Proc. Indian Nat. Sci. Acad. 38, 397–402.

Menon, N.R., Nair, N.B., 1972b. Indian species of the genus *Bugula* Oken. Proc. Ind. Nat. Sci. Acad. 38 (8), 403–413.

Menon, N.R., Nair, N.B., 1973. Species of the genus *Parasmittina* Osburn (Bryozoa Ascophora) from Indian waters – Distributional aspects. In: Zeitschel, B. (Ed.), The biology of the Indian Ocean Ecological studies. Springer Verlag, New York, pp. 405–406, 532.

Menon, N.R., Nair, N.B., 1974a. Two new records of *Hippopodina* (Bryozoan) from the Indian waters. J. Bombay Nat. Hist. Soc., 112–114.

Menon, N.R., Nair, N.B., 1974b. On the nature of tolerance to salinity in two euryhaline intertidal bryozoans *Victorella pavida* Kent and *Electra crustulenta* Pallas. Bull. Ind. Nat. Sci. Acad. 47, 414–424.

Menon, N.R., Nair, N.B., 1975. Indian species of Malacostega (Polyzoa, Ectoprocta). J. Mar. Biol. Assoc. India 17 (3), 553–579.

Menon, N.R., Katti, R.J., Shetty, H.P.C., 1977. Observations on the biology of fouling from Mangalore waters. Mar. Biol. (New York) 41, 127–140.

Nair, P.S.R., 1989. Studies in Bryozoa (Polyzoa) of the Southeast Coast of India. Ph.D. thesis. Annamalai University, Tamil Nadu. p. 165.

Nair, P.S.R., 1991. Occurrence of bryozoan in Vellar estuarine region, Southeast coast of India. Indian. Mar. Sci. 20 (4), 277–279.

Osburn, R.C., 1953. Bryozoans of the pacific coast of America. Part 2. Cheilostomata-Ascophora. Rep. Allan Hancock Pacific Exped. 14 (2), 271–611.

Pillai, S.R.M., 1978. A new species of *Hippoporina* from Bombay waters. Current Science 47, 61–63.

Pillai, S.R.M., 1981. A further report on taxonomy of fouling bryozoans from Bombay harbor and vicinity. J. Bombay Nat. Hist. Soc. 78, 317–329.

Pillai, S.R.M., Santhakumaran, L.N., 1972. Two new records of Bryozoans from Indian waters. J. Bombay Nat. Hist. Soc. 68 (3), 824–844.

Rao, S.N.V., Sastry, D.R.K., 2005. Fauna of Marine National Park Gulf of Kachchh (Gujarat). Zoological Survey of India, Kolkata, ISBN 10: 8181710614.

Rao, K.S., 1975. The Systematics and some aspects of the Ecology of Littoral bryozoan on the north-east coast of India, Ph.D. Thesis, Andhra University, Waltair. p. 235.

Rao Satyanarayan, K., Ganapati, P.N., 1972a. Some new and interesting bicellariellids (Polyzoacheilostomata) from Visakhapatnam coast, Bay of Bengal. Proc. Indian Nat. Sci. Acad. (B) 38, 3–4.

Rao, K.S., Ganapati, P.N., 1972b. On the common anascan genus *Electra* from Visakhapatnam and its vicinity. Proc. Indian Nat. Sci. Acad. 38, 220–224.

Rao, K.S., Ganapati, P.N., 1975. Littoral bryozoa in the Godavari Estuary. Bull. Oep. Mar. Sci. Uni. Cochin 7 (3), 591–600.

Rao, K.S., Ganapati, P.N., 1978. Ecology of fouling bryozoans at Visakhapatnam Harbour. Proc. Indian Acad. Sci. (Animal Sciences) 87, 63–75.

Rao, K.S., Ganapati, P.N., 1980. Epizoic fauna of *Thalamoporella var. indica* and *Pherusclla tubulosa* (Bryozoa). Bull. Marine Sci. 30, 34–44.

Rao, K.S., Viswanadham, 1984. First description of the ancestrula of *Thalamoporella stapifera* and preliminary observation on its early astogeny. Geobios New Reports 3, 90–92.

Raveendran, T.V., De Souza, A.P., Wagh, A.B., 1990. Fouling polyzans of Bombay off shore waters. Mahasagar 23 (2), 169–178.

Robertson, A., 1921. Report on a collection of Bryozoa from the Bay of Bengal and other eastern seas. Rec. Indian Mus. 22, 33–65.

Soja, L., 2006. Taxonomy, bionomics and biofouling of bryozoans from the coast of India and the Antarctic waters. Ph.D. Thesis. Cochin University of Science and Technology. p. 336.

Soja, L., Menon, N.R., 2009. *Biflustra perambulata n.* sp. (Cheilostomata: Bryozoa), a new alien species from Cochin Harbour, Kerala, India. Zootaxa 2066, 59–68.

Swami, B.S., Gaokar, S.N., 1998. Studies on the variability among macrofouling test coupons exposed in Bombay harbor. Indian Jour. Marine Sci. 27, 333–339.

Swami, B.S., Karande, A.A., 1987. Encrusting bryozoans in coastal waters of Bombay. Mahasagar—Bulletin National Institute of Oceanography 20 (4), 225–236.

Swami, B.S., Karande, A.A., 1994. Encrusting bryozoans in Karwar waters, central west coast of India. Indian J. Marine Sci. 23, 170–172.

Swami, B.S., Udhayakumar, M., 2010. Seasonal influence on settlement, distribution and diversity of organisms at Mumbai harbor. Indian J. Marine Sci. 39 (1), 57–67.

Thornely, L.R., 1905. Report on the Polyzoa collected by Prof. Herdman, at Ceylon. *Ceylon Pearl Oyster Fisheries*. Suppl. Rep. 26, 107–115.

Thornely, L.R., 1906. Additions and correction. Report Pearl Oyster Fisheries, Gulf of Mannar, IV. Suppl. Rep. 26, 449–450.

Thornely, L.R., 1907. Report on the marine Polyzoa in the collection of the Indian Museum. Rec. Indian Mus. 1, 179–196.

Thornely, L.R., 1912. Marine polyzoa of the Indian Ocean. Tran. Linn. Soc. London (Zool.) 15, 137–157.

Thornely, L.R., 1916. Report on the Polyzoa. In: Hornell (Ed.) Report to the Government of Baroda on the Marine Zoology of Okhamandal in Kattiawar, 2, 157–165.

Tilbrook, K.J., Hayward, P.J., Gordon, D.P., 2001. Cheilostomatous Bryozoa of Vanatu. Zool. J. Linn. Soc. 131, 35–109.

Winston, J.E., 1982. Marine Bryozoans (Ectoprocta) of the Indian River Area (Florida). Bull. American Mus. Nat. Hist. 73 (2), 99–176.

Chapter 8

Diversity, Distribution and Nesting Behavior of Sea Urchins along the Coast of Port Blair, South Andaman

J.K. Mishra, Yasmin, A. Mishra, J. Sinduja, D. Adhavan and N.P. Kumar
Department of Ocean Studies and Marine Biology, Pondicherry University, Port Blair, Andaman and Nicobar Islands, India

INTRODUCTION

Sea urchins are small, spiny oceanic animals representing a diverse group of marine deuterostomes (Smith *et al.*, 2004; Smith *et al.*, 2006) in the animal kingdom, which appeared around 540 million years ago in the Ordovician period and now comprise about one thousand living species (Kier, 1977; Smith, 1984). Studies on sea urchins have taken centre stage in recent years and gained significance on account of the animals' close association with humans, having a common origin in deuterostomes, with regard to the genome (Materna *et al.*, 2006; Materna and Cameron, 2008). The animal is also used as an excellent model organism for research in the areas of developmental biology under controlled conditions (Davidson *et al.*, 2002; Ben-Tabou de Leon and Davidson, 2007).

At the same time, the role of these spiny oceanic creatures in the food chain is very significant because of their influence on the community structure in the marine ecosystem (Quinn, 1965; Mills *et al.*, 2000; Lessios *et al.*, 2001; Nishizaki and Ackerman, 2007; Benitez-Villalobos *et al.*, 2008; Williams and Jorge, 2010). As reported the grazing behavior of sea urchins contributes significantly towards maintaining the rocky substrate from any kind of algal overgrowth, allowing other marine organism such as corals and sponges to colonize the rocky substratum (Calderon *et al.*, 2007). In addition, sea urchins also help in the release of nutrients into the water column from oceanic rocks, as they erode solid rocks by defacing them (Bak, 1993).

Apart from being a model in the study of developmental biology and its role in the marine food web, sea urchins also serve as an important component of the food industry, their gonad being considered a delicacy in Japan and several other countries (Kitamura *et al.*, 1993; Yur'eva *et al.*, 2003) and also as food among

Marine Faunal Diversity in India. DOI: 10.1016/B978-0-12-801948-1.00008-2

the traditional habitats, as in the case of the Nicobari population of the Andaman and Nicobar Islands (Tikadar and Das, 1985). Also, the colored and beautifully shaped sea urchin tests are used for ornamental purposes and have a good market potential (Mishra, unpublished), bringing sound economic benefits to the fishing community and aqua-farmers. All these economic components associated with this organism have resulted in increased sea urchin fishery in the world, both in terms of harvesting from natural holding areas and also aquaculture of the organism.

Because of this ecological and economic importance, scientists around the world are investigating the potential for aquaculture of these organisms (Kitamura *et al.*, 1993, 2003; Koh *et al.*, 1996; Kelly *et al.*, 2000). Several studies also have been carried out using larval sea urchins as biomarker for assessing environmental factors such as pH (Pagano *et al.*, 1985) and salinity (Dinnel *et al.*, 1981, 1983) as well as to detect the presence of xenobiotics and metals (Pagano *et al.*, 2000). Similarly, studies have been carried out involving sea urchins in the field of taxonomy, histology and morphology (Yokota, 2002), genome studies (Materna *et al.*, 2006; Materna and Cameron, 2008), immunological studies (Cooper and Alder, 2006; Hibino *et al.*, 2006), and larval settlement studies (Pearce and Scheibling, 1990; Kitamura *et al.*, 1993, 2003; Harris and Chester, 1996; Huggett *et al.*, 2006; Koh *et al.*, 1996; Miller and Emlet, 1999; Takahashi *et al.*, 2002; Swanson *et al.*, 2004). In India, studies on the early development and metamorphosis and anatomy of sea urchin species *Salmacis bicolor* from the Madras coast (southeast coast of India) was carried out by Aiyar (1936), followed by several other studies on different aspects of other species such as *Salmacis bicolor* and *Stomopneustes variolaris* (Aiyar and Menon, 1944), *S. variolaris* (Shetty, 1960; Giese *et al.*, 1964; Mary, 1979; Reuben *et al.,* 1980; Sastry, 1985) along the coast of India, and *Echinometra mathaei* (Jose *et al.*, 2007; Mishra *et al.*, 2012).

In the Andaman Sea, though about 80 species of *Echinoidea* are reported (Sastry, 2005, 2007), very little is known about the distribution pattern, habitat structure and population density of these sea urchin species at a local scale along the rocky coasts of these islands. The urchins are observed in their natural habitat extending from the shallow intertidal region to the deep abyssal plains. But in this zone, both the species diversity and their pattern of distribution depends on a number of local factors such as substratum type, type of substratum colonization and the structural morphology of the reef. However, the intertidal environment along the coast of South Andaman has been found to serve as a suitable habitat for the available sea urchin species. In the present chapter, the comprehensive distribution pattern of ten different sea urchin species and their habitat type along the coast of Port Blair, South Andaman Sea are discussed.

METHODS

For the present study, seven different stations were selected for sampling along the coast of Port Blair, South Andaman (Figure 8.1). The study locations were selected on the basis of the exposed rocky shore during the low tide period and

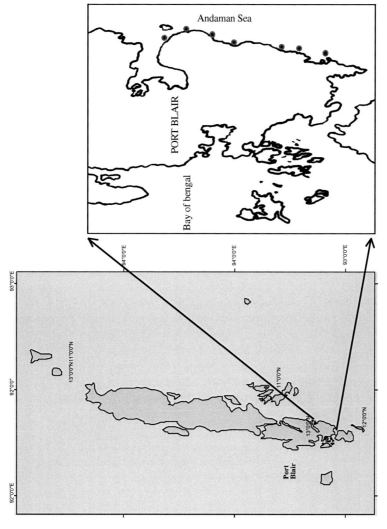

FIGURE 8.1   Study area along the coast of **Port Blair, South Andaman.**

accessibility for carrying out sampling throughout the period of investigation. These sampling areas were:

- S1: South Point (Lat. 11° 39.922′ N; Long. 92° 45.378′ E)
- S2: Science centre (Lat. 11° 50.894′ N; Long. 92° 51.097′ E)
- S3: Hornbill Nest Resort (Lat. 11° 39.054′ N; Long. 92° 45.439′ E)
- S4: Carbyns Cove (Lat. 11° 35.470′ N; Long. 92° 44.557′ E)
- S5: Rangachang Bay (Lat. 11° 33.226′ N; Long. 92° 44.066′ E)
- S6: Burmanalha (Lat. 11° 31.516′ N; Long. 92° 43.475′ E)
- S7: Kodiaghat (Lat. 11° 31.744′ N; Long. 92° 43.425′ E).

All the field investigations were carried out during the low tide period in coherence with the full moon and new moon phases extending over a period from March 2009 to April 2012. Distribution and population density of sea urchins were studied by employing the quadrant method, for which a quadrant of 1 m^2 area was marked at each sampling point and sea urchins within the marked area were counted and identified to species level. In this way, three random replications of quadrant sampling were taken for each station. During the investigation, both the size and shape of the crevices inhabited by different species were observed and recorded.

RESULTS

As observed during the study period, a number of sea urchins representing six genera and ten species were recorded from the seven sampling stations (Figures 8.2 and 8.3). A high degree of species diversity was found in the study area. As depicted in Table 8.1, species such as *Echinometra mathaei, E. oblonga* and *Echinostrephus molaris* were dominant at four sampling stations, i.e., South Point, Science Centre, Hornbill Nest Resort and Carbyns Cove. All these stations have a similar topography and rock type except Science Centre, which has a high degree of seaweed cover and with different corals. Similarly, *Diadema setosum* was dominant at Burmanalha, and the topography of this study area is different, being dominated by pebbles and small rock pools with seaweeds and small patches of coral.

Similarly, population density along the sampling stations exhibited a specific pattern on the basis of the species distribution and population structure. As explained in Table 8.2, species richness was greatest at South Point, followed by Carbyns Cove, Science Centre, Hornbill Nest and Rangachang Bay, and least at Burmanalha and Kodiaghat. Population density was observed to be maximum at Hornbill Nest (mean density of about 102 individuals/m^2) followed by Science Centre (81 individuals/m^2). Among the species encountered during the investigation, *Echinostrephus molaris* was found to be dominant (up to 335 individuals/m^2) at five stations; but at two places—Rangachang Bay and Kodiaghat—it was completely absent. *Diadema setosum* was found to occur at all the stations, with a mean density of 25 individuals/m^2, except that of Kodiaghat. Both

FIGURE 8.2 **Six species of sea urchin found in the study area.** (a) *Echinometra mathaei* (Blainville, 1825); (b) *Echinometra oblonga* (Blainville, 1825); (c) *Stomopneustes variolaris* (Lamarck, 1816); (d) *Echinostrephus molaris* (Blainville, 1825); (e) *Diadema setosum* (Leske, 1778); (f) *Diadema savigyni* (Audouin, 1829).

(a)   (b)

(c)   (d)

FIGURE 8.3 **Four species of sea urchin found in the study area.** (a) *Echinothrix calamaris* (Pallas, 1774); (b) *Echinothrix diadema* (Linnaeus, 1758); (c) *Tripneustes depressus* (A. Agassiz, 1863); (d) *Tripneustes gratilla* (Linnaeus).

the *Echinometra mathaei* and *E. oblonga* were found at all the stations except Burmanalha, with a respective population density of 34 and 52 individuals/m^2 Among the ten reported species in this study, *Tripneustes depressus* and *T. gratilla* were found to be least available, being confined to South Point area only with a minimum population density of four and two individuals/m^2 respectively.

Though the study area comprises mainly rocky coastline with different types of rocks, boulders and at times pebbles and sandy patches, the dominancy of some coral species such as *Acropora formosa, Acropora palifera, Stylophora pistillata, Pavona duerdeni, Porites solida, Pocilophora eydouxi, Favia speciosa, Galaxea fascicularis* and *Symphillia radians* was noted. In most cases, urchins such as *Echinometra mathaei* and *Echinometra oblonga* were found to remain associated with corals such as *Pavona duerdeni* and *Porites solida*.

Habitat pattern in sea urchins has been found to be specific to particular species, as was encountered in all ten species studied during this investigation. In the case of *Echinometra mathaei* (Figure 8.2a), they typically inhabit

TABLE 8.1 Distribution and Dominancy Pattern of Sea Urchin Species along the Coast of Port Blair, South Andaman

Sl. No.	Species	South Point	Science Centre	Hornbill Nest Resort	Carbyns Cove	Rangachang Bay	Burmanalha	Kodiaghat
1	Echinometra mathaei	+	+	+	+	+	-	+
2	Echinometra oblonga	+	++	+	+	+	+	+
3	Echinostrephus molaris	+++	+++	+++	+	-	+	-
4	Echinothrix calamaris	-	-	-	+	-	-	-
5	Echinothrix diadema	-	-	-	-	+	-	-
6	Diadema setosum	+	+	+	+	-	+	-
7	Diadema savignyi	+	-	-	+	-	-	-
8	Stomopneusteus variolaris	+	+	-	-	-	-	-
9	Tripneustes depressus	+	-	-	-	-	-	-
10	Tripneustes gratilla	+	-	-	-	-	-	-

-, Nil; +, 1–100; ++, 101–200; +++, >200

TABLE 8.2 Population Density of Sea Urchins per m² Area along the Coast of Port Blair, South Andaman

Sl. No.	Species	South Point	Science Centre	Hornbill Nest Resort	Carbyns Cove	Rangachang Bay	Burmanalha	Kodiaghat
1	Echinometra mathaei	39	59	39	52	10	-	6
2	Echinometra oblonga	24	128	28	47	49	-	34
3	Echinostrephus molaris	263	209	335	35	-	12	-
4	Echinothrix calamaris	-	-	-	6	-	-	-
5	Echinothrix diadema	-	-	-	-	6	-	-
6	Diadema setosum	25	4	5	8	18	32	-
7	Diadema savigyni	3	-	-	3	-	-	-
8	Stomopneusteus variolaris	6	3	-	-	-	-	-
9	Tripneustes depressus	4	-	-	-	-	-	-
10	Tripneustes gratilla	2	-	-	-	-	-	-

tunnel-shaped crevices made with the help of their spines and are often found hiding themselves deeper in these rock/coral (mainly porites) crevices. Along the study area, this species was found in both the sheltered and high wave action areas with boulders and hard corals. *E. mathaei* is readily identifiable by the prominent basal rings at the root of the spines. Their spines are olive green, grey or reddish brown with a light colored spinal tip. But *Echinometra oblonga* (Figure 8.2b) were found to have round cup-shaped crevices, and in most cases adults are found beneath the rocks or inside tunnel-shaped crevices. Here they were found in both sheltered and high wave action areas on algal turf and their nests were often seen around porites. This species has black test with a purple tint, mostly on the oral side, and the peristome is usually burgundy in color with monochrome black and thick spines.

In the case of *Stomopneustes variolaris* (Figure 8.2c) the nesting areas mostly were rock pools and rock crevices, where they were found to hide. They preferred shadowy zones with less wave action but with constant water circulation. The species has characteristic long, black and stout spines with a blue sheen, and the test is usually black in color. On the other hand, *Echinostrephus molaris* (Figure 8.2d) inhabit cylindrical depressions formed in rocks and only their aboral spines and tube feet are usually visible. These are abundant in rock pool areas and reef crest. This species have purple test with a light green tint. Spines are long, slender and purple with a sharp tip. The spines at the aboral side are long with a closed tip, but at the oral side spines are small and blunt, which help to anchor and fit them correctly inside their bore.

Diadema setosum (Figure 8.2e) were mostly associated with coral reefs and often found in the boulder zone, where they use the narrow elevated area for nesting. They have long, sharp and brittle black spines with a blue/green sheen. In juveniles, banded spines were observed with an orange anal ring, and they possess five prominent white spots on the test. *Diadema savignyi* (Figure 8.2f) mostly occurred in sand flats and coral reef areas in the fringe zone. It has bold iridescent blue or green lines along the interambulacra and around the periproct, with five pale white spots on the test. In juveniles, spines were banded with dark purple and white color.

The habitat of *Echinothrix calamaris* (Figure 8.3a) along the study area comprised reefs, dead corals and big boulders. They mostly hide beneath the rocks or they hang under the rock with the help of tube feet, and this species was mostly found in groups in one area. This species has a wide number of color morphs. It possesses a large periproct with white platelets and a bold green or blue band on the interambulacral region. Spines are blunt at the end. But *Echinothrix diadema* (Figure 8.3b) exists mostly on reef crest and boulders. Juveniles of this species are mostly found hiding under the rock crevices or under coral colonies. However, they are mostly found near seagrass and seaweed beds. In *E. diadema*, spines were black with a bluish green sheen. The periproct has a small black anal cone. Juveniles usually have banded spines, but this disappears in adults.

Tripneustes depressus (Figure 8.3c) were found on intermittent sandy shores with small pieces of dead coral and observed along with algal species such as *Padina* sp. and attached to dead corals such as *Acropora* sp. and *Porites solida*. This species is also known as the brown sea urchin. Spines are light brown and the test uniformly dark brown. Another species, *Tripneustes gratilla* (Figure 8.3d) occurred on the soft sand bottoms associated with patches of seagrass and seaweed beds. They are found covering themselves with seaweeds, observed in seaweed beds of *Padina* sp. and *Amphiroa rigida*. The species possesses light brown to mud white spines, and the test is dark brown in ambulacral and interambulacral areas and, along rows of tube feet, it is light brown in color. Spines are inclined towards adjacent ambulacra leaving a clear test surface over most interambulacra, and they exhibit a clear appearance of pentaradiate symmetry.

DISCUSSION

Sea urchins are mainly found to shelter in rocky shores or such habitats, where they can find easy access for predation to survive or they can protect themselves from other predators found in their vicinity, particularly in the subtidal zones (Andrew, 1993; Andrew and Underwood, 1993; Alves *et al.*, 2001). In our study of intertidal zones, we found the species availability and distribution are highly related to the habitat type, which was mainly dominated by rocks and boulders with intermittent sand beds. Though the rocky intertidal coast of South Andaman was suitable for the sea urchins, the two species *Tripneustes depressus* and *T. gratilla* exhibited the lowest population density. These two species were mainly confined to the patchy sandy substratum available, indicating their preference towards a specific habitat type.

As reported, habitat colonization by sea urchins is related to the availability of algae in any rocky substratum (Turon *et al.*, 1995; Falcon *et al.*, 1996; Alves *et al.*, 2001). But our study suggests that there is no major relationship between the sea urchin density and the availability of any particular seaweed species. However, seaweeds such as *Padina* sp., *Dictyospira* sp., *Amphiroa rigida*, *Halimida* sp., *Turbinaria ormnata* and Coralline algae were found to be dominant in the study areas. Some studies pertaining to the role of substrate structure suggest that sea urchins crave their own nest and inhabit rocky intertidal or subtidal areas, where there is a coating of algal or bacterial films (Hinegardner, 1969; Cameron and Schroeter, 1980). But the constraints on species richness and distribution are always influenced by the space availability in any habitat.

On the other hand, involvement of a species specific chemical cue in the environment in attracting the settlement stage larvae to a particular habitat cannot be ruled out. As suggested by Cameron and Schroeter (1980), larvae of the purple sea urchin settle in close proximity to their own adults, indicating the existence of a kind of chemical (cue) communication. It is reported that invertebrate larvae have affinity towards a particular substratum with the involvement of chemical cues such as pheromones emanating from the environment

(Crisp, 1974; Mishra *et al.,* 2001; Mishra, 2002). A study of sea urchin settlement on a substratum might be helpful in elucidating a cue specific orientation mechanism of settlement stage pluteus larvae in the environment.

ACKNOWLEDGEMENTS

The financial support by the University Grant Commission (UGC) as a Major Research Project on sea urchins from 2009 to 2012 to JM is highly acknowledged. Authors also acknowledge the support and encouragement provided by Pondicherry University in terms of logistics and infrastructure.

REFERENCES

Aiyar, R.G., 1936. Early development and metamorphosis of the tropical echinoid *Salmacis bicolor* (Agassiz). Proc. Indian Acad. Sci., 714–728, IB.

Aiyar, R.G., Menon, M., 1944. Observations on the spicules of *Salmacis bicolor* (Agassiz) and *Stomopneustes variolaris*. Ann. Mag. Nat. Hist. 10 (13), 468–473.

Alves, F.M.A., Chicharo, L.M., Serrao, E., Abreu, A.D., 2001. Algal cover and sea urchin spatial distribution at Madeira Island (NE Atlantic). Sci. Mar. 65 (4), 383–392.

Andrew, N.L., 1993. Spatial heterogeneity, sea urchin grazing, and habitat structure on reefs in temperate Australia. Ecol. 74, 292–302.

Andrew, N.L., Underwood, A.J., 1993. Density-dependent foraging in the sea urchin *Centrostephanus rodgersii* on shallow subtidal reefs in New South Wales, Australia. Mar. Ecol. Prog. Ser. 99, 89–98.

Bak, R.P.M., 1993. Sea urchin bioerosion on coral reefs: place in the carbonate budget and relevant variables. Coral Reefs 33, 99–103.

Benitez-Villalobos, F., Gomez, M.T.D., Lopez Perez, R.A., 2008. Temporal variation of the sea urchin *Diadema mexicanum* population density at Bahias de Huatulco, Western Mexico. Internat. J. Trop. Biol. 56 (3), 255–263.

Ben-Tabou de Leon, S., Davidson, E.H., 2007. Gene regulation: gene control network in development. Annu. Rev. Biophys. Biomol. Struct. 36, 191–212.

Calderon, E.N., Zilberberg, C., de Pavia, P.C., 2007. The possible role of *Echinometra lucunter* (Echinodermata: Echinoidea) in the local distribution of *Darwinella* sp. (Porifera: Dendroceratida) in Arraial do Cabo, Rio de Janeiro State, Brazil. In: Custódio, M.R., Lôbo-Hajdu, G., Hajdu, E., Muricy, G. (Eds.), Porifera Research: Biodiversity, Innovation and Sustainability. Série Livros 28. Museu Nacional, Rio de Janeiro, pp. 211–217.

Cameron, R.A., Schroeter, S.C., 1980. Sea urchin recruitment: Effect of substrate selection on juvenile distribution. Mar. Ecol. Prog. Ser. 2, 243–247.

Cooper, M.D., Alder, N., 2006. The evolution of adaptive immune systems. Cell. 124, 815–822.

Crisp, D., 1974. Factors influencing the settlement of marine invertebrate larvae. In: Grant, P., Mackie, A. (Eds.), Chemoreception in marine organisms. Academic Press, New York, pp. 177–265.

Davidson, E.H., Rast, J.P., Oliveri, P., Ransick, A., Calestani, C., Yuh, C.H., Minokawa, T., Amore, G., Hinman, V., Arenas-Mena, C., 2002. A provisional regulatory gene network for specification of endomesoderm in the sea urchin embryo. Dev. Biol. 246, 162–190.

Dinnel, P.A., Stober, Q.J., Dijulio, D.H., 1981. Sea urchin sperm bioassay for sewage and chlorinated sea water and its relation to fish bioassay. Mar. Environ. Res. 5, 29–39.

Dinnel, P.A., Stober, Q.J., Link, J.M., Letourneau, M.W., Robert, W.E., Felton, S.P., Nakatani, R.E., 1983. Methodology and validitation of a sperm cell toxicity test for testing toxic substances in marine waters., Univ. Washington Sea Grant Prog. in Coop. U.S. Environmental protection agency. p. 38.

Falcon, J.M., Bortone, S.A., Brito, A., Bundrick, C.M., 1996. Structure and relationships within and between the littoral, rock-substrate fish communities off four islands in the Canarian Archipelago. Mar. Biol. 125, 215–231.

Giese, A.C., Krishnaswamy, S.B., Vasu, S., Lawerence, J., 1964. Reproductive and biochemical studies on sea urchin, *Stomopneustes variolaris* from Madras Harbour. Comp. Biochem. Physiol. 13, 367–380.

Harris, L.G., Chester, C.M., 1996. Effects of location, exposure and physical structure on juvenile recruitment of the sea urchin *Strongilocentrotus droebachiensis* (Muller). J. Exp. Mar. Biol. Ecol. 176, 107–126.

Hibino, T., Loza, C.M., Messier, C., Majeske, A., Cohen, A., Terrewilliger, D., Buckley, K., Brockton, V., Nair, S., Berney, K., 2006. The immune gene repertoire encoded in the purple sea urchin genome. Dev. Biol. 300, 349–365.

Hinegardner, R.T., 1969. Growth and development of the laboratory cultured sea urchin. Biol. Bull. 137, 465–475.

Huggett, J.M., Williamson, E.J., De Nys, R., Kjelleberg, S., Steinberg, D.P., 2006. Larval settlement of the common sea urchin, *Heliocidaris erythrogramma* in response to bacteria from the surface of coralline algae. Oceanologia 149, 604–619.

Jose, J.J., Lipton, A.P., Anil, M.K., 2007. Induced spawning and larval rearing of sea urchin *Echinometra mathaei* (de Blainville, 1825). J. Mar. Biol. Assoc. India 49 (2), 230–233.

Kelly, M.S., Hunter, A.J., Scholfield, C.L., Mckenzie, J.D., 2000. Morphology and survivalship of larval *Psammaechinus miliaris* (Gmelin) (Echinodermata: Echinoidea) in response to varying food quantity and quality. Aquaculture 183, 223–240.

Kier, P.M., 1977. The poor fossil record of the regular Echinoid. Paleabiology 3, 168–174.

Kitamura, H., Kitahara, S., Koh, H.B., 1993. The induction of larval settlement and metamorphosis of two sea urchins, *Pseudocentrotus depressus* and *Anthocidaris crassispina*, by free fatty acids extracted from the coralline red alga *Corallina pilulifera*. Mar. Biol. 115, 387–392.

Kitamura, H., Katahara, S., Koh, H.B., 2003. The induction of larval settlement and metamorphosis of two sea urchins, *Pseudocentrotus depressus* and *Anthocidaris crassispina*, by free fatty acids extraction from the coralline red algae *Coralline pilulifera*. Mar. Biol. 115, 387–392.

Koh, H.B., Kitamura, H., Hirayama, K., 1996. Effect of water soluble substances extracted from the coralline red alga, *Corallina pilulifera* on the larval metamorphosis of the sea urchin, *Pseudocentrotus depressus*. Sessile Organisms 13 (1), 1–5.

Lessios, H.A., Kessing, B.D., Pearse, J.S., 2001. Population structure and speciation in tropical seas: global phylogeography of the sea urchin *Diadema*. Evolution 55, 955–975.

Mary, B.M. (1979). Occurrence of the sea urchin *Stomopneustes variolaris* (Lamarck, 1816) along the coasts of Kannyakumari, S. India. *Rec. Zool. Surv. India.* pp. 103–104.

Materna, S.C., Cameron, A.R., 2008. The sea urchin genome as a window on function. Biol. Bull. 214, 266–273.

Materna, S.C., Berney, K., Cameron, K.A., 2006. The *S. puprpuratus* genome: a comparative perspective. Dev. Biol. 300, 485–495.

Miller, A.B., Emlet, B.R., 1999. Development on newly metamorphosed juvenile sea urchins (*Strongylocentrotus franciscanus* and *S. purpuratus*): morphology, the effects of temperature and larval food ration, and a method for determining age. J. Exp. Mar. Biol. Ecol. 235 (199), 67–90.

Mills, S.C., Clausade, M.P., Fontaine, M.F., 2000. Ingestion and transformation of algal turf by *Echinometra mathaei* on Tiahura fringing reef (French Polynesia). J. Exp. Mar. Biol. Ecol. 254, 71–84.

Mishra, J.K., 2002. Pheromone, a novel mode of chemical communication for inducing settlement of the barnacle, *Balanus Amphitrite*. Advances in Marine, Antarctic Sciences. Published by APH Publishing Corporation, New Delhi, pp. 150–161.

Mishra, J.K., Kitamura, H., Ishibashi, F., Tomoda, K., 2001. Volatile substances from adult extracts induce larval settlement of the barnacle, *Balanus amphitrite*. Biofouling 17 (1), 23–28.

Mishra, A., Mishra, J.K., Yasmin, Mohan, P.M., 2012. Distribution pattern of Echinoderm larvae with special reference to the larvae of sea urchins in the Andaman Sea. J. Coast. Environ. 3 (1), 81–88.

Nishizaki, M.T., Ackerman, J.D., 2007. Juvenile–adult associations in sea urchins (*Strongilocentrotus franciscanus* and *S. droebachiensis*): protection from predation and hydrodynamics in *S. franciscanus*. Mar. Biol. 151, 135–145.

Pagano, G., Cipollaro, M., Corsale, G., Esposito, A., Ragucci, E., Giordano, G.G., 1985. pH-Induced changes in mitotic and developmental patterns in sea urchin embryogenesis. Exposure of embryos. Teratog. Carcinog. Mutagen. 5, 101–112.

Pagano, G., Korkina, L.G., Iacarrino, M., deBiase, A., Doronin, B., Guida, Y.K., Melluso, M., Meric, S., Oral, R., Trieff, N.M., Warnau, M., 2000. Development, cytogenetic and biochemical effects of spiked or environmentally polluted sediments in sea urchin bioassays. In: Garrigues et al., (Ed.), Biomarkres in Marine Ecosystems: a practical approach. Elsevier, Amsterdam, pp. 1–50.

Pearce, M.C., Scheibling, R.E., 1990. Induction of metamorphosis of larvae of the green sea urchin, *Strongilocetrotus droebachiensis*, by coralline red algae. Biol. Bull. 79, 304–311.

Quinn, B.G., 1965. Predation in sea urchins. Bull. Mar. Sci. 15 (1), 259–264.

Reuben, S., Apparo, T., Sampson, M.E., 1980. Sea urchin resources of Waltair coast (Abstract). Proceed. Symp. on Coastal Aquaculture. Mar. Biol. Assoc. India, Cochin. p. 113.

Sastry, D.R.K., 1985. Observations on the distribution of *Stomopneustes variolaris* (Lamarck) (Echinode mala: Echinoidea) along the Visakbapatnam coast. Second National Seminar on Marine Intertidal Ecology, February 14-16, 1985. Department of Zoology, Andhra University, Waltair. Abstract No. 42.

Sastry, D.R.K., 2005. Echinodermata of Andaman and Nicobar Islands, Bay of Bengal: An annotated list. Rec. Zool. Surv. India, Occ. Paper No. 233, 1–207.

Sastry, D.R.K., 2007. Echinodermata of Andaman and Nicobar Islands, Bay of Bengal: An annotated list. Rec. Zool. Surv. India, Occ. Paper No. 271, 1–387.

Shetty, H.P.C., 1960. Observations on the early development of *Stomopneustes variolaris* Agassiz. Proc. Indian Acad. Sci. 52B (3), 91–102.

Smith, A.B., 1984. Echinoid Paleobiology (Special topics in Palaeontology). George Allen & Unwin (Publ.), London, p. 199.

Smith, A.B., Peterson, K.J., Wray, G.A., Littlewood, D.T.J., 2004. From Bilateral symmetry to pentaradiality. The phylogeny of hemichordates and echinoderms. In: Craft, J., Donoghue, M.J. (Eds.), Assembling the tree of life. Oxford University Press, Oxford, pp. 365–383.

Smith, L.C., Rast, J.P., Brockton, V., Terwillieger, D.P., Nair, S.V., Buckley, K.M., Majeske, A.J., 2006. The sea urchin immune system. Invertebrate Survival J 3, 25–39.

Swanson, R.L., Williamson, J.E., De Nys, R., Kumar, N., Bucknall, M.P., Steinberg, P.D., 2004. Induction of settlement of larvae of sea urchin *Holopneustes purpurascens* by Histamine from a host algae. Biol. Bull. 206, 161–172.

Takahashi, Y., Itoh, K., Ishi, M., Suzuki, M., Itabashi, Y., 2002. Induction of larval settlement and metamorphosis of the sea urchin *Strongylocentrotus intermedius* by glycoglycerolipids from the green algae, *Ulvella lens*. Mar. Biol. 140, 763–771.

Tikadar, B.K., Das A.K., 1985. Glimpses of animal life of Andaman and Nicobar Islands. p. 97.

Turon, X., Giribet, G., Lopez, S., Palacin, C., 1995. Growth and population structure of Paracentrotus lividus (Echinodermata: Echinoidea) in two contrasting habitats. Mar. Ecol. Prog. Ser. 122, 193–204.

Williams, S.M., Jorge, G.S., 2010. Temporal and spatial distribution patterns of echinoderm larvae in La Parguera, Puerto Rico. Rev. Biologia Trop. 58 (3), 1–7.

Yokota, Y., 2002. The Sea Urchin Biology. In: Yokota, Y., Matranga, V., Smiolenicka, Z. (Eds.), The Sea Urchin: From Basic Biology to Aquaculture. A.A. Balkema Publ, The Netherlands, pp. 1–10.

Yur'eva, M.I., Lisakovskaya, O.V., Akulin, V.N., Kropotov, A.V., 2003. Gonads of sea urchins as the source of medication stimulating sexual behavior. Russ. J. Mar. Biol. 29 (3), 189–193.

Chapter 9

Coral Reef Associated Macrofaunal Communities of Rutland Island, Andaman and Nicobar Archipelago

C. Raghunathan* and K. Venkataraman[†]

*Andaman and Nicobar Regional Centre, Zoological Survey of India, Port Blair, Andaman and Nicobar Islands, India; [†]Zoological Survey of India, Kolkata, West Bengal, India

INTRODUCTION

The Andaman and Nicobar Islands comprise a 'beaded cluster' mountain chain of 572 islands, islets and outcrops, distributed over a 800 km long stretch. Andaman and Nicobar are the two major groups and are separated by the wide, turbulent Ten Degree Channel (Tikadar and Das, 1985). Rutland is one of the largest of the labyrinthine group of islands in South Andaman. It is situated in the southern region of the Andaman group, separated by Duncan Passage from Little Andaman Island. The island is located between Lat. 11° 20′ to 11° 28′ N and Long. 92° 35′ to 92° 45′E, with a total area of 137.2 km^2 and around 60 km of coastline, across the Macpherson Strait from South Andaman and also separated from Cinque Island by Manners Strait. Mount Ford is the highest peak on Rutland Island at 435 m in height. The orientation of the mountain and its formation has given rise to a convoluted coastline of inlets and bays around this island. The island is a treasure-trove of biodiversity in both terrestrial and marine aspects. The diversification and differentiation of physical and climatic variables contributed greatly to the development of the ecological niche for the sustainable development of biological diversity of these islands. Rutland Island is characterized by the presence of a variety of coastal habitats supported by a large bay, sandy and muddy beaches, creeks, rocky areas, mangrove swamps and a variety of coral reefs. Marine macrofauna include a wide range of faunal communities such as polychaetes, molluscs, echinoderms, sea anemones, corals, sponges, tunicates, etc. that are greater than 1 mm in size. The sea around the islands is home to many marine organisms, based on coral reefs as the building blocks.

Marine Faunal Diversity in India. DOI: 10.1016/B978-0-12-801948-1.00009-4
121

A living coral reef ecosystem is one of the most glorious and extraordinary sights on our planet (Rao, 2010). Coral reefs are popularly known as the baseline animals for the construction of undersea ecological pyramids of the Andaman Sea (Whitaker, 1985). The scleractinian corals of the Andaman and Nicobar Islands are well distributed as a fringing reef pattern (Venkataraman *et al.*, 2003). Benthic faunal communities are well known as biological indicators as they can provide information on environmental parameters and features, either because of the sensitivity of single indicator species or by integrating environmental signals over a long period of time (Tagliapietra and Sigovini, 2010). The fringing reefs of Rutland Island extensively sustain wide ecological avenues for the other reef associated faunal components such as molluscs, echinoderms, polyclads, crustaceans, sponges, fishes, etc. The present investigation was carried out to study the diversity and distribution of three major groups of the marine macrofauna of the island: scleractinian corals, molluscs and echinoderms.

METHODS

Extensive underwater surveys were conducted of the shelf region along reef areas of Rutland Island during 2008–2013 by self-contained underwater breathing apparatus (SCUBA) diving and snorkeling. Ten reef areas were selected randomly to carry out the faunal studies (Table 9.1 and Figure 9.1). The physicochemical parameters of the seawater were also recorded from the study area. Surface seawater temperature was measured using a standard mercury thermometer. The salinity was recorded at all the places of study by using a hand-held refractometer, model ERMA, Japan. The seawater pH was measured,

TABLE 9.1 Coordinates of the Study Areas

Sl. No.	Study area	Coordinates	
		Latitude N	Longitude E
1	Station 1	11° 22.520′	092° 35.110′
2	Station 2	11° 24.275′	092° 36.267′
3	Station 3	11° 26.109′	092° 36.449′
4	Station 4	11° 29.202′	092° 36.536′
5	Station 5	11° 30.180′	092° 39.026′
6	Station 6	11° 28.051′	092° 40.569′
7	Station 7	11° 25.510′	092° 41.214′
8	Station 8	11° 24.281′	092° 39.560′
9	Station 9	11° 22.434′	092° 41.498′
10	Station 10	11° 21.006′	092° 39.091′

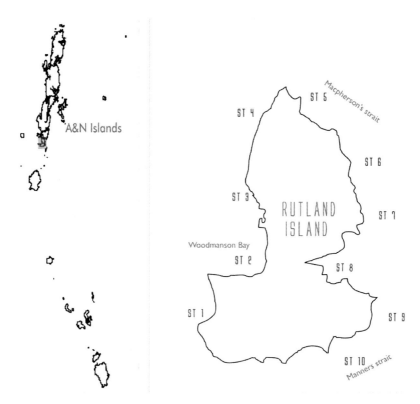

FIGURE 9.1 Map showing the areas surveyed in Rutland Island, South Andaman.

soon after collection of a water sample, using a portable water quality analyzer, model Systronics Water Analyzer 371. The transparency of the seawater column was measured by using a Secchi disc from the sea surface to assess the depth of light penetration. The seawater turbidity was measured by a turbidity meter, model Eutech Instruments ECTN100IR, Singapore.

Corals and Associated Faunal Communities

The diversity of corals and their associated faunal communities was assessed using a preliminary Manta Tow survey (Done *et al.*, 1982) in shallow reef areas. Photography and videography was used for identification in addition to specimen sampling. The assessment of live corals and associated fauna was done by randomly laying out 20 m long line intercept transects (LITs), following English *et al.* (1997). Digital photographs of individual species was taken by underwater camera (Sony-Cyber Shot, Model-T900, marine pack, 12.1 megapixels; Sony-Cyber Shot, Model-TX1, marine pack, 10.2 megapixels; and Canon Powershot G15). Species of corals were identified with reference to Veron and Pichon (1976, 1979, 1982), Veron *et al.* (1977), Veron and Wallace (1984),

Veron (2000) and Wallace (1999). Health of coral reefs was studied simultaneously, as bleaching was found during May and June 2010.

Data on echinoderms and molluscs were also collected from study sites during the study period. Identification of echinoderms was made in conjunction with the key characters of Clark and Rowe (1971) and Sastry (2005, 2007). Molluscs were identified and categorized following Subba Rao (2000, 2003) and other standard monographs.

Diversity indices such as Shannon-Wiener, Simpson, Pielou's Evenness, Berger-Parker, and Fisher Alpha were calculated and presented.

RESULTS

The diversity of marine faunal communities was assessed in ten selected reef areas at Rutland Island during the period of study.

Physico-Chemical Parameters

Physico-chemical parameters are the most important to depict the ecological status of any environment. These variables of seawater were measured and tabulated (Table 9.2). The mean surface seawater temperature of the study sites ranged from 26.5°C to 28.0°C, while salinity was in the range 32.90 ppt to 33.80 ppt. The concentration of hydrogen ions (pH) varied from 7.3 at Station 5 to 7.7 at Station 9. The transparency in terms of penetration of light in the seawater column ranged from 7 m to 10 m. The turbidity of seawater was also measured by Nudson Turbidity Unit (NTU) and was found to be minimum (730) at Station

TABLE 9.2 Physico-Chemical Parameters of Seawater at Study Areas in Rutland Island

Station	Temp. (°C)	Salinity (ppt)	pH	Transparency (m)	Turbidity (NTU)	Intertidal exposure (m)
1	27.5	33.41	7.5	9.2	740	30
2	27.5	33.18	7.6	10.0	760	40
3	27.0	33.70	7.5	7.5	800	35
4	26.5	32.90	7.4	7.5	810	30
5	26.5	33.52	7.3	9.0	730	25
6	27.0	33.46	7.5	8.0	750	20
7	27.5	33.61	7.4	9.5	850	10
8	28.0	33.48	7.6	10.0	870	30
9	28.0	33.80	7.7	7.0	810	25
10	28.0	33.16	7.6	7.0	790	20

5 and maximum (870) at Station 8. The intertidal exposure during low tide at the surveyed area ranged from 10 m at Station 7 to 40 m at Station 2.

Scleractinian Corals

A total of 327 species belong to 68 genera and 15 families of scleractinian corals were reported during the study period. See Figure 9.2 for examples. A maximum of 88 species were observed from the Acroporidae family, whereas

Favites abdita (Ehrenberg, 1834)   *Diploastrea heliopora* (Lamarck, 1816)

Favia speciosa (Dana, 1846)   *Goniastrea retiformis* (Lamarck, 1816)

Stylophora pistillata (Esper, 1797)   *Acropora cytherea* (Dana, 1846)

FIGURE 9.2   Scleractinian corals of Rutland Island. (Please see color plate at the back of the book.)

TABLE 9.3 Number of Species Under Different Families of Scleractinian Corals Reported

Sl. No.	Family	ANI total	Rutland Island	Percent of ANI total in RI
1	Acroporidae	160	88	55.00
2	Pocilloporidae	22	13	59.09
3	Oculinidae	5	2	40.00
4	Siderastreidae	15	9	60.00
5	Agariciidae	35	24	68.57
6	Astrocoeniidae	3	1	33.33
7	Fungiidae	52	29	55.76
8	Merulinidae	9	7	77.77
9	Mussidae	28	21	75.00
10	Faviidae	91	82	90.11
11	Pectinidae	15	9	60.00
12	Poritidae	55	21	38.18
13	Dendrophylliidae	18	13	72.22
14	Euphyllidae	8	7	87.50
15	Caryophylliidae	8	1	12.50
	Total	524	327	62.40

ANI, Andaman and Nicobar Islands; RI, Rutland Island.

the minimum of 1 species was recorded under each of the Astrocoeniidae and Caryophyllidae families (Table 9.3). Rutland Island accommodated 62.40 percent of the scleractinian corals of the Andaman and Nicobar Islands.

The extensive studies on scleractinian corals of Rutland Island showed a cumulative progressive status of ecological indices such as diversity, dominance, evenness, community and equitability indices (Figure 9.3). The recorded numerical values for these indices imply an optimal degree of health status for the scleractinian corals of Rutland Island. It is also pertinent to mention that the species diversity shows maximum value ($H' = 4.10$). Coral species were distributed at a depth between 2 m and 35 m. The mean density of corals was assessed as 12 to 23 colonies/10 m^2. The live coral cover was estimated as 57–63 percent, of which 50–65 percent of corals were reported bleached in 2010 due to an increase in sea surface temperature in the Andaman Sea. Probably because of the favorable environment, 28–87 percent recovery of corals was observed in subsequent surveys during 2011. It is important to note that new recruitment of scleractinian corals was also recorded (23–56%) during the study period.

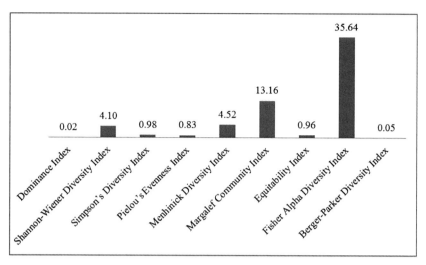

FIGURE 9.3   Ecological indices for scleractinian corals of Rutland Island.

Molluscs

A total of 130 species of molluscs under 69 genera, 39 families and 9 orders were reported during the study period in Rutland Island. See Figure 9.4 for examples. A maximum of 48 species were observed under the order Neogastropoda, whereas one species were recorded under each of the orders Architectonoidea and Dentaliida (Table 9.4). Rutland Island accommodated 11.33 percent of the total number of molluscan species of the Andaman and Nicobar Islands, with 9 orders out of 20.

The Shannon-Wiener diversity index showed 3.38, while Menhinick diversity index registered 4.20 for molluscs. Other calculated indices showed a wide range of values, however. Fisher alpha diversity index was the highest at 27.53. The results show that the diversity, dominance, equitability, community and evenness indices for molluscs were optimum at Rutland Island (Figure 9.5). The density of the molluscs was recorded at all the study areas, and it ranged from 14 to 25 individuals/10 m^2.

Echinoderms

One hundred and three species of echinoderms belonging to 60 genera, 35 families, 15 orders and 5 classes were reported during the study period. See Figure 9.6 for examples. A maximum of 31 species were observed under the class Holothuroidea, whereas the minimum of 9 species were identified under class Ophiuroidea (Table 9.5). Rutland Island accommodated 24.23 percent of the total number of echinoderm species of the Andaman and Nicobar Islands.

The cumulative data from all ten study areas were analyzed to assess the ecological status of echinoderms of Rutland Island. It was revealed that the

Chicoreus ramosus
(Linnaeus,1758)

Conus nobilis
(Linnaeus, 1758)

Conus straitus
(Hwass & Bruguiere, 1792)

Lambis (Lambis) lambis
(Linnaeus, 1758)

Nassa serta
(Bruguiere, 1789)

Rhinoclavis (Rhinoclavis)
sinensis (Gmelin,1791)

Strombus (Canarium) labiatus
(Roeding, 1798)

Strombus variabilis
(Swainson, 1820)

Terebralia palustrisa
(Linnaeus, 1758)

FIGURE 9.4   Molluscs of Rutland Island. (Please see color plate at the back of the book.)

TABLE 9.4 Number of Species, Under Different Orders of Molluscs Reported

Sl. No.	Order	ANI total	Rutland Island	Percent of ANI total in RI
1	Archeogastropoda	116	16	13.79
2	Mesogastropoda	273	37	13.55
3	Neogastropoda	421	48	11.40
4	Architectonoidea	19	1	5.26
5	Pulmonata	14	3	21.43
6	Cephalaspidea	32	3	9.38
7	Mytiloida	87	3	3.45
8	Veneroida	177	18	10.17
9	Dentaliida	8	1	12.50
	Total	1147	130	11.33

ANI, Andaman and Nicobar Islands; RI, Rutland Island.

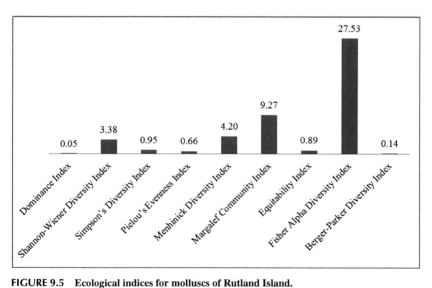

FIGURE 9.5   Ecological indices for molluscs of Rutland Island.

echinoderms are well distributed with optimum diversity, evenness and equitability indices (Figure 9.7). The species diversity was recorded as 3.53, 0.96 and 2.87 for Shannon-Wiener, Simpson's and Menhinick diversity indices, respectively. The density of echinoderms varied in the range 14–25 individuals/10 m^2.

Cenometra bella
(Hartlaub, 1890)

Culcita noveguineae
(Muller and Troschel, 1842)

Echinometra mathaei
(de Blainville, 1825)

Fromia indica
(Perrier, 1869)

Stephanometra indica
(Smith, 1876)

Thelenota ananas
(Jaeger, 1833)

Holothuria (Halodeima) edulis
(Lesson, 1830)

Acaudina molpadioides
(Semper, 1868)

Linckia laevigata
(Linneaus, 1758),

Pearsonothuria graeffei
(Semper, 1868)

Ophiomastix annulosa
(Lamarck, 1816)

Stichopus vastus
(Sluiter, 1887)

FIGURE 9.6   **Echinoderms of Rutland Island.** (Please see color plate at the back of the book.)

TABLE 9.5 Number of Species, Under Different Classes of Echinoderms Reported

Sl. No.	Class	ANI total	Rutland Island	Percent of ANI total in RI
1	Crinoidea	50	16	32.00
2	Asteroidea	104	25	24.04
3	Ophiuroidea	103	9	8.74
4	Echinoidea	80	22	27.50
5	Holothuroidea	88	31	35.23
	Total	425	103	24.23

ANI, Andaman and Nicobar Islands; RI, Rutland Island.

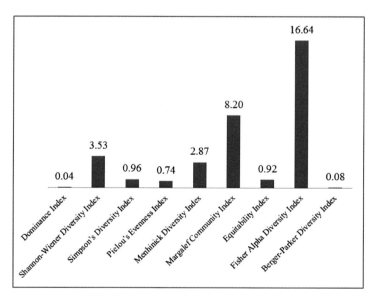

FIGURE 9.7   Ecological indices of echinoderms at Rutland Island.

DISCUSSION

Macrobenthic communities are the most important in the marine realm for the maintenance of aquatic ecosystems as primary and secondary consumers (Popchenko, 1971; Nassaj *et al.*, 2010). Yap *et al.* (2003) accepted the significance of macrobenthic faunal groups as biological indicators. The marine environment of Rutland Island and its shelf areas support an enormous wealth of marine faunal communities. Existing conducive environment variables such as temperature, pH, salinity, transparency and turbidity at optimal levels ensure the maintenance of the rich marine biodiversity of the island. The present study reports a healthy number of scleractinian corals, molluscs and echinoderms in the shelf region of the island during the period of study.

Of all ocean habitats, coral reefs seem to have the greatest development of complex symbiotic associations. Among India's four major reef areas, the Andaman and Nicobar Islands (ANI) are endowed with maximum diversity. A total of 479 species of scleractinian corals under 79 genera and 15 families (Tamal *et al.*, 2012a) have so far been reported from these islands, which contribute 60 percent of the global coral diversity. A total of 424 species of scleractinian corals (89 percent of India's coral diversity) are present at the Andaman group of islands and 242 species of scleractinians are distributed in the Nicobar group, which contributes 51 percent of India's coral diversity. All of the 15 families recorded from other reef regions of India are reported from ANI. Out of 89 genera reported from India, only the three genera *Stephanocyathus*, *Flabellum* and *Cladangia* have not been recorded so far from these islands. ANI have two endemic species—*Deltocyathus andamanensis* and *Polycyathus andamanensis*—which are azooxanthellate in nature. ANI reefs are dominated by the families Acroporidae, Faviidae, Poritidae, Fungidae and Agariciidae.

Rutland Island is one of the major islands in the Andaman group. The presence of a total of 327 species of scleractinian corals of this island represents 62.40 percent of scleractinian species of the entire ANI. The high values of species diversity indices indicate the rich scleractinian ecosystem of this island. A total of 91 species under the family Faviidae have been recorded until now from these groups of islands (Tamal *et al.*, 2012b,c), of which 90.11 percent are reported from this island. Of the coral species documented, 35 species of scleractinian corals were reported as new records to Indian waters (Tamal *et al.*, 2010, Tamal *et al.*, 2011a–d, Tamal *et al.*, 2012d–e, 2014). The diversity and status of scleractinian corals from this island indicates the importance of the biogenic habitat of Rutland Island for marine faunal communities. The value obtained from Shannon-Wiener diversity index (H′ = 4.10) proves the healthy status of scleractinian species diversity at Rutland Island.

Five of the seven major classes of the phylum Mollusca—Polyplacophora, Gastropoda, Scaphopoda, Bivalvia and Cephalopoda—are represented in India. Against a total of 586 families of molluscs in the world, an estimated 279 families occur in the Indian region. A total of 130 species of molluscs under 69 genera and 39 families were reported from Rutland Island, with high diversity (H′ = 3.38). Reported Mollusca of Rutland Island represent a total of 11.33 percent, with 9 orders out of 20, of species reported from the whole ANI.

Echinodermata is one of the best characterized and most distinct phyla of the animal kingdom. The echinoderms, being common and conspicuous marine animals, have been known since ancient times. They are found at every ocean depth, from the intertidal zone to the abyssal zone. The echinoderms are important both biologically and geologically: biologically because few other groupings are as abundant in the biotic desert of the deep sea, as well as the shallower oceans, and geologically as their ossified skeletons are major contributors to many limestone formations, and can provide valuable clues as to the geological environment. Further, it is held by some that the radiation of echinoderms was responsible for the Mesozoic revolution of marine life. The echinoderms comprise five classes: echinoids, crinoids, asteroids, ophiuroids and holothuroids. All five classes are represented in ANI. Bell (1887) was the pioneer of echinoderm study and listed the echinoderms from the Andaman. In 1983, James dealt with the sea cucumber and sea urchin resources of ANI and gave a list of echinoderms known from these islands. Later, Sastry (2005, 2007) compiled all the available published literature for 425 species of echinoderms of ANI against the 649 species reported from Indian waters. A total of 103 species of echinoderms belonging to 60 genera, 35 families and 15 orders were recorded from Rutland Island during the present study period. The values of species diversity indices are high (H' = 3.53), which implies a great deal of echinoderm diversity in Rutland Island.

Life history, recruitment, survival, development, diversity, distribution and abundance of any sort of macrobenthos exclusively depend on the characteristic features of the environment where they live, mediated by their physiological condition (Dahanayakar and Wijeyaratne, 2006; Perkins, 1974; Nassaj *et al.,* 2010). The enriched physico-chemical attributes of the marine environment of Rutland Island and surrounding areas provide sustainable ecological strata to macrobenthic organisms that allow them to flourish in great diversity and with a wide distributional pattern. The data collected from the present study prove that the shelf region of Rutland Island harbors rich macrofaunal diversity. The information gathered on species of different faunal groups will be useful for the conservation and protection of these living resources. Having found such enormous biodiversity in the marine habitat, it is tempting to recommend that Rutland Island be declared a Marine Sanctuary.

ACKNOWLEDGEMENTS

The authors are grateful to the authorities of the Department of Environment and Forests, Andaman and Nicobar Administration for logistic support in the conduct of the surveys.

REFERENCES

Bell, F.L., 1887. Report on a collection of echinodermata from the Andaman Islands. Proc. Zool. Soc. London 1, 130–145.

Clark, A.M., Rowe, F.W.E., 1971. Monograph of shallow-water Indo-West Pacific Echinoderms. Trustees of the British Museum (Natural History), London, p. 238.

Dahanayakar, D.D.G.L., Wijeyaratne, M.J.S., 2006. Diversity of macrobenthic community in the Negombo estuary, Sri Lanka, with special reference to environmental conditions. Sri Lanka J. Aquat. Sci. 11, 43–61.

Done, T.J., Kenchinton, R.A., Zell, L.D., 1982. Rapid, large area, reef resource surveys using a manta board. Proceedings of the Fourth International Coral Reef Symp. Manila 2, 597–600.

English, S., Wilkinson, C., Baker, V., 1997. Survey manual for tropical marine resources, 2nd Edition Australian Institute of Marine Science, p. 390.

James, D.B., 1983. Sea cucumber and sea urchin resources. Bull. Cent. Mar. Fish. Res. Inst. 34, 85–93.

Nassaj, S.M.S., Nabavi, S.M.B., Yavari, V., Savari, A., Maryamabadi, A., 2010. Species Diversity of Macrobenthic Communities in Salakh Region, Qeshm Island, Iran. World J. Fish Marine Sci. 2 (6), 539–544.

Perkins, E.J., 1974. The biology of estuarine and coastal waters. Academic Press, London, p. 678.

Popchenko, V.I., 1971. Consumption of oligochaeta by fish and invertebrates. J. Ichthyol. 11, 75–80.

Rao, D.V., 2010. Field Guide to Coral and Coral Associates of Andaman & Nicobar Islands. Zoological Survey of India, Kolkata, p. 283.

Sastry, D.R.K., 2005. Echinoderms of Andaman and Nicobar Islands., Bay of Bengal. An annotated list. Rec. Zool. Sur. India., Occ. Paper No. 233, 1–202.

Sastry, D.R.K., 2007. Echinodermata of India: An Annotated list. Rec. Zool. Surv. India. Occ. Paper No. 271, 1–387.

Subba Rao, N.V., 2000. Catalogue of Marine mollusks of Andaman and Nicobar Islands. Rec. Zool. Surv. India. Occ. Paper No. 187, 1–323.

Subba Rao, N.V., 2003. Indian Seashells (Part-I): Polyplacophora and Gastropoda. Rec. Zool. Surv. India, Occ. Paper No. 192, 1–416.

Tagliapietra, D., Sigovini, M., 2010. Benthic fauna: collection and identification of macrobenthic invertebrates. NEAR Curriculum in Natural Environmental Science. Terre et Environnement 88, 253–261.

Tamal, M., Raghunathan, C., Ramakrishna, 2011a. New Record of Five Scleractinian Corals from Rutland Island, South Andaman Archipelago. Asian J. Exp. Biol Sci. 2 (1), 114–118.

Tamal, M., Raghunathan, C., Ramakrishna, 2011b. Notes on three new records of scleractinian corals from Andaman Islands. J. Ocean. Marine Sci. 2 (5), 122–126.

Tamal, M., Raghunathan, C., Ramakrishna, 2011d. Addition of thirteen Scleractinians as New Record to Indian Water from Rutland Island, Andamans. Asian J. Exp. Biol. Sci. 2 (3), 383–390.

Tamal, M., Raghunathan, C., Venkataraman, K., 2011c. Five Scleractinian Corals as a New Record from Andaman Islands - A New Addition to Indian Marine Fauna. World J. Fish Marine Sci. 3 (5), 450–458.

Tamal, M., Raghunathan, C., Venkataraman, K., 2012a. Diversity and Distribution of Corals in Andaman and Nicobar Islands. Jour. Coast Env. 3 (2), 101–110.

Tamal, M., Raghunathan, C., Venkataraman, K., 2012b. An account of Faviid Corals of Andaman & Nicobar Islands. Res. J. Sci. Tech. 4 (2), 62–66.

Tamal, M., Raghunathan, C., Venkataraman, K., 2012c. New Record of five Scleractinian Corals to Indian Water from Andaman and Nicobar Islands. Res. J. Sci. Tech. 4 (6), 278–284.

Tamal, M., Raghunathan, C., Venkataraman, K., 2012d. New Distribution Report of Ten Scleractinian Corals to Indian Water from Andaman and Nicobar Islands. Res. J. Sci. Tech. 4 (4), 152–157.

Tamal, M., Raghunathan, C., Venkataraman, K., 2012e. New Record of Five Scleractinian Corals to Indian Water from Andaman and Nicobar Islands. IJABR 2 (2), 699–702.

Tamal, M., Raghunathan, C., Venkataraman, K., 2014. First Report of one Caryophylliid and two Dendrophylliid corals in Indian water from Andaman and Nicobar Islands. Indian J. Geo-Mar. Sci. 43 (4), 538–541.

Tamal, M., Raghunathan, C., Ramakrishna, 2010. New record of nine Scleractinian Corals from Rutland Island, Andaman. Int. J. Biol. Sci. 1, 155–170.

Tikadar, B.K., Das, A.K., 1985. Glimpses of animal life of Andaman & Nicobar Islands. Director, Zoological Survey of India, Kolkata, p. 170.

Venkataraman, K., Satyanarayan, Ch., Alfred, J.R.B., Wolstenholme, J., 2003. Handbook on Hard Corals of India. Zoological Survey of India, Kolkata, p. 266.

Veron, J.E.N., 2000. Corals of the World1-3Australian Institute of Marine Science.

Veron, J.E.N., Pichon, M., 1976. Scleractinia of Eastern Australia. *Part I.* Australian Institute of Marine Science, p. 86.

Veron, J.E.N., Pichon, M., 1979. Scleractinia of Eastern Australia. *Part III.* Australian Institute of Marine Science, p. 421.

Veron, J.E.N., Pichon, M., 1982. Scleractinia of Eastern Australia. *Part IV.* Australian Institute of Marine Science, p. 159.

Veron, J.E.N., Wallace, C.C., 1984. Scleractinia of Eastern Australia. *Part V.* Australian Institute of Marine Science, p. 485.

Veron, J.E.N., Pichon, M., Wijsman-Best, M., 1977. Scleractinia of Eastern Australia. *Part II.* Australian Institute of Marine Science, p. 233.

Wallace, C.C., 1999. Staghorn Corals of the world. CSIRO Publications, Melbourne, p. 421.

Whitaker, R., 1985. Endangered Andamans. Department of Environment, Government of India, p. 51.

Yap, G.K., Ismail, A.R., Ismail, A., Tan, S.G., 2003. Species diversity of macrobenthic invertebrates in Semenyih river. Pertanika J. Trop. Agric. Sci. 26 (2), 139–146.

Chapter 10

Diversity and Distribution of Sea Grass Associated Macrofauna in Gulf of Mannar Biosphere Reserve, Southern India

K. Paramasivam,* K. Venkataraman,†
C. Venkatraman,* R. Rajkumar* and S. Shrinivaasu*
*Marine Biology Regional Centre, Zoological Survey of India, Chennai, Tamil Nadu, India;
†Zoological Survey of India, Kolkata, West Bengal, India

INTRODUCTION

Sea grasses are one of the groups of flowering plants capable of completing their life cycle in a marine environment (Kuo and McComb, 1989). Sea grass ecosystems are highly productive, with immense ecological and socioeconomic importance. They support various kinds of biota, produce a considerable amount of organic matter, are a major energy source in the coastal marine food web, and play a significant role in nutrient regeneration and shore stabilization processes.

The sea grass ecosystem is considered to be very productive and is seen as a nursery and breeding ground for many marine organisms. It also acts as a sediment stabilizer, provides a suitable substratum for epiphytes and a good source of food for marine herbivores, and is a source of fodder and manure. Sea grasses are effective in removing nutrients from marine waters and surface sediments and are therefore, important in the control of water quality of coastal waters. Of the 13 genera and 58 species of sea grasses reported worldwide, 14 species belonging to 6 genera are in India, out of which 13 species are found in the Gulf of Mannar Biosphere Reserve (Kannan et al., 1999).

Despite the importance attached, sea grass meadows are experiencing high rates of loss in some parts of the world (Coles et al., 2011), and at a rate which may be as high as 7 percent of their global area per year (Orth et al., 2006; Waycott et al., 2009), with the subsequent loss of the fauna they support (Hughes et al., 2009). While most of the threats are anthropogenic, having both direct and indirect impacts (Coles and Lee Long, 1999), there are also significant impacts by climate change and local natural events (Kensworthy *et al.,* 2006). Therefore, given their role as a habitat for a wide range of species and the accelerated loss of many sea grass areas, it is imperative to make an inventory of the sea grass habitats for faunal associates, to develop baseline information, prepare checklists and assess status, in order to initiate effective management applications for conservation of biodiversity and management of resources.

The importance of sea grass ecosystems has been realized very lately in India, and only recently have studies of the extent of coverage of sea grass ecosystems been made (Jagtap and Inamdar, 1991; Jagtap, 1991; Parthasarathy *et al.,* 1991; Bahuguna and Nayak, 1994; Jagtap *et al.,* 2003; Umamaheswari *et al.,* 2009; Manikandan *et al.,* 2011a). The very few studies on faunal assemblages include a few selected localities on the Tamil Nadu coast and Lakshadweep Islands (Jagtap *et al.,* 2003; Susan *et al.,* 2012). However, detailed study on diversity, distribution and abundance of sea grass associated macrofauna is still lacking in India.

METHODS

Study Area

The Gulf of Mannar Biosphere Reserve extends from Rameswaram Island to Tuticorin and lies between 8° 45' N to 9° 25' N and 78° 5' E to 79° 30' E, extending for 140 km. There are 21 islands running almost parallel to the coastline of Mannar. These islands lie between 8° 47' N to 9° 15' N and 78° 12' E to 79° 14' E (Figure 10.1). These islands are situated at an average distance of about 8 km from the coastline of the Gulf of Mannar. The reserve is bounded by Palk Bay and Rameswaram Island in the north, by Ramanathapuram district in

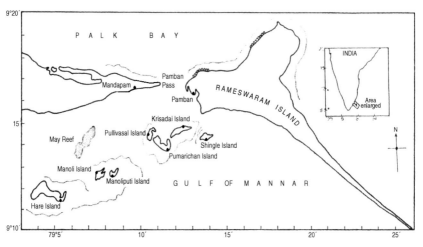

FIGURE 10.1   Map showing the study areas.

the northwest and west, by Tuticorin district in the south, and by Bay of Bengal in the east. It is endowed with three distinct ecosystems: corals, sea grass and mangroves.

The study was carried out at the following selected sea grass ecosystems located on fringe areas of the Mandapam group of Islands in the Gulf of Mannar Biosphere Reserve.

- **Krusadai Island** (Lat. 9° 14' N; Long. 79° 13' E) is situated near Mandapam and Pamban of Rameswaram Island and covers an area of 0.658 km^2 (Venkataraman *et al.*, 2004). The area of sea grass beds in the island is ~3 km^2 (Umamaheswari *et al.*, 2009).
- **Pullivasal Island and Poomarichan Island** (Lat. 9° 14' N; Long. 79° 11' E) and (Lat. 9° 14' N; Long. 79° 11' E) cover an area of 0.3 km^2 and 0.17 km^2, respectively (Venkataraman *et al.*, 2004). The area of sea grass beds in these islands is ~5.89 km^2 (Umamaheswari *et al.*, 2009).
- **Manoli and Manoliputti Island** (Lat. 9° 14' N; Long. 79° 7' E) and (Lat. 9° 13' N; Long. 79° 7' E) cover an area of 0.26 km^2 and 0.034 km^2 respectively (Venkataraman *et al.*, 2004). The area of sea grass beds in these islands is ~5.89 km^2 (Umamaheswari *et al.*, 2009).
- **Musal / Muyal theevu** (Lat. 9° 12' N; Long. 79° 5' E) is the largest island in the Gulf of Mannar and is also called Hare Island. The sea grass beds in this island of area 1.29 km^2 (Venkataraman *et al.*, 2004) cover an area of ~9.5 km^2 (Umamaheswari *et al.*, 2009).

Adequate spatial coverage of sampling in the sea grass ecosystems in each of these islands was made depending upon the area cover of the particular sea grass bed.

Sampling Methods

The sampling was done by a combination of methods, which included shore-based quadrat collections, collections from the landing sites of the sea grass fishery areas, trawl surveys (using prawn trawl nets), dredge trawls and push-nets. In the shore-based quadrat collections, quadrats were placed over sea grass beds during low-tide exposures, and the epibenthic and epifauna extracted to the maximum possible extent. Fishing trawlers with nets used by fishermen for prawn fishery were employed for the trawl surveys (Figure 10.2).

The swept area a is estimated as

$$a = D \, hr \, x_2$$

where D is the distance covered, hr is the length of the head-rope, and x_2 is that fraction of the head-rope length hr which is equal to the width of the path swept by the trawl. For Southeast Asian bottom trawls, values of x_2 from 0.4 (Shindo, 1973) to 0.66 (SCSP, 1987) are reported. Pauly (1980) suggests $x_2 = 0.5$ as the best compromise.

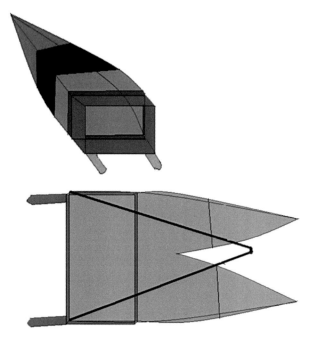

FIGURE 10.2   **Fishing trawlers.**

In the dive transects, two divers who swam along the transect line of 100 m laid over sea grass beds, counted, photographed and collected as many sessile and mobile epibenthos as possible. The dredge trawl and push-net were custom designed, based on Holme and McIntyre (1971) and Riley (1965) respectively, for operations at 2–4 m depth (Dredge-trawl) and at depths <2 m.

Preservation and Identification of Samples

Collected specimens were relaxed and preserved according to standard protocols. (Alfred and Ramakrishna, 2004). Preservation was in 75 percent or 95 percent ethanol, 5–10 percent seawater formalin, or Bouin's fixative, depending on the taxon. Photo documentation was done either in live condition or fixed form. The number of samples collected from each sampling were sorted taxonomically, weighed and counted for making further analysis.

Statistical Analysis

Species Richness and Abundance The species richness and abundance were calculated from the census data and field observations.

Diversity Indices Shannon index, Simpson index and Hill's diversity numbers N1 and N2 were calculated for different locations using the programme SPDIV-ERS.BAS developed by Ludwig and Reynolds (1988).

Similarity Measures Jaccard and Sorenson similarity index values between the different islands were calculated using the formula of Magurran (1988).

Dominance Index The dominance of the each species was calculated using the dominance index.

RESULTS

Species Composition and Abundance

Five major macrofaunal groups—Cnidaria, Crustacea, Mollusca, Echinodermata and Pisces—were found to associate with sea grass in the Mandapam group of islands in the Gulf of Mannar. A total of 2439 individuals belonging to 149 species were observed during the study period. The highest number of individuals—1119, belonging to 56 species—was recorded at Manoliputti Island, followed by 536 individuals belonging to 51 species at Manoli Island. The highest number of species, 60, was recorded at Krusadai Island (Table 10.1).

TABLE 10.1 Sea Grass Associated Macrofauna Encountered during the Study Period

Species	Localities					
	1	2	3	4	5	Total
Order: Decapoda						
Family: Alpheidae						
Alpheus rapax (Fabricius, 1798)				1		1
Order: Rajiformes						
Family: Dasyatidae						
Amphotistius imbricatus (Bloch & Schneider, 1801)			1			1
Order: Tetraodontiformes						
Family: Monacanthidae						
Anacanthus barbatus (Gray, 1830)			1			1
Order: Stomatopoda						
Family: Squillidae						
Anchisquilla fasciata (de Hann, 1844)	1	1				2
Order: Neogastropoda						
Family: Olividae						
Ancilla cinnamomea (Lamarck, 1801)				1	1	2

1, Krusadai Island; 2, Pullivasal Island; 3, Poomarichan Island; 4, Manoli Island; 5, Manoliputti Island.

(Continued)

TABLE 10.1 Sea Grass Associated Macrofauna Encountered during the Study Period *(cont.)*

Species	Localities					
	1	2	3	4	5	Total
Order: Archaeogastropoda						
Family: Angaridae						
Angaria delpbinus (Linnaeus, 1758)				2	6	8
Order: Pectinoida						
Family: Anomiidae						
Anomia achaeus (Gray, 1850)					1	1
Order: Perciformes						
Family: Apogonidae						
Apogon fasciatus (White,1790)	4					4
Apogon guamensis (Valenciennes,1832)	5	2		3	19	29
Order: Siluriformes						
Family: Ariidae						
Arius thalassinus (Ruppell, 1837)	4					4
Order: Tetraodontiformes						
Family: Tetraodontidae						
Arothron hispidus (Linnaeus, 1758)	2	1	1		1	5
Arothron immaculatus (Bloch & Schneider, 1801)	1	1	2			4
Order: Archaeogastropoda						
Family: Turbinidae						
Astralium semicostatum (Kiener, 1850)	1					1
Order: Cephalaspidea						
Family: Haminoeidae						
Atys elongatus (A. Adams, 1850)					1	1
Order: Arcoida						
Family: Arcidae						
Barbatia fusca (Bruguire, 1789)				2		2
Order: Batrachoidiformes						
Family: Batrachoididae						
Batrachomoeus trispinosus (Gunther, 1861)			1			1
Order: Decapoda						
Family: Scyllaridae						

1, Krusadai Island; 2, Pullivasal Island; 3, Poomarichan Island; 4, Manoli Island; 5, Manoliputti Island.

TABLE 10.1 Sea Grass Associated Macrofauna Encountered during the Study Period *(cont.)*

Species	Localities					
	1	2	3	4	5	Total
Biarctus sordidus (Stimpson, 1860)			1			1
Order: Neogastropoda						
Family: Turridae						
Brachytoma crenularis (Lamarck, 1801)					3	3
Order: Cephalaspidea						
Family: Bullidae						
Bulla ampulla (Linnaeus, 1758)					2	2
Order: Mesogastropoda						
Family: Bursidae						
Bursa crumena (Lamarck, 1816)	2					2
Order: Tetraodontiformes						
Family: Tetraodontidae						
Canthigaster solendri (Richardson, 1845)			3			3
Order: Perciformes						
Family: Carangidae						
Carangoides armatus (Ruppell, 1830)		1				1
Caranx praeustus (Anonymous (Bennett), 1830)	1				1	1
Order: Veneroida						
Family: Cardiidae						
Cardium asiaticum (Bruguire, 1789)	1				13	14
Cardium flavum (Linnaeus, 1758)				11		11
Order: Veneroida						
Family: Cardiidae						
Cerastoderma edule (Linnaeus, 1758)	4			1		5
Order: Mesogastropoda						
Family: Cerithidae						
Cerithium tenellum (Sowerby, 1855)	58		128	319	666	1171
Cerithium traillii (Sowerby, 1855)			17	14	2	33
Cerithidea cingulata (Gmelin, 1791)				31		31
Cerithium atratum (Born, 1778)					7	7
Cerithium balteatum (Philippi, 1848)	4			4		8

1, Krusadai Island; 2, Pullivasal Island; 3, Poomarichan Island; 4, Manoli Island; 5, Manoliputti Island.

(Continued)

TABLE 10.1 Sea Grass Associated Macrofauna Encountered during the Study Period *(cont.)*

Species	Localities					
	1	2	3	4	5	Total
Cerithium columna (Sowerby, 1834)	1	1	2	9	73	86
Cerithium coralium Kiener, 1841				7	22	29
Cerithium petrosa gennesi (Fischer & Vignal, 1901)				12	24	36
Cerithium scabridium (Philippi, 1848)					6	6
Family: Portunidae						
Charybdis anisodon (De Hann, 1850)	22	2				24
Charybdis natator (Herbst, 1794)			1			1
Order: Tetraodontiformes						
Family: Tetraodontidae						
Chelonodon patoca (Hamilton, 1822)	1					1
Order: Veneroida						
Family: Veneridae						
Circe scripta (Linnaeus, 1758)		1				1
Order: Archaeogastropoda						
Family: Trochidae						
Clangulus margaritarius (Philippi, 1846)					1	1
Order: Mesogastropoda						
Family: Cerithidae						
Colina macrostoma (Hinds, 1844)				2		2
Order: Batrachoidiformes						
Family: Batrachoididae						
Colletteichthys dussumieri (Valenciennes, 1837)			1			1
Order: Neogastropoda						
Family: Conidae						
Conus araneosus (Lightfoot, 1786)		1				1
Order: Syngnathiformes						
Family: Syngnathidae						
Corythoichthys haematopterus (Bleeker, 1851)		1	1			2
Order: Decapoda						
Family: Parthenopidae						

1, Krusadai Island; 2, Pullivasal Island; 3, Poomarichan Island; 4, Manoli Island; 5, Manoliputti Island.

TABLE 10.1 Sea Grass Associated Macrofauna Encountered during the Study Period *(cont.)*

Species	Localities					
	1	2	3	4	5	Total
Cryptopodia fornicata (Fabricius, 1787)				1		1
Order: Pleuronectiformes						
Family: Cynoglossidae						
Cynoglossus bilineatus (Lacepede, 1802)	1					1
Cynoglossus lida (Bleeker, 1851)			1			1
Order: Veneroida						
Family: Donacidae						
Donax cuneatus (Linnaeus, 1758)	1					1
Order: Perciformes						
Family: Serranidae						
Epinephelus coioides (Hamilton, 1822)	1			1	1	3
Order: Caenogastropoda						
Family: Epitoniidae						
Epitonium subauriculatum (Souverbie, 1866)					1	1
Order: Sepiolida						
Family: Sepiolidae						
Euprymna berryi (Sasaki, 1929)	4					4
Order: Veneroida						
Family: Veneridae						
Gafrarium tumidum (Roeding, 1798)				21	6	27
Order: Perciformes						
Family: Gerreidae						
Gerres oyena (Forsskal, 1775)	14	1		1	1	17
Order: Decapoda						
Family: Corystidae						
Gomeza bicornis (Gray, 1831)				1		1
Order: Perciformes						
Family: Labridae						
Halichoeres nigrescens (Bloch & Schneider, 1801)			1			1
Order: Syngnathiformes						
Family: Syngnathidae						

1, Krusadai Island; 2, Pullivasal Island; 3, Poomarichan Island; 4, Manoli Island; 5, Manoliputti Island.

(Continued)

TABLE 10.1 Sea Grass Associated Macrofauna Encountered during the Study Period *(cont.)*

Species	Localities					
	1	2	3	4	5	Total
Hippocampus spinosissimus (Weber, 1913)		1				1
Hippocampus trimaculatus (Leach, 1814)			2			2
Order: Aspidochirotida						
Family: Holothuridae						
Holothuria atra (Jaeger, 1833)					2	2
Holothuria leucospilata (Brandt, 1835)	1					1
Order: Perciformes						
Family: Gobiidae						
Istigobius ornatus (Ruppell, 1830)	1				2	3
Order: Archaeogastropoda						
Trochidae						
Jujubinus striatus (Linnaeus, 1758)	2		8	6	7	23
Order: Tetraodontiformes						
Family: Tetrodontidae						
Lagocephalus lunaris (Bloch & Schneider,1801)	1					1
Order: Mesogastropoda						
Family: Strombidae						
Lambis lambis (Linnaeus, 1758)				1		1
Order: Perciformes						
Family: Latidae						
Lates calcarifer (Bloch, 1790)		2				2
Order: Perciformes						
Family: Leiognathidae						
Leiognathus brevirostris (Valenciennes, 1835)	5					5
Family: Lethrinidae						
Lethrinus ornatus (Valenciennes, 1830)		1		2		3
Lethrinus variegatus (Valenciennes, 1830)			1	1		2
Order: Myopsida						
Family: Loliginidae						
Loligo uyii (Wakia & Ishikawa, 1921)			1			1
Order: Perciformes						

1, Krusadai Island; 2, Pullivasal Island; 3, Poomarichan Island; 4, Manoli Island; 5, Manoliputti Island.

TABLE 10.1 Sea Grass Associated Macrofauna Encountered during the Study Period *(cont.)*

Species	Localities					
	1	2	3	4	5	Total
Family: Lutjanidae						
Lutjanus russelli (Bleeker, 1849)				1		1
Lutjanus fulviflamma (Forsskal, 1775)	1	2		2		5
Lutjanus malabaricus (Bloch & Schneider, 1801)	3		1			4
Order: Veneroida						
Family: Mactridae						
Mactra maculata (Gmelin, 1791)				2	3	5
Mactrellona exoleta (Gray, 1837)	5	1				6
Order: Neogastropoda						
Family: Marginellidae						
Marginella angustata (G.B. Sowerby,1846)					1	1
Order: Decapoda						
Family: Epialtidae						
Menaethius monoceros (Latrielle, 1825)					5	5
Family: Penaeidae						
Metapenaeopsis palmensis (Haswell, 1879)		1	2			3
Metapenaeus affinis (H. Milne Edwards, 1837)			1			1
Metapenaeus dobsoni (Miers, 1878)	2					2
Order: Mytiloida						
Family: Mytilidae						
Modiolus philippinarium (Hanley, 1843)	1	1	9			11
Order: Mugiliformes						
Family: Mugilidae						
Mugil cephalus (Linnaeus, 1758)	6					6
Order: Neogastropoda						
Family: Muricidae						
Murex trapa (Roeding, 1798)	1					1
Order: Perciformes						
Family: Acanthuridae						
Naso lituratus (Forster, 1801)		1				1

1, Krusadai Island; 2, Pullivasal Island; 3, Poomarichan Island; 4, Manoli Island; 5, Manoliputti Island.

(Continued)

TABLE 10.1 Sea Grass Associated Macrofauna Encountered during the Study Period *(cont.)*

Species	Localities					
	1	2	3	4	5	Total
Order: Neogastropoda						
Family: Buccinidae						
Nassaria laevior (E.A.Smith, 1899)					1	1
Nassarius albescens gemmuliferus (A. Adams, 1852)	2		1			3
Family: Nassariidae						
Nassarius distortus (A. Adams, 1852)		1			10	11
Nassarius luridus (Gloud, 1850)				1		1
Nassarius pullus (Linnaeus, 1758)				1		1
Order: Mesogastropoda						
Family: Naticidae						
Natica vitellus (Linnaeus, 1758)				2	1	3
Order: Octopoda						
Family: Octopodidae						
Octopus aegina (Gray, 1849)		1				1
Order: Neogastropoda						
Family: Olividae						
Oliva annulata (Gmelin, 1791)			1			1
Order: Stomatopoda						
Family: Squillidae						
Oratosquilla oratoria (De Hann, 1844)			1			1
Order: Perciformes						
Family: Gobiidae						
Oxyurichthys papuensis (Valenciennes, 1837)	3					3
Order: Veneroida						
Family: Veneridae						
Paphia undulata (Born, 1778)	1					1
Family: Gobiidae						
Parachaeturichthys ocellatus (Day, 1873)	1		1			2
Order: Tetraodontiformes						
Family: Monacanthidae						

1, Krusadai Island; 2, Pullivasal Island; 3, Poomarichan Island; 4, Manoli Island; 5, Manoliputti Island.

TABLE 10.1 Sea Grass Associated Macrofauna Encountered during the Study Period *(cont.)*

Species	Localities					
	1	2	3	4	5	Total
Paramonacanthus choirocephalus (Bleeker, 1851)		2	3			5
Paramonacanthus frenatus (Peters, 1855)	1					1
Order: Pleuronectiformes						
Family: Soleidae						
Pardachirus pavoninus (Lacepede, 1802)		1	1		1	3
Order: Perciformes						
Family: Terapontidae						
Pelates quadrilineatus (Bloch, 1790)			1	1	5	8
Order: Decapoda						
Family: Penaeidae						
Penaeus indicus (H. Milne Edwards, 1837)	5					5
Penaeus semisulcatus (De Hann, 1844)	11	5	3	2	4	25
Order: Perciformes						
Family : Blennidae						
Petroscirtes breviceps (Valenciennes, 1836)	1	2	4		1	8
Petroscirtres mitratus (Ruppell, 1830)		1				1
Order: Archaeogastropoda						
Family: Phasianellidae						
Phasianella nivosa (Reeve, 1862)					2	2
Phasianella variegata (Lamarck, 1822)	5			6	8	19
Order: Neogastropoda						
Family: Buccinidae						
Phos senticosus (Linnaeus, 1758)				2		2
Order: Pteroida						
Family: Anomiidae						
Pinna bicolor (Gmelin, 1791)					1	1
Order: Siluriformes						
Family: Plotosidae						
Plotosus lineatus (Thunberg, 1787)		2				2
Order: Mesogastropoda						
Family: Naticidae						

1, Krusadai Island; 2, Pullivasal Island; 3, Poomarichan Island; 4, Manoli Island; 5, Manoliputti Island.

(Continued)

TABLE 10.1 Sea Grass Associated Macrofauna Encountered during the Study Period *(cont.)*

Species	Localities					
	1	2	3	4	5	Total
Polinices melanostoma (Gmelin, 1791)				5		5
Order: Decapoda						
Family: Crangonidae						
Pontophilus spinosus (Leach, 1816)					1	1
Order: Decapoda						
Family: Portunidae						
Portunus (portunus) pelagicus (Linnaeus, 1758)	5		5	6		16
Portunus haanii (Stimpson, 1858)			1			1
Portunus hastatoides (Weber, 1795)	8		1			9
Order: Neogastropoda						
Family: Columbellidae						
Pyrene flava (Bruguiere,1789)			1			1
Pyrene obscura (Sowerby, 1844)			1	1		2
Pyrene scripta (Lamarck, 1822)	51		42	12	96	201
Order: Neogastropoda						
Family: Muricidae						
Rapena rapiformes (Born, 1778)	1					1
Order: Decapoda						
Family: Parthenopidae						
Rhinolambrus pelagicus (Ruppell,1830)					3	3
Order: Camerodonta						
Family: Temnoplueridae						
Salmacis virgulata (L. Agassiz & Desor, 1846)			1			1
Order: Pennatulacea						
Family: Pennatulidae						
Sarcoptilus grandis (Gray, 1848)		1				1
Order: Cluepeiformes						
Family: Clupeidae						
Sardinella albella (Valenciennes, 1847)	1					1
Sardinella gibbosa (Bleeker, 1849)		1				1

1, Krusadai Island; 2, Pullivasal Island; 3, Poomarichan Island; 4, Manoli Island; 5, Manoliputti Island.

TABLE 10.1 Sea Grass Associated Macrofauna Encountered during the Study Period *(cont.)*

Species	Localities					
	1	2	3	4	5	Total
Order: Aulopiformes						
Family: Synodontidae						
Saurida gracilis (Quoy & Gaimard, 1824)		2		2	9	13
Order: Perciformes						
Family: Scaridae						
Scarus ghobban (Forsskal, 1775)		1		1	4	6
Scarus psittacus (Forsskal, 1775)					3	3
Order: Perciformes						
Family: Leiognathidae						
Secutor insidiator (Bloch, 1787)	23	1				24
Order: Cephalopoda						
Family: Sepiidae						
Sepia aculeata Van Hasselt, 1835	4					4
Sepiella inermis (Van Hasselt, 1835)	1		2			3
Order: Myopsida						
Family: Loliginidae						
Sepioteuthis lessoniana (Lesson, 1830)		2				2
Order: Perciformes						
Family: Siganidae						
Siganus canaliculatus (Park, 1797)		5	58	5	13	81
Siganus javus (Linnaeus, 1758)	11	1		3	4	19
Order: Perciformes						
Family: Sillaginidae						
Sillaho sihama (Forsskal,1775)	5					5
Order: Scorpaeniformes						
Family: Platycephalidae						
Sorsogona tuberculata (Cuvier, 1829)	5	1			1	7
Order: Actiniaria						
Family: Stichodactylidae						
Stichodactylus haddoni (Saville-Kent,1893)			1			1
Order: Mesogastropoda						
Family: Strombidae						

1, Krusadai Island; 2, Pullivasal Island; 3, Poomarichan Island; 4, Manoli Island; 5, Manoliputti Island.

(Continued)

TABLE 10.1 Sea Grass Associated Macrofauna Encountered during the Study Period *(cont.)*

Species	Localities					
	1	2	3	4	5	Total
Strombus (Canarium) labiatus labiatus (Roeding, 1798)			3	3		6
Order: Perciformes						
Family: Callionymidae						
Synchiropus postulus (Smith,1963)	1					1
Order: Syngnathiformes						
Family: Platycephalidae						
Syngnathoides biaculeatus (Bloch, 1785)	1	1	2	1	5	10
Order: Veneroida						
Family: Tellinidae						
Tellina staurella (Lamarck, 1818)		1		1		2
Tellina virgata Linnaeus, 1758	1					1
Order: Perciformes						
Family: Terapontidae						
Terapon puta (Cuvier, 1829)	2	1	1			4
Order: Mesogastropoda						
Family: Strombidae						
Terebellum terebellum (Linnaeus, 1758)				1		1
Order: Neogastropoda						
Family: Terebridae						
Terebra maculata (Linnaeus, 1758)				1		1
Order: Neogastropoda						
Family: Muricidae						
Thais rugosa (Born,1780)				3	1	4
Order: Decapoda						
Family: Portunidae						
Thalamita parvidens (Rathbun, 1907)				1	2	3
Thalamita crenata (Ruppell, 1830)			1	7		8
Thalamita integra (Dana, 1852)	73	4	8	8	57	150
Order: Perciformes						
Family: Trichiuridae						
Trichiurus savala (Cuvier, 1829)			1			1

1, Krusadai Island; 2, Pullivasal Island; 3, Poomarichan Island; 4, Manoli Island; 5, Manoliputti Island.

TABLE 10.1 Sea Grass Associated Macrofauna Encountered during the Study Period *(cont.)*

Species	Localities					
	1	2	3	4	5	Total
Order: Archaeogastropoda						
Family: Turbinidae						
Turbo brunneus (Roeding, 1798)				1	2	3
Order: Perciformes						
Family: Mullidae						
Upeneus tragula (Richardson, 1846)		4		2	3	9
Upeneus vittatus (Forsskal, 1775)	1					1
Order: Perciformes						
Family: Gobiidae						
Yongeichthys criniger (Valenciennes, 1837)	3		1			4
No. of individuals	389	64	331	536	1120	2440
Overall No. of species	60	41	47	52	56	150

1, Krusadai Island; 2, Pullivasal Island; 3, Poomarichan Island; 4, Manoli Island; 5, Manoliputti Island.

In terms of groups, the highest numbers (971 and 487) of Molluscan individuals, belonging to 31 species, was observed in Manoli and Manoliputti Island, respectively, followed by 10 individuals of 9 species in Pullivasal Island. Similarly, the highest number of individuals (111) of Pisces, belonging to 31 species, was recorded in Krusadai Island, followed by 88 individuals of 23 species at Poomarichan Island (Table 10.2).

A total of nine species of crab and one species of sea anemone predominantly occurred in sea grass habitat of the Mandapam group of islands. Their dominance index ranged from 59.52 to 5.09. Of these, *Thalamita integra* (Dana, 1852), *Portunus (portunus) pelagicus* (Linnaeus, 1758) and *Thalamitta crenata* (Ruppell, 1830) occurred at more than one island. Species *Sarcoptilus grandis* Gray, 1848 and *Charybdis anisodon* (De Hann, 1850) were restricted to Pullivasal island, and *Thalamita parvidens* (Rathbun, 1907), *Menaethius monoceros* (Latrielle, 1825) and *Rhinolambrus pelagicus* (Ruppell, 1830) were restricted to Manoliputti Island (Table 10.3).

Species Diversity

Among the macrofauna observed, Pisces and Mollusca show highest diversity in all islands. The diversity of Pisces and Mollusca was respectively 1.49 and

TABLE 10.2 Species Richness and Abundance of Faunal Groups in the Study Area

Island	Animal group	Richness	Abundance
Krusadai Island	Crustaceans	8	124
	Molluscans	21	151
	Pisces	31	111
	Echinodermata	1	1
Pullivasal Island	Cnidaria (sea pen)	1	1
	Crustaceans	5	13
	Molluscans	9	10
	Pisces	26	42
Poomarichan Island	Cnidaria	1	1
	Crabs	11	25
	Mollusca	11	214
	Echinodermata	1	1
	Pisces	23	88
Manoli Island	Crustaceans	7	26
	Mollusca	31	487
	Pisces	13	25
Manoliputti Island	Crab	7	73
	Mollusca	31	971
	Pisces	16	73
	Echinodermata	1	2

1.32 at Krusadai Island and 1.42 and 0.95 at Pullivasal Island. Highest diversity (1.20) was observed in Manoliputti followed by 1.11 in Manoli Island. Apart from Pisces and Mollusca the diversity of Crustaceans was observed to be high at Krusadai (0.90) and Manoli and Manoliputti Islands (0.95) (Table 10.4).

Similarity

The similarity value ranged from 11.14 to 60.35. Highest species similarity (60.35) was observed between Manoli and Manoliputti Islands, followed by 46.98 found between Krusadai and Manoli Islands. Least species similarity was found between Pullivasal and Manoliputti Islands (11.14), followed by 21.85 between Pullivasal and Manoli Islands (Table 10.5).

TABLE 10.3 Dominance Index of Sea Grass Macrofauna in the Study Area

Species	Localities				
	1	2	3	4	5
Thalamita integra (Dana, 1852)	73(18.72)		42(12.77)	319(59.29)	666(59.52)
Portunus (portunus) pelagicus (Linnaeus, 1758)	58(14.87)		17(5.17)		
Portunus hastatoides (Weber, 1795)	51(13.08)				
Sarcoptilus grandis (Gray, 1848)		5(7.58)			
Charybdis anisodon (De Hann, 1850)		5(7.58)			
Stichodactylus haddoni (Saville-Kent, 1893)			128(38.91)		
Thalamitta crenata (Ruppell, 1830)			58(17.63)	31(5.76)	
Thalamita parvidens (Rathbun, 1907)					96(8.58)
Menaethius monoceros (Latrielle, 1825)					73(6.52)
Rhinolambrus pelagicus (Ruppell, 1830)					57(5.09)

1, Krusadai Island; 2, Pullivasal Island; 3, Poomarichan Island; 4, Manoli Island; 5, Manoliputti Island.

DISCUSSION

This study clearly revealed that the sea grass ecosystem in the Mandapam group of islands, Gulf of Mannar Biosphere Reserve, supports five major macrofaunal groups: Cnidaria, Crustacea, Mollusca, Echinodermata and Pisces. Of these, Mollusca and Pisces predominantly occurred in sea grass. Sea grass beds serve as nursery and spawning grounds for diverse biota (Kemp, 1983; Dennison *et al.,* 1993). Sea grasses provide direct food for some organisms such as sirens, sea turtles, sea urchins, crustaceans and fishes (Short and Short, 1984; Stevenson, 1988; Fortes, 1989; Bell and Pollard, 1989). The diversity of fish species was high in all islands during the study period. Sea grass support large numbers

TABLE 10.4 Diversity Index of Faunal Groups in the Study Area

Animal group	Diversity Indices				
	Shannon	Simpsons	Alpha	Margaleff	Mackintosh
Krusadai Island					
Crustacea	0.90	0.37	1.897	14.26	1.092
Mollusca	1.32	0.26	6.629	13.768	1.083
Pisces	1.49	0.08	14.267	14.668	1.096
Echinodermata	0.48	0.17	5.451	49.829	1.475
Pullivasal Island					
Cnidaria (sea pen)	0.48	0.27	2.387	38.553	1.346
Crustacea	0.70	0.22	2.974	26.931	1.242
Mollusca	0.95	0.02	43.424	30	1.261
Pisces	1.42	0.03	29.105	18.481	1.143
Poomarichan Island					
Cnidaria	0.48	0.17	5.451	49.829	1.238
Crabs	1.04	0.14	7.502	21.46	1.171
Mollusca	1.00	0.29	2.931	15.508	1.099
Echinodermata	0.48	0.36	1.575	31.439	1.201
Pisces	1.36	0.44	10.125	15.428	1.095
Manoli Island					
Crustacea	0.85	0.20	3.143	21.202	1.149
Mollusca	1.49	0.44	7.375	11.163	1.043
Pisces	1.11	0.07	10.913	21.46	1.145
Manoliputti Island					
Crab	0.85	0.62	1.908	16.1	1.098
Mollusca	1.49	0.49	6.103	10.043	1.031
Pisces	1.20	0.12	6.326	16.1	1.096
Echinodermata	0.48	0.25	1.743	33.219	1.072

of fish species and individuals, and provide nursery habitats for juveniles of many species, as compared with adjacent unvegetated areas that have different fish assemblages, usually characterized by fewer species and fewer individuals (Kikuchi, 1974). The epiphytes also form an important component of the sea grass food web (Alcovera *et al.*, 1997). The importance of sea grass is that is are often correlated with the fish breeding ground since it can assist fish breeding (Patriquin, 1972). Major attractions of sea grass for fauna are the protection it offers especially from predators as well as the availability of food (Hale, 1976).

TABLE 10.5 Similarity Index between Islands

	Krusadai Island	Pullivasal Island	Poomarichan Island	Manoli Island	Manoliputti Island
Krusadai Island	0	28.95	31.13	46.98	30.48
Pullivasal Island		0	42.70	21.85	11.14
Poomarichan Island			0	25.71	16.06
Manoli Island				0	60.35
Manoliputti Island					0

The Manoli and Manoliputti islands were found to support the most organisms in terms of species richness, abundance and diversity, followed by Krusadai Island. This may be because most sampling points are on the shoreward side. Sea grass diversity, shoot density, biomass, epiphytic biomass and associated flora and fauna were significantly higher on the shoreward side than on the seaward side (Manikandan *et al.*, 2011b). The seaward side is sandy bottom with rock, and another important factor is that depth on the seaward side is high at more than 20 m. Depth on the shoreward side is 6–8 m or less, and depth plays a vital role in the occurrence of sea grass. Distribution of sea grass meadows in deep water habitats is particularly affected by light reduction from pulse turbidity events (Longstaff and Dennison, 1999).

Threats to Sea Grasses

The current threats to sea grass resources of the Gulf of Mannar are both anthropogenic and natural. The most common threats observed during the assessment were human disturbance due to fishing activity, in particular operation of trawlers, shore seine gill net and boat anchorage. Because of bottom trawling, tons of sea grasses are swept ashore every day. These disturbances are common in the Gulf of Mannar (Mathews *et al.*, 2013). Since the sea grass has been said to provide shallow marine areas with a "valuable benthic substratum" (Howard *et al.*, 1989) and forms a significant grazing ground for the sea cow, *Dugong dugon* (Muller, 1776) and sea turtle, there is a need for the continuous monitoring of the sea grass resources because of their importance to the marine environment. The present study suggested that exploitation of sea grasses from this environment should be restricted, to preserve resources in a sustainable way.

REFERENCES

Alcovera, T.C., Durate, C., Romero, S., 1997. The influence of herbivores on *Posidonia oceanica* epiphytes. Aquatic Botany 174, 247–256.

Alfred, J.R.B., Ramakrishna, 2004. Collection, Preservation and Identification of Animals. Zoological Survey of India, Kolkata, p. 310.

Bahuguna, A., Nayak, S., 1994. Coral reef mapping of Tamil Nadu using satellite data. *SAC (ISRO)* Ahamedabad. SAC/RSA/RSAG/DOD-COS/SN/07/94, p. 10.

Bell, J.D., Pollard, D.A., 1989. Ecology of fish assemblages and fisheries associated with sea grasses. In: Larkum, A.W.D., Macomb, A.J., Shepherd, S. (Eds.), Biology of Sea grasses: A treatise on the biology of sea grasses with special reference to the Australian region. Elsevier, Amsterdam, pp. 565–609.

Coels, R.G., Lee Long W. J., 1999. Seagrasses. In: Marine/Coastal Biodiversity in the Tropical Island Pacific Region: Volume 2. Population, Development and Conservation Priorities Workshop Proceedings. Pacific Science Association/ East West centre, Honolulu, Hawaii.

Coles, R., Grech, A., Rasheed, M., McKenzie, L., Unsworth, R., Short, F., 2011. Tropical Ecology and Threats in the Tropical Indo-Pacific Bioregion. In: Pirog, R.S. (Ed.), Sea grass Ecology, Uses and Threats. Nova Science, Publishers Inc, pp. 225–239.

Dennison, W.C., Orth, R.J., Moore, K.A., Stevenson, J.C., Carter, V., Kollar, S., Bergstron, P.W., Batiuk, R.A., 1993. Assessing water quality with submersed aquatic vegetation. Bio-Science 43, 86–94.

Fortes, M.D., 1989. Sea grass: a Resource Unknown in the Asian Region. ICLARM Education Series, Manila, Philippines, p. 46.

Hale, H.M., 1976. The crustaceans of South Australia, (Government Printer, Adelaide, South Marine animal standing stocks in the Philippine Australia.) p. 380.

Holme, N.A., McIntyre, A.D., 1971. Methods for the Study of Marine Benthos. Blackwell Scientific Publications, Oxford, p. 334.

Howard, R.K., Edgar, G.J., Hutchings, P.A., 1989. Faunal assemblages of sea grass beds. In: Larkum, A.W.D., Mc Comb, A.J., Shepherd, S.A. (Eds.), Biology of Sea grasses. Elsevier, Amsterdam, pp. 536–564.

Hughes, R.A., Williams, S.L., Duorte, C.M., Heck, J.K.L., Waycott, M., 2009. Associations of concern: Declining sea grass and threatened dependent species. Front. Ecol. Environ. 7 (5), 242–246.

Jagtap, T.G., 1991. Distribution of sea grasses along the Indian coast. Aquatic Botany 40 (4), 379–386.

Jagtap, T.G., Inamdar, S.N., 1991. Mapping of Sea grass Meadows from the Lakshadweep Islands (India) using aerial photographs. J. Indian Soc. Remote Sensing 19 (2), 77–81.

Jagtap, T.G., Deepali, S., Rouchelle, K., Rodrigues, S., 2003. Status of a sea grass ecosystem: an ecologically sensitive wetland habitat from India. Wetlands 23 (1), 161–170.

Kannan, L., Thangaradjou, T., Anantharaman, P., 1999. Status of sea grasses of India. Seaweed Res. Util. 21 (1&2), 25–33.

Kemp, W.M., 1983. Sea grass communities as a coastal resources. Marine Tech. Soc. J. 17, 3–5.

Kensworthy, W.J., Wyllie-Echeverria, S., Coles, R.G., Pergent, G., Pergent, C., 2006. Sea grass conservation Biology: An Interdisciplinary Science for Protection of the sea grass biome. In: Larkum, A.W.D., Orth, R.J., Duarte, C. (Eds.), Seagrasses: Biology, Ecology and Conservation. Springer, The Netherlands, pp. 595–623.

Kikuchi, T., 1974. Japanese contributions on consumer ecology in Sea grass (*Zostera marina*) beds, with special reference to trophic relationships and resources in inshore fisheries. Aquaculture 4, 145–160.

Kuo, J., McComb, A.J., 1989. Sea grass taxonomy, structure and development. In: Larkum, A.W.D., McComb, A.J., Shepherd, S.A. (Eds.), Biology of Sea grasses. Elsevier, Amsterdam, pp. 6–73.

Longstaff, B.J., Dennison, W.C., 1999. Sea grass survival during pulsed turbidity events: The effects of light deprivation on the sea grasses *Halodule pinifolia* and *Halophila ovalis*. Aquat. Bot. 65, 105–121.

Ludwig, J.A., Reynolds, J.F., 1988. Statistical Ecology: A Primer on Methods and Computing. Wiley, New York, p. 337.

Magurran, A.E., 1988. Ecological Diversity and its Measurement. Croom Helm, London, p. 179.

Manikandan, S., Ganesapandian, S., Parthiban, K., 2011a. Distribution and zonation of sea grasses in the Palk Bay, Southeastern India. J. Fish. Aquat. Sci. 6, 178–185.

Manikandan, S., Ganesapandian, S., Singh, M., Kumaraguru, A.K., 2011b. Sea grass Diversity and Associated Flora and Fauna in the Coral Reef Ecosystem of the Gulf of Mannar, Southeast Coast of India. Res. J. Environ. Earth Sci. 3 (4), 321–326.

Mathews, G.K., Diraviya Raj, Thinesh, T., Patterson, J., Patterson Edward, J.K., Wilhelmsson, D., 2013. Status of Sea grass Diversity, Distribution and Abundance in Gulf of Mannar Marine National Park and Palk Bay, Southeastern India. In: Ramachandran, P., Lakshmai, A., Ramachandran, R. (Eds.), Lagoons, Lives and Livelihood. National Centre for Coastal Management, Government of India, Chennai, pp. 50–66.

Orth, R.J., Carruthers, T.J.B., Dennison, W.C., Duarte, C.M., Fourqurean, J.W., Heck, Jr., K.L., Hughes, A.R., Kendrick, G.A., Kenworthy, W.J., Olyarnic, S., Short, F.T., Waycott, M., Williams, S.L., 2006. A Global contemporary crisis for sea grass ecosystems. Bioscience 56, 987–996.

Parthasarathy, N., Ravikumar, K., Ganesan, R., Ramamurthy, K., 1991. Distribution of sea grasses along the coast of Tamil Nadu, southern India. Aquatic Botany 40 (2), 145–153.

Patriquin, D.G., 1972. The origin of nitrogen and phosphorus for growth of the marine angiosperm *Thalassia testudinum*. Mar. Biol. 15, 35–46.

Pauly, D., 1980. A selection of simple methods for the assessment of tropical fish stocks. FAO Fish. Circ., (729): p. 54. Issued also in French. Superseded by Pauly. D., 1983. FAO Fish. Tech. Pub., (234), p. 52.

Riley, 1965. A Sublittoral survey of Port Erin Bay, particularly as an environment for young plaice. Rep. Mar. Biol. Stn. Port Erin 77, 49–53.

SCSP (South China Sea Development Programme), 1978. Report on the workshop on the demersal resources of the Sunda shelf, Part 1. Manalia, South China Fisheries Development and Co-ordinating Programme, SCS/GEN/77/12:44.

Shindo, S., 1973. General review of the trawl fishery and the demersal fish stocks of the South China Sea. FAO Fish. Tech. Pub. (120), 49.

Short, F.T., Short, C.A., 1984. The Sea grass filter: Purification of Estuarine and coastal water. In: Kennedy, V.S. (Ed.), The Estuary as a Filter. Academic Press, Orlando, pp. 395–413.

Stevenson, J.C., 1988. Comparative ecology of submersed sea grass beds in fresh water, estuarine and marine environments. Limnol. Oceanogr. 33, 867–893.

Susan, V.D., Pillai, N.G.K., Satheesh Kumar, P., 2012. A Checklist and Spatial Distribution of Molluscan Fauna in Minicoy Island, Lakshadweep. India World J. Fish Marine Sci. 4 (5), 449–453.

Umamaheswari, R., Sundararajan, R., Elavumkudi, P.N., 2009. Mapping the extend of sea grass meadows of Gulf of Mannar Biosphere Reserve, India using IRS ID satellite Imagery. Internat. J. Biodiv. Conserv. 1 (5), 187–193.

Venkataraman, K., Jeyabaskaran, R., Satyanarayana, Ch., Raghuram, K.P., 2004. Status of Coral reefs in Gulf of Mannar Biosphere Reserve. Rec. Zool. Surv. India 103 (Part 1-2), 1–15.

Waycott, M., Duante, C.M., Caruthers, T.J.B., Orth, R.J., Dennisson, W.C., Olyarnik, S., Calladine, A., Fourqurea, J.W., Heck, K.L., Hughes, A.R., Kendrick, G.A., Kenworth, W.J., Short, T., Williams, S.L., 2009. Accelerating loss of sea grass across the globe threatens coastal ecosystems. PNAS 106 (30), 12377–12381.

Chapter 11

Diversity and Ecology of Sedentary Ascidians of the Gulf of Mannar, Southeast Coast of India

G. Ananthan, S. Mohamed Hussain, A. Selva Prabhu and T. Balasubramanian
Centre of Advanced Study in Marine Biology, Faculty of Marine Sciences, Annamalai University, Parangipettai, Tamil Nadu, India

INTRODUCTION

The ascidians are sedentary tunicates occurring exclusively in marine environments and unable to survive in low salinity areas. But in rare instances, some ascidians are able to acclimatize to lower salinities. Ascidians can settle on all kinds of surfaces, such as hard rocks, stones, hulls of ships, mangroves, dead corals, algae, floating objects, hard sand and muddy surfaces. Ascidians are distributed over a wide range extending from the tropics to the polar regions, and majority of their forms occur in the littoral zone. They are major components of the fouling community occurring on the hulls of ship, piers, pilings, test panels, buoys, floats, cables and various other harbor installations. From an evolutionary point of view, ascidians occupy an interesting position between invertebrates and chordates.

The exceptional filtering capability of adult ascidians causes them to accumulate pollutants that may be toxic to embryos and larvae as well as impeding enzyme function in adult tissues. This property has made some species sensitive indicators of pollution. Over the last few hundred years, most of the world's harbors have been invaded by non-native sea squirts that have clung to ship hulls or to introduced organisms such as oysters and seaweeds. Several factors, including quick attainment of sexual maturity, tolerance of a wide range of environments, and fewer predators, allow sea squirt populations to grow rapidly. Unwanted populations on docks, ship hulls, and farmed shellfish cause significant economic problems.

Ascidians are the natural prey for some animals, including nudibranchs, flatworms, molluscs, rock crabs, sea stars and sea otters. They are also eaten by

Marine Faunal Diversity in India. DOI: 10.1016/B978-0-12-801948-1.00011-2

humans in many parts of the world, including Japan, Korea, Chile and Europe, where they are sold under the name "sea violet." As a chemical defence, many ascidians have an extremely high concentration of vanadium in the blood, very low pH in easily-ruptured bladder cells, and produce secondary metabolites harmful to predators and invaders. Some of these metabolites are toxic to cells and are of potential use in pharmaceuticals.

Since the advent of nuclear weapons and nuclear power stations, the presence of radionuclides in the sea and their uptake by marine organisms has received considerable attention. So far as ascidians are concerned, the practical importance of the uptake of radioactive material is the possibility that they will be concentrated in the tissues and passed on to man, either when ascidians are eaten or indirectly through the consumption of commercial fish which themselves feed on the ascidians.

An exhaustive survey of ascidians from the sea adjoining the Indian coast has not been done so far. Random collections have been made from Indian coasts, and the intervals between collections run into several years. The localities of some collections are also not recorded. Herdman (1906) collected some ascidians from the Gulf of Mannar. Das (1936) described the anatomy of *Herdmania pallida*, collected from the sea adjoining Tuticorin. Renganathan (1990) reported that ascidians form an important group of marine fouling organisms. Ascidians are the major foulants in Tuticorin harbor. Methylene-chloride extract of the ascidians *Syelapigmentata* and *Pyurapallida* was toxic to *Artemia salina* respectively at 8 and 4 mg/mL weight of the animal upon 96 hours of exposure. Antibacterial and cytotoxic activity has been previously reported from extracts of some tunicates (Thompson *et al.,* 1985; Hussain and Ananthan, 2009; Sivaperumal *et al.,* 2010; Ananthan *et al.,* 2011; Hussain *et al.,* 2011).

Three species of ascidians of the family Polyclinidae were reported for the first time from the southeast coast of India by Meenakshi (1998). Hussain *et al.* (2011) examined the cytotoxic potential of crude methanol and ethyl acetate extract of two colonial ascidians, using the brine shrimp lethality assay. Three species of ascidians were reported, of which *Distaplia nathensis* sp. nov. is new to science; the other two—*Phallusia nigra* (Savingy, 1816) and *Eusyntyelatincta* (Van Name, 1902)—are first records for Indian waters. From the above, it is evident that ascidians collections from the Indian coasts are too few, and at very wide intervals. Karthikeyan *et al.* (2010) worked on the diversity of ascidians from Palk Bay, southeast coast of India.

Ascidians are a human food source in many parts of the world (Monniot *et al.,* 1991). Only a few species are economically important and all are stolidobranchs. *Halocynthia roretzi* (Von Drasche, 1884) and *Halocynthia aurantium* (Pallas, 1787) are farmed in Japan; *Halocynthia roretzi* and *S. Clava* are farmed in Korea. Both countries export these species to Europe and North America. *Microcosmus sabatieri* (Roule, 1885) and *Microcosmus vulgaris* (Heller, 1877) are eaten in Europe (referred to in fish markets as the sea violet), while *P. chilensis* is an important food item in Chile (Davis, 1995). The Maoris of New Zealand are known to eat a native *Pyura*. Floating artificial structures in harbors provide a unique environment that results in the establishment of invertebrate assemblages not duplicated on natural benthic surfaces (Connell, 2000; Holloway and Connell, 2002; Lambert and Lambert, 2003). One important factor is that they are subject to less predation by benthic predators.

METHODS

The method suggested by Patricia Kott (1990), Queensland Museum, Australia was followed for the collection, preservation and identification.

Collection Samples were collected from intertidal to deep sea areas during low tide at a depth of 4–12 m, with the help of a SCUBA (self-contained underwater breathing apparatus) diver during the period February 2009 to January 2012. Specimens were carefully detached from the substratum without any damage. Before removing the animal, its color, appearance of the living colony and its habitat were noted.

Narcotization The collected specimens were transferred to a tray filled with fresh sea water. Care was taken that the samples were distributed uniformly with enough space between them. A few crystals of magnesium sulfate were placed in the four corners of the tray to induce defecation. After 2 hours, a few crystals of menthol were sprinkled over the water in the tray. The tray was covered with a glass plate and kept for 1–2 hours without any disturbance. Complete narcotization was tested by gentle touching of the extended apertures with a needle. When both or either one of the siphons contracted, the narcotization process was continued.

Fixation The collected specimens were preserved in 10% sea water formalin.

Study Area

The Gulf of Mannar Biosphere Reserve (08° to 09°N; 78° 12' to 70° 14'E) covers an area of 1,050,000 ha on the southeast coast of India. It is one of the world's richest regions from a marine biodiversity perspective. In the present study a total of 30 stations were selected from Rameshwaram to Tuticorin for study of ascidians diversity (Figure 11.1).

RESULTS

Ascidians were found attached in 16 different habitats: trawl, intertidal, deep sea, hull of ship, barge, pipeline, pearl oyster farm, pearl oyster cage, peal oyster bed, seaweed raft, seaweed rope, fishing harbor, fish landing center, cement block, rock pillars and dead coral. A total of 44 species of ascidians were identified during the study period. On completion of the study, all the specimens were deposited with the museum documentation database in the Marine Biology Reference Museum, Centre of Advanced Study in Marine Biology, Annamalai University, Parangipettai.

FIGURE 11.1   The study area on the Gulf of Mannar.

Species Composition and Distribution of Ascidians along the Gulf of Mannar

Most marine organisms have two phases in their life cycle: the pelagic larval stage and the benthic adult stage. The planktonic larval stage in their life cycle introduces the potential for considerable spatial and temporal variations. The spatial and temporal variability in patterns of settlement and recruitment of marine invertebrates can strongly influence the distribution and abundance of adult populations.

Ascidian Diversity

The diversity of ascidians was found to descend in the following order by location: Harbor green gate (3.802) > Habor green gate barge (3.703) >Tuticorin Roch park (3.355) >Vellapatti (2.858) >Near CECRI (Tuticorin harbor) (2.803). The lowest biodiversity (0.979) was observed in Inico Nagar. The richness (d) values were maximum (2.616) at Harbor green gate barge followed by Harbor green gate (2.44), Tuticorin Roch Park (2.119) and Near CECRI (Tuticorin harbor) (1.648), while the minimum (0.251) richness was observed in Annam Kovil. The evenness (J') values ranged between 0.876 and 1.000 in Mookkaiyur, Rameswaram and Tuticorin New Beach, respectively. The values of Simpson index varied from 0.5007 to 0.9274 in Annam Kovil and Harbor green gate, respectively (Table 11.1).

TABLE 11.1 Diversity Indices of Ascidians of Gulf of Mannar

Sl. No.	Sampling station	Diversity indices			
		Shannon (H')	Simpson (λ)	Species richness	Evenness
1	Rameswaram	1.00	0.54	0.38	1.00
2	Mandapam	2.17	0.77	0.88	0.94
3	Near CMFRI (Mandapam)	2.83	0.85	1.33	0.94
4	Pamban	1.45	0.61	0.45	0.91
5	Thonithurai	1.80	0.69	0.71	0.90
6	Annam Kovil	0.99	0.50	0.25	0.99
7	Vedhalai	1.56	0.68	0.58	0.99
8	Mookkaiyur	2.04	0.74	1.08	0.88
9	Narippaiyur	2.26	0.81	1.21	0.97
10	Seeniappa Dharga	2.06	0.73	0.92	0.89
11	Pudumadam	1.90	0.73	0.77	0.95
12	Barathinagar, Keelakkarai	1.49	0.64	0.55	0.94
13	Erandhurai algal farm	2.21	0.78	0.94	0.95
14	ChinnaErvadi	1.53	0.66	0.58	0.96
15	Valinokkam	0.00	0.00	0.00	0.00
16	Habor green gate	3.80	0.93	2.44	0.97
17	Harbor red gate	2.83	0.85	1.43	0.94
18	Tuticorin harbor pearl oyster farm	2.73	0.84	1.37	0.91
19	Van Island	2.44	0.81	0.94	0.94
20	Hare Island	1.00	0.54	0.38	1.00
21	Inico Nagar	2.17	0.77	0.88	0.94
22	Fishing harbor	2.83	0.85	1.33	0.94
23	Tharuvaikulam	1.45	0.61	0.45	0.91
24	Vellapatti	1.80	0.69	0.71	0.90
25	Harbor green gate barge	0.99	0.50	0.25	0.99
26	Tuticorin Roch park	1.56	0.68	0.58	0.99
27	Tuticorin new beach	2.04	0.74	1.08	0.88
28	Vembar	2.26	0.81	1.21	0.97
29	Near CECRI (Tuticorin harbor)	2.06	0.73	0.92	0.89
30	North breakwater (Tuticorin harbor)	1.90	0.73	0.77	0.95

k-Dominance Plot

A k-dominance curve, showing the cumulative dominance with species rank and stations, was plotted. It shows the highest diversity at Harbor green gate and the lowest diversity at Valinokkam.

Cluster

Cluster analysis was carried out to understand the similarity among the stations (Figure 11.2). A total of 8 groups were formed in this cluster. Among these, the highest similarity was observed in the first group between Narippaiyur and Tuticorin Roch Park; the lowest similarity was observed in the fifth group between Rameswaram and Inico Nagar.

Ellipse

An ellipse diagram combining delta+ and lambda+ values for number of species found showed that data for all stations are within the 95 percent confidence limit except for Valinokkam, Mandapam, Thonithurai, Erandhurai algal farm, Tuticorin new beach, Annam kovil, Van island, Hare island, Seeniappa Dharga and Inico Nagar. At most stations, at least 10 species of ascidians were found; Harbor green gate showed richest diversity: 20 species of ascidians.

Physicochemical Parameters

The salinity values ranged from 27.0 to 35.0 psu along this study area. The seasonal distribution of salinity was in general characterized by minimum (27.0 psu) in monsoon, due to heavy rainfall. In contrast, maximal (35.0 psu) salinity values were observed during summer (April–June). The pH value ranged from 7.50 in monsoonto a maximum of 8.50 in summer. The dissolved oxygen values varied from 3.20 to 4.89 mg/L. The maximum (4.89 mg/L) concentration was recorded during October–December as a result of the freshwater inflow. The water temperature in the Gulf of Mannar varied from 25.5 $^{\circ}$C (monsoon) to 34. 5 $^{\circ}$C (summer). The temperature was highest in April–June and lowest during October–December.

DISCUSSION

Ascidians belong to the subphylum Urochordata of the phylum Chordata. They are marine animals—most of the species are intertidal organisms found in all seas at all depths, extending from the littoral zone down to the abyssal depths of over three miles. The Ascidiacea, the largest class of the Tunicata, are fixed filter feeding organisms found in most marine habitats from intertidal to hadal depths. The class contains two orders: the Enterogona in which the atrial cavity (atrium) develops from paired dorsal invaginations, and the Pleurogona in

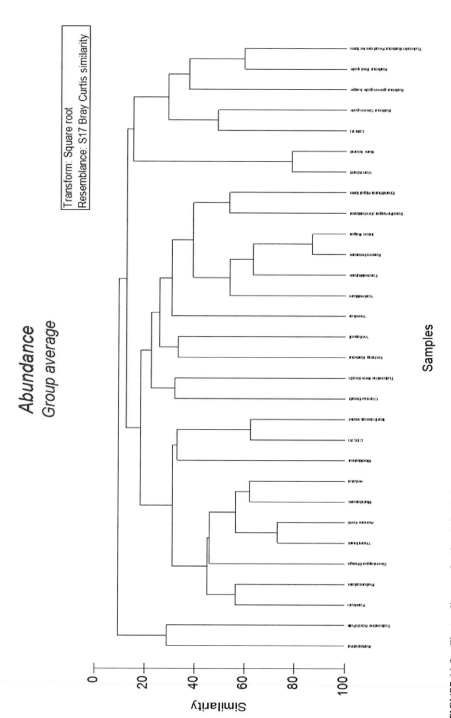

FIGURE 11.2 Cluster diagram for the study stations.

which it develops from a single median invagination. These ordinal characters are not present in adult organisms. Accordingly, the sub ordinal groupings, Aplousobranchia and Phlebobranchia (Enterogona) and Stolidobranchia (Pleurogona), are of more practical use at the higher taxon level. Reports reveal that certain groups of ascidians are endemic to this region, *viz. P. madrasensis* and *P. indicum*. Compared with other organisms, studies relating to taxonomy and systematics of ascidians are limited to Palk Bay. The present survey reports the occurrence of 2 orders, 3 suborders, 8 families, 18 genera and 44 species of ascidians from the Gulf of Mannar, southeast coast of India. A total of 303 species of ascidian fauna was reported in Indian waters by Meenakshi, (2008). The distribution of solitary and colonial forms displayed variability. The solitary forms are dominant in stable environments (especially at great depth and on overhangs) and colonial forms are relatively more abundant in shallow waters and on horizontal or vertical walls (Jackson, 1977; Buss, 1979). Similarly in this study, it was found that solitary ascidians are more abundant in offshore areas, whereas colonial ascidians are more abundant in intertidal or shallow waters.

Species diversity is a simple and useful measure of a biological system (Redding and Cory, 1985). Sanders (1968) found a high level of agreement between the species diversity and nature of the environment, and hence the measure of the species diversity is an ecologically powerful tool. The conservation of biological diversity (biodiversity) has become one of the major issues since the late twentieth century. There is a worldwide recognition that reductions in diversity of life will, sooner or later, affect us in some manner (WCMC, 1992). Quantification of biodiversity is fundamental to the identification of changes that may be taking place and the understanding of their possible consequences. Biodiversity can be measured at many levels and in numerous ways (Harper and Hawksworth, 1994; May, 1994; Hambler and Speight, 1995). Biodiversity simply relates to the number of species in an area. The number of species is, however, not the only measure of biodiversity; the relative abundance of the different species is important.

The prevailing salinity was also found to be the most important physical factor determining the presence or absence of ascidian species in different parts of two marine ponds in Norway (Dybern, 1969). In stable neritic waters, salinity is high during summer and post-monsoon seasons, providing an environment conducive to the support of acidians, and this might explain the higher diversity values observed during these seasons in the present study. Barrington (1965), Dybern (1967, 1965, 1969), Gunter and Hall (1963), Diehl (1957) and Monniot (1965) found a similar decrease in species diversity in areas of low salinity.

A wide range of variation of diversity of ascidian fauna in the intertidal areas has been observed (Millar, 1971).The variations in species diversity observed in the present study might be ascribed to the dynamic nature of the marine environment.

REFERENCES

Ananthan, G., Sivaperumal, P., Mohamed Hussain, S., 2011. Cytotoxicity of the crude extracts of marine ascidians (Tunicata: Ascidiacea) from Tuticorin, Southeast coast of India. Archives Applied Sci. Res. 3 (2), 139–142.

Barrington, E.J.W., 1965. The Biology of Hemichordata and Prochordata. Oliver and Boyd, Edinburgh and London, p. 176.

Buss, L.W., 1979. Habitat selection, directional growth and spatial refuges: why colonial animals have more hiding places. In: Coulson, J.C., London, D.F. (Eds.), Biology and systematics of colonial organisms. Academic Press, New York, pp. 459–497.

Connell, S.D., 2000. Floating pontoons create novel habitats for sub tidal epibiota. J. Exp. Mar. Biol. Ecol. 247, 183–194.

Das, S.M., 1936. *Herdmania*, the monascidian of the Indian Seas. Indian Zool. Mem. 5, 103.

Davis, A.R., 1995. Over-exploitation of *Pyurachilensis* (Ascidiacea) in southern Chile: the urgent need to establish marine reserves. Rev. Chil. Hist. Nat. 68, 107–116.

Diehl, M., 1957. Die Okologie der Ascidie Styelacoriacea in der KielerBucht. KielerMeereforsch 13, 59–68.

Dybern, B.I., 1965. The lifecycle of *Ciona intestinalis* (L.) *f. typicain* relation to the environmental temperature. Oikos 16, 109–131.

Dybern, B., 1967. The distribution and salinity tolerance of *Ciona intestinalis* (L.) *f. typica* with special reference to waters around southern Scandinavia. Ophelia 4, 27–36.

Dybern, B., 1969. Distribution and ecology of ascidians in Kviturdvikpollen and Vagsbopollen on the west coast of Norway. Sarsia 37, 21–40.

Gunter, G., Hall, G.E., 1963. Biological investigations of the St. Lucie estuary (Florida) in connection with Lake Okeechobee discharges through the St. Lucie canal. Gulf Res. Rep. 1, 189–307.

Hambler, C., Speight, M.R., 1995. Biodiversity conservation in Britain: science replacing tradition. British Wildlife 6, 137–147.

Harper, J.L., Hawksworth, D.L., 1994. Biodiversity: measurement and estimation. Phil. Trans. Res. Soc., Series B 345, 5–12.

Herdman, W.A., 1906. On the Tunicata. Report to the Government of Ceylon on the Ceylon pearl oyster fisheries of the Gulf of Manaar. Supplementary Reports 39, 295–348.

Holloway, M.G., Connell, S.D., 2002. Why do floating structures create novel habitats for subtidalepibiota? Mar. Ecol. Prog. Ser. 235, 43–52.

Hussain, S.M., Ananthan, G., 2009. Antimicrobial activity of thecrude extracts of compound Ascidians, *Didemnum Candidum* and *Didemnum Psammathodes* (Tunicata: Didemnidae) from Mandapam (Southeast coast of India). Curr.Res.J. Biol.Sci. 1 (3), 168–171.

Hussain, S.M., Ananthan, G., Sivaperumal, P., 2011. Exploration of cytotoxic potential from marine ascidians *Trididemnum clinids* and *Trididemnum savignii*. J. Pharmacy Res. 4 (7), 2032–2033.

Jackson, J., 1977. Competition on marine hard substrata: the adaptive significance of solitary and colonial strategies. Am. Nat. 111 (980), 743–767.

Karthikeyan, M.M., Ananthan. G., and Balasubramanian, T. (2010). A monograph on ascidians of Palk Bay southeast coast of India. Envis library of Marine Biology, Annamalai University. pp. 1–58.

Kott, P., 1990. The Australian Ascidiacea part 2, Aplousobranchia (1). Mem. Qld. Mus. 29, 1–266.

Lambert, C.C., Lambert, G., 2003. Persistence and differential distribution of nonindigenous ascidians in harbors of the Southern California Bight. Mar. Ecol. Prog. Ser. 259, 145–161.

May, R.M., 1994. Conceptual aspects of the quantification of the extent of biological diversity. Phil. Trans. Res. Soc., Ser. B 345, 13–20.

Meenakshi, V.K., 1998. Occurrence of a new ascidian species – *Distaplianathensis* sp. nov. and two species - *Eusynstyelatincta* (Van Name, 1902), *Phallusianigra* (Savingny, 1816) new records for Indian waters. Indian J. Mar. Sci. 27, 477–479.

Meenakshi, V.K., 2008. A report on the biodiversity of Indian ascidians. Glimpses of Aquatic Biodiversity—Rajiv Gandhi Chair Spl. Pub., 7: 213–219.

Millar, R.H., 1971. The biology of ascidians. Adv. Mar. Biol. 9, 1–100.

Monniot, C., 1965. Redescription de six ascidies du golfed'Elatrécoltées par H. Schumacher. Israel J. Zool. 22, 51–62.

Monniot, C., Monniot, F., Laboute, P., 1991. Coral Reef Ascidians of New Caledonia. ORSTOM, Paris, p. 247.

Redding, J.M., Cory, R.L., 1985. Macroscopic benthic fauna of three tidal creeks adjoining the Rhode River, Maryland. Water Resources Investigation Report. US Environmental Protection Agency, Washington, DC, pp. 39–75.

Renganathan, T.K., 1990. Systematics and ecology of Indian ascidians. In: Nair, K.V.K., Venugopalan, V.P. (Eds.), Marine biofouling and power plants. BARC, Bombay, India, pp. 263–271.

Sanders, H.L., 1968. Marine benthic diversity: A comparative study. American Naturalist 102, 243–282.

Sivaperumal, P., Ananthan, G., Mohamed Hussain, S., 2010. Exploration of antibacterial effects on the crude extract of marine ascidian *Aplidium multiplicatum* against clinical isolates. Int. J. Medicine and Medical Sciences 2 (12), 382–386.

Thompson, J.E., Walker, R.P., Faulkner, D.J., 1985. Screening and bioassays for biologically-active substances from forty marine sponge species from San Diego, California, USA. Mar. Biol. 88, 11–12.

WCMC (World Conservation Monitoring Centre), 1992. Global biodiversity: Status of the Earth's living resources. Chapman and Hall, London, p. 594.

Chapter 12

Diversity of Marine Fish of India

K.C. Gopi and S.S. Mishra
Fish Division, Zoological Survey of India, Kolkata, West Bengal, India

INTRODUCTION

Fish contribute more than one-half of the total number of approximately 54,711 valid vertebrate species (Nelson, 2006). There are descriptions of an estimated 33,059 valid species of fish known from around the world (Eschmeyer and Fong, 2014). Marine fishes are species that spend at least a part of their life cycle in the sea. India's biodiversity richness is truly reflected in the diversity of marine and freshwater fishes. Variety and variability of fish diversity have direct bearing on the diversity exemplified in the ecosystems/habitat assemblages in inland and marine waters.

India is a mega-diversity nation well known for its biodiversity richness. The Indian biogeographic territory (including the Lakshadweep and the Andaman and Nicobar Islands) is an integral part of the Central Indian Ocean Region which includes also the other countries Bangladesh, Indonesia, Maldives, Malaysia, Myanmar, Thailand and Sri Lanka. The Central Indian Ocean Marine Region consists of three distinct marine ecosystem areas or zones: the Arabian Sea, the Bay of Bengal and a large area of the Indian Ocean south of India and Sri Lanka. The Indian Peninsula flanks two Large Marine Ecosystems (LMEs)—the Arabian Sea and Bay of Bengal—bounding the west coast and east coast, respectively, both merging with the Indian Ocean at the peninsula's southernmost extremity.

India is endowed with a coastline of over 8000 km, an Exclusive Economic Zone (EEZ) of 2.02 million km^2 integral to its continental shelf and the offshore islands, and a variety of coastal ecosystems encompassing estuaries, lagoons, mangroves, backwaters, salt marshes, rocky coasts, sandy stretches and coral reefs (Venkataraman and Wafar, 2005). Four major coral reef island groups—Andaman and Nicobar, Gulf of Mannar, Gulf of Kachchh and Lakshadweep—also fall within the marine waters of India's EEZ.

But, as is the case in the world scenario, the biodiversity wealth of India, especially its marine fish diversity, is in ever-increasing danger, with depletion of living resources such as fish, despite the acknowledged notion that it is crucially important for the survival of humanity itself. We are not even aware of the full dimension of the potential of biodiversity. At the world level, altogether fewer than 2 million species of animals, plants and microorganisms have been identified.

Marine Faunal Diversity in India. DOI: 10.1016/B978-0-12-801948-1.00012-4
171

The number of species from all groups and all habitats of the seas could be of the order of several million, but only a small share of that wealth is now known to us.

In the case of the marine biodiversity of India, the number of species known could be of the order of 13,000 or higher (Ramakrishna and Venkataraman 2001). The inventory of species has been done in detail only in the case of commercially important groups such as fishes or molluscs, and is very weak as regards the other groups and minor phyla or microbial organisms. In terms of spatial coverage, probably only two-thirds of the total marine habitats have been covered, and the remote islands and other minor estuaries still remain virtually unexplored. The unknown diversity of life in coastal/marine ecosystems probably far exceeds what is already known. Mora *et al.* (2008) estimated that the global marine fish inventory is only about 79 percent completed, and the remaining 21 percent has yet to be discovered and named.

STUDIES ON INDIAN MARINE FISHES

The study of fish systematics in India has a long history. Although taxonomy and fishery science is of recent origin in India, references to the utility of fish as a source of food are reflected in Kautilya's *Arthasastra* as early as 300 BC (Hora, 1948). The edict on the second pillar of King Ashoka implies prohibition of consumption of fish during a certain lunar period, which Hora (1950) interpreted as being based on principles of conservation of natural resources.

As regards knowledge of fishes of the Indian region, we have had, to begin with, some species described by Linnaeus (1758) from Indian waters, though many of them are erroneously tagged with *'Habitat in India'*. The work of Bloch and Schneider (1801) and a few of the earlier works of M.E. Bloch also dealt with some fishes from Indian waters. A great credit goes first to Russell (1803), who made the pioneering work on the description of 200 fishes from one 'locality', Vizagapatnam, of India. Since his work on documenting species was not binomial in nature, most of his illustrations were later used by several other authors to describe many species. Hamilton (1822) made a monumental work on *'An account of the fishes found in the River Ganges and its branches'* describing 71 estuarine fishes.

The period from about 1825 to about 1895, especially the mid-1800s, was a time of the great marine expeditions by Europeans, by such ichthyologists as Cuvier and Valenciennes, Sir F. Day, and A.W. Alcock, which provided high points in new marine species descriptions. Cuvier and Valenciennes (1828–1849) described 70 nominal species of fish from the marine waters of Pondicherry, now known as Puduchery, as inferred by Krishnan and Mishra (2004). *'The Fishes of India'* by Day (1875–78, 1888), a monumental treatise on fishes of the Indian region, still continues to be a valuable reference work. Day (1889) was first to provide a consolidated report on the *'Fishes of British India'* that included 1418 species, under 342 genera, of both freshwater and marine fishes. Alcock (1889, 1890) through his publications described 162 species new to science from the seas of India.

A code of nomenclature was established in 1842. Extensive marine expeditions by explorers from several countries continued from 1890 to 1920, and many monographs on many groups of fishes and geographical areas were published. During the 1920s to 1950s, there was an increase in the number of taxonomic workers as well as in the publication of regional studies on marine fishes. The search for fishable commercial stocks in marine waters increased after 1950, as a result of which many taxonomic groups, especially commercially important ones, were relatively well studied. After 1950, ichthyological studies became more efficient, with technological advances and better tools for fish collections, molecular genetic studies, better communication, modern data processing and precision gadget facilities. The sophistication and accuracy of description of new taxa increased substantially with more workers entering icthyological studies.

In the twentieth century, several authors ventured to study the ichthyofaunal wealth of Indian coasts. Some notable works on fishes, the series on '*The Fishes of Indo-Australian Archipelago*' (Weber and de Beaufort, 1916–1936; de Beaufort, 1940; de Beaufort and Chapman, 1951; Koumans, 1953; de Beaufort and Briggs, 1962), accounting many Indian fishes have considerably strengthened our knowledge. '*Fishes of the Laccadive Archipelago*' by Jones and Kumaran (1980) is an outstanding work on fishes of the marine environment around these island waters. Talwar and Kacker (1984) gave an illustrated account of '*Commercial Sea Fishes of India*' contributing to our knowledge on many of the most important commercial sea fishes. The work dealt with 548 species under 89 families. The '*FAO Species-Identification sheets for Fishery Purposes—Western Indian Ocean*' (Fischer and Bianchi, 1984) is a very good resource manual providing valuable information for marine fish and fishery workers.

In later years, the Zoological Survey of India, based on its survey collections and review of scattered literature, published the *State Fauna Series* to list the fishes found in the marine and estuarine waters of all maritime states of India: from Lakshadweep (Rao, 1991), West Bengal (Talwar *et al.*, 1992), Gujarat (Barman *et al.*, 2000), Puduchery (Mishra and Krishnan, 2003), Andhra Pradesh (Barman *et al.*, 2004), Odisha (Barman *et al.*, 2007), Tamil Nadu (Barman *et al.*, 2011), Maharashtra (Barman *et al.*, 2012) and Karnataka (Barman *et al.*, 2013). Rajan *et al.* (2013) provided an updated checklist of fishes from the Andaman and Nicobar Islands.

India still lacks a complete, scientific database of its marine fish diversity. Even an authentic checklist of marine fishes of India has yet to be completed. In the process of preparing such a checklist, the authors attempted to analyze and assess the marine fish diversity so far known from Indian marine waters. A synopsis of the salient aspects of the fish diversity known from Indian marine waters, based on our effort, is presented in this chapter.

MARINE FISH DIVERSITY

Table 12.1 gives the estimated number of valid genera and species under 230 families. They are enumerated based on the validation of the latest taxonomic works, especially revisions of families and genera. Although subspecies is

TABLE 12.1 Marine Fishes of India: Systematic Index of Higher Taxa with Genera and Species Numbers

Sl. No.	Phylum: Chordata Subphylum: Craniata Superclass: Gnathostomata	India		World	
		No. of Genera	No. of Species	No. of Genera	No. of Species
	Class: Chondrichthyes				
	Order: Chimaeriformes				
1	Rhinochimaeridae	1	1	3	8
	Order: Orectolobiformes				
2	Hemiscyllidae	1	5	2	17
3	Stegostomatidae	1	1	1	1
4	Ginglymostomatidae	1	1	3	3
5	Rhinodontidae	1	1	1	1
	Order: Lamniformes				
6	Odontaspididae	1	2	2	4
7	Alopiidae	1	3	1	3
8	Lamnidae	2	2	3	5
9	Pseudocarchariidae	1	1	1	1
	Order: Carcharhiniformes				
10	Scyliorhinidae	6	9	17	150
11	Proscyllidae	1	1	3	7
12	Triakidae	2	3	9	46
13	Hemigaleidae	3	3	4	8
14	Carcharhinidae	10	26	12	56
15	Sphyrnidae	2	4	2	10
	Order: Hexanchiformes				
16	Hexanchidae	2	2	2	2
	Order: Echinorhoniformes				
17	Echinorhinidae	1	1	1	1
	Order: Squaliformes				
18	Etmopteridae	2	2	2	2
19	Squalidae	1	2	1	2
20	Centrophoridae	2	8	2	8
21	Somniosidae	2	2	2	2

*Number of freshwater species under these families omitted.

TABLE 12.1 Marine Fishes of India: Systematic Index of Higher Taxa with Genera and Species Numbers *(cont.)*

Sl. No.	Phylum: Chordata Subphylum: Craniata Superclass: Gnathostomata	India		World	
		No. of Genera	No. of Species	No. of Genera	No. of Species
	Order: Torpediniformes				
22	Torpedinidae	1	5	1	22
23	Narcinidae	2	6	4	30
24	Narkidae	2	4	6	12
	Order: Pristiformes				
25	Pristidae	2	4	2	7
	Order: Rajiformes				
26	Rajiidae	7	7	18	176
27	Rhinobatidae	4	10	11	60
	Order: Myliobatiformes				
28	Hexatrygonidae	1	1	1	1
29	Placiobatidae	1	1	1	1
30	Dasyatidae	7	23	8	88
31	Gymnuridae	2	4	1	14
32	Myliobatidae	5	16	7	39
	Class: Actinopterygii				
	Order: Elopiformes				
33	Elopidae	1	1	1	7
34	Megalopidae	1	1	1	2
	Order: Albuliformes				
35	Albulidae	1	2	2	11
	Order: Notacanthiformes				
36	Halosauridae	2	5	3	16
37	Notacanthidae	1	1	4	12
	Order: Anguilliformes				
38	Anguillidae	1	5	2	22
39	Moringuidae	1	6	2	15
40	Muraenidae	10	38	16	200
41	Synaphobranchidae	2	3	12	38

Number of freshwater species under these families omitted.

(Continued)

TABLE 12.1 Marine Fishes of India: Systematic Index of Higher Taxa with Genera and Species Numbers *(cont.)*

Sl. No.	Phylum: Chordata Subphylum: Craniata Superclass: Gnathostomata	India		World	
		No. of Genera	No. of Species	No. of Genera	No. of Species
42	Ophichthidae	17	24	59	308
43	Colocongridae	1	1	2	9
44	Muraenesocidae	4	6	6	15
45	Nemichthyidae	2	2	3	9
46	Congridae	12	17	30	194
47	Nettastomatidae	2	2	6	42
48	Serrivomeridae	1	1	2	9
	Order: Clupeiformes				
49	Dussumieriidae	1	2	2	9
50	Clupeidae*	12	26	54	200
51	Pristigasteridae	4	12	9	38
52	Engraulididae	5	34	17	145
53	Chirocentridae	1	2	1	2
	Order: Gonorynchiformes				
54	Chanidae	1	1	1	1
	Order: Siluriformes				
55	Plotosidae	1	3	10	40
56	Ariidae	10	25	30	153
	Order: Argentiniformes				
57	Platytroctidae	3	4	13	39
58	Alepocephalidae	9	14	19	98
	Order: Stomiiformes				
59	Gonostomatidae	4	6	8	31
60	Sternoptychidae	4	8	10	73
61	Phosichthyidae	2	3	7	24
62	Stomiidae	6	9	28	287
	Order: Aulopiformes				
63	Synodontidae	4	23	4	70
64	Chlorophthalmidae	1	3	2	18
65	Ipnopidae	2	4	6	30

Number of freshwater species under these families omitted.

TABLE 12.1 Marine Fishes of India: Systematic Index of Higher Taxa with Genera and Species Numbers *(cont.)*

Sl. No.	Phylum: Chordata Subphylum: Craniata Superclass: Gnathostomata	India		World	
		No. of Genera	No. of Species	No. of Genera	No. of Species
66	Evermannellidae	2	2	3	8
67	Alepisauridae	1	2	1	2
68	Paralepipidae	2	3	12	58
	Order: Myctophiformes				
69	Neoscopelidae	2	3	3	6
70	Myctophidae	11	41	33	248
	Order: Lampriformes				
71	Veliferidae	1	1	2	2
72	Lophotidae	1	1	2	4
73	Regalecidae	1	1	2	3
74	Ateleopodidae	2	3	4	13
	Order: Polymixiiformes				
75	Polymixiidae	1	3	1	10
	Order: Gadiformes				
76	Bregmacerotidae	1	1	1	14
77	Macrouridae	9	18	34	396
78	Moridae	1	2	18	108
	Order: Ophidiformes				
79	Carapidae	3	5	8	35
80	Ophidiidae	16	28	50	255
81	Bathytidae	6	7	52	208
82	Aphyonidae	1	1	6	23
	Order: Batrachoidiformes				
83	Batrachoididae	3	5	23	82
	Order: Lophiiformes				
84	Lophiidae	2	3	4	28
85	Antennaridae	2	9	13	47
86	Chaunacidae	1	1	2	19
87	Ogcocephalidae	5	11	10	73
88	Oneirodidae	1	1	16	64

Number of freshwater species under these families omitted.

(Continued)

TABLE 12.1 Marine Fishes of India: Systematic Index of Higher Taxa with Genera and Species Numbers *(cont.)*

Sl. No.	Phylum: Chordata Subphylum: Craniata Superclass: Gnathostomata	India		World	
		No. of Genera	No. of Species	No. of Genera	No. of Species
89	Diceratiidae	1	1	2	6
90	Ceratiidae	1	1	2	4
	Order: Mugiliformes				
91	Mugilidae*	7	18	25	75
	Order: Athriniformes				
92	Atherinidae	4	9	14	71
93	Notocheiridae	1	1	1	1
	Order: Beloniformes				
94	Exocoetidae	6	18	7	68
95	Hemiramphidae	5	16	8	63
96	Zenarchopteridae	2	8	5	61
97	Belonidae	4	8	10	47
	Order: Stephanoberyciformes				
98	Melamphaidae	1	1	5	61
	Order: Beryciformes				
99	Monocentridae	1	1	2	4
100	Trachichthyidae	2	3	8	51
101	Berycidae	2	4	2	10
102	Holocentridae	4	25	8	83
	Order: Zeiformes				
103	Parazenidae	1	1	3	4
104	Grammicolepididae	2	2	3	3
105	Zeidae	1	2	2	6
	Order: Gasterosteiformes				
106	Pegasidae	2	3	2	5
	Order: Syngnathiformes				
107	Aulostomidae	1	1	1	3
108	Fistulariidae	1	2	1	4
109	Macrorhamphosidae	1	1	3	8
110	Centriscidae	2	2	2	4

Number of freshwater species under these families omitted.

TABLE 12.1 Marine Fishes of India: Systematic Index of Higher Taxa with Genera and Species Numbers *(cont.)*

Sl. No.	Phylum: Chordata Subphylum: Craniata Superclass: Gnathostomata	India		World	
		No. of Genera	No. of Species	No. of Genera	No. of Species
111	Solenostomidae	1	2	1	6
112	Syngnathidae	14	42	56	298
	Order: Scorpaeniformes				
113	Dactylopteridae	1	5	2	7
114	Apistidae	1	1	3	3
115	Scorpaenidae	15	35	26	215
116	Setarchidae	2	3	3	7
117	Tetrarogidae	9	12	17	40
118	Synanceiidae	5	13	9	36
119	Aploactinidae	4	6	17	48
120	Triglidae	2	7	9	124
121	Peristediidae	5	7	6	44
122	Bembridae	1	1	4	9
123	Platycephalidae	11	16	18	79
	Order: Perciformes				
124	Caproidae	1	2	2	18
125	Ambassidae*	1	11	8	49
126	Latidae	2	2	3	13
127	Acropomatidae	2	5	8	31
128	Symphysanodontidae	1	3	1	12
129	Serranidae	19	85	75	532
130	Centrogenyidae	1	1	1	1
131	Ostracoberycidae	1	1	1	3
132	Pseudochromidae	4	9	24	152
133	Plesiopidae	3	5	12	49
134	Opistognathidae	1	7	3	79
135	Priacanthidae	3	9	4	19
136	Apogonidae	19	63	33	346
137	Sillaginidae	2	11	5	33
138	Malacanthidae	2	3	5	45

Number of freshwater species under these families omitted.

(Continued)

TABLE 12.1 Marine Fishes of India: Systematic Index of Higher Taxa with Genera and Species Numbers *(cont.)*

Sl. No.	Phylum: Chordata Subphylum: Craniata Superclass: Gnathostomata	India		World	
		No. of Genera	No. of Species	No. of Genera	No. of Species
139	Lactariidae	1	1	1	1
140	Pomatomidae	1	1	1	1
141	Coryphaenidae	1	2	1	2
142	Rachycentridae	1	1	1	1
143	Echeneidae	3	6	3	8
144	Carangidae	20	66	30	146
145	Menidae	1	1	1	1
146	Leiognathidae	9	22	9	48
147	Bramidae	3	3	7	20
148	Emmelichthyidae	1	1	3	17
149	Lutjanidae	10	45	17	110
150	Caesionidae	4	16	4	23
151	Datnioididae	1	1	1	5
152	Lobotidae	1	1	1	2
153	Gerreidae	2	11	7	54
154	Haemulidae	3	28	19	133
155	Hapalogenyidae	1	1	1	7
156	Nemipteridae	4	33	5	67
157	Lethrinidae	5	24	5	38
158	Sparidae	7	12	36	133
159	Polynemidae	5	11	8	43
160	Sciaenidae	19	43	66	283
161	Mullidae	3	27	6	82
162	Pempherididae	2	7	2	27
163	Bathyclupeidae	1	1	1	7
164	Monodactylidae	1	3	2	6
165	Toxotidae	1	2	1	7
166	Kyphosidae	1	3	14	52
167	Drepanidae	1	2	1	3
168	Chaetodontidae	8	48	12	129

Number of freshwater species under these families omitted.

TABLE 12.1 Marine Fishes of India: Systematic Index of Higher Taxa with Genera and Species Numbers *(cont.)*

Sl. No.	Phylum: Chordata Subphylum: Craniata Superclass: Gnathostomata	India		World	
		No. of Genera	No. of Species	No. of Genera	No. of Species
169	Pomacanthidae	6	21	8	88
170	Pentacerotidae	1	1	7	13
171	Terapontidae	2	4	16	52
172	Kuhliidae	1	3	1	14
173	Cirrhitidae	4	8	12	33
174	Cepolidae	2	4	5	23
175	Pomacentridae	19	92	29	387
176	Labridae	28	85	70	517
177	Scaridae	7	29	10	100
178	Chiasmodontidae	3	3	4	32
179	Champsodontidae	1	2	1	13
180	Trichonotidae	1	2	2	2
181	Pinguipedidae	1	12	7	42
182	Creediidae	1	1	8	18
183	Percophidae	2	3	11	48
184	Ammodytidae	1	2	7	28
185	Uranoscopidae	2	6	8	53
186	Pholidichthyidae	1	1	1	2
187	Tripterygiidae	3	8	29	171
188	Clinidae	1	1	26	88
189	Blenniidae	26	65	57	396
190	Callionymidae	4	21	20	188
191	Eleotrididae	11	18	33	177
192	Xenisthmidae	1	1	5	13
193	Kraemeriidae	1	1	2	9
194	Gobiidae*	71	190	250	1674
195	Microdesmidae	3	9	12	86
196	Schindleriidae	1	2	1	3
197	Kurtidae	1	1	1	2
198	Ephippidae	3	4	8	15

Number of freshwater species under these families omitted.

(Continued)

TABLE 12.1 Marine Fishes of India: Systematic Index of Higher Taxa with Genera and Species Numbers *(cont.)*

Sl. No.	Phylum: Chordata Subphylum: Craniata Superclass: Gnathostomata	India		World	
		No. of Genera	No. of Species	No. of Genera	No. of Species
199	Scatophagidae	1	1	2	4
200	Siganidae	1	17	1	28
201	Zanclidae	1	1	1	1
202	Acanthuridae	5	39	6	82
203	Sphyraenidae	1	10	1	27
204	Gempylidae	9	10	16	24
205	Trichiuridae	6	12	10	44
206	Scombrolabracidae	1	1	1	1
207	Scombridae	11	22	15	54
208	Xiphiidae	1	1	1	1
209	Istiophoridae	3	5	5	11
210	Centrolophidae	1	2	7	31
211	Nomeidae	2	3	3	16
212	Ariommatidae	1	1	1	7
213	Stromateidae	1	2	3	15
	Order: Pleuronectiformes				
214	Psettodidae	1	1	1	3
215	Citharidae	1	1	4	6
216	Paralichthyidae	2	9	14	111
217	Bothidae	9	21	20	162
218	Pleuronectidae	3	4	40	103
219	Samaridae	2	2	3	27
220	Soleidae	11	27	32	174
221	Cynoglossidae	3	21	3	143
	Order: Tetraodontiformes				
222	Triacanthodidae	6	6	11	23
223	Triacanthidae	3	5	4	7
224	Balistidae	11	22	12	42
225	Monacanthidae	14	22	28	107
226	Ostraciidae	4	7	8	25

Number of freshwater species under these families omitted.

TABLE 12.1 Marine Fishes of India: Systematic Index of Higher Taxa with Genera and Species Numbers *(cont.)*

Sl. No.	Phylum: Chordata Subphylum: Craniata Superclass: Gnathostomata	India		World	
		No. of Genera	No. of Species	No. of Genera	No. of Species
227	Triodontidae	1	1	1	1
228	Tetraodontidae	8	32	26	190
229	Diodontidae	3	6	7	18
230	Molidae	3	4	3	4
	Total	**927**	**2443**		

*Number of freshwater species under these families omitted.

considered as a valid category, as a taxon having its own evolutionary history, subspecies are not included in the species counts. While updating the species counts the authors have consulted all the relevant literature, including the web-based sources (Eschmeyer, 2014).

Fish diversity known from the fresh and marine waters of India constitutes 9.7 percent of the total number of about 33,059 species of fish known from the world (Eschmeyer and Fong, 2014), and of these the marine fishes alone account for 7.4 percent. An authentic assessment of species diversity of fishes of India recognizes an estimated 3231 valid species of both freshwater and marine fishes. Of this total fish diversity, marine fishes constitute 75.6 percent, comprising of 2443 species.

In India, as also the world over, fish groups of families and genera are expanding with newly described species. But as a corollary, species are being synonymized while new ones are described. However, a net increase in species every year suggests that there are still many new discoveries or records of taxa to be made from the diverse habitat environment of our marine and freshwater ecosystems.

The estimated number of 2443 species of marine fish recognized from India is taxonomically distributed in 230 families. Of these, the 12 families with the highest species richness (each with over 40 species) contain approximately 35.4 percent of all species (comprising altogether 865 species). These families in order of decreasing number of species are: Gobiidae (190), Pomacentridae (92), Labridae (85), Serranidae (85), Carangidae (66), Blenniidae (65), Apogonidae (63), Chaetodontidae (48), Lutjanidae (45), Sciaenidae (43), Syngnathidae (42) and Myctophidae (41).

Among the marine fishes of India, 9 families are monotypic, containing only one species, and 7 families have two species in one or two genera. In the Indian context, 35 families have 20 or more species, 12 of which have over 40 species.

In India, 96 families are represented by only one genus each, with a total of 205 species (Table 12.1). The family having the highest species richness with only one genus is the Siganidae, with 17 species. The average number of species in the Indian marine environment per family is 11. The most species-rich group of fishes known from Indian marine waters is the order Perciformes, comprising 1367 species and accounting for 56 percent of the total marine fishes recorded; this is followed by three orders— Scorpaeniformes, Anguilliformes and Tetra-odontiformes—each with a share of 4.3 percent of the total.

Talwar (1991) made an attempt to enumerate the estimated number of species of marine/freshwater fishes known from India. However, the current number of known species has increased to 2443 (total marine fish diversity). The number of species of fish recognized as valid (already known species after an assessment of the synonymy of the species, new species described and new records reported) has increased over the years.

DISTRIBUTION AND ENDEMISM

The highest number of marine fish species occurs in tropical areas of the Indian and western Pacific oceans east to the central Pacific, an area known as the Indo-Pacific faunal region. Randall's (1998) classic work on Indo-Pacific shore fishes richly contributes to the knowledge on endemism, disjunctive distributions, and species versus subspecies. Halas and Winterbottom (2009) provided information on the origin of the East Indies coral reef biota, and defined the "East Indies Triangle" as islands between Sumatra in the west, New Guinea in the east, and Luzon, Philippines in the north. This area has the highest marine biodiversity, including fishes, in the world. The number of marine fish species declines with distance from this area.

The Andaman and Nicobar archipelago shows the highest degree of species diversity in the Indian region, reportedly with 1431 species (Rajan *et al.*, 2013), possibly due to its proximity to and similarity with the Indonesian plate in zoogeography. The coral islands of the Lakshadweep, located offshore of the country along its western coast, exhibit a moderate diversity with 753 species (Rao, 1991). The east coast of India, with its peculiar estuary-like habitat-environment in the north and coral reefs of the Gulf of Mannar in the south, harbors 1121 species (Mishra, 2013). Current updating of Indian marine fish diversity lists the presence of 1071 species of fish along the west coast of India. Most of the shore fishes of India are derived from the centre of distribution in the "East Indies Triangle" (Menon, 1961; Krishnan and Mishra, 1993).

Endemism, as the concept used in the description of biodiversity, refers to a geographical area with taxa (species or genera) that occur only in that area. It is generally assumed that the taxa evolved there. Although the endemism of marine species is difficult to ascertain, many estuarine forms have restricted distribution, confined within the coastal waters of India, so as to be treated as endemic to India.

In marine fishes, the concept of endemism has received less attention than among freshwater fishes and other taxonomic groups. The presence of endemic genera or species involves many complex issues, such as geological age of the coastal area, its isolation and location, arrival of ancestral progenitors, competition, suitable habitat, and environmental stability over time (Eschmeyer *et al.*, 2010).

Mishra *et al.* (2013) have documented the endemic fishes of India, including 86 estuarine and marine fishes as endemic to Indian marine waters. A further 5 more species are described exclusively from Indian waters: *Aseraggodes martine* (Randall *et al.*, 2013); *Awaouichthys menoni* (Chatterjee and Mishra, 2013); *Chelidoperca maculipinna* (Bineesh *et al.*, 2013), *Hapalogenys bengalensis* (Mohapatra *et al.*, 2013) and *Ptereleotriscaeruleo marginata* (Allen *et al.*, 2012), thereby making the total number of coastal water endemics 91 species, under 74 genera, including an endemic genus *Awaouichthys* of 49 families (Chatterjee and Mishra, 2013).

MARINE FISHES AND ECOLOGICAL DIVERSITY

Fish inhabit diverse ecosystems/habitats within their vast environment of the extensive marine coastal waters of India. As is the nature of the environment, marine fishes of India show diverse morphology and behavior. While many species are territorial (e.g., Opistognathidae), some species congregate in schools (e.g., Clupeiformes). Some fishes show commensal relationships with other fish and other animals (e.g., Echeneidae, clown fishes, Carapidae). Adults of *Carapus acus* typically live as commensals in the gut of shallow-water holothurians.

Marine fishes found along the coastal and marine waters feed on a wide variety of food, including almost all classes of animals and some plants. A large number of species are specialized or adapted to feed on items such as zooplankton, snails, and coral and its associates. Some species show a parasitic mode of feeding (e.g., Antennariidae). There are also fishes producing venom (e.g., Scorpaeniformes), electricity (e.g., Torpediniformes), sound (e.g., Siganidae) or light (e.g., Myctophiformes), assisting them in their mode of feeding or enabling them to escape from their enemies using the adaptation as an offensive or defensive mechanism.

Most fishes in the Indian marine waters are ectotherms. Internal fertilization occurs in certain species, and females of some species provide nutrients to developing embryos (e.g., most of the elasmobranchs). Some fish exhibit parental care (e.g., Syngnathidae), and others release millions of eggs as a means to thrive against hazardous predation. While most of the species characteristically exhibit distinct sexual patterns in their reproductive life cycle, many species are hermaphroditic in nature (e.g., Labridae, Polynemidae, Scaridae, etc.).

Individuals of most of the species normally reproduce more than once in a season. Semelparity is found among the species of Anguillids; a few of them die relatively soon after a single spawning period. Anguillids are also diadromous.

Fishes in all types of aquatic environments, and especially in marine ecosystems, migrate long distances and use various homing mechanisms. Larvae and early juveniles of some oceanic species (e.g., Exocoetidae, Coryphaenidae) regularly inhabit shore waters, whereas the larvae of many shore fishes inhabit oceanic waters (e.g., Scombridae).

Marine fishes inhabit almost every conceivable habitat regime of the marine ecosystem. Some fish may tolerate a wide range of salinity (euryhaline) or a narrow range (stenohaline). Fishes in Indian marine waters range in length from 2.5 cm (*Brachygobius nunus*) to 20 m (*Rhincodon typus*). Fish body shape varies from elongated, cylindrical and snake-like (eels) to ball shaped (puffer fishes). A large number of species associated with corals and the environment alike are brilliantly colored (Chaetodontids) whereas many are drab looking (Batrachoidids). Some fishes are sleek and graceful in their movement, experiencing little water resistance (e.g., Scombridae), whereas some are ugly or grotesque and, if their livelihood does not depend on speed, remain virtually stationary in wait to pounce upon their prey (e.g., Antennaridae). The bodies of some are devoid of scales (e.g., eels, stonefish, sea catfish). In some fishes fins may be missing, particularly paired fins (e.g., eels), or be highly modified in to hold organs (remoras). Many bizarre specializations also exist in the internal anatomical diversity in hard and soft parts, exemplifying the morphological diversity among the fish of the Indian marine waters.

MARINE FISHES AND CONSERVATION

Fish are of immense value to humans, as they have long been a staple item of food. Unfortunately, overdependence on fish has led to overfishing, resulting in the dwindling of many species.

Fish and fisheries today form an important element in the economy of many nations, including India. Fish also provide incalculable recreational, psychological and aesthetic value to naturalists and fish hobbyists. Some fishes cause concerns to people as they are considered dangerous because of their poisons, stinging, shocking, or biting.

Fish constitute a resource that is a subject of international and domestic agreements and disagreements. In India, some institutions at the national or state level are devoted to the study of the propagation of some food fish species, based on their biology and reproductive potential. Intentional or clandestine introductions of alien species of marine ornamental fish are also in vogue in some important metropolitan commercial hubs in the country. Marine fish, as for other animal resources, constitute an important subject for scientific studies of behavior, ecology, evolution, genetics, physiology and medicine.

A healthy habitat presupposes or evokes the desirability of diversity and variability among species. The richer the diversity of life in an ecosystem, the greater is the ecosystem's strength and resilience, allowing it to maintain its equilibrium. Fish and their diversity are often used as general ecological indicators for assessing the health of the habitat, enabling humans to better realize the importance of their natural heritage and the sustainable livelihood of the people.

Systematists desire to observe diversity and variability of species for their studies. Their effort therefore has a leading role not only in realizing species diversity, but also in protecting its potential by recognizing diversity in the ecosystem and documenting it. Humans recognize fish, marine fish in particular, as an easily accessible source of food and appreciate their dependency upon that source. But some of our own actions that upset the integrity of the environment pose serious threats to fish stocks—for example, overfishing of marine waters, resulting in depletion of diversity and abundance of stocks.

There is a growing need to augment our efforts, through research and ecosystem restorative measures, to realize the actual and potential diversity of marine fish known from all the diverse habitats of the vast coasts of India and its offshore limits. Increasing efforts are needed to monitor and distinguish the status of species of marine fish to determine whether they are at risk of becoming vulnerable or endangered, possibly leading to their extinction as a result of human action or inaction.

Ever-increasing are the concerns over problems of overexploitation of marine resources, especially overfishing and habitat destruction that degrades the fisheries potential of many species, diminishing their populations even to the

TABLE 12.2 Threatened Fish Species in Indian Marine Waters

Sl. No.	Species	Common name
	Elasmobranchii (sharks, skates and rays)	
	Critically Endangered	
1	*Anoxypristis cuspidata* (Latham, 1794)	Knifetooth Sawfish
2	*Carcharhinus hemiodon* (Valenciennes, 1839)	Pondicherry Shark
3	*Glyphis gangeticus* (Müller & Henle, 1839)	Narrowsnout Sawfish
4	*Pristis microdon* (Latham, 1794)	Largetooth Sawfish
5	*Pristis pectinata* (Latham, 1794)	Wide Sawfish
6	*Pristis zijsron* (Bleeker, 1851)	Narrowsnout Sawfish
	Endangered	
7	*Aetobatus flagellum* (Bloch & Schneider, 1801)	Longheaded Eagle Ray
8	*Aetomylaeus maculatus* (Gray, 1834)	Mottled Eagle Ray
9	*Himantura fluviatilis* (Hamilton, 1822)	Ganges Stingray
10	*Lamiopsis temmincki* (Müller & Henle, 1839)	Broadfin Shark
11	*Sphyrna lewini* (Griffith & Smith, 1834)	Scalloped Hammerhead

(Continued)

TABLE 12.2 Threatened Fish Species in Indian Marine Waters *(cont.)*

Sl. No.	Species	Common name
12	*Sphyrna mokarran* (Ruppell, 1837)	Squat-headed Hammerhead Shark
	Vulnerable	
13	*Aetomylaeus nichofii* (Bloch & Schneider, 1801)	Banded Eagle Ray
14	*Alopias pelagicus* (Nakamura, 1935)	Pelagic Thresher
15	*Alopias superciliosus* (Lowe, 1841)	Bigeye Thresher Shark
16	*Alopias vulpinus* (Bonnaterre, 1788)	Common Thresher Shark
17	*Carcharhinus longimanus* (Poey, 1861)	Oceanic Whitetip Shark
18	*Carcharias taurus* (Rafinesque, 1810)	Sand Tiger
19	*Centrophorus granulosus* (Bloch and Schneider, 1801)	Smallfin Gulper Shark
20	*Centrophorus squamosus* (Bonnaterre, 1788)	Deepwater Spiny Dogfish
21	*Chaenogaleus macrostoma* (Bleeker, 1852)	Hooktooth Shark
22	*Gymnura zonura* (Bleeker, 1852)	Zonetail Butterfly Ray
23	*Hemipristis elongata* (Klunzinger, 1871)	Snaggletooth Shark
24	*Heteronarce garmani* (Regan, 1921)	Natal Electric Ray
25	*Himantura gerrardi* (Gray, 1851)	Whitespotted Whipray
26	*Himantura uarnak* (Forsskal, 1775)	Reticulate Whipra
27	*Isurus oxyrinchus* (Rafinesque, 1810)	Shortfin Mako
28	*Nebrius ferrugineus* (Lesson, 1830)	Tawny Nurse Shark
29	*Negaprion acutidens* (Ruppell, 1837)	Sharptooth Lemon Shark
30	*Rhina ancylostoma* (Bloch & Schneider, 1801)	Bowmouth Guitarfish
31	*Rhincodon typus* (Smith, 1829)	Whale Shark
32	*Rhinobatos granulatus* (Cuvier, 1829)	Sharpnose Guitarfish
33	*Rhinobatos obtusus* (Müller & Henle, 1841)	Widenose Guitarfish
34	*Rhinobatos typus* (Bennett, 1830)	Common Shovel Nose Ray
35	*Rhinoptera javanica* (Müller & Henle, 1841)	Javanese Cow Nose Ray
36	*Rhynchobatus djiddensis* (Forsskal, 1775)	Whitespotted Wedge Fish
37	*Sphyrna zygaena* (Linnaeus, 1758)	Smooth Hammerhead
38	*Stegostoma fasciatum* (Herman, 1783)	Leopard Shark
39	*Taeniura meyeni* (Müller & Henle, 1841)	Black-blotched Stingray
40	*Urogymnus asperrimus* (Bloch & Schneider, 1801)	Porcupine Ray

TABLE 12.2 Threatened Fish Species in Indian Marine Waters *(cont.)*

Sl. No.	Species	Common name
	Near Threatened	
41	*Aetobatus narinari* (Euphrasen, 1790)	Spotted Eagle Ray
42	*Atelomycterus marmoratus* (Anonymous [Bennett], 1830)	Coral Cat Shark
43	*Carcharhinus albimarginatus* (Ruppell, 1837)	Silvertip Shark
44	*Carcharhinus amblyrhynchoides* (Whitley, 1934)	Graceful Shark
45	*Carcharhinus amblyrhynchos* (Bleeker, 1856)	Grey Reef Shark
46	*Carcharhinus brevipinna* (Müller & Henle, 1839)	Spinner Shark
47	*Carcharhinus dussumieri* (Müller & Henle, 1839)	Widemouth Blackspot Shark
48	*Carcharhinus falciformis* (Bibron, 1839)	Silky Shark
49	*Carcharhinus leucas* (Valenciennes, 1839)	Bull Shark
50	*Carcharhinus limbatus* (Müller & Henle, 1839)	Blacktip Shark
51	*Carcharhinus macloti* (Müller & Henle, 1839)	Hardnose Shark
52	*Carcharhinus melanopterus* (Quoy & Gaimard, 1824)	Blacktip Reef Shark
53	*Carcharhinus sealei* (Pietschmann, 1913)	Blackspot Shark
54	*Carcharhinus sorrah* (Müller &Henle, 1839)	Spottail Shark
55	*Centrophorus acus* (Garman, 1906)	Needle dogfish
56	*Chiloscyllium arabicum* (Gubanov, 1980)	Arabian Carpet Shark
57	*Chiloscyllium griseum* (Müller & Henle, 1838)	Grey Bamboo Shark
58	*Chiloscyllium indicum* (Gmelin, 1789)	Ridgebacked Bamboo Shark
59	*Chiloscyllium plagiosum* (Bennett, 1830)	Whitespotted Bamboo Shark
60	*Chiloscyllium punctatum* (Müller & Henle, 1838)	Brownbanded Bamboo Shark
61	*Dasyatis zugei* (Müller & Henle, 1841)	Pale-edged Stingray
62	*Eusphyra blochii* (Cuvier, 1816)	Slender Hammerhead
63	*Galeocerdo cuvier* (Peron & LeSueur, 1839)	Tiger Shark
64	*Gymnura poecilura* (Shaw, 1804)	Longtail Butterfly Ray
65	*Heptranchias perlo* (Bonnaterre, 1788)	Sharpnose Sevengill Shark
66	*Hexanchus griseus* (Bonnaterre, 1788)	Bluntnose Sixgill Shark
67	*Manta birostris* (Donndorff, 1792)	Prince Alfred's Ray
68	*Mobula eregoodootenkee* (Bleeker, 1859)	Pygmy Devilray
69	*Mobula japanica* (Müller & Henle, 1841)	Spinetail Mobula
70	*Mobula thurstoni* (Lloyd, 1908)	Smoothtail Devil Ray

(Continued)

TABLE 12.2 Threatened Fish Species in Indian Marine Waters *(cont.)*

Sl. No.	Species	Common name
71	*Prionace glauca* (Linnaeus, 1758)	Blue Shark
72	*Pseudocarcharias kamoharai* (Matsubara, 1936)	Crocodile Shark
73	*Scoliodon laticaudus* (Müller & Henle, 1838)	Spadenose Shark
74	*Taeniura lymma* (Forsskal, 1775)	Ribbontailed Stingray
75	*Triaenodon obesus* (Ruppell, 1837)	Whitetip Reef Shark
	Actinopterygii (Bony Fishes)	
	Endangered	
76	*Cheilinus undulatus* (Rüppell, 1835)	Humphead Wrasse
	Vulnerable	
77	*Cromileptes altivelis* (Valenciennes, 1828)	Hump-back Rock-cod
78	*Epinephelus lanceolatus* (Bloch, 1790)	Queensland Grouper
79	*Hippocampus comes* (Cantor 1849)	Tiger Tail Seahorse
80	*Hippocampus kuda* (Bleeker, 1852)	Estuary Seahorse
81	*Hippocampus trimaculatus* (Leach, 1814)	Low-crowned Seahorse
82	*Hyporhamphus xanthopterus* (Valenciennes, 1847)	Red-tipped halfbeak
83	*Plectropomus areolatus* (Rüppell, 1830)	Squaretail Leopard Grouper
84	*Plectropomus laevis* (Lacepede, 1801)	Blacksaddled Coral Grouper
85	*Thunnus obesus* (Lowe, 1839)	Big Eye Tuna
	Near Threatened	
86	*Arius gagora* (Hamilton, 1822)	Gagora Catfish
87	*Chaetodon trifascialis* (Quoy & Gaimard, 1825)	Triangulate Butterflyfish
88	*Epinephelus bleekeri* (Vaillant, 1878)	Duskytail Grouper
89	*Epinephelus coioides* (Hamilton, 1822)	Orange-spotted Grouper
90	*Epinephelus diacanthus* (Valenciennes, 1828)	Spinycheek Grouper
91	*Epinephelus fuscoguttatus* (Forsskål, 1775)	Brown-marbled Grouper
92	*Epinephelus malabaricus* (Bloch & Schneider, 1801)	Malabar Grouper
93	*Epinephelus polylepis* (Randall & Heemstra, 1991)	Smallscaled Grouper
94	*Epinephelus polyphekadion* (Bleeker, 1849)	Camouflage Grouper
95	*Plectropomus pessuliferus* (Fowler, 1904)	Roving Coral Grouper

extent of possible extinction of some. As of today, as many as 50 species of marine fishes are threatened (IUCN, 2014), 6 of them Critically Endangered, 7 Endangered and 37 Vulnerable (Table 12.2). Of the 50 threatened species, most of them (40) are cartilaginous fishes or elasmobranchs—sharks, skates and rays—and the remaining 10 are teleost, i.e. bony, fishes. In addition, 45 more species are Near-Threatened, being already on the path to vulnerability.

The Wildlife (Protection) Act, 1972 of the Government of India as amended up to 1993 includes only 10 species of sharks and rays in Schedule I Part IIA Fishes. Those fishes are: *Rhincodon typus, Anoxypristis cuspidata, Carcharhinus hemiodon, Glyphis gangeticus, Glyphis glyphis, Himantura fluviatilis, Pristis microdon, Pristis zijsron, Rhynchobatus djiddensis, Urogymnus asperrimus.* Subsequently, seahorses (Sygnathidians), about 10 species, and the Giant Grouper, *Epinephelus lanceolatus*, were added to the list. Capture, killing and trade of these fishes attract punishment under the Act.

Hence conservation of species and fish stocks is, needless to say, of paramount importance. Appropriate and suitable natural conservation areas such as fish sanctuaries and Marine Protected Areas (MPAs) in the biogeographically optimal marine interface zones should be designated wherever required along the extensive marine coastal waters of India in order to safeguard the vulnerable or endangered species and their populations, creating a cascading effect on the improvement of the protection and safety of marine living resources in general.

Wanton exploitation of fish resources, especially overfishing of large fish, which are mostly top predators in the food web of the marine ecosystem, needs to be checked by implementing a moratorium on selective catching and killing of top predators, such as sharks and other forms, among the fast-dwindling species of cartilaginous and bony fishes, through effective policy decisions.

REFERENCES

Alcock, A.W., 1889. Natural history notes from H.M.S. Indian marine survey steamer 'Investigator' – No. 12. Descriptions of some new and rare species of fishes from the Bay of Bengal, obtained during the season 1888–89. J. Asiatic Soc. Bengal 58 (2), 296–305.

Alcock, A.W., 1890. Natural history notes from H.M.S. Indian marine survey steamer 'Investigator' Commander R.F. Hoskyn, R.N., commanding – No 16. On the bathybial fishes collected in the Bay of Bengal during the season 1889-90. Ann. Mag. Nat. Hist. Series 6, 197–222.

Allen, G.R., Erdmann, M.V., Cahyani, D., 2012. Description of *Ptereleotris caeruleomarginata* new species (Pisces: Ptereleotridae). In: Allen, G.R., Erdmann, M.V. (Eds.), Reef fishes of the East Indies, volume III, Tropical Reef Research, Perth, Australia, pp. 1190–1193.

Barman, R.P., Kar, S., Mukherjee, P., 2004. Marine and Estuarine fishes. Fauna of Andhra Pradesh, State Fauna Series. Zoological Survey of India, Kolkata 5 (2), 97–311.

Barman, R.P., Mishra, S.S., Kar, S., Mukherjee, P., Saren, S.C., 2007. Marine and estuarine fish fauna of Orissa. Rec. Zool. Surv. India. Occ. Paper 260, 186.

Barman, R.P., Mishra, S.S., Kar, S., Mukherjee, P., Saren, S.C., 2011. Marine and Estuarine Fish. Fauna of Tamil Nadu, State Fauna Series. Zoological Survey of India, Kolkata 17 (2), 293–418.

Barman, R.P., Mishra, S.S., Kar, S., Mukherjee, P., Saren, S.C., 2012. Marine and estuarine fish. Fauna of Maharastra, State Fauna Series. Zoological Survey of India, Kolkata 20 (1), 369–480.

Barman, R.P., Mishra, S.S., Kar, S., Saren, S.C., 2013. Marine and estuarine fishes. Fauna of Karnataka, State Fauna Series. Zoological Survey of India, Kolkata 21, 277–388.

Barman, R.P., Mukherjee, P., Kar, S., 2000. Marine and Estuarine fishes. Fauna of Gujarat, State Fauna Series. Zoological Survey of India, Kolkata 8 (1), 311–411.

Bineesh, K.K., Akhilesh, K.V., Abdussamad, E.M., Pillai, N.G.K., 2013. *Chelidoperca maculicauda*, a new species of perchlet (Teleostei: Serranidae) from the Arabian Sea. Aqua. Int. J. Ichthyology 19 (2), 71–78.

Bloch, M.E., Schneider, J.G., 1801. Systema Ichthyologiae. Iconibus e Illustratum. Berlin, p. 584.

Chatterjee, T.K., Mishra, S.S., 2013. A new genus and new species of Gobioid fish (Gobiidae: Gobionellinae) from Sunderbans, India. Rec. Zool. Surv. India 112 (4), 85–88.

Cuvier, G., Valenciennes, A., 1828–1849. *Histoire naturelle des poissons.* F.G. Levrault, Paris. 22 vols, 11,030p, 621 pls.

Day, F., 1875–78 (1888). The fishes of India, being a Natural History of the fishes known to inhabit the seas and freshwater of India, Burma and Ceylon. London. Part 1, 1875: 1-168, 1-40 pls.; Part 2, 1876: 169-368, 41-78 pls. (+ 51 A-C); Part 3, 1877: 369-552, 70-138 pls.; Part 4, 1878: i-xx + 553-778, 139-195 pls.; Suppl., 1888: 779-816, 7 figs.

Day, F., 1889. The fauna of British India: Fishes. Today and Tomorrow Printers and Publishers. New Delhi, 1 & 2, 548, p. 509.

de Beaufort, L.F., 1940. The Fishes of Indo-Australian Archipelago. E. J. Brill, Leiden 8, 508.

de Beaufort, L.F., Briggs, J.C., 1962. The Fishes of Indo-Australian Archipelago. E. J. Brill, Leiden, 11, p. 481.

de Beaufort, L.F., Chapman, W.M., 1951. The Fishes of Indo-Australian Archipelago. E. J. Brill, Leiden, 9, p. 484.

Eschmeyer, W.N. (Ed). 2014. Catalog of Fishes: Genera, Species, References. (hhttp://research. calacademy.org/research/ichthyology/catalog/fishcatmain.asp). Electronic version accessed 01 April 2014.

Eschmeyer, W.N., Fong, J.D., 2014. Species By Family/Subfamily (http://research.calacademy. org/research/ichthyology/catalog/SpeciesByFamily.asp). Electronic version accessed 01 April 2014.

Eschmeyer, W.N., Fricke, R., Fong, J.D., Polack, D.A., 2010. Marine fish diversity: history of knowledge and discovery (Pisces). Zootaxa 2525, 19–50.

Fischer, W., Bianchi, G., 1984. FAO species identification sheets for fishery purposes. Western Indian Ocean (Fishing Area 51). FAO, Rome. Vol. 1-5.

Halas, D., Winterbottom, R., 2009. A phylogenetic test of multiple proposals for the origins of the East Indies coral reef biota. J. Biogeography 36, 1847–1860.

Hamilton, F., 1822. An account of the fishes found in the River Ganges and its branches. Archibald Constable and Company, Edinburgh and London: 405 pp, 39 pls.

Hora, S.L., 1948. Knowledge of the ancient Hindus concerning fish and fisheries of India - 1. References to fish in Arthasastra (ca. 300 B. C.). J. R. Asiat. Soc. Bengal 14 (1), 7–10.

Hora, S.L., 1950. Knowledge of the ancient Hindus concerning fish and fisheries of India - 2. Fishery legislation in Ashoka's Pillar Edict V (246 B. C.). J. R. Asiat. Soc. Bengal 16 (1), 43–56.

IUCN (2014). The IUCN;1; Red List of Threatened Species. Version 2013.2. <http://www. iucnredlist.org>.

Jones, S., Kumaran, K., 1980. Fishes of the Laccadive Archipelago. Natural Conservation and Aquatic Sciences Service Cornell University, p. 760.

Koumans, F.P., 1953. Gobioidea. In: Weber, M., de Beaufort, L. (Eds.), The Fishes of Indo-Australian Archipelago, 10, E. J. Brill, Leiden, p. 423.

Krishnan, S., Mishra, S.S., 1993. On a collection of fish from Kakinada-Gopalpur sector of the east coast of India. Rec. Zool. Surv. India 93 (1-2), 201–240.

Krishnan, S., Mishra, S.S., 2004. An inventory of fish species described originally from fresh and coastal marine waters of Pondicherry. Rec. Zool. Surv. India 102 (3–4), 65–87.

Menon, A.G.K., 1961. On a collection of fish from the Coromandel Coast of India including Pondicherry and Karaikal areas. Rec. Indian Mus. 59 (4), 369–404.

Mishra, S.S., 2013. Coastal Marine Fish Fauna of East Coast of India. In : Venkataraman, K., Sivaperuman, C., Raghunathan, C., (Eds.), *Ecology and conservation of tropical marine faunal communities*, Springer-Verlag, Berlin, Heidelberg. 245–260.

Mishra, S.S., Krishnan, S., 2003. Fish fauna of Pondicherry and Karaikal. Rec. Zool. Surv. India. Occ. Paper 216, 53.

Mishra, S.S., Kosygin, L., Rajan, P.T. and Gopi, K.C. 2013. Pisces. In : Venkataraman, K., Chattopadhyay, A., Subramanian, K.A., (Eds.), *Endemic animals of India (Vertebrates)*. Zoological Survey of India, Kolkata. 7–107.

Mohapatra, A., Ray, D., Kumar, V., 2013. A new fish species of the genus *Hapalogenys* (Perciformes: Hapalogenyidae) from the Bay of Bengal. India. Zootaxa 3718 (4), 367–377.

Mora, C., Tittensor, D.P., Myers, R.A., 2008. The completeness of taxonomic inventories for describing the global diversity and distribution of marine fishes. Proc. Royal Soc. (B) 275, 149–155.

Nelson, J.S., 2006. Fishes of the world. John Wiley & Sons, Inc, Hoboken, New Jersey, p. 601.

Rajan, P.T., Sreeraj, C.R., Immanuel, T., 2013. Fishes of Andaman and Nicobar Islands: A Check-list. J. Andaman Sci. Assoc. 17 (1), 47–87.

Ramakrishna, Venkataraman, K., 2001. Marine. In : Ecosystems of India, (Eds.) Alfred, J.R.B., Das, A.K. and Sanyal, A.K., ENVIS Centre, Zoological Survey of India, Kolkata: 291–315.

Randall, J.E., 1998. Zoogeography of shore fishes of the Indo-Pacific region. Zoological Studies 37, 227–268.

Randall, J.E., Bogorodsky, S.V., Mal, A.O., 2013. Four new soles (Pleuronectiformes: Soleidae) of the genus *Aseraggodes* from the western Indian Ocean. J. Ocean Sci. Foundation 8, 1–17.

Rao, G.C., 1991. Lakshadweep: General features, Fauna of Lakshadweep, State Fauna Series. Zool. Surv. India 2, 5–40.

Russell, P., 1803. Descriptions and Figures of Two Hundred Fishes; collected at Vizagapatam on the coast of Coromandel. Vol. 1-2. London, 78 p. + 85 p.

Talwar, P.K., 1991. Pisces. In, Animal Resources of India. Zoological Survey of India, Kolkata, 577–630.

Talwar, P.K., Kacker, R.K., 1984. Commercial Sea Fishes of India. Hand Book. Zoological Survey of India 4, 997.

Talwar, P.K., Mukherjee, P., Saha, D., Paul, S.N., Kar, S., 1992. Marine and estuarine fishes, Fauna of West Bengal, State Fauna Series. Zoological Survey of India, Kolkata 3 (2), 243–342.

Venkataraman, K., Wafar, M., 2005. Coastal and marine biodiversity of India. Indian J. Mar. Sci. 34 (1), 57–75.

Weber, M., de Beaufort, L.F., 1916–36. The Fishes of Indo-Australian Archipelago. E. J. Brill, Leiden, 3 (1916): 455 p; 4 (1922): 410 p; 5 (1929): 458 p; 6 (1931): 448 p; and 7 (1936): p. 607.

Chapter 13

Fish and Shellfish Fauna of Chilika Lagoon: An Updated Checklist

A. Mohapatra,* S.K. Mohanty[†] and S.S. Mishra**
*Marine Aquarium and Regional Centre, Zoological Survey of India, Digha, West Bengal, India;
[†]Chilika Development Authority, Bhubaneswar, Odisha, India; **Marine Fish Section, Zoological
Survey of India, Kolkata, India

INTRODUCTION

Chilika lagoon is the largest coastal wetland in India and an internationally recognized Ramsar Site. It is regarded as a rich storehouse of living aquatic resources. With unique ecological characteristics resulting from two antagonistic hydrological processes—fresh water inflow and sea water influx—Chilika has no parallel in the tropical world. Chilika lagoon, lying on the east coast of India, is situated between 19° 28′ to 19° 54′ N latitude and 85° 05′ to 85° 38′ E longitude; it fluctuates in area from a monsoon maximum of 1165 km^2 to a dry season minimum of 906 km^2 (annual average 923 km^2); its linear axis is 64.3 km in length, and it has an average mean width of 20.1 km (Pattnaik, 2002; Ghosh and Pattnaik, 2005; Ghosh et al., 2006). The lagoon is separated from the Bay of Bengal by a sand bar whose width varies from 100 m to 1.5 km; a 30 km outer (inlet) channel connects the main lagoon with the Bay of Bengal. The 14 km long Palur canal connects the southern end of the lagoon to the sea through Rushikulya river mouth. The lagoon was critically threatened during the last few decades because of natural changes coupled with anthropogenic pressure. In the process of degradation of the ecosystem, the lagoon fishery became the major victim, causing misery to fishing communities. Hence, it was imperative to restore the fragile ecosystem of Chilika lagoon to recover *inter alia* the fisheries and biodiversity for the greater benefit of the wetland communities. After the opening of a new artificial lagoon mouth for restoration of the lagoon by Chilika Development Authority (CDA), the fishery production has increased many fold. The lagoon had witnessed multidimensional problems: increased sediment load, decreased salinity, Ghery prawn culture (Pens), excessive fishing pressure, invasive weed growth and several other natural changes

Marine Faunal Diversity in India. DOI: 10.1016/B978-0-12-801948-1.00013-6

coupled with incessant anthropogenic pressure/activities, which together altered the natural attributes of the lagoon.

The lagoon, being of estuarine character, is shallow throughout. The major part of the lagoon (the northern sector) has a depth of less than 0.3 m, while the maximum of 4.2 m was seen in the central sector during the pre-restoration period. The Magarmukh, being very shallow (20–30 cm deep) pre-intervention, could not be navigated even by flat-bottomed country boats.

Numerous islands are present in the lagoon, especially near the channel. Best known among them is Nalaban, a low, flat marshy island of 15.53 km^2, covered with low vegetation. Nalaban Island was designated as a Sanctuary in 1987 in consideration of its unique features as a habitat for avifauna. The Nalaban Bird Sanctuary is the only protected part of the Chilika lagoon. Several rocky islands in the southern sector, such as Kalijai, Somolo, Dumkudi, Honeymoon, Breakfast and Bird Island, represent the inundated remains of the Eastern Ghats (Bandyopadhyay and Gopal, 1991). Along the channel and between the channel and the lagoon proper are many islands made up of entrenched sand dunes, known collectively as Garh Krushnaprasad Block. These islands cover an area of about 34.4 km^2 (Bandyopadhyay and Gopal, 1991).

Hydrologically, Chilika lagoon is influenced by three subsystems: the Mahanadi distributaries (Delta Rivers), 52 rivulets and streams draining into the lagoon from the western catchment, and the sea (Bay of Bengal). The lagoon is situated at the southern margin of the Mahanadi delta; it receives less than 6 percent of the total Mahanadi flow, but this volume is close to half the total fresh water inflow into the lagoon. The lagoon receives inflows from its western catchments (1560 km^2), and runoff and irrigation drainage from the delta region (2250 km^2). The total Chilika drainage basin, including the lagoon itself, the contributing islands and coastal strip has an area of 4300 km^2 (World Bank, 2005). The Chilika drainage basin is estimated to contribute about 1760 million cubic metres (mcm) of water into the lagoon, while direct precipitation has been estimated at about 870 mcm, and total evaporation losses are estimated at about 1286 mcm (ORSAC, 1988). The freshwater inflows influence the biogeochemistry of Chilika lagoon in several ways, although few of these are well quantified. Firstly, and most importantly it is the freshwater inflows that drive the spatial and temporal salinity dynamics, that contribute to the temporal and spatial mosaic of different aquatic habitats for plants and animal species, and their varying lifecycle requirements (World Bank, 2005). It is primarily this dynamic salinity regime that enables the lagoon to support high biodiversity and productive fishery.

Ecologically, Chilika lagoon is an assemblage of shallow to very shallow marine, brackish and freshwater ecosystems. Salinity is the dominant factor determining the lagoon's ecology, and salinity dynamics are controlled jointly by the nature of the connection to the sea, associated tidal fluctuation, and the volume and timing of freshwater inflows to the lagoon from the Delta Rivers and western catchments (Mohapatra et al., 2007a). Both of these controlling factors are subjected to natural variability, and have been affected by anthropo-genic pressure. The lagoon is broadly divided into four ecological sectors based on differences in ecological features. These sectors are called northern, cen-tral, southern, and outer channel sectors. Magarmukh is the gateway between the main lagoon and the outer channel (Mohapatra *et al.*, 2007a). The lagoon undergoes a cyclical variation in salinity throughout the year, with different patterns seen in different ecological sectors. It is the periodicity in salinity that allows freshwater as well as marine and estuarine species to thrive in the lagoon. No mangrove vegetation is found in the lagoon except for two species (few in number confined to only one rocky island in southern sector): *Cassipourea cey-lanica* (rare and endemic to Chilika) belonging to the Rhizophoraceae family, which is only reported from Chilika lagoon (Panda and Patnaik, 2002), and *Ae-giceras corniculatum* of the Myrsinaceae family. This small patch of mangrove vegetation has no relevance to the mud crab habitat in the lagoon. The sea grass beds, which were almost lost during the eco-degradation phase, reappeared after the opening of the new lagoon mouth and have covered extensive areas in the central, southern and outer channel sectors. An extensive area in the northwest part of the lagoon is dominated by *Phragmites karka*.

The physical configuration and limnology of the lagoon is greatly influenced by different oceanographic processes, mainly wave action, littoral drift and tidal influence. Along Chilika coast, maximum wave height (3–4 m) occurs during the southwest monsoon period. The wave period of that season ranges from 6 to 14 s, whereas it varies in the range 6–18 s during the northeast monsoon. The Bay of Bengal in general experiences two surface currents: a clockwise northerly current during January to April/May and an anticlockwise southerly current from October to December (La Fond, 1958). This in turn affects the lagoon–seawater exchange and shoal formation along the mouth region. High salinity, greater transparency and low temperature prevail during the clockwise circulation of the summer season, whereas during the anticlockwise circulation the water mass is characterized by low salinity, less transparency and high nu-trients and temperature (Ganapati and Ramasarma, 1958).

Shrinkage of the old mouth has been attributed primarily to the littoral drift. It has been estimated that one million tonnes of sediment move in a northeastern direction every year during March to October, and such net transport is believed to be responsible for shrinkage of the old mouth. The tide of this region is a mixed semi-diurnal type. The semidiurnal tidal fluctuations control the ingress of seawater into the lagoon. Tidal range near Chilika old mouth has been esti-mated at 85 cm (Rajan, 1971).

The unique and fragile ecosystem of Chilika lagoon, with its estuarine char-acter and rich fishery resources, gradually lost its ecological characteristics as a result of changing coastal processes, a degraded drainage basin and anthro-pogenic pressure over the last few decades. Being a coastal wetland, the la-goon is quite sensitive to changes in hydrological regimes. During the last few decades, the lagoon gradually moved towards freshwater systems because of significant decrease in salinity. Several natural changes were the major factors contributing to the rapid deterioration of the lagoon ecosystem: shifting of the lagoon mouth by more than 30 km from the lagoon proper, rapid siltation at the rate of 7.53×10^5 m^3 annually (World Bank, 2005), explosive spread of invasive weeds, shoal formations in the inlet channel, choking of the Magarmukh and Palur canals, etc. (Pattnaik, 2000, 2001; Mohanty and Behera, 2002; Nayak et al., 2002). The cross section of the outer channel was significantly reduced by shoal formation along the channel, which resulted in considerable hydraulic head loss and flushing action. Various studies on the coastal process indicated that about one mcm of littoral drift prevails in this region of the coast, which led to shifting of the mouth in the northeastern direction (Chandramohan et al., 2002). On average, the shifting was about 350 m per annum. The depth of the water at Magarmukh, which is considered as the gateway between the main lagoon and outer channel, was observed to be alarmingly low, at only about 30 cm in December 1997. This was identified as the most critical zone preventing the discharge of freshwater sediment during monsoon to the sea through the outer channel and the ingress of saline water to the lagoon during post-monsoon months. These natural changes distinctively contributed to the eco-degradation in the lagoon, which was further aggravated by incessant anthropogenic activities such as carving out extensive fringe areas with earthen dikes for agriculture, prawn culture pond development along fringe areas, unabated expansion of eco-inimical and illegal shrimp pen culture (Prawn gheries) inside the lagoon covering more than 10,000 ha (11% of the lagoon area) (Mohanty et al., 2004a,b), and excessive fishing activity including destructive fishing.

Analysis of satellite data revealed that during the years 1973, 1977, 1985 and 1993 the weed-covered areas in the lake were 20, 60, 200 and 398 km^2, respectively, as reported by Ghosh (2002). The northern sector exhibited dominance of freshwater weeds. Annual invasion of weeds by 1998 was calculated to be 15 km^2.

The lagoon is one of the hotspots of biodiversity and shelters a number of endangered or threatened species listed in the IUCN Red List of threatened species. The lagoon has profound socio-economic significance, since its rich fishery resources support the livelihood and nutritional security of about 0.2 million fisherfolk living in and around the lagoon. The fishery resources of the lagoon suffered serious setback from the later part of the 1980s when the salinity level sharply decreased and recruitment routes (outer channel and Palur canal) gradually became silted up, adversely affecting the recruitment of fish and shellfish seed from the sea into the lagoon. Jhingran (1963) estimated that 63–75 percent of the annual fish production was contributed by migratory species. Annual fish landing, on an average, was 7200 tons during the 1980s but decreased to 1737 tons during 1999–2000, a decrease of about 76 percent. Significant loss of overall biodiversity of the lagoon took place during this period (Ghosh, 2002).

In the aftermath of the gradual closure of the old mouth and Palur canal, the lagoon had turned more towards a freshwater ecosystem, resulting in substantial changes in species composition with significant increase in freshwater forms. The Zoological Survey of India (ZSI) after completing the Chilika Expedition (1985–87) opined that the lake ecosystem was tending towards a freshwater ecosystem and warranted urgent restoration measures. Owing to the degraded state of the lagoon's ecosystem with drastic changes in the ecological characteristics and overall loss of biodiversity, the Chilika lagoon was included in the Montreux Record (Threatened list of Ramsar Sites) in 1993.

From the 1950s to 2000 the lagoon was in continual decline, with increasing sediment loads and decreasing salinity. The fishery showed a major decline, invasive weeds began to take hold, and the entire lagoon progressively shrunk in area and volume (World Bank, 2005). The eco-degradation reached a critical stage with loss of ecological characteristics, change in ecosystem function and overall loss of biodiversity which brought miseries to about 0.2 million local people directly and indirectly depending on the lagoon's resources (goods) and services. Being a *bona fide* signatory to the Ramsar Convention, and since Chilika is listed as a Ramsar Site, the Government of India took urgent measures to restore Chilika lagoon through the State Government and provided the necessary funds from the Tenth and Eleventh Finance Commission Grants. The State Government constituted an Authority called Chilika Development Authority (CDA) which implemented the eco-restoration programme in Chilika lagoon during 2000–01 and 2001–02 following an ecosystem approach. The restoration programme included four major activity components:

1. Hydrological intervention
2. Palur canal renovation
3. Catchments treatment
4. Operation of the Naraj Barrage for preferential environmental flow.

Hydrology is the most important factor for the maintenance of a wetland's structure and function. Hydrologic conditions affect many vital biotic and abiotic factors including the salinity regime of coastal wetlands. Chilika lagoon is a highly productive coastal wetland. The lagoon was in a degraded state and tending towards a freshwater ecosystem due to siltation and choking of the mouth (inlet) resulting in the poor exchange of water. This had resulted in decline in productivity and loss of biodiversity. Against this backdrop the CDA initiated an integrated adaptive management process to address the complex ecological and socio-economic issues of the Chilika lagoon with an ecosystem approach. There was an assessment of the principal causes of degradation, with the objective to implement appropriate and effective methods to restore the lagoon to its former state through targeted scientific studies and wide stakeholders' consultations. Intensive studies of the coastal processes showed that the tidal influx into the lagoon was adversely affected by the shoal formation along the lead channel and continuous shifting of the mouth as a result of littoral drift. This also adversely affected the auto-recruitment and breeding migration of fish species through the mouth opening into the sea. The services of the Central Water and Power

Research Station (CWPRS, Pune) were commissioned for the creation of a two-dimensional numerical model to establish the optimum salinity gradient. From the findings of the two-dimensional model studies, the CWPRS recommended a 'straight cut' to improve the salinity gradient of the lagoon to the desired level. Accordingly, a new mouth was opened on September 2000 at a distance of 11 km from Magarmukh. To improve the circulation and the exchange of water, a lead channel of 3.2 km was dredged at Magarmukh, which formed the link between the lagoon and the outer channel, connecting to the sea. Environment Impact Assessment (EIA) was carried out by the National Institute of Oceanography (NIO), Goa, to assess the impact of the hydrological intervention. The lead channel at Magarmukh was extended towards the river outfall point over a length of 22.6 km for better propagation of salinity and flushing out of sediment from this sector. The dredged channel also facilitates the dispersal of fish and shellfish juveniles to this sector. After this intervention, there has been a significant improvement of the lagoon ecosystem, as is evident from the enhancement of fishery resources, improvement of the tidal and salinity flux into the lagoon, flushing out of sediment, the decrease of freshwater invasive species, increase in the Irrawaddy dolphin population, expansion of the sea grass meadows and decreased water logging.

The hydrological intervention not only rejuvenated the ecosystem of the lagoon but also immensely benefited the community depending on the lagoon, whose average annual income increased due to enhancement of the fishery resources. There have been significant improvements of the salinity gradient after the opening of the mouth. Before the opening, the salinity level of the northern sector of the lagoon was almost zero throughout the year. There was an abrupt change in salinity level of the central sector and outer channel sector at the onset of monsoon. After the opening of the new mouth, less fluctuation of the salinity level was observed. The gradual reduction in the salinity from the lagoon mouth to the lagoon interior after the opening of the mouth is providing the desirable sense of direction for the euryhaline forms to enter into the lagoon from the sea. This facilitates the auto-recruitment of fish, prawn and crab into the lagoon. After the hydrological intervention, the average fish landing increased by many fold (Mohapatra et al., 2007a; Mohanty et al., 2009).

Palur canal (14 km) connects the Palur Bay of Chilika lagoon in the southern sector to the Rushikulya estuary. Because of heavy siltation in the canal the exchange of water between the sea and the southern sector of the lagoon was not taking place. Palur canal plays a vital role in maintaining the salinity gradient and the auto-recruitment of economic species from the sea to enrich the fishery resources of the southern sector. The canal was renovated by CDA with an objective to enhance the fishery resources of the southern sector. It became functional from the year 2004. This renovated canal is providing an effective recruitment and migration route for fish and shellfish to the southern sector. After renovation of the Palur canal, there has been a significant improvement of the fishery resources of this sector.

Although hydrological intervention was a key to the restoration of Chilika lagoon, some other ameliorative measures outside the lagoon, particularly the vast catchments (2250 km^2) under Mahanadi river basin and 1560 km^2 western catchments are also important. More than 48 catchment streams drain freshwater into the lagoon during rains, with an estimated sediment load of 1.06 million tons to the lagoon per year. The hilly areas in the western catchment were once thickly forested, but were mostly deforested by the local communities. This caused increased silt/sediment loading in the lagoon. Hence, with a view to arresting silt loading in the lagoon and preventing rapid soil erosion, a planned plantation programme through participatory micro-watershed implementation was taken up in the degraded catchments. The micro-watershed projects with community participation were undertaken on a priority basis to demonstrate regeneration of the highly degraded catchment ecosystem and sharing of benefits of water and land resources. The programme was initiated in 2001–02 with a very successful implementation of the 'Dengei Pahada' participatory micro-watershed project in the central part of the western catchment.

Coastal wetlands like Chilika are especially sensitive to the hydrologic regime, as this controls the depth variations and influences the water quality conditions that determine the range and character of aquatic habitats. Salinity is the most dominant factor determining the lagoon's ecology, and the salinity dynamics are controlled jointly by two factors:

- the nature of the connection to the sea, associated tidal fluctuations; and
- the volume and timing of freshwater inflows to the lagoon from the delta (Mahanadi) distributaries and western catchments.

Both the controlling factors are subjected to natural variability and have been affected by anthropogenic activities. The freshwater inflows into the lagoon have long been affected by upstream water management for irrigation, drainage and flood mitigation.

The World Bank-assisted Orissa Water Resources Consolidation Project (OWRCP) considered the water management in Chilika lagoon very critically for sustainability of the lagoon's ecological function and bio-resources, particularly fisheries, avifauna and the dolphin population, since the major freshwater inflow into the lagoon is affected by Mahanadi distributaries. An operation schedule was formulated for flow control at the new Naraj Barrage on Mahanadi to allow preferred flow of freshwater into Chilika for maintenance of the lagoon's eco-function at its optimal condition.

DIVERSITY STATUS OF FINFISH AND SHELLFISH

Pre-intervention Period

During 1914–24 the Zoological Survey of India (ZSI) carried out pioneering work on faunal diversity and recorded 112 fish species, 24 prawn and 26 brachyuran crabs from Chilika lagoon (Kemp, 1915; Chaudhuri, 1916a–c;

TABLE 13.1 Diversity Status (Number of Species) of Fish and Shellfish in Chilika Lagoon during Pre-restoration Phase

Status parameter	Pre-restoration survey (1914–2000)
Species recorded by ZSI, CIFRI and individual workers	
Fish	225 (G/149, F/72, O/16)
Shrimp & prawn	24 (G/13,F/9,SO/2)
Lobster	Not recorded
Crab	28 (G/22, F/9, SO/1)
Inventory of recorded species—last survey by ZSI (1985–87)	
Fish	71 (G/60, F/14, O/13)
Shrimp & prawn	13 (G/6, F/5, SO/2)
Lobster	Not recorded
Crab	11 (G/11, F/9, SO/1)

G, Genus; F, Family; O, Order; SO, Suborder; ZSI, Zoological Survey of India; CIFRI, Central Inland Fisheries Research Institute; CDA, Chilika Development Authority.

Hora, 1923). The Central Inland Fisheries Research Institute (CIFRI), while investigating the fisheries of Chilika lagoon, recorded 55 additional fish species during 1957–65, and some individual workers during 1954–86 added a further collection of 46 fish species (Kaumans, 1941; Jones and Sujansinghani, 1954; Menon, 1961; Misra, 1969, 1976a,b; Jhingran and Natrajan, 1966, 1969; Rajan *et al.*, 1968; Mohanty, 1973; Talwar and Jhingran, 1991). During the Chilika expedition (1985–87), the ZSI recorded another 4 species of fish and 2 species of crab. Thus by 1987, 217 fish species (Rama Rao, 1995), 28 crab species (Maya Deb, 1995) and 24 prawn species (Reddy, 1995) were listed. Later, Bhatta *et al.* (2001) added 8 new records of fish, bringing the total to 225 species of fishes, 24 species of prawns and 28 crab species before opening of the new mouth under hydrological intervention in September 2000. During the eco-degradation phase in the 1990s when the ecological characteristics were changed, there may have been loss of faunal diversity to a certain extent. Status of fish and shellfish biodiversity in Chilika lagoon during the pre-restoration period is given in Table 13.1.

Post-intervention Period

During the post-restoration phase (2000–01 to 2003–04), CDA made inventory in Chilika lagoon of fish, prawn and crab fauna since 2000–01. A total of 221 species have been collected from the lagoon up to March 2004, comprising 187 species of fish, 18 species of prawn, 14 species of crab and 2 species of lobster (Mohanty *et al.*, 2006; Mohapatra *et al.*, 2007a). The faunal inventory included 56 new records (43 species of fish, 4 prawns, 7 crabs and 2 lobsters) during the period. From the post-restoration records, 69.78 percent fish, 64.28 percent prawn and 40 percent crab species were recovered, and the overall recovery was 66.36 percent up to March 2004. Freshwater elements in the catches have decreased significantly by 45.3 percent during the post-restoration period as compared with the pre-restoration period. Species richness in the Outer Channel sector considerably increased after opening of the new mouth. Some further fish species were added by Wetlands International-South Asia (2011), and the total list of fish comprised 314 species from the pre-restoration to the post-restoration period. During the recent survey carried out by the Marine Aquarium and Regional Centre, Zoological Survey of India (MARC, ZSI) on the "Ornamental Fauna of East coast of India," one species of fish was collected and identified as *Acanthurus triostegus* (Linnaeus, 1758), Registration No. MARC/ZSI/F2516, which was not reported to date from Chilika lagoon; this brings the total of ichthyofauna to 315 species. A detail account of brachyuran crab fauna of Chilika lagoon from the pre-restoration to post-restoration period has been given by Mohapatra *et al.*, 2007b in which the total crab fauna reported from Chilika was 35 species. Thereafter, one species *Charybdis feriata,* was wrongly reported as a new record to Chilika lagoon by Sahoo *et al.*, 2008. The same species was reported from Chilika lagoon as *Charybdis cruciata* (Mohapatra *et al.*, 2007b), which is synonymous to *Charybdis feriata*; thus, in total, 35 species of crabs were reported from the Chilika lagoon to date. A detailed checklist of the finfish and shellfish of the lagoon is given in Table 13.2.

TABLE 13.2 Inventory and Updated Checklist of Fish and Shellfish of Chilika Lagoon (1916–2012)

Sl. No.	Family	Species
Finfish		
1.	**Hemiscylliidae** (Bamboo sharks)	*Chiloscyllium indicum* (Gmelin, 1789)**
2.	**Carcharhininidae** (Requiem sharks)	*Carcharhinus leucas* (Müller & Henle, 1839)**
3.		*Carcharhinus limbatus* (Müller & Henle, 1839)
4.		*Carcharhinus melanopterus* (Quoy & Gaimard, 1824)
5.		*Glyphis gangeticus* (Müller & Henle, 1839)
6.		*Scoliodon laticaudus* (Müller & Henle, 1838)*
7.	**Sphyrnidae** (Hammerheaded shark)	*Eusphyra blochii* (Cuvier, 1816)** [Previously recorded as *Sphyrna blochii* (Cuvier, 1817)]
8.		*Sphyrna lewini* (Griffith & Smith, 1834)**

*Collection under post-restoration inventorial survey;
**New records during post-restoration period.

(Continued)

TABLE 13.2 Inventory and Updated Checklist of Fish and Shellfish of Chilika Lagoon (1916–2012) *(cont.)*

Sl. No.	Family	Species
9.	**Pristidae** (Sawfish)	*Pristis clavata* (Garman, 1906)
		[Previously recorded as *Pristis pectinata* (Latham, 1794)]
	Rhinobatidae (Guitar fishes)	
10.		*Rhynchobatus djiddensis* (Forsskål, 1776)**
11.	**Dasyatidae** (Stingrays)	*Himantura imbricata* (Bloch & Schneider, 1801)
12.		*Himantura marginata* (Blyth, 1860)**
		[Previously recorded as *Dasyatis marginatus* (Blyth, 1860)]
13.		*Himantura uarnak* (Forsskål, 1775)*
14.		*Himantura walga* (Müller & Henle, 1841)*
15.		*Pastinachus sephen* (Forsskål, 1775)*
		[Previously recorded as *Hypolophus sephen* (Forsskål, 1775)]
16.	**Myliobatidae** (Eaglerays)	*Aetobatus flagellum* (Bloch & Schneider, 1801)*
17.		*Aetobatus narinari* (Euphrasen, 1790)*
18.		*Aetomylaeus nichofii* (Bloch & Schneider, 1801)*
19.	**Notopteridae** (Featherbacks)	*Chitala chitala* (Hamilton, 1822)*
		[Previously recorded as *Notopterus chitala* (Hamilton, 1822)]
20.		*Notopterus notopterus* (Pallas, 1769)*
21.	**Elopidae** (Tenpounders)	*Elops machnata* (Forsskål, 1775)*
22.	**Megalopidae** (Tarpons)	*Megalops cyprinoides* (Broussonet, 1782)*
23.	**Anguillidae** (Freshwater eels)	*Anguilla bengalensis bengalensis* (Gray, 1831)*
24.		*Anguilla bicolor bicolor* (McClelland, 1844)*
25.	**Muraenidae** (Moray eels)	*Strophidon sathete* (Hamilton, 1822)*
		[Previously recorded as *Thyrosoidea macrura* (Blecker, 1854)]
26.	**Ophichthidae** (Snake eels)	*Lamnostoma orientalis* (McClelland, 1844)
27.		*Pisodonophis boro* (Hamilton, 1822)*

Collection under post-restoration inventorial survey;
***New records during post-restoration period.*

TABLE 13.2 Inventory and Updated Checklist of Fish and Shellfish of Chilika Lagoon (1916–2012) *(cont.)*

Sl. No.	Family	Species
28.		*Pisodonophis cancrivorus* (Richardson, 1848)
29.	**Muraenesocidae** (Pike congers)	*Congresox talabonoides* (Bleeker, 1853)*
30.		*Muraenesox bagio* (Hamilton,1822)**
31.		*Muraenesox cinereus* (Forsskål, 1775)*
32.	**Clupeidae** (Herrings, shads, sprats, sardines, pilchards)	*Amblygaster leiogaster* (Valenciennes, 1847)**
33.		*Amblygaster sirm* (Walbaum, 1792) [Previously recorded as *Sardinella sirm* (Walbaum,1792)]
34.		*Anodontostoma chacunda* (Hamilton, 1822)*
35.		*Corica soborna* (Hamilton, 1822)*
36.		*Dussumieria acuta* (Valenciennes, 1847)
37.		*Dussumieria elopsoides* (Bleeker, 1849)**
38.		*Ehirava fluviatilis* (Deraniyagala, 1929)**
39.		*Escualosa thoracata* (Valenciennes, 1847)*
40.		*Gonialosa manmina* (Hamilton, 1822)*
41.		*Gudusia chapra* (Hamilton, 1822)*
42.		*Hilsa kelee* (Cuvier, 1829)*
43.		*Nematalosa nasus* (Bloch, 1795)*
44.		*Sardinella fimbriata* (Valenciennes, 1847)**
45.		*Sardinella longiceps* (Valenciennes, 1847)**
46.		*Sardinella melanura* (Cuvier, 1829)
47.		*Tenualosa ilisha* (Hamilton, 1822)* [Previously recorded as *Hilsa ilisha* (Hamilton, 1822)]
48.		*Tenualosa toli* (Valenciennes, 1847)**
49.	**Engraulidae** (Anchovies)	*Setipinna phasa* (Hamilton, 1822)
50.		*Stolephorus baganensis* (Hardenberg, 1933)*
51.		*Stolephorus commersonnii* (Lacepède, 1803)*
52.		*Stolephorus dubiosus* (Wongratana, 1883)*
53.		*Stolephorus indicus* (Van Hasselt, 1823)*

*Collection under post-restoration inventorial survey;
**New records during post-restoration period.

(Continued)

TABLE 13.2 Inventory and Updated Checklist of Fish and Shellfish of Chilika Lagoon (1916–2012) *(cont.)*

Sl. No.	Family	Species
54.		*Thryssa gautamiensis* (Babu Rao, 1971)**
55.		*Thryssa hamiltonii* (Gray, 1835)*
56.		*Thryssa kammalensoides* (Wongratana, 1883) [Previously recorded as *Thryssa kammaleneis* (Bleeker, 1849)]
57.		*Thryssa malabarica* (Bloch, 1795)*
58.		*Thryssa mystax* (Bloch & Schneider, 1801)*
59.		*Thryssa polybranchialis* Wongratana, 1983*
60.		*Thryssa purava* (Hamilton, 1822)*
61.		*Thryssa setirostris* (Broussonet, 1782)**
62.		*Thryssa vitrirostris* (Gilchrist & Thompson, 1908)**
63.	**Chirocentridae** (Wolf herrings)	*Chirocentrus dorab* (Forsskål, 1775)
64.	**Pristigasteridae** (Pellonas)	*Ilisha elongata* (Bennett, 1830)**
65.		*Ilisha megaloptera* (Swainson, 1839)*
66.		*Ilisha melastoma* (Bloch & Schneider, 1801)
67.		*Opisthopterus tardoore* (Cuvier, 1829)**
68.	**Chanidae** (Milkfish)	*Chanos chanos* (Forsskål, 1775)*
69.	**Cyprinidae** (Carps & minnows)	*Amblypharyngodon mola* (Hamilton, 1822)*
70.		*Chela cachius* (Hamilton, 1822)*
71.		*Laubuca laubuca* (Hamilton, 1822)* [Previously recorded as *Chela laubuca* (Hamilton, 1822)]
72.		*Cirrhinus mrigala* (Hamilton, 1822)*
73.		*Cirrhinus reba* (Hamilton, 1822)*
74.		*Crossocheilus latius* (Hamilton, 1822)
75.		*Danio rerio* (Hamilton, 1822) [Previously recorded as *Brachydanio rerio* (Hamilton, 1822)]
76.		*Esomus danricus* (Hamilton, 1822)*
77.		*Catla catla* (Hamiltton, 1822)*

*Collection under post-restoration inventorial survey;
**New records during post-restoration period.*

TABLE 13.2 Inventory and Updated Checklist of Fish and Shellfish of Chilika Lagoon (1916–2012) *(cont.)*

Sl. No.	Family	Species
78.		*Labeo boga* (Hamilton, 1822)**
79.		*Labeo calbasu* (Hamilton, 1822)*
80.		*Labeo gonius* (Hamilton, 1822)**
81.		*Labeo rohita* (Hamilton, 1822)*
82.		*Osteobrama cotio peninsularis* (Silas, 1952)**
83.		*Osteobrama vigorsii* (Sykes, 1839)
84.		*Puntius chola* (Hamilton, 1822)*
85.		*Systomus sarana* (Hamilton, 1822)*
86.		*Puntius sophore* (Hamilton, 1822)*
87.		*Pethia ticto* (Hamilton, 1822)*
88.		*Puntius vittatus* (Day, 1865)
89.		*Rasbora daniconius* (Hamilton, 1822)* [Previously recorded as *Parluciosoma daniconius* (Hamilton, 1822)]
90.		*Rasbora rasbora* (Hamilton, 1822).*
91.		*Salmophasia bacaila* (Hamilton, 1822)*
92.	**Cobitidae** (Loaches)	*Lepidocephalichthys guntea* (Hamilton, 1822) [Previously recorded as *Lepidocephalus guntea* (Hamilton, 1822)]
93.	**Bagridae** (Bagrid catfishes)	*Mystus cavasius* (Hamilton, 1822)*
94.		*Mystus gulio* (Hamilton, 1822)*
95.		*Mystus vittatus* (Bloch, 1794)*
96.		*Sperata seenghala* (Sykes, 1839)* [Previously recorded as *Aorichthys seenghala* (Sykes, 1839)]
97.	**Siluridae** (Eurasian catfishes)	*Ompok bimaculatus* (Bloch, 1794)*
98.		*Ompok pabda* (Hamilton, 1822)*
99.		*Wallago attu* (Bloch & Schneider, 1801)*
100.	**Schilbeidae** (Schilbid catfishes)	*Ailia coila* (Hamilton, 1822)*
101.		*Eutropiichthys vacha* (Hamilton, 1822)
102.		*Silonia silondia* (Hamilton, 1822)

*Collection under post-restoration inventorial survey;
**New records during post-restoration period.

(Continued)

TABLE 13.2 Inventory and Updated Checklist of Fish and Shellfish of Chilika Lagoon (1916–2012) *(cont.)*

Sl. No.	Family	Species
103.	**Pangasiidae** (Shark catfish)	*Pangasius pangasius* (Hamilton, 1822)*
104.	**Sisoridae** (Sisorid catfish)	*Bagarius bagarius* (Hamilton, 1822)*
105.		*Bagarius yarrelli* (Sykes, 1839)**
106.	**Clariidae** (Air-breathing catfish)	*Clarias magur* (Hamilton, 1822)* [Previously recorded as *Clarias batracacus* (Linnaeus, 1758)]
107.	**Heteropneustidae** (Airsac catfish)	*Heteropneustes fossilis* (Bloch, 1794)*
108.	**Ariidae** (Sea catfishes)	*Arius arius* (Hamilton, 1822)*
109.		*Arius maculatus* (Thunberg, 1792)
110.		*Nemapteryx caelata* (Valenciennes, 1840)* [Previously recorded as *Arius caelatus* Valenciennes, 1840]
111.		*Osteogeneiosus militaris* (Linnaeus, 1758)*
112.		*Plicofollis argyropleuron* (Valenciennes, 1840) [Previously recorded as *Arius satparanus* Chaudhuri, 1916]
113.		*Plicofollis tenuispinis* (Day, 1877)
114.	**Plotosidae** (Stinging catfishes)	*Plotosus canius* (Hamilton, 1822)*
115.		*Plotosus lineatus* (Thunberg, 1787)*
	Synodontidae (Lizardfishes)	
116.		*Saurida tumbil* (Bloch, 1795)**
117.		*Trachinocephalus myops* (Forster, 1801)**
118.	**Mugilidae** (Mullets)	*Liza macrolepis* (Smith, 1846)*
119.		*Liza melinoptera* (Valenciennes, 1836)*
120.		*Liza parsia* (Hamilton, 1822)*
121.		*Liza planiceps* (Valenciennes, 1836)* [Previously recorded as *Liza tade* (Forsskal, 1775)]
122.		*Liza subviridis* (Valenciennes, 1836)*

Collection under post-restoration inventorial survey;
**New records during post-restoration period.*

TABLE 13.2 Inventory and Updated Checklist of Fish and Shellfish of Chilika Lagoon (1916–2012) *(cont.)*

Sl. No.	Family	Species
123.		*Liza vaigiensis* (Quoy & Gaimard, 1825)
124.		*Mugil cephalus* Linnaeus, 1758*
125.		*Rhinomugil corsula* (Hamilton, 1822)*
126.		*Valamugil cunnesius* (Valenciennes, 1836)*
127.		*Valamugil seheli* (Forsskål, 1775)
128.		*Valamugil speigleri* (Bleeker, 1858–59)*
129.	**Atherinidae** (Oldworld silversides)	*Atherinomorus duodecimalis* (Valenciennes, 1835)**
130.		*Atherinomorus lacunosus* (Forster, 1801)**
131.	**Belonidae** (Needlefishes)	*Strongylura leiura* (Bleeker, 1850)*
132.		*Strongylura strongylura* (Van Hasselt, 1823)*
133.		*Xenentodon cancila* (Hamilton, 1822)*
134.	**Hemiramphidae** (Halfbeaks)	*Hemiramphus far* (Forsskål, 1775)**
135.		*Hyporhamphus limbatus* (Valenciennes, 1847)*
136.	**Adrianichthyidae** (Adrianichthyids)	*Oryzias dancena* (Hamilon, 1822)*
137.	**Aplocheilidae** (Asian rivulines)	*Aplocheilus panchax* (Hamilton, 1822)*
138.	**Syngnathidae** (Pipefishes and Seahorses)	*Hippocampus fuscus* (Rüppell, 1838)* [Previously recorded as *Hippocampas brachyrhynchus* (Duncker, 1940)]
139.		*Ichthyocampus carce* (Hamilton, 1822)*
140.		*Hippichthys cyanospilos* (Bleeker, 1854)** [Previously recorded as *Syngnathus cyanospilos* (Bleeker, 1854)]
141.	**Synbranchidae** (Swamp eels)	*Ophisternon bengalense* McClelland, 1844**
142.	**Mastacembelidae** (Spiny eels)	*Macrognathus aral* (Bloch & Schneider, 1801)*
143.		*Macrognathus pancalus* (Hamilton, 1822)*
144.		*Mastacembelus armatus* (Lacepède, 1800)*

*Collection under post-restoration inventorial survey;
**New records during post-restoration period.

(Continued)

TABLE 13.2 Inventory and Updated Checklist of Fish and Shellfish of Chilika Lagoon (1916–2012) *(cont.)*

Sl. No.	Family	Species
145.	**Scorpaenidae** (Scorpionfishes)	*Pterois radiata* (Cuvier, 1829)* [Previously recorded as *Pteropterus radiata* (Cuvier, 1829)]
146.	**Tetrarogidae** (Waspfishes)	*Tetraroge niger* (Cuvier, 1829)**
147.	**Platycephalidae** (Flatheads)	*Cociella crocodilus* (Cuvier, 1829)**
148.		*Kumococius rodericensis* (Cuvier, 1829)** [Previously recorded as *Suggrundus rodericensis* (Cuvier, 1829)]
149.		*Platycephalus indicus* (Linnaeus, 1758)*
150.	**Ambassidae** (Perchlets, glass fishes)	*Ambassis ambassis* (Lacepede, 1802)* [Previously recorded as *Ambassis commersoni* (Cuvier, 1828)]
151.		*Ambassis gymnocephalus* (Lacepède, 1802)*
152.		*Chanda nama* Hamilton, 1822*
153.		*Parambassis ranga* (Hamilton, 1822)* [Previously recorded as *Pseudoambassis ranga* (Hamilton, 1822)]
154.	**Latidae** (Lates perches)	*Lates calcarifer* (Bloch, 1790)*
155.	**Serranidae** (Groupers, Rock-cods)	*Epinephelus coioides* (Hamilton, 1822)**
156.		*Epinephelus lanceolatus* (Bloch, 1790) [Previously recorded as *Promicrops lanceolatus* (Bloch, 1790)]
157.		*Epinephelus malabaricus* (Bloch & Schneider, 1801)**
158.		*Epinephelus tauvina* (Forsskål, 1775)*
159.	**Sillaginidae** (Smelt Whitings)	*Sillaginopsis panijus* (Hamilton, 1822)
160.		*Sillago sihama* (Forsskål, 1775)*
161.		*Sillago vincenti* (McKay, 1880)**
162.	**Lactariidae** (False trevallies)	*Lactarius lactarius* (Bloch & Schneider, 1801)**
163.	**Rachycentridae** (Cobias)	*Rachycentron canadum* (Linnaeus, 1766)*

*Collection under post-restoration inventorial survey;
**New records during post-restoration period.

TABLE 13.2 Inventory and Updated Checklist of Fish and Shellfish of Chilika Lagoon (1916–2012) *(cont.)*

Sl. No.	Family	Species
164.	**Echeneidae** (Sharksuckers, Discfishes)	*Echeneis naucrates* (Linnaeus, 1758)*
165.	**Carangidae** (Jacks, Trevallies, Pompanos & Scads)	*Alectis indicus* (Rüppell, 1830)*
166.		*Alepes djedaba* (Forsskål, 1775)*
167.		*Atule mate* (Cuvier, 1833)
168.		*Carangoides gymnostethus* (Cuvier, 1833)
169.		*Carangoides praeustus* (Bennett, 1830)*
170.		*Caranx ignobilis* (Forsskål, 1775)
171.		*Caranx melampygus* (Cuvier, 1833)
172.		*Caranx sexfasciatus* (Quoy & Gaimard, 1825)*
173.		*Megalaspis cordyla* (Linnaeus, 1758)*
174.		*Parastromateus niger* (Bloch, 1795) [Previously recorded as *Apolectus niger* (Bloch, 1795)]
175.		*Scomberoides commersonnianus* (Lacepède, 1801)**
176.		*Scomberoides lysan* (Forsskål, 1775)
177.		*Scomberoides tala* (Cuvier, 1832)*
178.		*Scomberoides tol* (Cuvier, 1832)**
179.		*Selar boops* (Cuvier, 1833)**
180.		*Selar crumenophthalmus* (Bloch, 1793)**
181.		*Selaroides leptolepis* (Cuvier, 1833)*
182.		*Trachinotus blochii* (Lacepède, 1801)
183.		*Trachinotus mookalee* (Cuvier, 1832)**
184.	**Leiognathidae** (Pony fishes)	*Eubleekeria splendens* (Cuvier, 1829) [Previously recorded as *Leiognathus splendens* (Cuvier, 1829)]
185.		*Gazza minuta* (Bloch, 1795)*
186.		*Leiognathus daura* (Cuvier, 1829)
187.		*Leiognathus dussumieri* (Valenciennes, 1835)*
188.		*Leiognathus equulus* (Forsskål, 1775)*

*Collection under post-restoration inventorial survey;
**New records during post-restoration period.

(Continued)

TABLE 13.2 Inventory and Updated Checklist of Fish and Shellfish of Chilika Lagoon (1916–2012) *(cont.)*

Sl. No.	Family	Species
189.		*Leiognathus fasciatus* (Lacepède, 1803)**
190.		*Nuchequula blochii* (Valenciennes, 1835)* [Previously recorded as *Leiognathus blochii* (Valenciennes, 1835)]
191.		*Nuchequula gerreoides* (Bleeker, 1851**
192.		*Photopectoralis bindus* (Valenciennes, 1835)** [Previously recorded as *Leiognathus bindus* (Valenciennes, 1835)]
193.		*Secutor insidiator* (Bloch, 1787)*
194.		*Secutor ruconius* (Hamilton, 1822)**
195.	**Lutjanidae** (Snappers)	*Lutjanus argentimaculatus* (Forsskål, 1775)*
196.		*Lutjanus johnii* (Bloch, 1792)*
197.		*Lutjanus kasmira* (Forsskål, 1775)*
198.		*Lutjanus russellii* (Bleeker, 1849)*
199.	**Datnioididae** (Freshwater tripletails)	*Datnioides polota* (Hamilton, 1822)* [Previously recorded as *Datnioides quadrifasciatus* (Sevastianov, 1809)]
200.	**Gerreidae** (Silver biddies)	*Gerres erythrourus* (Bloch, 1791)** [Previously recorded as *Gerres abbreviatus* (Bleeker, 1850)]
201.		*Gerres filamentosus* Cuvier, 1829*
202.		*Gerres limbatus* Cuvier, 1830*
203.		*Gerres macracanthus* Bleeker, 1854
204.		*Gerres oyena* (Forsskål, 1775)*
205.		*Gerres phaiya* Iwatsuki & Hampstra, 2001 [Previously recorded as *Gerres poieti* Cuvier, 1830)]
206.		*Gerres setifer* (Hamilton, 1822)* [Previously recorded as *Gerreomorpha setifer* (Hamilton, 1822)]
207.	**Haemulidae** (Grunts & Rubberlips)	*Plectorhinchus nigrus* (Cuvier, 1830) [Previously recorded as *Plectorhinchus nigers* (Cuvier, 1830)]

*Collection under post-restoration inventorial survey;
**New records during post-restoration period.

TABLE 13.2 Inventory and Updated Checklist of Fish and Shellfish of Chilika Lagoon (1916–2012) *(cont.)*

Sl. No.	Family	Species
208.		*Pomadasys argenteus* (Forsskål, 1775)*
209.		*Pomadasys kaakan* (Cuvier, 1830)**
210.		*Pomadasys multimaculatus* (Playfair, 1867)**
211.	**Sparidae** (Seabreams)	*Acanthopagrus berda* (Forsskål, 1775)*
212.		*Acanthopagrus latus* (Houttuyn, 1782)
213.		*Argyrops spinifer* (Forsskål, 1775)
214.		*Crenidens crenidens* (Forsskål, 1775)*
215.		*Rhabdosargus sarba* (Forsskål, 1775)*
216.	**Nemipteridae** (Threadfin breams)	*Nemipterus japonicus* (Bloch, 1791)**
217.	**Sciaenidae** (Croakers)	*Daysciaena albida* (Cuvier, 1830)*
218.		*Dendrophysa russelii* (Cuvier, 1829)*
219.		*Johnius carruta* (Bloch, 1793)**
220.		*Johnius dussumieri* (Valenciennes, 1837)
221.		*Johnius belangerii* (Cuvier, 1830)*
222.		*Johnius coitor* (Hamilton, 1822)
223.		*Johnius macropterus* (Bleeker, 1853)
224.		*Nibea maculata* (Bloch & Schneider, 1801)**
225.		*Otolithes ruber* (Bloch & Schneider, 1801)**
226.		*Otolithoides biauritus* (Cantor, 1849)
227.		*Otolithoides pama* (Hamilton, 1822)* [Previously recorded as *Pama pama* (Hamilton, 1822)]
228.		*Paranibea semiluctuosa* (Cuvier, 1830)*
229.		*Protonibea diacanthus* (Lacepède, 1802)*
230.	**Polynemidae** (Threadfinfishes)	*Eleutheronema tetradactylum* (Shaw, 1804)*
231.		*Leptomelanosoma indicum* (Shaw, 1804)* [Previously recorded as *Polydactylus indicus* (Shaw, 1804) & *Polynemus indicus* (Shaw, 1804)]
232.		*Polydactylus plebeius* (Broussonet, 1782)**
233.		*Polydactylus sextarius* (Bloch & Schneider, 1801)*
234.	**Mullidae** (Goatfishes)	*Upeneus sulphureus* (Cuvier, 1829)**

Collection under post-restoration inventorial survey;
***New records during post-restoration period.**

(Continued)

TABLE 13.2 Inventory and Updated Checklist of Fish and Shellfish of
Chilika Lagoon (1916–2012) *(cont.)*

Sl. No.	Family	Species
235.	**Drepaneidae** (Sicklefishes)	*Drepane punctata* (Linnaeus, 1758)*
236.	**Monodactylidae** (Moonies)	*Monodactylus argenteus* (Linnaeus, 1758)*
237.	**Nandidae** (Leaf fishes)	*Nandus nandus* (Hamilton, 1822)*
238.	**Terapontidae** (Terapon perches)	*Pelates quadrilineatus* (Bloch, 1790)
239.		*Terapon jarbua* (Forsskål, 1775)*
240.		*Terapon puta* (Cuvier, 1829)*
241.		*Terapon theraps* (Cuvier, 1829)*
242.	**Cichlidae** (Cichlids)	*Etroplus suratensis* (Bloch, 1790)*
243.		*Oreochromis mossambicus* (Peters, 1852)**
244.	**Uranoscopidae** (Stargazers)	*Ichthyoscopus lebeck* (Bloch & Schneider, 1801)* [Previously recorded as *Ichthyoscopus inermis* (Cuvier, 1829)]
245.	**Blenniidae** (Blennies & allies)	*Omobranchus zebra* (Bleeker, 1868)
246.	**Eleotridae** (Gudgeons)	*Butis butis* (Hamilton, 1822)
247.		*Eleotris fusca* (Forster, 1801)
248.		*Eleotris melanosoma* (Bleeker, 1852)**
249.	**Gobiidae** (Gobies)	*Acentrogobius cyanomos* (Bleeker, 1849)*
250.		*Acentrogobius griseus* (Day, 1876)
251.		*Acentrogobius masoni* (Day, 1873)
252.		*Acentrogobius viridipunctatus* (Valenciennes, 1837)
253.		*Amoya madraspatensis* (Day, 1868) [Previously recorded as *Acentrogobius madraspatensis* (Day, 1868)]
254.		*Bathygobius fuscus* (Ruppell, 1830)
255.		*Bathygobius ostreicola* (Chaudhuri, 1916)
256.		*Brachygobius nunus* (Hamilton, 1822)
257.		*Drombus globiceps* (Hora, 1923)*
258.		*Eugnathogobius mas* (Hora, 1923)

*Collection under post-restoration inventorial survey;
**New records during post-restoration period.

TABLE 13.2 Inventory and Updated Checklist of Fish and Shellfish of Chilika Lagoon (1916–2012) *(cont.)*

Sl. No.	Family	Species
		[Previously recorded as *Glossogobius mas* Hora, 1923]
259.		*Glossogobius giuris* (Hamilton, 1822)*
260.		*Gobiopterus chuno* (Hamilton, 1822)
261.		*Oligolepis acutipennis* (Valenciennes, 1837)
262.		*Oligolepis cylindriceps* (Hora, 1923)*
263.		*Oxyurichthys microlepis* (Bleeker, 1849)*
264.		*Oxyurichthys tentacularis* (Vallenciennes, 1837)
265.		*Parapocryptes rictuosus* (Valenciennes, 1837)
266.		*Periophthalmus kalolo* (Lesson, 1831)* [Previously recorded as *Periophthalmus koelreuteri* (Pallas, 1770)]
267.		*Psammogobius biocellatus* (Valenciennes, 1837)* [Previously recorded as *Glossogobius biocellatus* (Vallenciennes, 1837)]
268.		*Pseudapocryptes elongatus* (Cuvier,1816) [Previously recorded as *Pseudapocryptes lanceolatus* (Bloch & Schneider, 1801)]
269.		*Pseudogobius javanicus* (Bleeker, 1856) [Previously recorded as *Stigmatgobius javanicus* (Bleeker, 1856)]
270.		*Stigmatogobius minima* (Hora, 1923)
271.		*Taenioides buchanani* (Day, 1873)
272.		*Trypauchen vagina* (Bloch & Schneider, 1801)*
273.		*Yongeichthys nebulosus* (Forsskal, 1775)**
274.	**Ephippidae** (Spadefishes)	*Ephippus orbis* (Bloch, 1787)**
275.		*Platax orbicularies* (Forsskål, 1775)**
276.	**Scatophagidae** (Scats)	*Scatophagus argus* (Linnaeus, 1766)*
277.	**Siganidae** (Spinsfoots, Rabbitfishes)	*Siganus canaliculatus* (Park, 1797)**
278.		*Siganus javus* (Linnaeus, 1766)*
279.		*Siganus vermiculatus* (Valenciennes, 1835)*
280.	**Acanthuridae** (Surgeon fishes)	*Acanthurus mata* (Cuvier, 1829)**

*Collection under post-restoration inventorial survey;
**New records during post-restoration period.

(Continued)

TABLE 13.2 Inventory and Updated Checklist of Fish and Shellfish of Chilika Lagoon (1916–2012) *(cont.)*

Sl. No.	Family	Species
281.		*Acanthurus triostegus* (Linnaeus, 1758)**
282.	**Sphyraenidae** (Barracudas)	*Sphyraena jello* (Cuvier, 1829)**
283.		*Sphyraena putnamae* (Jordan & Seale, 1905)**
284.	**Trichiuridae** (Hairtail fishes)	*Eupleurogrammus glossodon* (Bleeker, 1860)**
285.		*Trichiurus lepturus* (Linnaeus, 1758)**
286.		*Lepturacanthus savala* (Cuvier, 1829)**
287.	**Scombridae** (Mackerels, Seerfishes, Tunas, Albacores)	*Euthynnus affinis* (Cantor, 1849)**
288.		*Rastrelliger kanagurta* (Cuvier, 1816)**
289.		*Scomberomorus lineolatus* (Cuvier, 1829)*
290.	**Anabantidae** (Climbing perches)	*Anabas cobojius* (Hamilton, 1822)*
291.		*Anabas testudineus* (Bloch, 1792)*
292.	**Osphronemidae** (Gouramies)	*Trichogaster fasciata* (Bloch & Schneider, 1801)* [Previously recorded as *Colisa fasciata* (Bloch & Schneider, 1801)]
293.		*Trichogaster lalius* (Hamilton, 1822)* [Previously recorded as *Colisa lalia* (Hamilton, 1822)]
294.	**Channidae** (Snakeheads, Murrels)	*Channa gachua* (Hamilton, 1822)**
295.		*Channa marulius* (Hamilton, 1822)**
296.		*Channa punctata* (Bloch, 1793)*
297.		*Channa striata* (Bloch, 1793)*
298.	**Paralichthyidae** (Lefteye flounders)	*Pseudorhombus arsius* (Hamilton, 1822)*
299.		*Pseudorhombus micrognathus* (Norman, 1927)**
300.		*Pseudorhombus triocellatus* (Bloch & Schneider, 1801)**
301.	**Soleidae** (Soles)	*Brachirus orientalis* (Bloch & Schneider, 1801)* [Previously recorded as *Eyriglossa orientalis* (Bloch & Schneider, 1801)]

*Collection under post-restoration inventorial survey;
**New records during post-restoration period.

TABLE 13.2 Inventory and Updated Checklist of Fish and Shellfish of Chilika Lagoon (1916–2012) *(cont.)*

Sl. No.	Family	Species
302.		*Solea ovata* (Richardson, 1846)
303.	**Cynoglossidae** (Tongue soles)	*Cynoglossus lida* (Bleeker, 1851)**
304.		*Cynoglossus lingua* (Hamilton, 1822)*
305.		*Cynoglossus puncticeps* (Richardson, 1846)*
306.	**Triacanthidae** (Tripod fishes)	*Triacanthus biaculeatus* (Bloch, 1786)*
307.	**Balistidae** (Triggerfishes)	*Abalistes stellaris* (Bloch & Schneider, 1801)**
308.	**Tetraodontidae** (Puffers)	*Arothron reticularis* (Bloch & Schneider, 1801)
309.		*Arothron stellatus* (Bloch & Schneider, 1801)
310.		*Chelonodon patoca* (Hamilton, 1822)*
311.		*Lagocephalus lunaris* (Bloch & Schneider, 1801)
312.		*Takifugu oblongus* (Bloch, 1786)*
313.		*Tetraodon cutcutia* Hamilton, 1822*
314.		*Tetraodon fluviatilis* Hamilton, 1822* [Previously recorded as *Chelonodon fluviailis* (Hamilton, 1822)]
315.	**Diodontidae** (Porcupinefishes)	*Diodon hystrix* (Linnaeus, 1758)**
Shellfish		
Crabs		
316.	**Majidae** (Spider crabs)	*Doclea muricata* (Fabricius, 1787) [Previously reported as *Doclea hybrida* Edwards, 1834]
317.	**Calappidae** (Box crabs)	*Matuta planipes* (Fabricius, 1798)*
318.		*Ashtoret lunaris* (Forsskål, 1775)** [Previously reported as *Matuta lunaris* (Forsskål, 1775)]
319.	**Leucosiidae** (Nut crabs)	*Philyra malefactrix* (Kemp, 1915) [Previously reported as *Ebalia malefactrix* (Kemp, 1915)]
320.		*Philyra alcocki* (Kemp, 1915)*

Collection under post-restoration inventorial survey;
**New records during post-restoration period.*

(Continued)

TABLE 13.2 Inventory and Updated Checklist of Fish and Shellfish of Chilika Lagoon (1916–2012) *(cont.)*

Sl. No.	Family	Species
321.	**Hymenosomatidae** (False spider crabs)	*Elamina (Trigonoplax) cimex* (Kemp, 1915)
322.	**Ocypodidae** (Ghost crabs and Fiddler crabs)	*Ocypode ceratophthalma* (Pallas, 1772)
323.		*Ocypode macrocera* (Edwards, 1852)*
324.		*Ocypode platytarsis* (Edwards, 1852)
325.		*Uca (Austruca) annulipes* (Edwards, 1837) [Previously reported as *Uca annulipes* (Edwards, 1837)]
326.		*Dotilla pertinax* (Kemp, 1915)
327.		*Dotilla intermedia* (de Man, 1888)
328.		*Dotilla myctiroides* (Edwards, 1852)
329.		*Euplax leptophthalmus* (Edwards, 1852) [Previously reported as *Macrophalmus gastrodes* (Kemp, 1915)]
330.		*Camptandrium sexdentatum* (Stimpson, 1858)
331.		*Baruna socialis* (Stebbing, 1904) [Previously reported as *Leipocten sardidulum* (Kemp, 1915)]
332.	**Grapsidae** (Marsh crabs)	*Pachygrapsus propinquus* (de Man, 1908)
333.		*Varuna litterata* (Fabricius, 1798)*
334.		*Ptychognathus onyx* (Alcock,1900)
335.		*Sesarma plicatum* (Latreille, 1806)
336.		*Neosarmatium meinerti* (De Man, 1887) [Previously reported as *Sesarma tetragonum* (Miers, 1879)]
337.		*Nanosesarma batavicum* (Moreira, 1903) [Previously reported as *Sesarma batavicum* (Moreira, 1903)]
338.		*Sesarma quadrata* (Fabricius, 1798)**
339.		*Plagusia squamosa* (Herbst, 1790) [Previously reported as *Plagusia depressa tuberculata* (Lamarck, 1818)]
340.		*Metopograpsus messor* (Forskål, 1775)

*Collection under post-restoration inventorial survey;
**New records during post-restoration period.

TABLE 13.2 Inventory and Updated Checklist of Fish and Shellfish of Chilika Lagoon (1916–2012) *(cont.)*

Sl. No.	Family	Species
341.	**Gecarcinidae** (Land crabs)	*Cardisoma carnifex* (Herbst, 1796)
342.	**Xanthidae** (Pebble crabs or Rubble crabs)	*Benthopanope indica* (de Man, 1887) [Previously reported as *Heteropanope indica* (de Man, 1887)]
343.	**Portunidae** (Swimming crabs)	*Portunus (Portunus) pelagicus* (Linnaeus, 1758)* [Previously reported as *Portunus pelagicus* (Linnaeus, 1758)]
344.		*Portunus (Portunus) sanguinolentus* (Herbst, 1783)** [Previously reported as *Portunus sanguinolentus* (Herbst, 1783)]
345.		*Thalamita crenata* (Rüppell, 1830)*
346.		*Charybdis (Charybdis) feriata* (Linnaeus, 1758)** [Previously reported as *Charybdis cruciata* (Herbst, 1794)]
347.		*Charybdis (Charybdis) callianassa* (Herbst, 1789)** [Previously reported as *Charybdis callianassa* (Herbst, 1789)]
348.		*Scylla serrata* (Forskål, 1775)*
349.		*Scylla tranquebarica* (Fabricius, 1798)**
350.		*Podophthalmus vigil* (Fabricius, 1798)**
Shrimp and Prawns		
351.	**Penaeidae** (Penaeid shrimp)	*Penaeus monodon* (Fabricius, 1798)*
352.		*Penaeus semisulcatus* (De Haan, 1844)*
353.		*Fenneropenaeus indicus* (H. Milne Edwards, 1837)* [Previously reported as *Penaeus indicus* (Edwards, 1837)]
354.		*Melicertus canaliculatus* (Olivier, 1811)** [Previously reported as *Penaeus canaliculatus* (Olivier, 1811)]

*Collection under post-restoration inventorial survey;
**New records during post-restoration period.

(Continued)

TABLE 13.2 Inventory and Updated Checklist of Fish and Shellfish of
Chilika Lagoon (1916–2012) *(cont.)*

Sl. No.	Family	Species
355.		*Metapenaeus monoceros* (Fabricius, 1798)*
356.		*Metapenaeus affinis* (H. Milne Edwards, 1837)*
357.		*Metapenaeus dobsoni* (Miers, 1878)*
358.		*Metapenaeus ensis* (De Haan, 1844)**
359.	**Sergestidae** (Sergestid shrimps)	*Lucifer hanseni* (Nobili, 1905)
360.		*Philocheras hendersoni* (Kemp, 1915) [Previously reported as *Pontophilus hendersoni* (Kemp, 1915)]
361.	**Palaemonidae** (Palaemonid shrimps)	*Macrobrachium lamarrei lamarrei* (Edwards, 1837)*
362.		*Macrobrachium malcolmsonii malcolmsonii* (H. Milne Edwards, 1844)* [Previously reported as *Macrobrachium malcolmsoni* (Edwards, 1844)]
363.		*Macrobrachium rude* (Heller, 1862)*
364.		*Macrobrachium scabriculum* (Heller, 1862)
365.		*Macrobrachium rosenbergii* (De Man, 1879)**
366.		*Macrobrachium equidens* (Dana, 1852)**
367.		*Exopalaemon styliferus* (H. Milne Edwards, 1840)*
368.		*Phycomenes indicus* (Kemp, 1915) [Previously reported as *Periclimenes (Periclimenes) indicus* (Kemp, 1915)]
369.		*Cuapetes demani* (Kemp, 1915)* [Previously reported as *Periclimenes (Harpilius) demani* (Kemp, 1915)]
370.	**Alpheidae** (Snapping shrimps)	*Ogyrides striaticauda* (Kemp, 1915)
371.		*Athanas polymorphus* (Kemp, 1915)
372.		*Alpheus lobidens* (De Haan, 1849) [Previously reported as *Periclimenes (Harpilius) demani* (Kemp, 1915) & *Alpheus crassimanus* (Heller, 1865)]
373.		*Alpheus malabaricus* (Fabricius, 1775)

*Collection under post-restoration inventorial survey;
**New records during post-restoration period.*

TABLE 13.2 Inventory and Updated Checklist of Fish and Shellfish of Chilika Lagoon (1916–2012) *(cont.)*

Sl. No.	Family	Species
374.		*Alpheus paludicola* (Kemp, 1915)
375.	**Atyidae** (Basket shrimp)	*Caridina nilotica* (Roux, 1833)
376.		*Caridina propinqua* (De Man, 1908)*
377.	**Pasiphaeidae** (Glass shrimp)	*Leptochela (Leptochela) aculeocaudata* (Paul'son, 1875)
378.	**Callianassidae** (Ghost shrimp)	*Neocallichirus maxima* (A. Milne-Edwards, 1870)* [Previously reported as *Callianassa (Callichirus) maxima* (Milne-Edwards, 1870)]
379.	**Upogebidae** (Mud shrimps)	*Wolffogebia heterocheir* (Kemp, 1915)* [Previously reported as *Upogebia (Upogebia) heterocheir* (Kemp, 1915).]
Lobsters		
380.	**Palinuridae** (Spiny lobsters)	*Panulirus polyphagus* (Herbst, 1793)**
381.		*Panulirus ornatus* (Fabricius, 1798)**

Collection under post-restoration inventorial survey;
**New records during post-restoration period.

CONCLUSION

Chirocentridae and Cobitidae could not be found, though 14 families were represented the first time after restoration. Out of the 315 species presently known from this lake, 65 species of fish could not be collected during the post-restoration period, most of them small gobioid fishes (16 species); that may possibly be due to change of habitat. *Eugnathogobius mas* was never reported after it was described. *Pristis pectinata*, once a widely distributed sawfish, has been almost eliminated from large areas of its former range, and so, its absence from Chilika lagoon nowadays indicates its elimination from these environs. *Glyphis gangeticus* is presently known only from the Ganges river system, Hooghly river mouth, West Bengal, India. Records of this species from Chilika lagoon may probably be based on some other similar species, such as *Carcharhinus leucas* (Valenciennes). Absence of several commercial fishes also points to overexploitation causing depletion of their population from coastal waters and so not entering the lagoon. Among the 35 species of crab species reported to date, only 14 species were reported during the post-restoration period, and only 18 prawn species out of the total reported 29 were recorded during the post-restoration phase. In total, 315 species of fishes (24 orders, 87 families and 196 genera), 35 crab species (9 families and 24 genera), 29 prawn species (8 families and 18 genera) and 2 lobsters (1 family and 1 genera) were reported so far from the Chilika lagoon.

ACKNOWLEDGEMENTS

We are thankful to Dr K. Venkataraman, Director, Zoological Survey of India, Kolkata for providing the necessary facilities.

REFERENCES

Bandyopadhyay, S., Gopal, B., 1991. Ecosystem studies and management problems of a coastal lagoon, the lake Chilika. In: Gopal, B., Asthana, V. (Eds.), Aquatic Sciences in India. Indian Association for Limnology and Oceanography, India, pp. 117–172.

Bhatta, K.S., Pattnaik, A.K., Behera, B.P., 2001. Further contribution to the fish fauna of Chilika lagoon – A coastal wetland of Orissa. GIOBIOS 28 (2-3), 97–100.

Chandramohan, P., Pattnaik, A.K., Jena, B.K., 2002. Sediment dynamics at Chilika outer channel. Proc. International workshop on Sustainable Dev. of Chilika lagoon, Bhubaneswar, 12–14 December 1998, 22–30.

Chaudhuri, B.L., 1916a. Description of two new fishes from Chilika Lake. Rec. Indian Mus. 12 (3), 97–100.

Chaudhuri, B.L., 1916b. Fauna of Chilika Lake: Part I. Mem. Indian Mus 5 (4), 403–440.

Chaudhuri, B.L., 1916c. Fauna of Chilika Lake: Part II. Mem. Indian Mus. 5 (5), 441–458.

Ganapati, P.N., Ramasarma, D.V., 1958. Hydrography in relation to the production of plankton off Waltair coast. Andhra Univ. Mem.Oceanogr. 2, 168–192.

Ghosh, A.K., 2002. Integrated management strategy for Chilika. Proc. International Workshop on Sustainable Development of Chilika lagoon 12–14 December 1998. Chilika Development Authority, Bhubaneswar, pp. 274–280.

Ghosh, A.K., Pattnaik, A.K., 2005. Chilika lagoon. Experience and lessons learned brief, prepared for the GEF Lake Basin Management initiative. See www.worldlakes.org.

Ghosh, A.K., Pattnaik, A.K., Ballatore, T., 2006. Chilika lagoon: Restoring ecological balance and livelihoods through re-salinization. Lakes and Reservoir: Res. Manage 11, 239–255.

Hora, S.L., 1923. Fauna of Chilika Lake: Fish Part V. Mem. Indian Mus 5 (11), 737–770.

Jhingran, V.G., 1963. Report on the fisheries of the Chilika Lake (1957-60). Bull. Central Inland Fish. Res. Inst., Barrackpore, India. 1: 1–113.

Jhingran, V.G., Natarajan, A.V., 1966. Final Report on the Fisheries of the Chilika Lake (1957–1965). Central Inland Fisheries Research Institute. Bull No 8, 1–12.

Jhingran, V.G., Natarajan, A.V., 1969. Study of the fishery and fish populations of the Chilika lake during the period 1957-65. J. Inland Fish. Soc. India 1, 47–126.

Jones, S., Sujansinghani, K.H., 1954. Fish and fisheries of Chilika lake with statistics of fish catches for the years 1948-1950. Indian J. Fish (1-2), 256–344.

Kaumans, F.P., 1941. Goboid fishes of India. Mem. Indian Mus. 13 (3), 205–313.

Kemp, S., 1915. Crustacean : Decapoda Fauna of Chilika Lake. Mem. Indian Mus. 5, 199–325.

La Fond, E.C., 1958. On the circulation of surface layers off the east coast of India. Andhra Univ. Mem. Oceanogr. 2, 1–11.

Maya Deb, 1995. Crustacea: Brachyra. In: Ghosh, A.K. (Ed.), Wetland Ecosystem Series: Fauna of Chilika Lake. Zoological Survey of India, Kolkata, pp. 345–366.

Menon, M.A.S., 1961. On a collection of fish from lake Chilika. Orissa. Rec. Indian Mus. 59 (1&2), 41–69.

Misra, K.S., 1969, Pisces: Fauna of India and adjacent countries, 2nd ed. IManager Publication, New Delhi, p. 276.

Misra, K.S., 1976a, Pisces: Fauna of India and adjacent countries, 2nd ed. IIManager Publication, New Delhi, p. 438.

Misra, K.S., 1976b, Pisces: Fauna of India and adjacent countries, 2nd ed. IIIManager Publication, New Delhi, p. 367.

Mohanty, S.K., 1973. Further additions to the fish fauna of the Chilika lake. J Bombay Nat. Hist. Soc. 72 (3), 863–866.

Mohanty, N.D., Behera, G., 2002. Studies on shifting of inlet, variations of water level and its effect on salinity concentrations of Chilika lagoon. Proc. International Workshop on Sustainable Development of Chilika lagoon, Bhubaneswar, 12-14 Dec. 1948, 48–59.

Mohanty, S.K., Mohanty, R.K., Mohanty, S., 2004a. Evaluation of Ghery prawn culture in Chilika lagoon with special reference to social issues. J. Indian Soc. Coast Agric. Res. 22, 293–297.

Mohanty, S.K., Bhatta, K.S., Badapanda, H.S., 2004b. On shrimp ghery of Chilika lake. Fishing Chimes 24 (8), 41–44.

Mohanty, R.K., Mohanty, S.K., Mohapatra, A., Bhatta, K.S., Pattanaik, A.K., 2006. Postecorestoration impact on fish and shellfish biodiversity in Chilika lake. Indian J. Fish. 53 (4), 397–407.

Mohanty, R.K., Mohapatra, A., Mohanty, S.K., 2009. Assessment of the impacts of a new artificial lake mouth on the hydrology and fisheries of Chilika lake, India. Lakes and Reservoirs: Res. Manage 14, 231–245.

Mohapatra, A., Mohanty, R.K., Mohanty, S.K., Bhatta, K.S., Das, N.R., 2007a. Fisheries enhancement and biodiversity assessment of fish prawn and mud crab in Chilika lagoon through hydrological intervention. Wetlands Ecol. Manage 15, 229–252.

Mohapatra, A., Dey, S.K., Mohanty, S.K., Bhatta, K.S., Mohanty, R.K., 2007b. Brachyuran crab fauna of Chilika lagoon, Orissa, after opening of the new mouth. Fishing chimes 27 (2), 33–34.

Nayak, B.U., Ghosh, L.K., Roy, S.K., Kankara, R.S., 2002. A study of hydrodynamics and salinity in the Chilika lagoon. Proc. International Workshop on Sustainable Development of Chilika lagoon 12-14 December 1998, Chilika Development Authority, Bhubaneswar, 31–47.

ORSAC, 1988. Interim report on study of Chilika lake resources and environment. Orissa Remote Sensing Application Centre, Bhubaneswar.

Panda, P.C., Patnaik, S.N., 2002. An enumeration of the flowering plants of Chilika lagoon and its immediate neighborhood. Proc. International workshop on Sustainable Dev. of Chilika lagoon, Bhubaneswar, 12-14 December, 1998, 122–141.

Pattnaik, A.K., July 2000. Conservation of Chilika – An overview. Chilika Newsl. 1, 3–5.

Pattnaik, A.K., May 2001. Hydrological Intervention for Restoration of Chilika lagoon. Chilika Newsl. 1, 3–5.

Pattnaik, A.K., 2002. Chilika Lake – An overview. Proc. International workshop on Sustainable Dev. of Chilika lagoon, Bhubaneswar, 12-14 December, 1998. pp.12–21.

Rajan, S., 1971. Environmental studies of Chilika lake. 2. Benthic animal communities. Indian J. Fish. 12, 492–499.

Rajan, S., Pattanaik, S., Bose, N.C., 1968. New records of fishes from Chilika lake. J. Zool. Soc. India 20 (1), 80–83.

Rama Rao, K.V., 1995. Pisces. In: Ghosh, A.K. (Ed.), Wetland Ecosystem Series: Fauna of Chilika Lake. Zoological Survey of India, Kolkata, pp. 483–506.

Reddy, K.N., 1995. Crustacea: Decapoda (Prawns and Shrimps). In: Ghosh, A.K. (Ed.), Wetland Ecosystem Series 1. Fauna of Chilika Lake (ZSI), pp. 367–389.

Sahoo, D., Panada, S., Guru, B.C., Bhatta, K.S., 2008. A new record of Indo-Pacific crab *Charybdis ferriata* (Linn. Brachyura: portunidae) from Chilika lagoon, Orissa, India. The Ecoscan 2 (2), 177–179.

Talwar, P.K., Jhingran, A.G., 1991. Inland fishes of India and adjacent countries I & II Oxford & IBH Publ Cp, New Delhi, p. 1077.

World Bank, 2005. Scenario assessment of the provision environmental flows to Lake Chilika from Naraj Barrage, Orissa, India. Report from the Environmental Flows window of the Bank, Netherlands Water Partnership Programme (World Bank) to the Government of Orissa, India: p. 40.

Chapter 14

New Records of Reef Fishes from the Andaman and Nicobar Islands

K. Devi, K. Sadhukhan, J.S. Yogesh Kumar and S. Kumar Shah

Andaman and Nicobar Regional Centre, Zoological Survey of India, Port Blair, Andaman and Nicobar Islands, India

INTRODUCTION

Studies on ichthyofauna of the Andaman and Nicobar Islands (ANI) are known from the earlier works of Talwar (1990), Devi (1991), and subsequent reports by several researchers (Rao *et al.*, 1992a,b, 1993a,b, 1994, 2000; Krishnan and Mishra, 1992a,b; Devi *et al.*, 1993; Dhandapani and Mishra, 1993; Rajan *et al.* 1993; Rao, 2003). Rao (2009) reported a checklist of fishes with 1369 species under 586 genera belonging to 175 families recorded from ANI, based on field survey and available literature. Dam Roy *et al.* (2009) added 11 more species to the ichthyofauna of these islands. Recently, Ramakrishna *et al.* (2010) reported 83 new records with an updated checklist containing 1463 species from these Islands.

During the recent survey on taxonomic studies of fishes carried out using underwater photography in Mahatma Gandhi Marine National Park (MGMNP), nine new records of fishes from ANI were made. In this chapter a detailed description of morphological features of the newly recorded fish species with their habitat and distribution are presented.

METHODS

Undersea surveys were carried out by SCUBA diving to study fish from five sites in Mahatma Gandhi Marine National Park (MGMNP) at a maximum depth of 25 m from April 2011 to June 2012. Underwater photography was taken by Sony Cyber Shot (DSC-T900) camera with underwater housing facility. The fishes were identified based on the standard systematic taxonomic keys (Weber and De Beaufort, 1913–1953).

Marine Faunal Diversity in India. DOI: 10.1016/B978-0-12-801948-1.00014-8

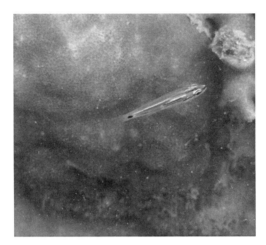

FIGURE 14.1   *Apogon neotes.*

RESULTS AND DISCUSSION

Systematic Account (Order Perciformes)

Family Apogonidae

1   ***Apogon neotes*** (Allen, Kuiter & Randall, 1994) (Figure 14.1)

Common Name   Larval cardinalfish

Material Observed   Tarmugli Island; depth 15–20 m.

Diagnostic Characters   The specimen was approximately 3–4 cm. Body oblong and compressed; head large, snout short; eye big in size; caudal emarginated. Body silvery-blue, translucent; black stripe intermittently bordered with silver stripes runs from above eye to back spot on mid-tail base.

Distribution   Indonesia, Palau, Philippines, Papua New Guinea, and Solomon Islands.

Habitat & Ecology   Found around coral reefs at varied depths in range 15–30 m in clear waters of the outer reef and lagoons. It is typically seen in areas of Gorgonian fans and soft corals. Solitary, in pairs or loose clusters and using crevices in the reef for shelter. Feeds on zooplankton and small, bottom-dwelling crustaceans.

Remarks   This species exhibits distinct pairing during courtship and spawning. The males are known for the unusual behavior of incubating egg masses inside their mouth.

FIGURE 14.2 *Monotaxis heterodon* (juvenile).

Family Lethrinidae

2 *Monotaxis heterodon* (Bleeker, 1854) (Figure 14.2)

Common Name Redfin Bream

Material Observed Grub Island and Tarmugli Island; depth 5–7 m.

Diagnostic Characters Attains 8–10 cm. Body oblong and compressed later-ally; eyes large, constituting less than one third of head length. The operculum, cheek and inner surface of pectoral fin base are scaled. Body color is brownish on the dorsal surface gradating to pale underside with 3–4 narrow white bars; black spot in axil of pectoral fin; yellow tail lobes.

Distribution Seychelles to Marshall Islands and New Caledonia

Habitat & Ecology Tropical carnivorous fish. Found around coral reefs, la-goons and outer reef slopes up to 100 m.

Family Pomacentridae

3 *Pomacentrus pavo* (Bloch, 1787) (Figure 14.3)

Common Name Blue Damsel/ Sapphire Damsel.

Material Observed Chester Island, Grub Island, Jolly Buoy and Tarmugli Islands; depth 8–10 m.

Diagnostic Characters Small fish; attains 6–9 cm. Body oblong and slightly compressed laterally; head profile convex. Body light blue to light green with

FIGURE 14.3 *Pomacentrus pavo.*

vertical blue to grey streaks on each scale; scattered blue spots on head; dark 'ear' spot present; spinous dorsal pale with blue margin, remainder of fin pale to dusky.

Distribution East Africa to Micronesia and French Polynesia. Taiwan to E. Australia.

Habitat & Ecology Good aquarium fish seen in loose groups on sandy areas down to 15 m around coral reefs. Feeds on plankton, filamentous algae and small invertebrates.

4   *Pomacentrus Polyspinus* (Allen,1991) (Figure 14.4)

Common Name   Thai Damsel.

Material Observed   Red Skin Island; depth 3–5 m.

Diagnostic Characters Attains 7–8 cm. Body compressed. Color greenish-grey with dark scales; small blue lines on snout and forehead; a blue-edged dark spot on rear dorsal fin; dark 'ear' spot present; anal fin margin blue; caudal and lobes of adjacent fins translucent.

Habitat & Ecology Found in shallow to moderate depth in coral reef areas. They usually swim around the reef in pairs. Feeds on algae and small benthic animals.

5   *Pomacentrus proteus* (Allen, 1991) (Figure 14.5)

Common Name   Colombo Damsel.

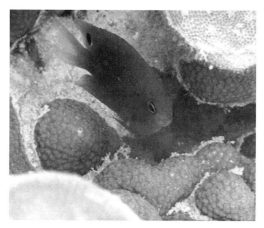

FIGURE 14.4 *Pomacentrus polyspinus.*

FIGURE 14.5 *Pomacentrus proteus.*

Material Observed Tarmugli Island; depth 5–7 m.

Diagnostic Characters Attains 6–8 cm. Body yellowish brown with blue streaks on scales; a blue line from snout to origin of spinous dorsal; a blue-edged dark spot on rear dorsal fin; In juvenile, dorsal half blue gradating to yellow below; light blue lines marking on upper head; rear dorsal fin with blue edged dark spot.

Distribution Sri Lanka.

FIGURE 14.6   *Pteragogus cryptus.*

Habitat & Ecology They are encountered in coral reef flats at depth of 5–10 m. The juveniles are seen to stay inside the shelter offered by branching corals. Adults are usually found on silty shoreline on mixed coral rubble. Feeds on plankton, filamentous algae and small invertebrates.

Family: Labridae

6   *Pteragogus cryptus* (Randall, 1981) (Figure 14.6)

Common Name Cryptic Wrasse.

Material Observed Tarmugli Island; depth 20–25 m.

Diagnostic Characters Attains 10–15 cm. Head pointed; caudal fin round. Body mottled shades of brown-red with scattered dark brown spots; white line passes from snout above eye to caudal; an oval dark spot, edged in pale on gill cover.

Distribution Red Sea, E. Africa to Samoa, Philippines and Micronesia to Australia.

Habitat & Ecology Found solitary beyond 60 m; usually swim around the reefs, sea grass beds and algal flats.

Family Gobiidae

7   *Eviota sigillata* (Jewett and Lachner, 1983) (Figure 14.7)

Common Name Seven-figure Pygmy Goby.

FIGURE 14.7 *Eviota sigillata.* (Please see color plate at the back of the book.)

Material Observed Tarmugli Island; depth 8–9 m.

Diagnostic Characters Very small fish; attains 2.5–3 cm. Head compressed; ventral separate; caudal fin rounded. Body translucent with white and brown dash marks alternate on spinal column; pectoral fin base white.

Distribution Seychelles to Papua New Guinea, Micronesia, Solomon Islands and Australia.

Habitat & Ecology Occur in sheltered areas of the coral reef at depths of 2–20 m. Mostly resting on live corals and on coral rock. Very small fish; because of its size and transparency is often difficult to notice.

8 *Koumansetta hectori* (Smith, 1957) (Figure 14.8)
Common Name Yellowstripe Goby.

Material Observed Tarmugli, Grub and Red Skin Islands; depth 10–12 m.

Diagnostic Characters Body bluish-black with 4 yellow longitudinal stripes from snout to back; 3 dark spots, one on first dorsal fin, yellow edged second spot on second dorsal fin, and third spot on upper tail base. Caudal fin pale.

Distribution Red Sea, Indonesia, Palau and S.W. Japan.

Habitat & Ecology Found in sand bottom or rubble close to the reef edge at depth of 3–30 m. Observed solitary or in pairs.

FIGURE 14.8 *Koumansetta hectori.* (Please see color plate at the back of the book.)

9 *Valenciennea limicola* Hoese & Larson, 1994 (Figure 14.9)

Common Name Mud Goby.

Material Observed Tarmugli Island; depth 5–8 m.

Diagnostic Characters Attains 9–10 cm. Snout elongate; ventral separate; caudal fin rounded. Body grey with pair of bright orange lines: one behind eye to caudal and other from snout to caudal; bright light-blue stripe below eye.

FIGURE 14.9 *Valenciennea limicola.* (Please see color plate at the back of the book.)

Distribution Gulf of Thailand and Indonesia to Fiji.

Habitat & Ecology Solitary or in pairs. Found around rubble on coral reefs or mud bottom of coastal reefs in 5–30 m.

ACKNOWLEDGEMENTS

The authors wish to thank the Director, Zoological survey of India, Kolkata and Officer-in-Charge, Andaman and Nicobar Regional Centre, Port Blair for providing necessary facilities and encouragement to carry out this study. Our Special thanks to Chief Wildlife Warden, Department of Environment and Forests, Andaman and Nicobar Administration Islands, Deputy Conservator of Forests (Wildlife), and Forest Range Officer, MGMNP, Wandoor for their necessary permission and cooperation to carry out this work.

REFERENCES

Allen, G., Steene, R., Humann. P., and Deloach. N., 2003. Reef fish Identification Tropical Pacific. Florida: New World Publications. p. 457.

Dam Roy, S., Krishnan, P., Grinson George, Srivastava, R.C., Kaliyamoorthy, M., Raghuraman, R., and Sreeraj, C.R., 2009. Reef Biodiversity of North Bay. Central Agricultural Research Institute, Andaman and Nicobar Islands, Port Blair. p. 132.

Devi, K., 1991. Supplementary list to the fishes of Bay Islands. J. Andaman Sci. Assoc. 7, 102–103.

Devi, K., Rao, D.V., Rajan, P.T., 1993. Additions to the gobioid fauna of Andaman Islands. Environ. Ecol. 11, 812–815.

Dhandapani, P., Mishra, S.S., 1993. New records of marine fishes from Great Nicobar. J. Andaman Sci. Assoc. 9 (1&2), 58–62.

Krishnan, S., Mishra, S.S., 1992a. New records of fishes from Andaman and Nicobar Islands. J. Andaman Sci. Assoc. 8 (1), 82–84.

Krishnan, S., Mishra, S.S., 1992b. Further new records of fishes from Andaman Islands. J. Andaman Sci. Assoc. 8 (2), 175–177.

Rajan, P.T., Rao, D.V., Devi, K., Day, S., 1993. New records of rare fishes from Andaman Islands. J. Andaman Sci. Assoc. 9 (1&2), 103–106.

Ramakrishna Titus Immanuel Sreeraj, C.R., Raghunathan, C., Raghuraman, R., Rajan, P.T., Yogesh Kumar, J.S., 2010. An account of additions to the Ichthyofauna of Andaman and Nicobar Islands. Rec. Zool. Surv. India. Occ. Paper 326, 1–40.

Rao, D.V., 2003. Guide to Reef Fishes of Andaman and Nicobar Islands. Zoological Survey of India, Kolkata, p. 555.

Rao, D.V., 2009. Checklist of fishes of Andaman and Nicobar Islands. Bay Bengal. Enviorn. Ecol. 27 (19), 334–353.

Rao, D.V., Devi, K., Rajan, P.T., 1992a. Some new records of Wrasses (Family: Libridae) from Andaman and Nicobar Islands. J. Andaman Sci. Assoc. 8 (1), 43–46.

Rao, D.V., Devi, K., Rajan, P.T., 1992b. New records of groupers (Family: Serranidae) and cardinalfishes (Family: Apogonidae) from Andaman and Nicobar Islands. J. Andaman Sci. Assoc 8 (1), 47–52.

Rao, D.V., Devi, K., Rajan, P.T., 1993a. Additions to the fish fauna of Andaman and Nicobar Islands. Environ. Ecol. 11, 882–887.

Rao, D.V., Devi, K., Rajan, P.T., 1993b. Further new records of fishes from Bay Islands. J. Andaman Sci. Assoc. 9 (1&2), 50–57.

Rao, D.V., Devi, K., Rajan, P.T., 1994. Additions to the Ichthyofauna of Bay Islands. J. Andaman Sci. Assoc. 10 (1&2), 28–31.

Rao, D.V., Devi, K., Rajan, P.T., 2000. An account of Ichthyofauna of Andaman and Nicobar Islands. Bay of Bengal. Rec. Zool. Surv. India. Occ. Paper 178, 434.

Smith, M.M., Heemstra, P.C., 1986. Smith's Sea Fishes. Springer-Verlag, New York.

Talwar, P.K., 1990. Fishes of Andaman and Nicobar Islands: A synoptic Analysis. J. Andaman Sci. Assoc. 6 (2), 71–102.

Talwar, P.K., Chatterjee, T.K., Devroy, M.K., 1982. Oxyurichthys dasi a new gobioid (Pisces: Goboiidae) from the Andaman Islands. Rec. Zool. Surv. India 79 (3-4), 483–487.

Weber, M., De Beaufort, L.F., 1913. Fishes of the Indo-Australian Archipelago. E. J. Brill, Leiden vol 2, p. 404.

Weber, M., De Beaufort, L.F., 1916. Fishes of the Indo-Australian Archipelago. E. J. Brill, Leiden vol 3, p. 455.

Weber, M., De Beaufort, L.F., 1922. Fishes of the Indo-Australian Archipelago. E. J. Brill, Leiden vol 4, p. 410.

Weber, M., De Beaufort, L.F., 1929. Fishes of the Indo-Australian Archipelago. E. J. Brill, Leiden vol 5, p. 458.

Weber, M., De Beaufort, L.F., 1931. Fishes of the Indo-Australian Archipelago. E. J. Brill, Leiden vol 6, p. 338.

Weber, M., De Beaufort, L.F., 1936. Fishes of the Indo-Australian Archipelago. E. J. Brill, Leiden vol 7, p. 607.

Weber, M., De Beaufort, L.F., 1940. Fishes of the Indo-Australian Archipelago. E. J. Brill, Leiden vol 8, p. 508.

Weber, M., De Beaufort, L.F., 1951. Fishes of the Indo-Australian Archipelago. E. J. Brill, Leiden vol 9, p. 484.

Weber, M., De Beaufort, L.F., 1953. Fishes of the Indo-Australian Archipelago. E. J. Brill, Leiden vol 10, p. 423.

Chapter 15

Ichthyofauna of Digha Coast, India

P. Yennawar,[*][†] A. Mohapatra,[*] D. Ray[*] and P. Tudu[*]
*Marine Aquarium & Regional Centre, Zoological Survey of India, Digha, West Bengal, India;
†Present address: Freshwater Biology Regional Centre, Zoological Survey of India,
Hyderabad, India

INTRODUCTION

Digha (Lat. 21° 36′ N; Long. 87° 30′ E), located near the Gangetic mouth, on the northern east coast of India, provides an unusual habitat of shallow muddy beach with high sedimentation load coming from the Ganga riverine system. It is also a major marine fish landing station as well as a famous beach destination of the east cost of India. Some efforts had already been made to make an inventory of the ichthyological faunal diversity of the Digha coast (Manna and Goswami, 1985; Goswami, 1992; Talwar *et al.,* 1992; Chatterjee *et al.* 2000). Recently, the coast was again monitored and the record of the species diversity there enriched (Yennawar and Tudu, 2010; Yennawar *et al.*, 2011; Yennawar *et al.*, 2012). During routine efforts to maintain diversity of the public aquarium in the Marine Aquarium and Regional Centre of the Zoological Survey of India, Digha, around 110 species were added to the existing species list of the area. The previous compilations of ichthyofauna of the region contributed 212 species from 145 genera and 88 families (Chatterjee *et al.*, 2000) and 238 species from 72 families (Manna and Goswami, 1985; Goswami, 1992). This chapter provides an updated list of ichthyofaunal diversity of the Digha coast (Table 15.1). Digha, being a famous tourist destination among northeastern states of India, the area is consistently receiving increasing pressure of tourism and related activities. The present study helps in conservation measures and in improving awareness of the fauna among tourists visiting Digha.

METHODS

Fish were collected from the fish-landing centre or from beach trawling sites on the Digha coast, West Bengal, India. All the freshly collected specimens were photographed after collection, and the preliminary observation of morphometric characters were carried out. After identification, specimens were preserved in

Marine Faunal Diversity in India. DOI: 10.1016/B978-0-12-801948-1.00015-X

TABLE 15.1 Checklist of Fish Species of Digha Coast, India

	Family	Species
Class Chondrichthyes		
Order Orectolobiformes		
1.	**Hemiscylliidae** (Bamboo sharks)	*Chiloscyllium griseum* (Müller & Henle, 1838)
2.		*Chiloscyllium indicum* (Gmelin, 1789)
3.	**Stegostomatidae** (Zebra shark)	*Stegostoma fasciatum* (Hermann, 1783)
4.	**Rhinocontidae** (Whale shark)	*Rhincodon typus* (Smith, 1828)
Order Carcharhiniformes		
5.	**Proscyllidae** (Finback shark)	*Eridancis radcliffei* (Smith, 1913)
6.	**Carcharhinidae** (Requiem sharks)	*Carcharhinus dussumieiri* (Valenciences, 1839)
7.		*Carcharhinus limbatus* (Müller & Henle, 1839)
8.		*Glyphis gangeticus* (Müller & Henle 1839)
9.		*Rhizoprionodon acutus* (Ruppell, 1837)
10.		*Scoliodon laticaudatus* (Müller & Henle, 1838)
11.		*Galeocerdo cuvier* (Peron & le Sueur, 1822)
Order Rajiformes		
12.	**Pristidae** (Sawfish)	*Anoxipristis cuspidatus* (Latham, 1794)
13.		*Pristis microdon* (Latham, 1794)
14.	**Narcinicae** (Electric rays)	*Narcine brunnea* (Annandale, 1909)
15.		*Naeke diptrigia* (Bloch & Schneider, 1801)
16.	**Rhinobaticae** (Guitar fish)	*Rhina ancylostoma* (Bloch & Schneider, 1801)
17.		*Rhinobatos annandalei* (Norman, 1926)
18.		*Rhina grannulatus* (Cauvier)
19.		*Rhina lionotus* (Norman)
20.		*Rhina obtusus* (Müller & Henle)
21.		*Rhynchobatus djeddensis* (Forsskål, 1775)
22.	**Dasyatoidae** (Sting rays)	*Dasyatis zugei* (Müller & Henle, 1841)
23.		*Himantura imbricata* (Bloch & Schneider, 1801)
24.		*Himantura uarnak* (Gmelin, 1789)
25.		*Himantura bleekeri* (Blyth, 1861)
26.		*Himantura gerrardi* (Gray, 1851)
27.	**Gymnuridae** (Butterfly rays)	*Aetoplaea tenaclata* (Valenciennes)

TABLE 15.1 Checklist of Fish Species of Digha Coast, India *(cont.)*

	Family	Species
28.		*Gymnura japonica* (Temminck & Schlegel, 1850)
29.		*Gymnura poecilura* (Shaw, 1804)
30.	**Myliobatidae** (Eagle rays)	*Aetobatus narinari* (Euphrasen, 1790)
31.		*Aetomylaeus nichofill* (Bloch & Schneider, 1801)
32.	**Sphyrnidae** (Hammerhead sharks)	*Sphyrna lewini* (Griffith & Smith, 1834)
33.		*Eusphyra blochii* (Cuvier, 1816)
Class Actinopterigii **Order Elopiformes**		
34.	**Megalopidae** (Tarpons)	*Megalops cypricides* (Broussonet, 1782)
Order Anguilliformes		
35.	**Anguillidae**	*Anguilla bengalensis bengalensis* (Gray, 1831)
36.		*Anguilla bicolor bicolor* (McClelland, 1844)
37.	**Moringuidae** (Worm eel)	*Moringua raitaborua* (Hamilton, 1822)
38.	**Muraenidae** (Moray eels)	*Echidna nebulosa* (Ahl, 1789)
39.		*Echidna zebra* (Shaw, 1797)
40.		*Gymnothorax melegris* (Shaw, 1795)
41.		*Strophidon sathete* (Hamilton, 1822)
42.		*Gymnothorax tile* (Hamilton, 1822)
43.		*Gymnothorax favegiatus* (Bloch & Schneider, 1801)
44.		*Gymnothorax javanicus* (Bleeker, 1859)
45.		*Gymnothorax reticularis* Bloch, 1795
46.		*Sideria picta* (Ahl, 1789)
47.		*Thyrsoidea macrura* (Bleeker, 1854)
48.	**Ophichthidae** (Snake eels)	*Limnostoma orientalis* (McClelland)
49.		*Psiodonophis boro* (Hamilton, 1822)
50.		*Psiodonophis cancrivorus* (Richardson, 1848)
51.	**Muraenesocidae** (Pike congers)	*Corgresox talabon* (Cuvier, 1829)
52.		*Corgresox talabooides* (Bleeker, 1853)
53.		*Muraenesox bagio* (Hamilton, 1822)
54.		*Muraenesox cinereus* (Forsskål, 1775)

(Continued)

TABLE 15.1 Checklist of Fish Species of Digha Coast, India *(cont.)*

	Family	Species
Order Clupeiformes		
55.	**Clupeidae** (Shads & Herrings)	*Herklotsichtys quadrimaculatus* (Ruppell, 1837)
56.		*Sardinella brachysoma* (Bleeker, 1852)
57.		*Anodontostoma chacunda* (Hamilton & Buchanan, 1822)
58.		*Escumosa thoracata* (Valenciennes, 1847)
59.		*Hilsa kelee* (Cuvier, 1829)
60.		*Nernatalosa nasus* (Bloch, 1795)
61.		*Sardinella fimbriata* (Valenciennes, 1847)
62.		*Sardinella longiceps* Valenciennes, 1847
63.		*Sardinella gibbosa* (Bleeker, 1849)
64.		*Tenualasa ilisha* (Hamilton & Buchanan, 1822)
65.		*Tenualosa toli* (Valenciennes, 1847)
66.		*Ophisthopterus tardoore* (Cuvier, 1829)
67.		*Raconda russeliana* (Gray, 1831)
68.	**Pristigasteridae** (Ilishas, Pellonas)	*Ilisha kampeni* (Weber & de Beaufort, 1913)
69.		*Ilisha megaloptera* (Swainson, 1839)
70.		*Ilisha melastoma* (Bloch & Schneider, 1801)
71.		*Pellona ditchela* (Valenciennes, 1847)
72.	**Engraulidae** (Anchovies)	*Coilia ramcarati* (Hamilton, 1822)
73.		*Coilia dussumieri* (Valenciennes, 1848)
74.		*Coilia neglecta* (Whitehead, 1968)
75.		*Coilia reynaldi* (Valenciennes, 1848)
76.		*Setipinna phasa* (Hamilton, 1822)
77.		*Setipinna taty* (Valenciennes, 1848)
78.		*Setipinna tenuifilis* (Valenciennes, 1848)
79.		*Thryssa dussumieri* (Valenciennes, 1848)
80.		*Thryssa hamiltonii* (Gray, 1835)
81.		*Thryssa malabarica* (Bloch, 1795)
82.		*Thryssa purava* (Hamilton, 1822)
83.		*Stolephorus commersoni* (Lacepede 1803)
84.		*Stolephorus indicus* (van Hasselt, 1823)
85.		*Stolephorus heterolobus* (Ruppell)
86.	**Chirocentridae** (Wolf Herrings)	*Chirocentrus nudus* (Swainson, 1839)

TABLE 15.1 Checklist of Fish Species of Digha Coast, India *(cont.)*

	Family	Species
Order Siluriformes		
87.	**Ariidae** (Sea catfishes)	*Arius jella* (Day, 1877)
88.		*Arius arius* (Hamilton, 1822)
89.		*Arius sagor* (Hamilton, 1822)
90.		*Arius tenuispinis* (Day, 1877)
91.		*Arius sona* (Hamilton, 1822)
92.		*Arius thalassinus* (Ruppell)
93.		*Arius maculatus* (Thunberg, 1792)
94.	**Plosidae** (Eeltail catfish)	*Plotossus canius* (Hamilton-Buchanan, 1822)
95.		*Plotossus lineatus* (Thunberg, 1787)
Order Aulopiformes		
96.	**Synodontidae** (Lizardfish)	*Saurida tumbil* (Bloch, 1745)
97.		*Saurida undosquamis* (Richardson, 1848)
98.		*Trachinocephalus myops* (Forster, 1801)
99.	**Harpadontidae** (Bombay Duck)	*Harpadon nehereus* (Hamilton, 1822)
Order Gadiformes		
100.	**Bregmacerotidae** (Codlets, codlings)	*Bregmaceros mcclellandi* (Thompson, 1840)
Order Batrachoidiformes		
101.	**Batrachoididae** (Toadfish)	*Batrichthys grunniens* (Linnaeus, 1758)
Order Bericiformes		
102.	**Holocentridae** (Soldierfish, Squirrelfish)	*Myripristis murdjan* (Forsskål, 1757)
103.		*Myripristis botche* (Cuvier, 1829)
104.		*Sargocentron praslin* (Lacepede, 1801)
Order Cyprinodontiformes		
105.	**Exocoetidae** (Flying fish)	*Exocoetus volitans* (Linnaeus, 1758)
106.		*Cypselurus poecilopterus* (Valenciennes, 1846)
107.	**Hemiramphidae** (Halfbeaks)	*Hemiramphus far* (Forsskål, 1775)
108.		*Hemiramphus lutkei* (Valenciennes, 1846)
109.		*Hyporamphus limbatus* (Valenciennes, 1846)
110.		*Rhynchorhamphus georgii* (Valenciennes, 1847)
111.	**Beloniidae** (Needlefish)	*Strongylura leiura* (Bleeker, 1850)
112.		*Strongylura strongylura* (Van Hasselt, 1823)
113.		*Tylosurus crocodiles* (Peron & Le Sueur, 1821)

(Continued)

TABLE 15.1 Checklist of Fish Species of Digha Coast, India *(cont.)*

	Family	Species
Order Syngnathiformes		
114.	**Syngnathidea** (Seahorses, Pipe fish)	*Hippichthys specifer* (Ruppell, 1838)
115.		*Ichthyocampus carce* (Hamilton, 1822)
116.	**Fistularidae** (Cornetfish)	*Fistularia petimba* (Lacepede, 1803)
Order Lophiformes		
117.	**Antennariidae** (Frogfish)	*Antennarius hispidus* (Bloch, 1801)
118.	**Ogcocephalidae** (Batfish)	*Halieutaea stellata* (Vahl, 1797)
Order Scorpaeniformes		
119.	**Scorpaenidae** (Scorpion fish)	*Pterois russelli* (Bennett, 1831)
120.		*Pterois volitans* (Linnaeus, 1758)
121.		*Brachypterois serrulata* (Richardson, 1846)
122.	**Synanceiidae** (Flatheads)	*Trachicephalus uranoscopus* (Bloch & Schneider, 1801)
123.	**Platycephalidae** (Spiny flatheads)	*Grammoplites scaber* (Linnaeus, 1758)
124.		*Platycephalus indicus* (Linnaeus, 1758)
125.		*Sorsogona tuberculata* (Cuvier, 1829)
Order Perciformes		
126.	**Ambassidae** (Perchlets)	*Ambassis nalua* (Hamilton, 1822)
127.	**Centropomidae** (Sea Perches)	*Lates calcariifer* (Bloch, 1790)
128.	**Serraniidae** (Grouper)	*Epinephelus tauvina* (Forsskål, 1775)
129.		*Epinephelus coioides* (Hamilton,1822)
130.		*Epinephelus lanceolatus* (Bloch,1790)
131.		*Epinephelus latifasciatus* (Temminck & Schlegel, 1842)
132.		*Epinephelus malabaricus* (Bloch & Schneider, 1801)
133.	**Terapontidae** (Terapon perches, terapons)	*Terapon jarbua* (Forsskål, 1775)
134.		*Terapon puta* (Cuvier, 1829)
135.		*Terapon threaps* (Cuvier, 1829)
136.	**Priacanthidae** (Bigeyes)	*Priacanthus tayenus* (Richardson, 1846)
137.	**Apogonidae** (Cardinal fish)	*Apogon lateralis* (Valenciennes, 1832)

TABLE 15.1 Checklist of Fish Species of Digha Coast, India *(cont.)*

	Family	Species
138.	**Sillaginidae** (Sillagos/ Whitings)	*Sillaginopsis panijus* (Hamilton-Buchanan, 1822)
139.		*Sillago sihama* (Forsskål, 1775)
140.	**Lactariidae** (False trevallies)	*Lactarius lactarius* (Bloch & Schneider, 1801)
141.	**Echeneidae** (Remoras, Sharksucker)	*Echeneis naucrates* (Linnaeus, 1758)
142.		*Remora remora* (Linnaeus, 1758)
143.	**Carangidae** (Jacks, Trevallies, Dart, Pompano)	*Alectis indicus* (Ruppell, 1830)
144.		*Alectes cilliaris* (Bloch, 1788)
145.		*Atropus atropos* (Bloch & Schneider, 1801)
146.		*Alepes djedaba* (Forsskål, 1775)
147.		*Carangoides ferdau* (Forsskål, 1775)
148.		*Caranx sexfasciatus* (Quoy & Gaimard, 1824)
149.		*Carangoides malabaricus* (Bloch & Schneider, 1801)
150.		*Decapterus russelli* (Ruppell, 1830)
151.		*Elagatis bipinnaulata* (Quoy & Gaimard, 1824)
152.		*Megalaspis cordyla* (Linnaeus, 1758)
153.		*Naucrates doctor* (Linnaeus, 1758)
154.		*Parastromateus niger* (Bloch, 1795)
155.		*Scomberoides commersonianus* (Lacepede, 1802)
156.		*Selar crumenophthalmus* (Bloch, 1793)
157.		*Selaroides leptolepis* (Cuvier, 1833)
158.		*Trachinotus blochi* (Lacepede, 1801)
159.		*Caranx caragus* (Bloch, 1793)
160.	**Coryphaenidae** (Dolphinfish)	*Coryphaena hippurus* Linnaeus, 1758
161.	**Rachycentridae** (Cobias)	*Rachycentron canadum* (Linnaeus, 1766)
162.	**Parastromateidae** (Black pomfrets)	*Parastromatus niger* (Bloch, 1795)
163.	**Menidae** (Moonfish)	*Mene maculate* (Bloch & Schneider, 1801)
164.	**Leiognathidae** (Ponyfish, Slipmouths, Toothponies)	*Photopectoralis bindus* (Valenciennes, 1835)

(Continued)

TABLE 15.1 Checklist of Fish Species of Digha Coast, India *(cont.)*

	Family	Species
165.		*Nauchequula blochii* (Valenciennes, 1835)
166.		*Leiognathus daura* (Cuvier, 1829)
167.		*Leiognathus equulus* (Forsskål, 1775)
168.		*Leiognathus fasciatus* (Lacepede, 1803)
169.		*Leiognathus splendens* (Cuvier, 1829)
170.		*Secutor insidiator* (Bloch, 1797)
171.		*Secutor ruconius* (Hamilton-Buchanan, 1822)
172.	**Lutjanidae** (Snappers)	*Lutjanus argentimaculatus* (Forsskål, 1775)
173.		*Lutjanus johnii* (Bloch, 1792)
174.		*Lutjanus lutjanus* Bloch, 1790
175.		*Lutjanus malabaricus* Bloch & Schneider, 1801
176.		*Lutjanus bengalensis* (Bloch, 1790)
177.		*Lutjanus russelli* (Bleeker, 1849)
178.	**Lobotidae** (Tripletails)	*Lobotes surinamensis* (Bloch, 1790)
179.		*Datnioides quadrifasciatus* (Sebastianov)
180.	**Gerreidae** (Silverbiddies, Mojarras)	*Gerres abbreviatus* (Bleeker, 1850)
181.		*Gerres filamentosus* (Cuvier, 1829)
182.		*Gerres oyena* (Cuvier, 1829)
183.		*Gerres poieti* (Cuvier, 1829)
184.		*Gerres setifer* (Hamilton, 1822)
185.	**Haemulidae** (Sweetlips & Grunts)	*Pomadasys maculatum* (Bloch, 1797)
186.	**Sparidae** (Sea breams)	*Acanthopagrus berda* (Forsskål, 1775)
187.		*Acanthopagrus latus* (Houttuyn, 1782)
188.		*Argyrops spinifer* (Forsskål, 1775)
189.		*Argyrops bleekeri* (Oshima, 1927)
190.		*Rhabdosargus sarba* (Forsskål, 1775)
191.	**Nemipteridae** (Threadfin breams)	*Nemipterus bipunctatus* (Valenciennes, 1830)
192.		*Nemipterus japonicas* (Bloch, 1791)
193.		*Nemipterus nematophorus* (Bleeker, 1853)
194.		*Nemipterus peronii* (Valenciennes, 1830)
195.		*Nemipterus randalli* (Russell, 1986)
196.		*Parascolopsis aspinosa* (Rao & Rao, 1981)

TABLE 15.1 Checklist of Fish Species of Digha Coast, India *(cont.)*

	Family	Species
197.		*Sclopsis vosmeri* (Bloch, 1792)
198.	**Sciaenidae** (Crokers, Drums)	*Johnius belangerii* (Cuvier, 1830)
199.		*Johnius carouna* (Cuvier, 1830)
200.		*Johnius sina* (Cuvier, 1830)
201.		*Johnius carutta* (Bloch,1793)
202.		*Johnius vogleri* (Bleeker, 1853)
203.		*Macrospinosa cuja* (Hamilton, 1822)
204.		*Nibea maculata* (Schneider, 1801)
205.		*Otolithoides pama* (Hamilton, 1822)
206.		*Otolithoides ruber* (Bloch, 1801)
207.		*Panna microdon* (Bleeker, 1849)
208.		*Pennahia macrophthalmus* (Bleaker, 1850)
209.		*Protonibea diacanthus* (Lacepéde,1802)
210.		*Pterotolithus maculatus* (Kuhl & Van Hassle, 1830)
211.	**Mullidae** (Goatfish, Red mullet)	*Parupeneus heptacanthus* (Lacepede, 1802)
212.		*Parupeneus indicus* (Shaw, 1803)
213.		*Upeneus guttatus* (Day, 1868)
214.		*Upeneus luzonius* (Jordon & Seale, 1907)
215.		*Upeneus moluccensis* (Bleeker, 1855)
216.		*Upeneus sulphureus* Cuvier, 1829
217.		*Upeneus sundaicus* (Bleeker,1855)
218.		*Upeneus taeniopterus* (Cuvier, 1829)
219.		*Upeneus tragula*, (Richardson, 1846)
220.		*Upeneus vittatus* (Forsskål, 1775)
221.	**Monodactylidae** (Silver batfish)	*Monodactylus argenteus* (Linnaeus)
222.	**Taxotidae** (Archer fish)	*Toxotes chatareus* (Hamilton, 1822)
223.		*Toxotes jaculator* (Pallas, 1767)
224.	**Ephippidae** (Spadefish)	*Ephippus orbis* (Bloch, 1787)
225.	**Drepaenidae** (Sicklefish)	*Drepane longimana* (Bloch & Schneider, 1801)
226.		*Drepane punctata* (Linnaeus, 1758)

(Continued)

TABLE 15.1 Checklist of Fish Species of Digha Coast, India *(cont.)*

	Family	Species
227.	**Ephippidae** (Batfish)	*Platax pinnatus* (Linaeus, 1758)
228.		*Platax tiera* (Forsskål, 1775)
229.	**Scatophagidae** (Spotted scat)	*Scatophagus argus* (Bloch, 1758)
230.	**Mugilidae** (Mullets)	*Liza macrolepis* (Smith, 1849)
231.		*Liza subviridis* (Valenciennes, 1836)
232.		*Liza parsia* (Hamilton, 1822)
233.		*Liza tade* (Forsskål, 1775)
234.		*Mugil cephalus* (Linnaeus, 1755)
235.		*Rhinomugil corsula* (Hamilton, 1822)
236.		*Valamugil cunnesius* (Valenciennes, 1836)
237.	**Sphyraenidae** (Barracudas)	*Sphyraena forsteri* (Cuvier, 1829)
238.		*Sphyraena jello* (Cuvier, 1829)
239.		*Sphyraena obtusata* (Cuvier, 1829)
240.	**Polynemidae** (Threadfins)	*Eleutheronema tetradactylum* (Shaw, 1804)
241.		*Polynemus indicus* (Shaw, 1804)
242.		*Polynemus plebeius* (Broussonet, 1782)
243.		*Polynemus paradiseus* (Linnaeus, 1758)
244.		*Polynemus sextarius* (Bloch & Schneider, 1801)
245.		*Polynemus longipectoralis* (Weber & de Beaufort, 1922)
246.	**Congrogadidae** (Eel-like blennies)	*Halidesmus thomaseni* (Nielson, 1961)
247.	**Uranoscopidae** (Stargazer)	*Ichthyscopus lebeck* (Bloch & Schneider, 1801)
248.		*Uranoscopus cognatus* (Cantor, 1849)
249.	**Callynymidae** (Dragnets)	*Callionymus belcheri* (Richardson, 1844)
250.		*Callionymus sagitta* (Pallas, 1770)
251.		*Eleutherochir opercularis* (Valenciennes, 1837)
252.	**Eleotridae** (Sleepers)	*Eleotris fusca* (Forster, 1801)
253.		*Eleotris melanosoma* (Bleeker, 1852)
254.		*Butis melanostigma* (Bleeker, 1849)
255.	**Gobiidae** (Gobies)	*Apocryptes bato* (Hamilton, 1822)
256.		*Apocryptodon madurensis* (Bleeker, 1849)
257.		*Parachaeturichthys polynema* (Bleeker, 1853)
258.		*Bathygobius fuscus* (Ruppell, 1830)
259.		*Boleopthalus boddarti* (Pallas, 1770)

TABLE 15.1 Checklist of Fish Species of Digha Coast, India *(cont.)*

	Family	Species
260.		*Glossogobius giuris* (Hamilton, 1822)
261.		*Periopthalmus pearsei* (Eggert, 1935)
262.		*Pseudapocrypts lanceolatus* (Bloch & Schneider, 1801)
263.		*Scartelaos histophorus* (Valenciennes, 1837)
264.	**Gobioididae**	*Odontamblyopus rubicundus*(Hamilton, 1822)
265.	**Trypauchinidae**	*Trypauchen vagina* (Bloch & Schneider, 1801)
266.	**Kurtidae** (Humphead)	*Kurtus indicus* (Bloch, 1786)
267.	**Siganidae** (Spinefoots, Rabbitfish)	*Siganus canaliculatus* (Park, 1797)
268.		*Siganus javus* (Linnaeus, 1766)
269.		*Siganus vermiculatus* (Valenciennes, 1835)
270.	**Trichiuridae** (Ribbonfish)	*Eupleurogrammus muticus* (Gray, 1831)
271.		*Eupleurogrammus glossodon* (Bleeker, 1860)
272.		*Eupleurogrammus pantuluvi* (Gupta)
273.		*Lepturacanthus savale* (Cuvier, 1829)
274.		*Trichiurus gangeticus* (Gupta, 1966)
275.		*Trichiurus lepturus* (Linnaeus, 1758)
276.	**Scombridae** (Mackerels)	*Rastrelliger kanagurta* (Cuvier, 1817)
277.		*Scomberomorus commerson* (Lacepède, 1801)
278.		*Scomberomorus guttatus* (Bloch & Schneider, 1801)
279.	**Istiophoridae** (Billfish)	*Istiophurus platypterus* (Shaw, 1792)
280.	**Stromateidae** (Pomfrets)	*Pampus argenteus* (Euphrasen, 1788)
281.		*Pampus chinensis* (Euphrasen, 1788)
282.	**Acanthuridae** (Surgeonfish, Tang, Unicorn-fish)	*Acanthurus lineatus* (Linnaeus, 1758)
283.		*Acanthurus mata* (Cuvier, 1829)
284.		*Acanthurus tristis* (Tickell, 1888)
285.	**Pomacanthidae** (Angelfish)	*Pomacanthus annularis* (Bloch, 1787)
286.		*Pomacanthus imperator* (Bloch, 1787)
287.		*Pomacnthus semicirculatus* (Cuvier, 1831)
288.	**Pomacentridae** (Damselfish, Sergeantfish)	*Abudefduf vaigiensis* (Quoy & Gaimard, 1825)

(Continued)

TABLE 15.1 Checklist of Fish Species of Digha Coast, India *(cont.)*

	Family	Species
289.	**Chaetodontidae** (Butterflyfish, Bannerfish)	*Haeniochus acuminatus* (Linnaeus, 1758)
290.	**Scaridae** (Parrotfish)	*Scarus ghobban* (Forsskål, 1775)
291.	**Ariommatidae** (Driftfish)	*Ariomma indica* (Day, 1870)
292.	**Acropomatidae** (Lanternbellies)	*Acropoma argentistigma* (Okamoto & Ida, 2002)
Order Puronectiformes		
293.	**Paralichthyidae** (Large tooth flounders)	*Pseudorhombus arsius* (Hamilton, 1822)
294.		*Pseudorhombus javanicus* (Bleeker, 1853)
295.	**Cynoglossidae** (Tongue fish)	*Cynoglossus cynoglossus* (Hamilton-Buchanan, 1822)
296.		*Cynoglossus lingua* (Hamilton-Buchanan, 1822)
297.		*Cynoglossus macrostomus* (Norman, 1928)
298.		*Cynoglossus puncticeps* (Richardson, 1846)
299.		*Cynoglossus semifasciatus* (Day, 1877)
300.		*Paraplagusia bilineata* (Bloch, 1784)
301.	**Citharidae**	*Brachypleura novae-zeelandiaae* (Gunther, 1862)
302.	**Soleidae** (Soles)	*Euryglossa orientalis* (Bloch & Schneider, 1801)
303.		*Solea elongata* (Day, 1877)
304.		*Synaptura commersoniana* (Lacepéde, 1802)
305.		*Zebrias altipinnis* (Alcock, 1890)
306.		*Zebrias quqgga* (Kaup, 1858)
307.	**Psettoodidae** (Indian Halibut)	*Psettodes erumei* (Bloch & Schneider, 1801)
Order Tetraodontiformes		
308.	**Triacanthidae** (Triplespines, Tripodfish)	*Triacanthus biaculeatus* (Bloch, 1786)
309.	**Tetraodontidae** (Puffers)	*Arothron stellatus* (Blach & Schneider, 1801)
310.		*Arthrodon immaculatus* (Bloch & Schneider, 1801)
311.		*Arthrodon nigropuctatus* (Bloch & Schneider, 1801)
312.		*Chelonodon patoca* (Hamilton, 1822)
313.		*Chelonodon fluviatilis* (Hamilton, 1822)
314.		*Lagocephalus lunaris* (Bloch & Schneider, 1801)
315.		*Lagocephalus sceleratus* (Gemlin, 1789)

TABLE 15.1 Checklist of Fish Species of Digha Coast, India *(cont.)*

	Family	Species
316.		*Lagocephalus inermis* (Temminck & Schlegel, 1847)
317.		*Takifugu oblongus* (Bloch, 1786)
318.	**Balistidae** (Triggerfish)	*Abalistes stellatus* (Lacepède, 1798)
319.	**Ostraciidae** (Boxfish)	*Ostraceon cubicus* (Linnaeus, 1758)
320.	**Monacanthidae** (Filefish, Leatherjackets)	*Alutera monoceros* (Linnaeus,1758)
321.	**Diodontidae** (Porcupinefish)	*Diodon hystrix* (Linnaeus, 1758)
322.	**Molidae** (Ocean Sunfish)	*Mola mola* (Linnaeus, 1758)

10 percent formaldehyde and kept in the museum of the Marine Aquarium and Regional Centre of the Zoological Survey of India, Digha. The present study was conducted at five different locations: Talsari, Udaipore, Digha, Mohana and Shankarpur (Figure 15.1).

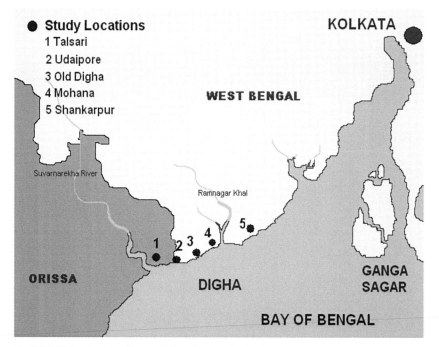

FIGURE 15.1   Map showing study locations on the Digha Coast.

ACKNOWLEDGEMENTS

We are thankful to Dr K. Venkataraman, Director, Zoological Survey of India, Kolkata for providing facilities and guidance throughout the study. DR is thankful to the Zoological Survey of India for award of a Senior Research Fellowship.

REFERENCES

Chatterjee, T.K., Ramakrishna, Talukdar, S., Mukherjee, A.K., 2000. Fish and fisheries of Digha coast of West Bengal. Rec. Zool. Surv. India. Occ. Paper 188 (1), 1–74.

Goswami, B.C.B., 1992. Marine fauna of Digha Coast of West Bengal, India. J. Mar. Biol. Ass. India 34 (1&2), 115–137.

Manna, B., Goswami, B.C.B., 1985. A check list of marine & estuarine fishes of Digha, West Bengal, India. Mahasagar 18 (4), 489–499.

Talwar, P.K., Mukherjee, P., Saha, D., Paul, S.N., Kar, S., 1992. Marine and estuarine fishes. Fauna of West Bengal, State Fauna Series, 3(Part 2): 243–364.

Yennawar, P., Tudu, P., 2010. New record of occurrence of Indian Yello Boxfish: *Ostracion cubicus* (Linnaeus, 1758) from Digha, Northern East Coast of India. Rec. Zool. Sur. India 110 (1), 115–118.

Yennawar, P., Tudu, P., Mohapatra, A., 2011. New record of two species of eels of the genus *Gymnothorax* (Muraenidae) in Digha coast of India. J. Bombay Nat. Hist. Soc. 108 (3), 232–233.

Yennawar, P., Ray, D., Mohapatra, A., 2012. First record of *Acropoma argentistigma* from Indian waters. Marine Biodiv. Rec. 5, e65, 1–3.

Chapter 16

Seasonal Abundance of Sea Snakes on the Chennai Coast, Southern India

C. Venkatraman,* P. Padmanaban* and C. Sivaperuman†
*Zoological Survey of India, Marine Biology Regional Centre, Chennai, Tamil Nadu, India;
†Andaman and Nicobar Regional Centre, Zoological Survey of India, Port Blair, Andaman
and Nicobar Islands, India

INTRODUCTION

Sea snakes are among the most unusually adapted serpents, spending most of their active lives at sea. They are commonly found in tropical and subtropical coastal waters of the Indian and Pacific Oceans (Heatwole, 1999). They are cold blooded, highly venomous and successful marine reptiles inhabiting the warm tropical waters of the world and not found in the Atlantic Ocean, Red Sea, or Mediterranean Sea. They comprise about 86 percent of living marine reptile species. Sea snakes are adapted to live in the sea by having a flattened body and oar-shaped tail.

Sea snakes are one of the most successful groups of reptiles among marine animals. Of the fifteen living families of snakes, four are marine. Sea snakes belong to the family Hydrophiidae, comprising 12 genera. The Hydrophids are the most widely distributed group of sea snakes, with 52 species. They occupy a variety of habitats such as mangroves, swamps, estuaries and lower reaches of rivers (Heatwole, 1999). Smith (1943) described 29 species of sea snakes in India, including the coast of Pakistan, Bangladesh, Sri Lanka and Myanmar. Ahmed (1978) also described 29 species from the Indian Ocean, of which 19 species are found in the collections of the Zoological Survey of India, Kolkata. Murthy (1992) and Whitaker (2004) listed 20 species of sea snakes from Indian coasts. Lobo *et al.* (2004) studied the morphometric relationships of marine snakes along the Goa coast. Significant differences in the proportion of the fish families represented in the diet with the size or sex of *Lapemis curtus* along the Goa coast were studied by Lobo *et al.* (2005), and their role in the trophic structure of coastal communities was described by Voris (1972).

The distribution patterns of 12 valid species that occur in the Coramantal coast region were reported by Karthikeyan and Balasubramanian (2007), and

Marine Faunal Diversity in India. DOI: 10.1016/B978-0-12-801948-1.00016-1
249

the distribution of eight species of sea snake along the Arabian Sea, Bay of Bengal, and Andaman and Nicobar Islands was recorded by Kannan and Rajagopalan (2008). Feeding behavior and parturition of the female Annulated sea snake *Hydrophis cyanocinctus* was observed in captivity by Karthikeyan *et al.* (2008). The diversity, biology and ecology of 10 species of sea snake that occur on the Parangipettai coast, southeast India were described by Damotharan *et al.* (2010). Eleven species of sea snake were caught by prawn trawl in the Gulf of Mannar Biosphere Reserve (Lobo, 2006). Sea snakes of the Chennai coast have been reported by Aiyar (1907), MacKenzie (1820), Wall (1918), Murthy (1977a,b, 1992), Kalaiarasan and Kanagasabai (1994) and Venkatraman *et al.* (2007).

Indian sea snakes have not been studied sufficiently, and there is no systematic study on sea snakes of India. Knowledge on taxonomy and biology of sea snakes is fragmentary, especially as regards seasonality. Sea snakes are commonly encountered, for example as by-catch in various fishing activities, yet there is little published information. Hence, the present study was intended to collect additional information on the incidental catches of sea snakes in fishing gear along the Chennai coast.

METHODS

Study Area

The study was conducted on the Chennai coast from January 2009 to December 2013 (Figure 16.1). The Chennai coast lies at 13° 06' N and 80° 18' E and is the location of more than 60 coastal villages involved in active fishing. The coast extends from Pulicat Lake to Kalpakkam and stretches for about 120 km. Kasimedu is the largest fish-landing centre on the coast. Chennai harbor is a protected area with placid waters. Effluents from the ships and docks pollute the seawater in and around the harbor area.

Procedure

Standard prawn trawl fishing nets modified with added sinkers/bobbins were used for the present survey. Trawl tracks were tentatively planned to make a complete representation of the study area. Each trawl tow was for a maximum duration of 1 h. The trawl tracks were saved in a GPS (GARMIN Oregon 550), and the distances of each trawl were then calculated using Google Earth Ver. 6.1.

Information was collected from local fisherfolk on the frequency of finding sea snakes trapped in their nets. The fish-landing centres Kasimedu, Nochikuppam and Kovalam were frequently visited during the early and late hours to examine the species caught in the nets. The species brought alive to the Marine Aquarium of Marine Biology Regional Centre, Zoological Survey of India, Chennai by local fisherfolk were also included for this study. Periodical visits

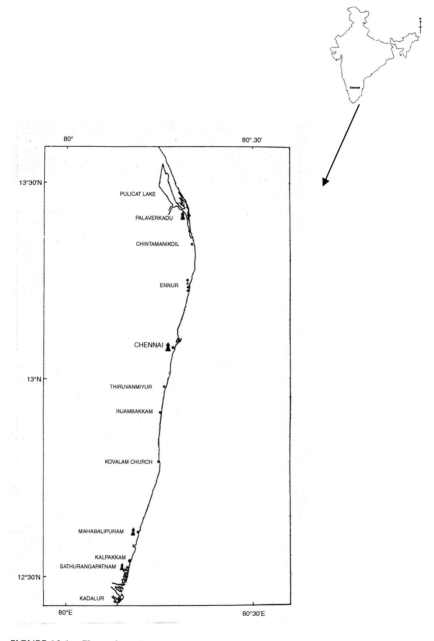

FIGURE 16.1 Chennai coast.

were undertaken all along the coast in order to collect dead specimens washed ashore. Photographs were also shown to fisherfolk, and enquiries made on their availability.

Morphometry

Morphometric measurements were made on *Enhydrina schistosa* (31), *Hydrophis cyanocinctus* (15), *Hydrophis spiralis* (14) and *Pelamis platurus* (7) that were captured in trawl nets. Snout to vent length (SVL) and total length (TL) were measured to the nearest 0.5 cm with a steel Freeman tape, and weight (Wt) of all snakes was measured to the nearest gram using a Pesola scale following the methods of Lobo *et al.* (2004).

Analysis

The morphometric variables were first subject to log normal transformation, and the relationships between SVL and Wt, and between TL and Wt, were examined for all the species. The statistical analyses were done using SPSS 16.0.

RESULTS AND DISCUSSION

During the present study, out of 12 species reported from previous studies, 10 species of sea snake belonging to five genera were recorded (Figures 16.2 and 16.3). The occurrences of sea snakes along the Chennai coast as reported by various authors from 1975 are given in Table 16.1. Fewer species were recorded in 1975, 1990 and 2005. Ten species were recorded in 1977 and also in the present study. *Hydrophis ornatus* was recorded in the present study for the first time on the Chennai coast. *Kerilia jerdoni* and *Thalassophina viperina* have not been recorded since 1977.

Out of 209 dead specimens collected, 70 were *Enhydrina schistosa* and 62 were *H. spiralis*. *Enhydrina schistosa* seems to be the most common species, and *Pelamis platurus* and *Lapemis curtus* are rare. Another two species, *H. cyanocinctus* and *H. gracilis,* were represented by 39 and 11 specimens, respectively. A single specimen of *Pelamis platurus* was recorded during January, and three specimens of *Lapemis curtus* were recorded in the period from April to June. The maximum number (68) was recorded during June, followed by July. There were two peaks in abundance: one during June to August and another during October to November. Similarly the minimum number of species was recorded during September, followed by December and February (Table 16.2).

The highest number (45) of sea snakes, belonging to nine species, was recorded during 2011, and the lowest number (38), belonging to 8 species, was recorded during 2009 (Table 16.3). There was a strong correlation ($p < 0.01$) between snout–vent length (SVL) and weight (Wt), and between total length

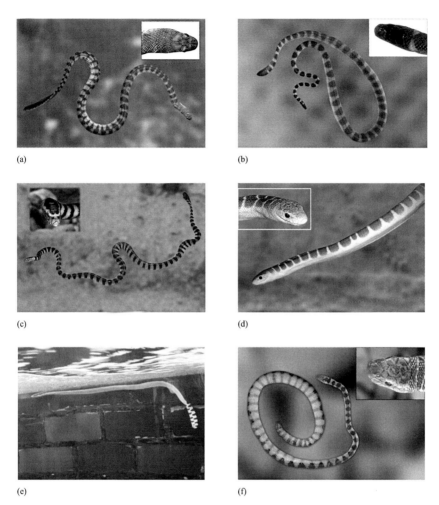

FIGURE 16.2   Sea snakes of Chennai coast. (a) *Enhydrina schistosa*, Hook-nosed sea snake; (b) *Hydrophis gracilis,* Small-headed sea snake; (c) *Hydrophis caerulescens*, Many-toothed sea snake; (d) *Hydrophis* ornatus, Ornate sea snake; (e) *Pelamis platurus*, Yellow-belly sea snake; (f) *Hydrophis cyanocinctus*, Annulated sea snake.

(TL) and weight, in the cases of *Enhydrirna schistosa* and *Hydrophis spiralis* (Table 16.4). That no significant correlation was found for some of the remaining species could be due to small sample size.

The frequency, abundance and species composition of sea snakes varied with the seasons. A similar observation was made by Heithaus (2001). The distribution and seasonality of sea snakes is influenced by parameters such as salinity, temperature and depth. Some species of marine snakes are nocturnal and others are diurnal. Other cycles may be of more importance than day–night.

(a)

(b)

(c)

FIGURE 16.3 Sea snakes of Chennai coast. (a) *Hydrophis spiralis*, Yellow sea snake; (b) *Hydrophis fasciatus*, Banded sea snake; (c) *Cerberus rynchops*, Dog-faced water snake.

TABLE 16.1 Sea Snakes Reported from the Chennai Coast by Various Authors

		Year				
Sl. No.	Species	1975	1977	1990	2005	Present study
1.	Hydrophis gracilis	-	+	+	+	+
2.	Enhydrina schistosa	-	+	+	+	+
3.	Hydrophis spiralis	-	+	+	+	+
4.	Hydrophis cyanocinctus	-	+	+	+	+
5.	Hydrophis caerulescens	-	+	+	-	+
6.	Hydrophis ornatus	-	-	-	-	+
7.	Hydrophis fasciatus	+	+	+	-	+
8.	Lapemis curtus	-	+	-	-	+
9.	Kerilia jerdoni	-	+	-	-	-
10.	Pelamis platurus	-	+	-	-	+
11.	Thalassophina viperina	-	+	-	-	-
12.	Cerberus rynchops	-	-	-	-	+
	Total	**1**	**10**	**6**	**4**	**10**

In some cases, a combination of day–night and tidal cycle may influence the distribution and availability of sea snakes. A number of species of *Hydrophis* are found only in deep waters whereas some others prefer to live in shallow waters, as reported by Das (2003).

Our data confirm the broad relationship of body size to weight observed in many other snake species (Das, 1991; Kaufman and Gibbons, 1975; Guyer and Donnelly, 1990; Lobo et al., 2004). Since TL, SVL and Wt are strongly influenced by growth rate, it would be very difficult to draw a conclusion on their sexual dimorphism without long term studies, since it is known that growth at maturity stabilizes on these characters (Madsen and Shine, 2000). Growth at maturity is also influenced by annual variation in the habitat and availability of prey (Madsen and Shine, 2000; Lobo et al., 2004). There is a need for the development of a well-designed awareness programme to educate fishermen on the potential dangers of sea snakes, and the setting up of proper treatment facilities in the event of casualties, which at present has not been given due attention (Lobo, 2006)

TABLE 16.2 Monthly Abundance of Sea Snakes During the Present Study Period

Sl. No.	Species	J	F	M	A	M	J	J	A	S	O	N	D	Total
1.	Enhydrina schistosa	5	2	4	3	2	26	13	1		5	8	1	70
3.	Hydrophis spiralis			4		3	19	9	12		6	7	2	62
2.	Hydrophis cyanocinctus	2	1	2	1		11	4	5		7	6		39
4.	Pelamis platurus	1												1
5.	Lapemis curtus				1		2							3
6.	Hydrophis ornatus				1		3	2	2	1	1			9
7.	Hydrophis gracilis	1		1			4	1			2	2		11
8.	Hydrophis fasciatus						2				1	1		5
9.	Hydrophis caerulescens						1		1					2
10.	Cerberus rynchops	1			2	1					1	2		7
Total		**10**	**3**	**11**	**8**	**6**	**68**	**29**	**21**	**1**	**23**	**26**	**3**	**209**

TABLE 16.3 Annual Abundance of Sea Snake Species during the Study Period

Sl. No.	Species	2009	2010	2011	2012	2013	Total
1.	Enhydrina schistosa	13	14	12	15	16	70
2.	Hydrophis spiralis	11	14	12	12	13	62
2.	Hydrophis cyanocinctus	8	7	9	6	9	39
4.	Pelamis platurus	0	1	0	0	0	1
5.	Lapemis curtus	0	1	2	0	0	3
6.	Hydrophis ornatus	2	1	3	2	1	9
7.	Hydrophis gracilis	1	4	2	2	2	11
8.	Hydrophis fasciatus	1	0	1	1	2	5
9.	Hydrophis caerulescens	1	0	1	0	0	2
10.	Cerberus rynchops	1	1	3	1	1	7
	No. of individuals	38	43	45	39	44	209
	No. of species	8	8	9	7	7	

TABLE 16.4 Correlation between Snout–Vent Length and Weight, and between Total Length and Weight, for Sea Snake Species Found during the Study Period

Species	Correlation		n
	SVL (cm) vs Wt (g)	TL (cm) vs Wt (g)	
Enhydrina schistosa	0.578**	0.563**	31
Hydrophis cyanocinctus	0.588*	0.636*	15
Hydrophis spiralis	0.909**	0.875**	14
Pelamis platurus	0.834*	0.806*	7

(Continued)

TABLE 16.4 Correlation between Snout–Vent Length and Weight, and between Total Length and Weight, for Sea Snake Species Found during the Study Period *(cont.)*

Species	Correlation		n
	SVL (cm) vs Wt (g)	*TL (cm) vs Wt (g)*	
Lapemis curtus	0.738^{NS}	0.797^{NS}	8
Hydrophis ornatus	0.457^{NS}	0.396^{NS}	9
Hydrophis gracilis	0.223^{NS}	0.180^{NS}	9
Hydrophis fasciatus	0.507^{NS}	0.706^{NS}	5
Hydrophis caerulescens	0.080^{NS}	0.091^{NS}	3

SVL, snout–vent length; TL, total length; Wt, weight.
*(p < 0.01); **(p < 0.05); NS, Not significant.

ACKNOWLEDGEMENTS

The authors are highly thankful to Dr. K. Venkataraman, Director, Zoological Survey of India, Kolkata for his encouragement and facilities provided.

REFERENCES

Ahmed, S., 1978. Sea snakes of the Indian Ocean in the collections of the Zoological Survey of India together with remarks on the geographical distribution of all Indian Ocean species. J. Mar. Biol.Assoc. India 17 (1), 73–81.

Aiyar, T.V.R., 1907. Notes on sea snakes caught at Madras. J. Asiatic Soc. Bengal 2, 69–72.

Damotharan, P., Arumugam, M., Vijayalakshmi, S., Balasubramanian, T., 2010. Diversity, biology and ecology of sea snakes (Hydrophiidae), distributed along the Parangipettai coast, southeast coast of India. Int. J. Curr. Res. 4, 062–069.

Das, I., 1991. Morphometrics of eryxconicus (Schneider) at a locality in south India (Squamata: Boidae). Hamadryad 16 (1&2), 21–24.

Das, I., 2003. Growth of knowledge on the reptiles of India, with an introduction to systematics, taxonomy and nomenclature. J. Bombay Nat. Hist. Soc. 100, 446–501.

Guyer, C., Donnelly, M.A., 1990. Length-mass relationships among an assemblage of tropical snakes in Costa Rica. J. Trop. Ecol. 6, 65–76.

Heatwole, H., 1999. Sea Snakes. Australian Natural History Series. University of New South Wales, Sydney, Australia, p. 85.

Heithaus, M.R., 2001. The biology of tiger sharks, *Galeocerdo cuvier*, in Shark Bay, Western Australia: sex ratio, size distribution, diet and seasonal changes in catch rates. Environ. Biol. Fishes 61, 25–36.

Kalaiarasan, V., Kanagasabai, R., 1994. Seasonal availability of sea snakes (Family: Hydrophiidae) in the Madras waters. Cobra 16, 18–19.

Kannan, P., Rajagopalan, M., 2008. Distribution of Sea Snakes in the Indian Coastal Waters. J. Sci. Trans. Environ. Tech. 1 (4), 218–223.

Karthikeyan, R., Balasubramanian, T., 2007. Species Diversity of the Sea Snake (Hydrophiidae) Distributed in the Coramantal Coast (East Coast of India). Internat. J. Zool. Res. 3 (3), 107–131.

Karthikeyan, R., Vijayalaksmi, S., Balasubramanian, T., 2008. Feeding and parturition of female annulated sea snake *Hydrophis cyanocinctus* in captivity. Curr. Sci. 94, 660–664.

Kaufman, G.A., Gibbons, J.W., 1975. Weight-length relationships in thirteen species of snakes in the southern united states. Herpetologica 31, 31–37.

Lobo, A., 2006. Sea snakes of Gulf of Mannar Marine National Park. The species and their conservation. Technical report submitted to the Rufford foundation.

Lobo, A., Pandav, B., Vasudevan, K., 2004. Weight-length relationships in two species of marine snakes along the coast of Goa, Western India. Hamadryad 29 (1), 89–93.

Lobo, A., Vasudevan, K., Pandav, B., 2005. Trophic ecology of *Lapemis curtus* (Hydrophiinae) along the Western coast of India. Copeia 3, 636–640.

MacKenzie, 1820. An account of venomous sea snakes on the coast of Madras. Asiatic. Res. 13, 329–336.

Madsen, T., Shine, R., 2000. Rain, fish and snakes: climatically driven population dynamics of Arafura file snakes in tropical Australia. Oecologia 124, 208–215.

Murthy, T.S.N., 1977a. Systematic index, distribution and bibliography of the sea snakes of India. Indian J. Zootomy 18 (2), 131–134.

Murthy, T.S.N., 1977b. On Sea snakes occurring in Madras waters. J. Mar. Biol. Assoc. India 9 (1), 68–72.

Murthy, T.S.N., 1992. Marine Reptiles of India: An overview. In: Strimple, P. (Ed.), Contributions in Herpetology, Greater Cincinnati. Herpetological Society, pp. 35–38.

Smith, M.A., 1943. The Fauna of British India. Vol III, Snakes. Taylor and Francis, London, p. 583.

Venkatraman, C., Jothinayagam, J.T., Raghunathan, C., 2007. Seas Snakes of Chennai coast, southern India. In: Editor-Director, National Symposium on Conservation and Valuation of Marine Biodiversity, Director, Zoological Survey of India, Kolkata. pp. 417–421.

Voris, H.K., 1972. The role of sea snakes (Hydrophiidae) in the trophic structure of coastal ocean communities. J. Mar. Biol. Assoc. India 14 (2), 429–441.

Wall, F., 1918. Notes on a collection of sea snakes from Madras. J. Bombay. Nat. Hist. Soc. 26, 599–607.

Whitaker, R., Captain, A., 2004. Snakes of India, The field guide. Draco books, Chengalpat (Tamil Nadu), p. 479.

Chapter 17

Coastal and Marine Bird Communities of India

C. Sivaperuman* and C. Venkatraman†

*Andaman and Nicobar Regional Centre, Zoological Survey of India, Port Blair, Andaman and Nicobar Islands, India; †Zoological Survey of India, Marine Biology Regional Centre, Chennai, Tamil Nadu, India

INTRODUCTION

Coastal wetlands include seasonal and relatively permanent coastal plain freshwater swamps and marshes, coastal beaches, rocky shorelines, estuarine salt marshes, mangrove swamps, seagrass beds, mud flats and sand bars. The Convention on Wetlands of International Importance Especially as Waterfowl Habitat, the so-called Rasmar Convention, broadly defines coastal wetlands to include "the areas of marine water the depth of which at low tide does not exceed six metres." The marine and coastal wetlands provide habitat to an enormous number of marine and coastal species, as do open sea ecosystems. India has a coastline of 7516 km of which the mainland accounts for 5422 km, the Lakshadweep coast 132 km and the Andaman and Nicobar Islands 1962 km (Venkataraman, 2008). The coasts are perhaps the most neglected biogeographic zone of India, mainly because charismatic species are not found there. Nonetheless, the coasts do have fabulous bird concentrations, as seen in Chilika Lake, important bird areas (IBA) and Bhitarkanika in Orissa, Point Calimere Wildlife Sanctuary (IBA) in Tamil Nadu, Sunderbans (IBA) in West Bengal, Sewri mudflats (IBA) in Maharashtra, and Kori Creek in Gujarat. Besides the sand beaches and rocky outcrops which are important as foraging sites for many waders, the mangroves serve as breeding ground for many species of birds: e.g. egrets, herons, storks, kingfishers and raptors. According to Rodgers *et al.* (2000), the Coasts Biogeographic Zone covers about 83,000 km^2, which is 2.52 percent of India's geographical area.

Coastal wetlands are special types of wetlands that are influenced by the fluctuating water levels to provide a habitat for a vast array of organisms, including many endangered species. These critically important features act as water purifier, fish spawning area and feeding grounds and habitat for many animal species. Some birds depend on wetlands almost totally for breeding, nesting, feeding, or shelter during their annual cycles. Birds that need functional access

Marine Faunal Diversity in India. DOI: 10.1016/B978-0-12-801948-1.00017-3

to a wetland or wetland products during their life cycle can be called "wetland dependent." The important migratory birds utilizing the coastal wetlands are ducks, shorebirds, gulls, terns and flamingos. Many birds that inhabit intertidal habitats are migrants and travel annually along the Central Asian Flyway, which extends from central Siberia through the Himalayas to the Indian subcontinent. During peak annual migration periods, hundreds of thousands of birds migrating along the Central Asian Flyway descend upon the coastal wetlands of India in search of refuge and food. Some species of shorebirds weighing as little as 25 g fly as far as 9000 km from the arctic breeding grounds to south Indian wintering grounds. Prior to breeding, they again fly northwards to their nesting grounds, thus, in one year they may fly 18,000 km (Balachandran, 2012). Even though many such records were available in the past, no effort was made to compile the avifauna of the coastal wetlands of India. An attempt is made here to compile the avifauna of the coastal wetlands on the basis of field surveys and published information.

METHODS

This chapter has been prepared on the basis of field studies conducted by the authors (CS & CV) on the east coast, west coast and in the Andaman and Nicobar Islands (Sivaperuman and Jayson, 2000, 2009; Jayson and Sivaperuman, 2003, 2004, 2005; Sivaperuman, 2011, 2013; Sivaperuman et al., 2010; Venkatraman, 2008; Venkatraman and Gokula 2009). In addition, the available important literature was also consulted (Sampath, 1989; Sampath and Krishnamurthy, 1989; Balachandran, 1990, 1995, 2006; Kurup, 1991, 1996; Anon. 1992, 1993; Venkatraman and Muthukrishnan, 1993; Jayson and Easa, 2000; Acharya and Kar 1996; Nagarajan and Thiyagesan 1996; Oswin, 1999; Verma et al., 2002; Biju Kumar, 2006; Gopi and Pandav, 2007; Pawar, 2011; Kannan and Pandiyan, 2012; Ramamurthy and Rajakumar, 2014; Jisha Kurian, 2014). Birds were classified as migratory or resident species based on Ali and Ripley (1983). The common and scientific names are after Manakadan and Pittie (2001).

RESULTS AND DISCUSSION

Two hundred and twenty three taxa of birds were recorded from the coastal wetlands of India, belonging to 30 families under nine orders (Tables 17.1 and 17.2). Out of these, 91 species were residents, 90 were trans-continental migrants, 37 were resident migrants, four species were vagrant and one species is a straggler. The highest number of species was recorded from the east coast (164), followed by the west coast (163) and the Andaman and Nicobar Islands (110). The order Charadriiformes was highest in dominance, followed by Falconiformes and Ciconiiformes (Figure 17.1). Out of 223 bird species, 74 were found in all three regions, these belonging to six orders and 12 families.

TABLE 17.1 List of Coastal Wetland Birds

Sl. No.	Common name	Scientific name	Status*	East coast	West coast	A & N Islands
Podicipediformes						
Podicipedidae						
1.	Little Grebe	*Tachybaptus ruficollis* (Pallas)	R, LC	√	√	
2.	Great Crested Grebe	*Podiceps cristatus* (Linnaeus)	M, LC		√	
Procellariiformes						
Procellariidae						
3.	Wilson's Storm-Petrel	*Oceanites oceanicus* (Kuhl)	M, LC			√
4.	Black-bellied Storm-Petrel	*Fretetta tropica* (Gould)	V, LC			√
Pelecaniformes						
Phaethontidae						
5.	Grey-backed Tropicbird	*Phaethon aethereus* (Linnaeus)	M, LC			√
6.	Red-tailed Tropicbird	*Phaethon rubricauda* (Boddaert)	R, LC			√
7.	Yellow-billed Tropicbird	*Phaethon lepturus* (Daudin)	R, LC			√
Pelecanidae						
8.	Great White Pelican	*Pelecanus onocrotalus* (Linnaeus)	RM, LC		√	
9.	Spot-billed Pelican	*Pelecanus philippensis* (Gmelin)	RM, NT	√	√	
10.	Dalmatian Pelican	*Pelecanus crispus* (Bruch)	M, VU		√	
Sulidae						
11.	Masked Booby	*Sula dactylatra* (Lesson)	R, LC		√	
12.	Red-footed Booby	*Sula sula* (Linnaeus)	R, LC			√
Phalacrocoracidae						
13.	Little Cormorant	*Phalacrocorax niger* (Vieillot)	R, LC	√	√	√

(Continued)

TABLE 17.1 List of Coastal Wetland Birds *(cont.)*

Sl. No.	Common name	Scientific name	Status*	East coast	West coast	A & N Islands
14.	Great Cormorant	*Phalacrocorax carpo* (Linnaeus)	R, LC	√	√	
15.	Indian Shag	*Phalacrocorax fuscicollis* (Stephens)	R, LC	√	√	
Anhingidae						
16.	Darter	*Anhinga melanogaster* (Pennant)	R, NT	√	√	
Fregatidae						
17.	Lesser Frigatebird	*Fregata ariel* (G.R. Gray)	M, LC	√	√	
18.	Christmas Island Frigatebird	*Fregata andrewsi* (Mathews)	M, CR			√
Ciconiiformes						
Ardeidae						
19.	Little Egret	*Egretta garzetta* (Linnaeus)	R, LC	√	√	√
20.	Western Reef Egret	*Egretta gularis* (Bosc)	RM, LC	√	√	
21.	Pacific Reef-Egret	*Egretta sacra* (Gmelin)	RM, LC	√		√
22.	Grey Heron	*Ardea cinerea* (Linnaeus)	RM, LC	√	√	√
23.	Goliath Heron	*Ardea goliath* (Cretzschmar)	V, LC	√		
24.	Great-billed Heron	*Ardea sumatarna* (Raffles)	RM, LC			√
25.	Purple Heron	*Ardea purpurea* (Linnaeus)	RM, LC	√	√	√
26.	Large Egret	*Casmerodius albus* (Linnaeus)	RM, LC	√	√	√
27.	Median Egret	*Mesophoyx intermedia* (Wagler)	RM, LC	√	√	√
28.	Cattle Egret	*Bubulcus ibis* (Linnaeus)	RM, LC	√	√	√
29.	Indian Pond Heron	*Ardeola grayii* (Sykes)	R, LC	√	√	√

TABLE 17.1 List of Coastal Wetland Birds *(cont.)*

Sl. No.	Common name	Scientific name	Status*	East coast	West coast	A & N Islands
30.	Chinese Pond-Heron	*Ardeola bacchus* (Bonaparte)	RM, LC	√		√
31.	Little Green Heron	*Butroides striatus* (Linnaeus)	R, NR	√	√	√
32.	Black-crowned Night-Heron	*Nycticorax nycticorax* (Linnaeus)	R, LC	√	√	√
33.	Malayan Night-Heron	*Goraschius melanolophus* (Raffles)	RM, LC			√
34.	Yellow Bittern	*Ixobrychus sinensis* (Gmelin)	RM, LC	√	√	√
35.	Chestnut Bittern	*Ixobrychus cinnamomeus* (Gmelin)	RM, LC	√	√	√
36.	Black Bittern	*Dupetor flavicollis* (Latham)	RM, LC	√	√	√
37.	Great Bittern	*Botaurus stellaris* (Linnaeus)	M, LC	√		
Ciconiidae						
38.	Painted Stork	*Mycteria leucocephala* (Pennant)	R, NT	√	√	
39.	Asian Openbill-Stork	*Anastomus osciatans* (Boddaert)	R, LC	√	√	
40.	Black Stork	*Ciconia nigra* (Linnaeus)	M, LC		√	
41.	White-necked Stork	*Ciconia epscopus* (Boddaert)	R, LC		√	
42.	European White Stork	*Ciconia ciconia* (Linnaeus)	M, LC	√	√	
43.	Black-necked Stork	*Ephippiorhyncus asiaticus* (Latham)	R, NT	√	√	
44.	Lesser Adjutant Stork	*Leptoptilys javanicus* (Horsfield)	R, VU	√		
Threskiornithidae						
45.	Glossy Ibis	*Plegadis falcinellus* (Linnaeus)	R, LC	√	√	

(Continued)

TABLE 17.1 List of Coastal Wetland Birds *(cont.)*

Sl. No.	Common name	Scientific name	Status*	East coast	West coast	A & N Islands
46.	Oriental White Ibis	*Threskiornis melanocephalus* (Latham)	R, LC	√	√	
47.	Black Ibis	*Pseudibis papillosa* (Temminck)	R, LC	√	√	
48.	Eurasian Spoonbill	*Platalea leucorodia* (Linnaeus)	R, LC	√	√	
Phoenicopteriformes						
Phoenicopteridae						
49.	Greater Flamingo	*Phoenicopterus ruber* (Linnaeus)	R, LC	√	√	
50.	Lesser Flamingo	*Pheonicopterus minor* (Geoffroy)	R, NT	√	√	
Anseriformes						
Anatidae						
51.	Large Whistling-Duck	*Dendrocygna bicolor* (Vieillot)	R, LC	√		
52.	Lesser Whistling-Duck	*Dendrocygna javanica* (Horsfield)	R, LC	√	√	√
53.	Greylag Goose	*Anser anser* (Linnaeus)	R, LC		√	
54.	Bar-headed Goose	*Anser indicus* (Latham)	R, LC	√		
55.	Brahminy Shelduck	*Tadorna ferruginea* (Pallas)	R, LC	√	√	√
56.	Comb Duck	*Sarkidiornis melanotus* (Pennant)	R, LC	√	√	
57.	Cotton Teal	*Nettapus coromandelianus* (Gmelin)	R, LC	√	√	√
58.	Gadwall	*Anas strepera* (Linnaeus)	M, LC	√	√	
59.	Eurasian Wigeon	*Anas penelope* (Linnaeus)	M, LC	√	√	√
60.	Mallard	*Anas platyrhynchos* (Linnaeus)	RM, LC	√	√	

TABLE 17.1 List of Coastal Wetland Birds *(cont.)*

Sl. No.	Common name	Scientific name	Status*	East coast	West coast	A & N Islands
61.	Andaman Teal	*Anas gibberifrons* (Muller)	R, LC			√
62.	Spot-billed Duck	*Anas poecilorhyncha* (J.R. Forester)	R, LC	√	√	√
63.	Northern Shoveller	*Anas clypeata* (Linnaeus)	M, LC	√	√	
64.	Northern Pintail	*Anas acuta* (Linnaeus)	M, LC	√	√	
65.	Garganey	*Anas querquedula* (Linnaeus)	M, LC	√	√	√
66.	Common Teal	*Anas crecca* (Linnaeus)	M, LC	√	√	√
67.	Red-crested Pochard	*Rhodonessa rufina* (Pallas)	M, LC	√		
68.	Common Pochard	*Aythya ferina* (Linnaeus)	M, LC	√	√	
69.	Ferruginous Pochard	*Aythya nyroca* (Guldenstadt)	R, NT		√	
70.	Baer's Pochard	*Aythya baeri* (Radde)	M, CR	√		
71.	Tufted Pochard	*Aythya fuligula* (Linnaeus)	RM, LC	√	√	
Falconiformes						
Accipitridae						
72.	Black Baza	*Aviceda leuphotes* (Dumont)	RM, LC		√	
73.	Andaman Blackcrested Baza	*Aviceda leuphotes andamanica* (Abdulali)	R, LC			√
74.	Oriental Honey-Buzzard	*Pernis ptilorhynchus* (Temminck)	RM, LC	√	√	√
75.	Black-shouldered Kite	*Elanus caeruleus* (Desfontaines)	RM, LC	√	√	
76.	Black Kite	*Milvus migrans* (Boddaert)	R, LC	√	√	√
77.	Brahminy Kite	*Haliastur indus* (Boddaert)	R, LC	√	√	√

(Continued)

TABLE 17.1 List of Coastal Wetland Birds *(cont.)*

Sl. No.	Common name	Scientific name	Status*	East coast	West coast	A & N Islands
78.	White-bellied Sea Eagle	*Haliaeetus leucogaster* (Gmelin)	R, LC	√		√
79.	Pallas's Fish-Eagle	*Haliaeetus leucoryphus* (Pallas)	M, VU	√		
80.	Greater Grey-headed Fish-Eagle	*Ichthyophaga ichthyaetus* (Horsfield)	R, NT		√	√
81.	Egyptian Vulture	*Neophron percnopterus* (Linnaeus)	R, EN	√		
82.	Indian White-backed Vulture	*Gyps bengalensis* (Gmelin)	R, CR	√	√	
83.	Short-toed Snake-Eagle	*Circaetus gallicus* (Gmelin)	R, LC	√		
84.	Crested Serpent-Eagle	*Spilornis cheela* (Latham)	R, LC	√	√	
85.	Andaman Serpent-Eagle	*Spilornis elgini* (Blyth)	R, NT			√
86.	Nicobar Serpent-Eagle	*Spilornis cheela minimus* (Hume)	R, NR			√
87.	Western Marsh Harrier	*Circus aeruginosus* (Linnaeus)	M, LC	√	√	√
88.	Pallid Harrier	*Circus macrourus* (S.G. Gmelin)	M, NT	√	√	√
89.	Pied Harrier	*Circus melanoleucos* (Pennant)	M, LC	√	√	
90.	Montagu's Harrier	*Circus phygargus* (Linnaeus)	M, LC	√		√
91.	Crested Goshawk	*Accipiter trivirgatus* (Temminck)	R, LC		√	
92.	Shikra	*Accipiter badius* (Gmelin)	R, LC	√	√	
93.	Katchal Shikra	*Accipiter badius obsoletus* (Richmond)	R, LC			√
94.	Besra Sparrowhawk	*Accipiter virgatus* (Temminck)	RM, LC	√	√	√

TABLE 17.1 List of Coastal Wetland Birds *(cont.)*

Sl. No.	Common name	Scientific name	Status*	East coast	West coast	A & N Islands
95.	Nicobar Sparrowhawk	*Accipiter butleri* (Gurney)	R, VU			√
96.	Car Nicobar Sparrowhawk	*Accipiter badius* (Gmelin)	R, LC			√
97.	Chinese Sparrowhawk	*Accipiter soloensis* (Horsfield)	M, LC			√
98.	Eurasian Sparrowhawk	*Accipiter nisus* (Linnaeus)	M, LC	√	√	√
99.	White-eyed Buzzard	*Butastur teesa* (Franklin)	R, LC		√	
100.	Long-legged Buzzard	*Buteo rufinus* (Cretzschmar)	M, LC		√	
101.	Black Eagle	*Ictinaetus malayensis* (Temminck)	R, LC	√	√	√
102.	Lesser Spotted Eagle	*Aquila pomarina* (Brehm)	RM, NR	√		
103.	Greater Spotted Eagle	*Aquila clanga* (Pallas)	M, VU	√	√	
104.	Tawny Eagle	*Aquila rapax* (Temminck)	R, LC	√		
105.	Steppe Eagle	*Aquila niplalensis* (Hodgson)	M, LC		√	
106.	Eastern Imperial Eagle	*Aquila heliaca* (Savigny)	M, VU		√	
107.	Booted Eagle	*Hieraaetus pennatus* (Gmelin)	R, LC	√	√	
108.	Rufous-bellied Eagle	*Hieraaetus kienerii* (E. Geoffroy)	M, LC		√	
109.	Changeable Hawk Eagle	*Spizaetus cirrhatus* (Gmelin)	R, LC	√	√	√
Pandionidae						
110.	Osprey	*Pandion haliaetus* (Linnaeus)	RM, LC	√	√	√
Falconidae						
111.	Common Kestrel	*Falco tinnunculus* (Linnaeus)	RM, LC	√	√	√
112.	Lesser Kestrel	*Falco naumanni* (Fleischer)	M, LC			√

(Continued)

TABLE 17.1 List of Coastal Wetland Birds *(cont.)*

Sl. No.	Common name	Scientific name	Status*	East coast	West coast	A & N Islands
113.	Saker	*Falco cherrug* (J.E. Gray)	R, EN			√
114.	Red-headed Falcon	*Falco chicquera* (Daudin)	R, LC	√		
115.	Eurasian Hobby	*Falco subbuteo* (Linnaeus)	M, LC		√	
116.	Laggar Falcon	*Falco juggar* (J.E. Gray)	R, NT		√	
117.	Peregrine Falcon	*Falco perigrinus* (Tunstall)	M, LC	√	√	√
118.	Shaheen Falcon	*Falco peregrinus peregrinator* (Sundevall)	R, LC			√
Gruiformes						
Gruidae						
119.	Common Crane	*Grus grus* (Linnaeus)	M, LC		√	
Rallidae						
120.	Andaman Crake	*Rallina canningi* (Blyth)	R, NT			√
121.	Slaty-legged Crake	*Rallina eurizonoides* (Lafresnaye)	RM, LC	√	√	
122.	Blue-breasted Rail	*Gallirallus striatus* (Linnaeus)	R, LC	√	√	√
123.	Nicobar Blue-breasted Rail	*Gallirallus striatus nicobariensis* (Abdulali)	R, LC			√
124.	Water Rail	*Rallus aquaticus* (Linnaeus)	R, LC		√	
125.	White-breasted Waterhen	*Amaurornis phoenicurus* (Pennant)	R, LC	√	√	
126.	Andaman White-breasted Waterhen	*Amaurornis phoenicurus leucocephalus* (Abdulali)	R, LC			√
127.	Baillon's Crake	*Porzana pusilla* (Pallas)	RM, LC		√	√

TABLE 17.1 List of Coastal Wetland Birds *(cont.)*

Sl. No.	Common name	Scientific name	Status*	East coast	West coast	A & N Islands
128.	Ruddy-breasted Crake	*Porzana fusca* (Linnaeus)	R, LC	√	√	√
129.	Watercock	*Gallicrex cinerea* (Gmelin)	RM	√	√	√
130.	Purple Moorhen	*Porphyrio porphyrio* (Linnaeus)	R, LC	√	√	√
131.	Common Moorhen	*Gallinula chloropus* (Linnaeus)	RM, LC	√	√	√
132.	Common Coot	*Fulica atra* (Linnaeus)	V, LC	√	√	√
Charadriiformes						
Jacanidae						
133.	Pheasant tailed Jacana	*Hydrophasianus chirurgus* (Scopoli)	M, LC	√	√	√
134.	Bronze-winged Jacana	*Metopidius indicus* (Latham)	R, LC	√	√	
Rostratulidae						
135.	Greater Painted Snipe	*Rostratula benghalensis* (Linnaeus)	R, NR	√	√	
Haematopodidae						
136.	Eurasian Oystercatcher	*Haematopus ostralegus* (Linnaeus)	M, LC	√	√	
Charadriidae						
137.	Pacific Golden-Plover	*Pluvialis fulva* (Gmelin)	M, LC	√	√	√
138.	Grey Plover	*Pluvialis squatarola* (Linnaeus)	M, LC	√	√	√
139.	Common Ringed Plover	*Charadrius hiaticula* (Linnaeus)	M, LC	√	√	
140.	Long-billed Plover	*Charadrius placidus* (J.E. Gray)	M, LC	√		
141.	Little Ringed Plover	*Charadrius dubius* (Scopoli)	RM, LC	√	√	√
142.	Kentish Plover	*Charadrius alexandrinus* (Linnaeus)	RM, LC	√	√	√

(Continued)

TABLE 17.1 List of Coastal Wetland Birds *(cont.)*

Sl. No.	Common name	Scientific name	Status*	East coast	West coast	A & N Islands
143.	Lesser Sand Plover	*Charadrius mongolus* (Pallas)	RM, LC	√	√	√
144.	Greater Sand Plover	*Charadrius leschenaultii* (Lesson)	M, LC	√	√	√
145.	Caspian Plover	*Charadrius asiatucus* (Pallas)	V, LC			√
146.	Black-fronted Dotterel	*Elseyornis melanops* (Vieillot)	V, LC	√		
147.	Yellow-wattled Lapwing	*Vanellus malabaricus* (Boddaert)	R, LC	√	√	
148.	Grey-headed Lapwing	*Vanellus cinereus* (Linnaeus)	M, LC	√		√
149.	Red-wattled Lapwing	*Vanellus indicus* (Boddaert)	R, LC	√	√	
150.	White-tailed Lapwing	*Vanellus leucurus* (Lichtenstein)	M, LC	√	√	
Scolopacidae						
151.	Eurasian Woodcock	*Scolopax rusticola* (Linnaeus)	R, LC		√	
152.	Wood Snipe	*Gallinago nemoricola* (Hodgson)	R, VU		√	
153.	Pintail Snipe	*Gallinago stenura* (Bonaparte)	M, LC	√	√	√
154.	Great Snipe	*Gallinago media* (Latham)	M, NT			√
155.	Swinhoe's Snipe	*Gallinago megala* (Swinhoe)	M, LC	√		√
156.	Common Snipe	*Gallinago gallinago* (Linnaeus)	RM, LC	√	√	√
157.	Jack Snipe	*Lymnocryptes minimus* (Brunnich)	M, LC	√	√	√
158.	Black-tailed Godwit	*Limosa limosa* (Linnaeus)	M, NT	√	√	√
159.	Bar-tailed Godwit	*Limosa lapponica* (Linnaeus)	M, LC	√	√	√
160.	Whimbrel	*Numenius phaeopus* (Linnaeus)	M, LC	√	√	√

TABLE 17.1 List of Coastal Wetland Birds *(cont.)*

Sl. No.	Common name	Scientific name	Status*	East coast	West coast	A & N Islands
161.	Eurasian Curlew	*Numenius arquata* (Linnaeus)	M, NT	√	√	√
162.	Spotted Redshank	*Tringa erythropus* (Pallas)	M, LC	√	√	
163.	Common Redshank	*Tringa totanus* (Linnaeus)	RM, LC	√	√	√
164.	Marsh Sandpiper	*Tringa stagnatilis* (Bechstein)	M, LC	√	√	√
165.	Common Greenshank	*Tringa nebularia* (Gunner)	M, LC	√	√	√
166.	Spotted Greenshank	*Tringa guttifer* (Nordmann)	M, EN	√		
167.	Green Sandpiper	*Tringa ochropus* (Linnaeus)	M, LC	√	√	√
168.	Wood Sandpiper	*Tringa glareola* (Linnaeus)	M, LC	√	√	√
169.	Terek Sandpiper	*Xenus cinereus* (Guldenstadt)	M, LC	√	√	√
170.	Common Sandpiper	*Actitis hypoleucos* (Linnaeus)	M, LC	√	√	√
171.	Ruddy Turnstone	*Arenaria interpres* (Linnaeus)	M, LC	√	√	√
172.	Asian Dowitcher	*Limnodromus semipalmatus* (Blyth)	M, NT	√		
173.	Great Knot	*Calidris tenuirostris* (Horsfield)	M, VU	√	√	√
174.	Red Knot	*Calidris canutus* (Linnaeus)	M, LC	√		
175.	Sanderling	*Calidris alba* (Pallas)	M, LC	√	√	√
176.	Spoonbill Sandpiper	*Calidris pygmeus* (Linnaeus)	M, CR		√	
177.	Little Stint	*Calidris minuta* (Leisler)	M, LC	√	√	√
178.	Rufous-necked Stint	*Calidris ruficollis* (Pallas)	M, LC	√		√
179.	Temminck's Stint	*Calidris temminckii* (Leisler)	M, LC	√	√	√

(Continued)

TABLE 17.1 List of Coastal Wetland Birds *(cont.)*

Sl. No.	Common name	Scientific name	Status*	East coast	West coast	A & N Islands
180.	Long-toed Stint	*Calidris subminuta* (Middendorff)	M, LC	√	√	√
181.	Dunlin	*Calidris alpina* (Linnaeus)	M, LC	√	√	
182.	Curlew Sandpiper	*Calidris ferruginea* (Pontoppidan)	M, LC	√	√	√
183.	Broad-billed Sandpiper	*Limicola falcinellus* (Pontoppidan)	M, LC	√	√	√
184.	Ruff	*Philomachus pugnax* (Linnaeus)	M, LC	√	√	
Recurvirostridae						
185.	Black winged Stilt	*Himantopus himantopus* (Linnaeus)	RM, LC	√	√	
186.	Pied Avocet	*Recurvirostra avosetta* (Linnaeus)	M, LC	√	√	
Phalaropidae						
187.	Red-necked Phalarope	*Phalaropus lobatus* (Linnaeus)	M, LC	√		
188.	Red Phalarope	*Phalaropus fulicaria* (Linnaeus)	M, LC		√	
Dromadidae						
189.	Crab-Plover	*Dromas ardeola* (Paykull)	M, LC	√		√
Burhinidae						
190.	Stone-Curlew	*Burhinus oedicnemus* (Linnaeus)	M, LC	√	√	
191.	Great Stone-Plover	*Esacus recurvirostris* (Cuvier)	R, LC	√	√	
192.	Beach Stone-Plover	*Esacus magnirostris* (Vieillot)	R, LC			√
Glareolidae						
193.	Indian Courser	*Cursorius coromandelicus* (Gmelin)	RM, LC		√	
194.	Oriental Pranticole	*Glareola maldivarum* (J.R. Forster)	RM, LC	√	√	√

TABLE 17.1 List of Coastal Wetland Birds *(cont.)*

Sl. No.	Common name	Scientific name	Status*	East coast	West coast	A & N Islands
195.	Collared Pratincole	*Glareola pratincola* (Linnaeus)	M, LC			√
196.	Small Pratincole	*Glareola lactea* (Temminck)	RM, LC	√	√	
Laridae						
197.	Heuglin's Gull	*Larus heuglini* (Bree)	M, NR	√	√	
198.	Yellow-legged Gull	*Larus cachinnans* (Pallas)	M, NR		√	
199.	Lesser Black-backed Gull	*Larus fuscus* (Linnaeus)	M, LC		√	
200.	Palla's Gull	*Larus ichthyaetus* (Pallas)	R, LC	√	√	
201.	Brown-headed Gull	*Larus brunnicephalus* (Jerdon)	M, LC	√	√	
202.	Black-headed Gull	*Larus ridibundus* (Linnaeus)	M, LC	√	√	
203.	Slender-billed Gull	*Larus genei* (Breme)	R, LC		√	
204.	Gull-billed Tern	*Gelochelidon nilotica* (Gmelin)	RM, LC	√	√	√
205.	Caspian Tern	*Sterna caspia* (Pallas)	M, LC	√	√	
206.	River Tern	*Sterna aurantia* (J.E. Gray)	R, NT	√	√	
207.	Lesser Crested Tern	*Sterna bengalensis* (Lesson)	R, LC	√	√	√
208.	Large Crested Tern	*Sterna bergii* (Lichtenstein)	R, LC	√	√	√
209.	Common Tern	*Sterna hirundo* (Linnaeus)	M, LC	√	√	
210.	Little Tern	*Sterna albifrons* (Pallas)	M, LC	√	√	√
211.	Saunders's Tern	*Sterna saundersi* (Hume)	R, LC		√	
212.	Roseate Tern	*Sterna duogalli* (Montagu)	R, LC	√		√
213.	Black-naped Tern	*Stern sumatrana* (Raffles)	R, LC			√

(Continued)

TABLE 17.1 List of Coastal Wetland Birds *(cont.)*

Sl. No.	Common name	Scientific name	Status*	East coast	West coast	A & N Islands
214.	Black-bellied Tern	*Sterna acuticauda* (J.E. Gray)	R, EN	√	√	
215.	Bridled Tern	*Sterna anaethetus* (Scopoli)	R, LC	√		√
216.	Sooty Tern	*Sterna. fuscata* (Linnaeus)	R, LC	√	√	√
217.	Sandwich Tern	*Sterna sandvicensis* (Latham)	M, LC	√	√	
218.	Whiskered Tern	*Chlidonias hybridus* (Pallas)	R, LC	√	√	
219.	White-winged Black Tern	*Chlidonias leucopterus* (Temminck)	M, LC	√	√	√
220.	Black Tern	*Chlidonias niger* (Linnaeus)	M, LC	√		
221.	Brown Noddy	*Anous stolidus* (Linnaeus)	R, LC	√		√
222.	Lesser Noddy	*Anous tenuurostris* (Temminck)	S, LC			√
Rynchopidae						
223.	Indian Skimmer	*Rynchops albicollis* (Swainson)	R, VU	√		

M, transcontinental migrant; R, resident; RM, resident migrant; S, straggler; V, vagrant.
* CR, Critically Endangered; EN, Endangered; LC, Least Concern; NR, Not Recognized; NT, Near Threatened; VU, Vulnerable.
* BirdLife International (2012).

Twenty species were found only from the east coast, and 28 each from the west coast and the Andaman and Nicobar Islands (Table 17.1).

A comparison of number of bird species recorded with those from the Indian subcontinent as a whole and worldwide is given in Table 17.3.

Globally Threatened Bird Species

Of the recorded species, 31 were listed under IUCN threated categories (Table 17.1). Four species—Christmas Island Frigatebird (*Fregata andrewsi*), Baer's Pochard (*Aythya baeri*), Indian White-backed Vulture (*Gyps bengalensis*), and Spoonbill Sandpiper (*Calidris pygmeus*)—were listed as Critically Engangered. The following species were listed as Endangered: Egyptian Vulture (*Neophron percnopterus*), Saker (*Falco cherrug*), Spotted Greenshank (*Tringa*

TABLE 17.2 Order and Status of Birds Recorded from Indian Coastal and Marine Ecosystems

Sl. No.	Order	Resident	Resident migrant	Migrant	Total
1	Podicipediformes	1		1	2
2	Procellariiformes			2	2
3	Pelecaniformes	8	2	4	14
4	Ciconiiformes	13	13	4	30
5	Phoenicopteriformes	2			2
6	Anseriformes	10	2	9	21
7	Falconiformes	25	7	16	48
8	Gruiformes	8	4	2	14
9	Charadriiformes	22	9	59	90

guttifer), and Black-bellied Tern (*Sterna acuticauda*). Fifteen species are Near Threatened and eight are Vulnerable.

Out of more than 9000 species of birds of the world, the Indian subcontinent supports about 1300, or over 13 percent of the world's birds (Grimmett *et al.*, 1998). This subcontinent, rich in avifauna, also boasts 48 bird families out of the total of 75 families in the world. The coastal wetands of the east coast, west coast and Andaman and Nicobar Isands support 17 percent of Indian avifauna. The high avian species richness recorded from the east and west coast arises from the presence of diverse microhabitats and extensive surveys carried

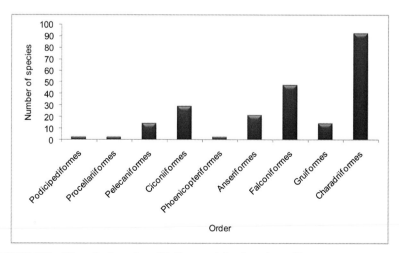

FIGURE 17.1   Diversity, by order, of Indian coastal and marine avifauna.

TABLE 17.3 Occurrence of Coastal and Marine Bird Species

Sl. No.	Order	World[1]	Indian subcontinent[2]	East Coast	West Coast	A & N Islands
1	Podicipediformes	22	05	1	2	-
2	Procellariiformes	32	21	-	-	2
3	Pelecaniformes	45	17	6	9	6
4	Ciconiiformes	88	34	25	23	14
5	Phoenifopteriformes	06	02	2	2	-
6	Anseriformes	145	44	18	16	8
7	Falconiformes	285	71	27	29	25
8	Gruiformes	186	33	8	11	9
9	Charadriiformes	199	121	77	71	46
	Total	**1008**	**348**	**164**	**163**	**110**

[1]Harrison (1978)
[2]Ali and Ripley (1983), Ali (1969)

out in the past by various ornithologists and amateurs in the coastal wetlands. Among the species recorded, 90 were transcontinental migrants, which showed the high influx of migratory birds during the migration season. The coastal wetland avifauna is dominated, at least numerically, by large numbers of Anseriforms, Ciconiiforms, and Charadriiforms. The avian populations increase considerably during migratory periods in various sites on the east coast, when large numbers of waterfowl and shorebirds congregate to feed and rest. Many coastal wetlands in the east and west coasts annually host significant portions of the world population of migratory species.

The coastal wetlands of India are the major wintering grounds for migratory shorebirds in the three important flyways: the Central Asian Flyway, East Asian–Australasian Flyway and Western Pacific Flyway. The coastal wetlands of Point Calimere, Chilika Lake, and Pulicate Lake on the east coast support a significant number of winter vistors. On the west coast, the Kadalundy estuary and the Vembanad-Kole Ramsar site are the major feeding and roosting grounds. The tsunami-inundated wetlands of the Andaman and Nicobar Islands also provide feeding and resting sites for many migratory shore birds in the East Asian–Australisian Flyway.

Though many checklists are available on the avifauna of coastal wetlands, still various aspects of ecological studies are lacking. Much important research needs to be carried out to better understand the ecology, particularly the movements within the ecosystem, and studies on different aspects of the ecology and habitat modeling of migratory shorebirds with satellite tracking needs to be initiated in the coastal wetlands of India. Studies also should be focused in the mangrove ecosystem of the east coast and the Andaman and Nicobar Islands. Information on phenological pattens of migratory birds and behavioral ecophysiology of migrating shorebirds are also important in this region. Foraging ecology studies should be carried out with the ultimate goal of understanding the consequences of habitat selection by wintering shorebirds in terms of meeting energy demands. Specifically, the information on shorebirds' diet, foraging rates among habitats and seasons, and food availability are lacking in this region, especially for the Andaman and Nicobar Islands. This study confirms the biodiversity value of these coastal wetlands as suitable habitats for feeding and as breeding grounds for wintering threatened waterbirds.

ACKNOWLEDGEMENT

The authors are grateful to the Director, Zoological Survey of India, Kolkata for his valuable support, encouragement and provision of necessary facilities.

REFERENCES

Acharya, S., Kar, S.K., 1996. Checklist of Waders (Charadriiformes) in Chilika Lake, Orissa. Newsl. Birdwatchers 36 (5), 89–90.

Ali, S., 1969. Birds Of Kerala. Oxford University Press, Oxford, p. 444.

Ali, S., Ripley, S.D., 1983. Hand Book of the Birds of India and Pakistan. Oxford University Press, Oxford, p. 737.

Anon., 1992. Birds of Kole wetlands: A survey report – I. Nature Education Society Thrissur (NEST), in collaboration with Kerala Forest Research Institute (KFRI) and Kerala Forest Department. p. 16.

Anon., 1993. Birds of Kole wetlands: A Survey Report – II. Nature Education Society Thrissur (NEST), in collaboration with Kerala Forest Research Institute (KFRI) and Kerala Forest Department. p. 18.

Balachandran, S., 1990. Studies on the coastal birds of Mandapam and the neighbouring islands (Peninsular India). Ph.D. Thesis, Annamalai University, Tamil Nadu. p. 229.

Balachandran, S., 1995. Shore birds of the Marine National Park in the Gulf of Mannar. Tamil Nadu. J. Bombay Nat. Hist. Soc. 92 (3), 303–313.

Balachandran, S., 2006. The decline in wader population along the east coast of India with special reference to Point Calimere, South-east India. pp. 296–301. In : Waterbirds around the, world., (Eds). Boere, G.C., Galbraith, C.A. and Stroud, D.A. The Stationery Office, Edinburgh, UK.

Balachandran, S., 2012. Avian diversity in coastal wetlands of India and their conservation needs. pp. 155–163. In: Marine Biodiversity, International Day for Biological Diversity, Uttar Pradesh Biodiversity Board, UP.

Biju Kumar, A., 2006. A checklist of avifauna of the Bharathapuzha River Basin. Kerala. Zoos' Print J. 21 (8), 2350–2355.

BirdLife International., 2012. The BirdLife checklist of the birds of the world, with conservation status and taxonomic sources. Version 5. Downloaded from http://www.birdlife.info/im/species/checklist.zip.

Gopi, G.V., Pandav, B., 2007. Avifauna of Bhitarkanika Mangroves. India. Zoos' Print J. 229 (10), 2839–2847.

Grimmett, R., Inskipp, C., Inskipp, T., 1998. Birds of Indian Subcontinent. Oxford, Oxford University Press, p. 888.

Harrison, C.J.O., 1978. Bird Families of the World. Elsevier Phaidon, Oxford, p. 264.

Jayson, E.A., Easa, P.S., 2000. Documentation of vertebrate fauna in Mangalavanam mangrove area. KFRI Research Report No. 183. Kerala Forest Research Institute, Peechi. p. 42.

Jayson, E.A., Sivaperuman, C., 2003. Avifauna of Kol wetlands of Thrissur, a newly declared Ramsar Site in Kerala. Regional Seminar on Ramsar sites of Kerala, 21st Feb. 2003, Kozhikode, Kerala.

Jayson, E.A., Sivaperuman, C., 2004. Avifauna of Kole wetlands of Thrissur, a newly declared Ramsar Site in Kerala. pp. 27–47. In: Proceeding of the National Seminar on Wetland Avian Ecology, 15th Jan. 2004, St. Aloysius College, Elthuruth, Thrissur.

Jayson, E.A., Sivaperuman, C., 2005. Avifauna of Thrissur district. Kerala. Zoos' Print J. 20 (2), 1774–1783.

Jisha Kurian, K., 2014. Spatio-temporal variation of waterbirds in Kachchh, Gujarat, Western India. Ph.D., Thesis. University of Madras, Chennai (Unpublished) p. 327.

Kannan, V., Pandiyan, J., 2012. Shorebirds (Charadriidae) of Pulicat Lake. India with special reference to conservation. World J. Zool. 7 (3), 178–191.

Kurup, D.N., 1991. Ecology of the birds of Malabar Coast and Lakshadweep. Ph.D. Dissertation, University of Calicut, Calicut.

Kurup, D.N., 1996. Ecology of the birds of Barathapuzha estuary and survey of the coastal wetlands of Kerala. Final Report submitted to Kerala Forest Department, Trivandrum. p. 59.

Manakadan, R., Pittie, A., 2001. Standardised Common and Scientific Names of the Birds of the Indian Subcontinent. Buceros 6 (1), 1–37.

Nagarajan, R., Thiyagesan, K., 1996. Waterbirds and substrate quality of the Pichavaram wetlands. Southern India. Ibis 138, 710–721.

Oswin, D.S., 1999. Avifaunal diversity of Muthpet Mangrove Forest. Zoos' Print J. 14 (6), 47–53.

Pawar, P.R., 2011. Species diversity of birds in mangroves of Uran (Raigad), Navi Mumbai, Maharashtra. West coast of India. J. Experim. Sci. 2 (10), 73–77.

Ramamurthy, V., Rajakumar, R., 2014. A study of avifaunal diversity and influences of water quality in the Udhayamarthandauram Bird Sanctuary, Thiruvarur District. Tamil Nadu. Int. J. Innovative Res. Sci. Engg. Tech. 3 (1), 8851–8858.

Rodgers, W.A., Panwar, H.S. and Mathur, V.B., 2000. Wildlife protected area network in India: A review. Wildlife Institute of India, Dehra Dun, India. p. 44.

Sampath, K., 1989. Studies on the ecology of Shorebirds (Aves: Charadriiformes) of the Great Vedaranyam Swamp and the Pichavaram Mangroves of India. Ph.D. Dissertation, Annamalai University, Tamil Nadu. p. 202.

Sampath, K., Krishnamurthy, K., 1989. Shore birds of the salt ponds at Great Vedaranyam Salt Swamp. Stilt 15, 20–23.

Sivaperuman, C., 2011. Avian communities of Ritchie's Archipelago, Andaman and Nicobar Islands. pp. 238. In: Proceeding of the International conference on Tropical Island Ecosystem: Issues related to livelihood, sustainable development and climate change. Central Agricultural Research Institute, Port Blair.

Sivaperuman, C., 2013. Diversity and species-abundance distribution of birds in the Ritchie's archipelago. Andaman and Nicobar Islands. Ann. Forestry 21 (1), 101–112.

Sivaperuman, C., Jayson, E.A., 2000. Birds of Kole Wetlands, Thrissur. Kerala. Zoos' Print J. 15 (10), 344–349.

Sivaperuman, C., Jayson, E.A., 2009. Population dynamics of wetland birds in the Kole wetlands of Kerala. India. J. Sci. Trans. Environ. Tech. 2 (3), 152–162.

Sivaperuman, C., Venkatraman, C., Raghunathan, C., 2010. Avifauna of Andaman and Nicobar Islands: An Overview. pp. 399–412. In : Recent Trends in Biodiversity of Andaman, Nicobar, Islands, (Eds.) Ramakrishna, C. Raghunathan and C. Sivaperuman. Zoological Survey of India.

Venkataraman, K., 2008. Coastal and Marine wetlands in India. pp. 392–400. In : Proceedings of Taal 2007 The 12th World Lake Conference, (Eds.) Sengupta, M. and Dalwani, R. Ministry of Environment and Forests, Govenment of India, New Delhi.

Venkatraman, C., 2008. Diversity of Coastal birds in Gulf of Mannar Biosphere Reserve, Southern India. (Ed) M.V. Reddy. *Wildlife Biodiversity Conservation in India*. Daya publishing house, New Delhi. pp 104–113.

Venkatraman, C., Gokula, V., 2009. Coastal birds of Tamil Nadu, Rec. Zool. Surv. India. Occ. Paper 303, 1–64.

Venkatraman, C., Muthukrishnan, S., 1993. Density of waterbirds at Vedanthangal Bird Sanctuary, Tamil Nadu. pp. 55–60. In : Bird Conservation: Strategies for the Nineties, Beyond, (Eds.) Verghese, A., S. Sridhar and A.K. Chakravarthy. Ornithological Society of India, Bangalore.

Verma, A., Chaturvedi, N., Balachandran, S., Kehimkar, I., 2002. Avian diversity in and around Mangroves and Mahul Creeks Mumbai, India. Proc. The National Seminar on Creeks, Estuaries and Mangroves – pollution and conservation. pp. 266–275.

Chapter 18

Diversity of Marine Mammals of India—Status, Threats, Conservation Strategies and Future Scope of Research

S. Venu and B. Malakar

Department of Ocean Studies and Marine Biology, Pondicherry University, Port Blair, Andaman and Nicobar Islands, India

INTRODUCTION

Marine mammals are some of the most amazing living creatures on earth. Their elegance and intelligence have been observed and studied by man from time immemorial. Close to 130 marine species are known from the world's oceans (Jefferson *et al.*, 2008), belonging to three major orders: Cetacea (whales, dolphins, porpoises), Sirenia (manatees and dugong) and Carnivora (sea otters, polar bears and pinnipeds). Their origin on earth can be traced back to 50 million years ago when the first whale-like animal, called "Archaeocetes," evolved from land mammals with four legs (Carwardine, 1995). Today they are known from all over the world's oceans and seas, from estuarine to coastal and oceanic forms. They are widely distributed from poles to the tropics. Though they are strikingly similar to marine fish such as sharks, yet their most common differentiation is the tail: marine mammals have a horizontal tail which moves up and down, whereas fish have a vertical tail which moves from side to side. Other differentiations are the skin, as fish have scales, and marine mammals do not have gills like fishes but breathe through their blowholes on the dorsal side of their heads, for which they need to surface regularly to breathe in air. There are few other morphological differences but many anatomical differences between fish and mammals. Behavior of marine mammals has been an interesting area of study for marine researchers over the years. Their social life is amongst the most prominent in the marine world. Pods of dolphins can be as large as two thousand to three thousand individuals, which are mostly offshore varieties, and as small as one to ten individuals, which are mostly near-shore forms; for whales the group size may vary from one to 50 individuals or more

Marine Faunal Diversity in India. DOI: 10.1016/B978-0-12-801948-1.00018-5

depending on their feeding or breeding behavior. Organized hunting and playful events are observed regularly amongst marine mammals. General behavior includes breaching, flipper-slapping, lobtailing, spyhopping, bow-riding, etc. Some of these behaviors are believed to be used to communicate with other individuals or groups, but much is still a mystery for man. One of the most commonly used techniques by all marine mammals to find direction and food is echolocation. Songs or other typical sounds are used by some marine mammals to communicate with other individuals and pods. Studies of some baleen whales' hearing apparatus suggest that their hearing is best adapted for low sound frequencies (McCauley, 1994; Richardson *et al.*, 1995). Songs of humpback whales are being studied to unravel the mystery behind their use and also as a method to detect the whales when not visually possible (Norris *et al.,* 1999; Swartz *et al.*, 2003).

Mammals play an important role at the top of the food chain in marine ecosystems. Most mammals are elusive creatures spending a major portion of their lives underwater and offshore in open seas, making them difficult to study (Carwardine, 1995). Coastal dolphins (*Tursiops* spp.) have been the most widely seen dolphins and are closely associated to fishing activities as they readily approach any fishermen's boat. Leatherwood and Reeves (1982) reported that some fishermen kill dolphins to prevent damage to fishing gear and to prevent their stealing catch or bait. The animals' inquisitive nature in the vast seas has been an interesting characteristic leading them to approach any boat or ship and ultimately making them easy prey for the whaling industry. Bannister (2002) estimated that in the 1930s the annual average catch by whaling was around 30,000 individuals per year, mostly consisting of blue whale, sei whale and minke whale, and by the 1970s the catch decreased to around ten thousand to fifteen thousand individuals per year. It is an interesting fact that, though whaling decreased the stock of most whales drastically, including the blue whale (the largest animal on planet earth), yet to date not a single species has become extinct through whaling (Bannister, 2002). The general awareness by the end of the last century helped in this regard when most of the countries turned towards conservation of marine mammals. By the end of the 20th century, whaling had decreased considerably as a result of the worldwide ban by the International Whaling Commission (IWC), and a new industry of whale watching began to thrive (Hoyt, 2002). By 1998 the value of this industry was estimated to be around US$1 billion, with 9 million people taking part in 87 countries worldwide (Hoyt, 2002). By 2009, it was estimated that 13 million tourists had undertaken tours to see whales, dolphins and porpoises in their natural habitat, generating US$2.1 billion and employing close to 13,000 people in 119 countries (O'Connor *et al.*, 2009; Cisneros-Montemayor *et al.,* 2010).

The habitat and distribution of marine mammals may vary from the tropics to the poles depending on their basic physiology, breeding and feeding patterns. For example, bottlenose dolphins are reported from the tropical waters around the globe whereas mammals like beluga and narwhal are known from polar and subpolar regions of the northern hemisphere only. Some larger whales such as the blue whale, fin whale, sei whale, minke whale and humpback whale are known to migrate thousands of kilometres every year in the Pacific and Atlantic Ocean from their feeding to breeding grounds and vice versa (Carwardine, 1995; Kasamatsu and Joyce, 1995; Smith *et al.*, 1999; Kasamatsu *et al.*, 2000). Generally the abundance of food in the polar regions makes them good feeding grounds for larger whales. These whales migrate to tropical and subtropical regions for breeding, as confirmed for species such as humpback whales. A total of 21 species (6 Pinnipeds, 8 Baleens and 7 Odontocetes) are known from the Antarctic region, and these were heavily exploited until the late 20th century (Boyd, 2002). The presence of large whales in the Antarctic region signifies the availability of food in the marine food chain in the region, though the environment is much more extreme than in tropical and subtropical regions. Food preferences vary between baleen whales, on one hand, and toothed whales and dolphins on the other. While baleen whales prefer krill and other crustaceans, toothed whales prefer squids and smaller fishes. Deep water giant squids are preferred by sperm whales and hence these whales are known for their deep dives. Dugongs prefer seagrass and rarely algae when seagrass is not abundant.

SPECIES COMPOSITION AND DISTRIBUTION IN INDIAN WATERS

In India all species of marine mammals are protected under the Wildlife (Protection) Act, 1972 (Rajagopalan and Menon, 2003). The International Whaling Commission created the Indian Ocean Sanctuary (IWC, 1980) for marine mammals, especially for whales (Kannan and Rajagopalan, 2013). This Sanctuary consists of those waters of the northern hemisphere from the coast of Africa to 100°E (including the Red and Arabian Seas and the Gulf of Oman) and those waters of the southern hemisphere between 20°E and 130°E from the equator to 55°S (IWC, 1980). Whales are given protection from commercial whaling here. Twenty-five species of cetaceans and one species of sirenian are known from Indian waters (Kumaran, 2002; Vivekanandan and Rajagopalan, 2011). The Central Marine Fisheries Research Institute (CMFRI) has played a pioneer role in marine research and conservation in India since the early 1980s. Close to 85 percent of works including sightings of marine mammals from India are published by CMFRI.

According to Kumaran (2012) India's diversity of marine mammals is one of the highest in the Indian Ocean Sanctuary as declared by IWC. Accordingly, out of 40 species of cetaceans that have been recorded thus far from the Indian Ocean region, 25 species have been reported in Indian waters (Kumaran, 2002). According to IUCN (www.iucnredlist.org), four of these species are endangered, three are vulnerable, eight are least concern and ten are data deficient species.

The Indian scenario on marine mammalian studies can be mainly subdivided into four zones: eastern coast of India (major part Bay of Bengal), western coast of India (Arabian Sea), southern Indian Ocean (including off Sri Lanka and Maldives) and finally Andaman and Nicobar Islands (southeastern part of Bay of Bengal and Andaman Sea). Mammalian research in India gathered momentum with the dawn of the new millennium, and CMFRI was the pioneer in this regard. In the later part of the last century various authors (Silas and Fernando, 1985; Lal Mohan, 1991; Dhandapani, 1992; Bensam and Menon, 1996; Das, 1996) proposed that marine mammalian research in India be given a proper direction, but this was lacking until the beginning of the 21st century.

Marine mammalian research in India is more sighting oriented than actual long term research on a particular group or individual of any species. This database of sightings has been strengthened by the numerous reports of accidental catch and occasional stranding reports along the east and west coast of India and very rarely from the Andaman and Nicobar Islands and Lakshadweep Islands. In general the Arabian Sea is more productive than the Bay of Bengal, because of the regular upwelling of nutrients in the Arabian Sea (McCreary *et al.*, 1993) and thereby is more productive in relation to marine resources and food availability for marine mammals. Rough-bottom topography with deep submarine canyons facilitates the aggregation of cetaceans in some parts of the Bay of Bengal (Kumaran, 1989).The Arabian marine mammals (Red Sea, Gulf of Oman, Gulf of Aden, Persian Gulf and Gulf of Aden) are some of the most well-studied cetaceans in the world, by contrast with those in Indian waters. Some 25 species of marine mammals are known from the Arabian region (Baldwin and Minton, 2000).

From the list of species in Table 18.1, only a few research studies are available, for selected species (six species such as Dugong and Irrawaddy dolphin), on their behavior and habitat, while for most other species (20 species such as blue whale, fin whale and pygmy sperm whale) details are either unavailable or insufficient (Vivekanandan and Jeyabaskaran, 2012) and have been confirmed only by rare sighting, occasional stranding or accidental catch in any fishing gear. This shows the neglected study of marine mammals in the context of Indian waters.

CONSERVATION MEASURES BY VARIOUS ORGANIZATIONS

Studies on marine mammals are mainly concentrated in international waters, and various institutes and NGOs have made a great breakthrough in studies related to life in the sea. The International Whaling Commission (IWC), United Nations Environmental Programme (UNEP), International Union for Conservation of Nature and Natural Resources (IUCN), etc. have played pioneer roles in protection, conservation and management of marine mammals in the world's oceans.

The IWC (www.iwc.int) is an international body set up by the terms of the International Convention for the Regulation of Whaling (ICRW), which was signed in Washington, DC, USA, on 2 December 1946. The main objective of the IWC is to provide for the proper conservation of whale stocks around the

TABLE 18.1 List of Marine Mammals Reported from India

Sl. No.	Common name	Scientific name
1.	Sea cow	*Dugong dugon* (Müller, 1776)
2.	Finless porpoise	*Neophocaena phocaenoides* (Cuvier, 1829)
3.	Killer whale	*Orcinus orca* (Linnaeus, 1758)
4.	False killer whale	*Pseudorca crassidens* (Owen, 1846)
5.	Pygmy killer whale	*Feresa attenuata* (Gray, 1874)
6.	Short finned pilot whale	*Globicephala macrorhynchus* (Gray, 1846)
7.	Melon headed whale	*Peponocephala electra* (Gray, 1846)
8.	Cuvier's beaked whale	*Ziphius cavirostris* (Cuvier, 1823)
9.	Indo-Pacific beaked whale	*Indopacetus pacificus* (Longman, 1926)
10.	Irrawaddy dolphin	*Orcaella brevirostris* (Owen, 1866)
11.	Rough toothed dolphin	*Steno bredanensis* (Cuvier, 1828)
12.	Risso's dolphin	*Grampus griseus* (Cuvier, 1812)
13.	Pantropical spotted dolphin	*Stenella attenuate* (Gray, 1846)
14.	Pantropical spinner dolphin	*Stenella longirostris* (Gray, 1828)
15.	Striped dolphin	*Stenella coeruleoalba* (Meyen, 1833)
16.	Long-beaked common dolphin	*Delphinus capensis* (Gray, 1828)
17.	Indo-Pacific humpbacked dolphin	*Sousa chinensis* (Osbeck, 1765)
18.	Common bottlenose dolphin	*Tursiops truncates* (Montagu, 1821)[1]
19.	Indo-Pacific bottlenose dolphin	*Tursiops aduncus* (Ehrenberg, 1832)
20.	Blue whale	*Balaenoptera musculus* (Linnaeus, 1758)
21.	Fin whale	*Balaenoptera physalus* (Linnaeus, 1758)
22.	Bryde's whale	*Balaenoptera edeni* (Anderson, 1879)
23.	Minke whale	*Balaenoptera acutorostrata* (Lacépède, 1804)
24.	Humpback whale	*Megaptera novaeangliae* (Borowski, 1781)
25.	Sperm whale	*Physeter macrocephalus* (Linnaeus, 1758)
26.	Pygmy sperm whale	*Kogia breviceps* (de Blainville, 1838)
27.	Dwarf sperm whale	*Kogia sima* (Owen, 1866)

[1]*The presence of* Tursiops truncatus *(common bottlenose dolphin) in Indian waters is a debatable issue.*

world and to monitor the whaling industry. In 1982 the IWC adopted a moratorium on commercial whaling. Countries such as Japan, Norway and Iceland have their own quotas as sanctioned by IWC every year. In 1994, the Southern Ocean Whale Sanctuary was created by the IWC to be similar to the Indian Ocean Sanctuary as declared in 1980. As of June 2013 there were 88 members signatory to the IWC.

Since 1978–79, the IWC has been conducting shipboard line transect surveys of whales in the Southern Ocean. By 2003–04, three circumpolar sets of surveys had been completed. It was found that abundance of Antarctic Minke whales (*Balaenoptera bonaerensis*) was 645,000, 786,000 and 338,000 calculated using conventional line transect estimation (Branch 2006). Being the most abundant of all the large whales, Minke whales are still killed in thousands every year by countries such as Japan and Norway. More than 330,000 blue whales *Balaenoptera musculus* were slaughtered for meat, skin, oil, etc. from the Antarctic between 1904 and 1978 (Tonnessen and Johnsen, 1982). Removal of large whales (blue whale, fin whale, humpback whale and sperm whale) can significantly affect marine ecosystems (Laws, 1985; Bengston and Laws, 1985; Kasamatsu, 2000). Kasamatsu (2000) described the species diversity of the whale community in the Antarctic after years of conservation efforts.

Worldwide there are a number of NGOs and institutes who are playing a major role in research and conservation of marine mammals. They include: the Whale and Dolphin Conservation Society, UK; Western Whale Research, Australia; Centre for Whale Research, Australia; The Wildlife Conservation Society, USA; Dolphin Research Institute, Australia, etc.

Interaction of marine mammals with fishery activities has been a subject of study over the years, as in most cases it causes loss of revenue for fisherfolk as well as death or injury to marine mammals. Gill nets and purse seines among all other fishing gear have been identified as the main cause of marine mammal mortality worldwide (Perrin *et al.*, 1994; Wise *et al.*, 2001; Read *et al.*, 2006). Purse seine nets have been the main cause of dolphin mortality worldwide (Archer *et al.*, 2001; Wise *et al.*, 2001).

Research over the years has revealed that individuals of the same species in any population can have distinct markings (scarring and pigmentation pattern) or minor variations in the shape of their dorsal and tail fin (Würsig and Jefferson, 1990; Carwardine, 1995; Dalla Rosa *et al.*, 2001). This in turn can help researchers study the behavioral pattern of particular individual(s) of a pod of dolphins or whales over a period of time. For example it has been confirmed that each humpback whale has distinctive black and white markings that acts like fingerprints in humans. Migration patterns have been studied for a number of species.

Detailed study of marine mammals worldwide over the years has solved some problems of identification of a number of species. For example, the close similarity between two *Tursiops* species—*Tursiops aduncus* and *Tursiops truncates*—has created difficulty in identification, and individuals have been misidentified on many occasions. Commonly known as bottlenose dolphins, these are the best known and most common of all cetaceans (Rice, 1998). Previously *Tursiops* was thought to be a mono-specific genus (Vivekanandan and Jeyabaskaran, 2012) with only *T. truncates* as the reported species. But recently, through genetic studies, it is confirmed that the genus consists of two species (Hale *et al.*, 2000), and the bottlenose dolphin from Indian seas is *T. aduncus*

(Jayasankar *et al.*, 2008). It has also been found through genetic studies that *T. aduncus* is more closely related to pelagic *Stenella* and *Delphinus* species and in particular to *S. frontalis*, than to *T. Truncates* (LeDuc *et al.*, 1999). There are minor variations in morphological characters: for example, *T. truncates* is larger than *T. aduncus* and has a shorter beak. Body coloration in more or less uniform grey in *T. truncates* (offshore ones) (Carwardine, 1995) whereas in *T. aduncus* coloration dorsally is dark grey and light pale grey ventrally and may have dark spotting with sexual maturity (Vivekanandan and Jeyabaskaran, 2012).

The gain in momentum of marine mammalian research and conservation since the 1980s has resulted in a number of success stories as well as greater understanding of some species. For example, years of conservation efforts have led to the encouraging reversals of stock of Pacific Grey whales and Atlantic Humpback whales after they were hunted to near extinction in the last century, as revealed by the IWC Scientific Committee (Bannister, 2002). But the committee is still concerned about the stock of North Atlantic Right whales, which has shown little improvement. Stocks of Humpback, Fin, Bowhead, and Grey whales are estimated to be increasing at a rate of 3–7 percent per year (George *et al.*, 2004; Zerbini *et al.*, 2006; Calambokidis *et al.*, 2008; Punt and Wade, 2010). The IWC releases yearly reports on the status of whaling and conservation of marine mammals, which are invaluable for mammalian research throughout the world.

RESEARCH ON MARINE MAMMALS IN INDIA

The Indian subcontinent had over 2.5 million fishermen and an estimated l,216,000 passive gill nets in the early 1990s (Lal Mohan, 1994). An estimated thousand to fifteen hundred cetaceans were caught annually in India, of which most were dolphins (Lal Mohan, 1994). Jayaprakash *et al.* (1995) studied by-caught dolphins landed at Fisheries Harbour, Cochin and noted that the magnitude of such mortality along the Indian coast can be compared to that of the eastern Pacific, which is very alarming. Recent estimates suggest that close to nine thousand to ten thousand cetaceans are killed by gill nets every year in India (Yousuf *et al.*, 2008). According to a marine fisheries census in 2006, there are 14,183 motorized gill netters operating all along the Indian coast (CMFRI, 2006). Interviewing fishermen, Yousuf *et al.* (2008) concluded that 95% were aware that marine mammals are protected under law in India. With the increase in the length of gill nets and the extension of fishing grounds to oceanic waters in search of more catch in recent years, the possibility of entanglement of more and more cetaceans has increased considerably (Yousuf *et al.*, 2008). Kumaran (2002) reviewed the status of mammalian research in India and suggested measures to enhance research on marine mammals, which has not been tapped to its potential. Prior to that, mammalian research in India mostly consisted of numerous reports of stranding and accidental catch during fishery activities (data mostly compiled by CMFRI), especially on the east and west coast of India.

Until 2003 marine mammal research in India was primarily passive in nature and was based on numerous reports of stranding (Silas and Pillay 1960; Balan 1976; Lal Mohan 1992; Baby 1996; Balasubramanium 2000, Anoop *et al.* 2004; Afsal and Rajagopalan 2007; Raghunathan *et al.*, 2013), accidental catch (Sivaprakasam, 1980; Shantha *et al.*, 1987; Venkataramana and Achayya, 1998; Chandrakumar, 1998; Thiagarajan *et al.*, 2000; Yousuf *et al.*, 2008) and occasional sightings (Rao *et al.*, 1989; Parsons, 1998; Sutaria and Jefferson, 2004; Sinha, 2004; Afsal *et al.*, 2008, D'Souza and Patankar, 2009, Abhilash *et al.*, 2011, Kannan and Rajagopalan, 2013). Thereafter, dedicated surveys were carried out by CMFRI over the last decade and collected a good amount of information on marine mammals of Indian seas. In recent years, sighting of killer whales near Nicobar (Abhilash *et al.*, 2011), stranding of 40 pilot whales in North Andaman (Raghunathan *et al.*, 2013) and a mammalian survey on board FORV *Sagar Sampada* in 2005 in the waters of the Andaman and Nicobar region (Kannan and Rajagopalan, 2013) have been three major reports from these islands. The most comprehensive study of marine mammalian fauna in India in recent times has been carried out on board FORV *Sagar Sampada* between October 2003 and November 2011 (Vivekanandan and Jeyabaskaran, 2012). During this period of survey, 626 sightings were recorded of 8674 individuals from 18 species of marine mammals. This project was part of the CMFRI initiated research project entitled "Studies on Marine Mammals of the Indian EEZ and the Contiguous Seas" which was funded by Centre for Marine Living Resources and Ecology (CMLRE), Ministry of Earth Sciences. The sightings carried out on board FORV *Sagar Sampada* found that the spinner dolphin (*Stenella longirostris*) was the most abundant species of marine mammal in the Indian EEZ (CMFRI, 2007). In recent times, studies on molecular aspects have been undertaken in CMFRI to confirm species of marine mammals and understand their phylogeny (Jayasankar *et al.*, 2006, 2008, 2011).

The study of dugong is an interesting case in India; once this animal was abundant throughout the coastal states of India including the islands of Lakshadweep and Andaman and Nicobar. Today it is an endangered mammal in India and immediately requires protection. Dugongs have already disappeared from countries such as Mauritius, Maldives, Cambodia, and Vietnam (UNEP). Dugongs are long-living mammals and have low reproductive rate and longer generation time (Marsh, 1999). According to Bryden *et al.* (1998), dugongs' low reproductive rate means that a very high proportion (more than 95%) of adult animals have to survive each year for a population to continue in any area. Studies on dugong have been carried out by D'souza and Patankar (2009) and D'Souza *et al.* (2011) in recent years from the Andaman and Nicobar Islands. It is now estimated that only 200 individual dugong are present in small pockets along the vast coastline of India (Sivakumar, 2013). As a result of hunting, accidental catch during fishery activities and rapid habitat destruction, this animal is endangered today in India (Vivekanandan and Jeyabaskaran, 2012). Dugongs are listed as 'Vulnerable to Extinction' in the IUCN Red List of Threatened Species, and in India it is listed under Schedule I of the Wildlife (Protection) Act, 1972 (D'souza and Patankar, 2009). At present in India dugongs are found primarily in four locations: Gulf of Mannar, Palk Bay, Gulf of Kachchh and Andaman and Nicobar Islands (Vivekanandan and Jeyabaskaran, 2012). Pandey *et al.* (2010) estimated the dugong population of India, using an interview-based survey, and found the population to be between 131 and 254 individuals. Lack of proper data and research on dugongs in India is evident, and the result is that very soon dugongs may become extinct if proper conservation measures are not taken immediately.

MODERN TECHNIQUES USED IN MAMMALIAN RESEARCH

Various methods are applied worldwide to study marine mammals, in countries such as the USA, Australia, Japan, the UK, South Africa and also in some Arabian nations. These same methods can be applied in India to further understand the resident as well as seasonal population of marine mammals. Some of the methods are described below.

Methods

Photo-Identification This is one of the most commonly used techniques to identify dolphins and whales over a geographical area over a period of time. This technique has been very successful in many countries. The data obtained from photo-identification, in conjunction with other data, can provide greater insight into the life history and behavior of individuals or pods (Haase and Schneider, 2001; Lusseau, 2005; Merriman, 2007). By photographing dorsal fins and tail fins, researchers can, with a non-invasive method, identify and follow individuals or pods over time. During the course of inter- and intra-specific interactions and contact with environmental (e.g. sand, rocks and corals) and anthropogenic factors (e.g. entanglement, ship/boat strikes), the thin posterior edge of the dorsal fin of dolphins may become irregular, resulting in recognizable patterns of notches and scars (Würsig and Würsig, 1977). These patterns are analogous to human fingerprints and are unique to each individual (Würsig and Würsig 1977; Wilson *et al.*, 1999). Dalla Rosa *et al.* (2001) catalogued humpback whale photo-id from the Antarctic Peninsula, which was organized according to decreasing amounts of white on the underside of the flukes, which act as 'fingerprints'. Similarly, grey whales were identified and studied in detail in the waters of Washington State by Calambokidis and Schelnder (1998).

Satellite Tagging One of the most reliable techniques to track movements of organisms, tagging has helped in tracking the movement of humpback whales in the Pacific Ocean. It is estimated humpback whales travel thousands of kilometres annually during their north–south migrations in the Pacific and elsewhere (Dawbin, 1966; Carwardine, 1995; Dalla Rosa *et al.*, 2001; Anon, 2010). Satellite tracking has revealed distinct movement patterns for Type B and Type C killer whales in the southern Ross Sea, Antarctica (Andrews *et al.*, 2008). Similarly, movements of satellite-monitored humpback whales *Megaptera novaeangliae* have been carried out by Mate *et al.* (1998), Lagerquist *et al.* (2008) and others over the years.

Biopsy Sampling Biopsy technique is used to collect tissue samples for DNA analysis and for molecular taxonomy of marine mammals. In this technique, small darts are fired upon individual(s) of a pod of dolphins to collect tissue samples to be analyzed later in labs. It is a harmless technique as only minute proportions of tissue are collected. It has been successfully used in the Gulf of Mexico (Sellas *et al.*, 2005; Mullin *et al.*, 2007) and Australia (Bilgmann *et al.*, 2007) to understand the genomics of bottlenose dolphins (*Tursiops* sp.) along with their abundance, stock structure, estimated survivorship, fecundity or breeding patterns and residency patterns over the years. Yearly or seasonal biopsy sampling can provide invaluable data on any given stock of mammals over a geographical area.

Aerial Survey Aerial survey using the line transects method is an efficient way to study marine mammals (Treacy, 1996; Rughet *et al.*, 2005; Hedley *et al.*, 2007). Aerial survey using a plane or helicopter has the advantages of covering great distance over a short time but with low disturbance to the mammals and having clear visual clues of the species being studied. Aerial survey might potentially link with a shipboard-based survey (Hedley *et al.*, 2007) and hence help in verification of data. In the Arctic, aerial surveys of bowhead whales (*Balaena mysticetus*), white whales (*Delphinapterus leucas*) and narwhals (*Monodon monoceros*) have been conducted (Treacy, 1996; Heide-Jørgensen and Acquarone, 2002; Rughet *et al.* 2005) over the years with great success.

Acoustic Surveys Most whales and dolphins sing songs to communicate. For example, male humpbacks sing for prolonged periods and can be good subjects for detection with passive acoustic methods (Swartz *et al.*, 2003). Recent attempts have been made to augment visual surveys with acoustic methods, and this has been found to be effective for blue whales (*Balaenoptera musculus*) and fin whales (*Balaenoptera physalus*) (Clark and Fristrup 1997), bowhead whales (*Balaenamys ticetus*) (Zeh *et al.*, 1993; Clark and Ellison, 1989), sperm whales (*Physeter macrocephalus*) (Barlow and Taylor, 1998; Leaper *et al.*, 2000), and humpback whales (Norris *et al.*, 1999).

Application

The application of these methods in India will not only help in understanding the marine mammals in the Indian Ocean region but also create an invaluable database which will definitely help in their protection and conservation in future.

Whale watching can develop as a by-product of accumulation of knowledge of the distribution and abundance of some species of dolphins and whales in particular arcas where they gather seasonally or annually for feeding or breeding. Such areas will hold enormous potential for mammal-watching programmes which can collect substantial revenues from the tourism industry for India, as in other countries such as the USA, Australia, New Zealand and South Africa.

THREATS TO MARINE MAMMALS

Present-day threats to marine mammals are multi-spectral and mostly related to anthropogenic activities. Leaving aside natural causes of death of marine mammals such as disease, predation, large-scale Unusual Mortality Events (UMEs) often associated with harmful algal blooms (HAB) (Mullin *et al.*, 2007), or stranding, man-made fatalities, direct or indirect, have increased over the years.

Whaling Whaling was a major industry from the 19th century until the latter half of the 20th century, when whales were hunted for meat, skin, oil and other products (Boyd, 2002). Whaling has reduced drastically over the years but not stopped completely. Countries such as Japan, Norway and Iceland continue to kill whales in the name of research and for food. IWC every year sanctions how many whales can be killed by these countries. But illegal killings, often crossing these limits, have ben reported over the years. The dolphin slaughters of Taiji, Japan and of the Faroe Islands are well known worldwide. Every year thousands of dolphins are slaughtered in Japan alone for meat, skin and oil. (Vail and Risch, 2006; Hemmi, 2011).

Habitat Loss Man's ever-increasing needs and ever-growing dependence on marine resources has led to rapid development of coastal areas, thereby degrading the delicate marine ecosystem. The most obvious example is the degrading of the coastal habitat of dugongs in India which has led to their near extinction (Silas and Fernando, 1985; D'souza and Patankar, 2009; Pandey *et al.*, 2010; Sivakumar, 2013). Human competition for the same fishery resources, such as fish and shrimps, often has a negative impact on mammals.

Pollution Marine pollution has been a key issue in recent years: oil pollution, heavy metals, solid waste, ghost nets, etc. Plastics and other marine debris have been found in the viscera of many mammals over the years (Simmonds, 2012). Laist (1997) has reviewed in detail the impact of entanglement and ingestion of marine debris by marine mammals. Wells and Scott (2002) reported that accumulations of chlorinated hydrocarbons like DDT and PCB can result in decline in immune system response in bottlenose dolphins.

Boat Strikes The use of more and more fishing vessels and huge ships for travel and transport has been a major threat to marine mammals, with numerous reports of ship strikes every year. Neilson *et al.* (2012) summarized the whale and vessel collisions in Alaska from 1978 to 2011; they found documentation of collisions to be a challenging task, yet there were an estimated 108 reports of collisions, of which 25 resulted in the death of a whale. The IWC has set up a Vessel Strike Data Standardization Group (VSDG) in 2005 to examine the issue of ship strikes with cetaceans and hence to create a global database and measurements to avoid the same. Other similar reports of ship strikes have been made by authors such as Wiley *et al.* (1995), Best *et al.* (2001), Capella and Falk (2001), Knowlton and Kraus (2001), Laist *et al.* (2001), Jensen and Silber (2003) and others. In most of the cases, collision with a large ship results in severe injury or death of the marine mammal. The effect of sonic devices used by boats, ships and submarines has been discussed by Gordon *et al.* (1998) but is still not clearly understood. Lately, the impact of boat traffic has also been an area of study, especially in relation to growing tourism of mammal-watching programs (Constantine, 2001; Stensland and Berggren, 2007).

Accidental Catch Accidental catch of marine mammals in fishing gear has been reported extensively (Sivaprakasam, 1980; Shantha *et al.*, 1987; Venkata-ramana and Achayya, 1998; Thiagarajan *et al.*, 2000; Yousuf *et al.*, 2008) from around the world. With ever-increasing fishing activities, more efficient and larger gear, the chances of marine mammals becoming entangled is increased. India's growing dependence on marine living resources will only complicate this danger further in future.

Climate Change Unknown impacts of climate change on habitat may directly or indirectly affect the food availability of the organisms. Besides, it may also make these wonderful animals susceptible to infections caused by microorganisms, parasites, etc. This issue is still being studied, and much is yet to be understood. But some evidence of climate change and its impact on mammalian fauna can be seen in Antarctica. Smetacek (2008) studied the declining trend of krill stock over the last century due to climate change and the decrease in abundance of whales. Understanding these complex phenomena might take some time, but the impacts are starting to appear. According to Burns (2002), climate change can have either positive or negative impact on marine mammals, which will vary depending on species and their geographical locations.

CONCLUSION AND RECOMMENDATIONS

The presence of mammals in the marine environment is vital for the sustainability of the complex food web in the seas and oceans of the world. Man's constant interference might one day result in the catastrophe of complete extinction of many marine mammals. India can play a leading role in future study and conservation of marine mammals, considering its vast coastline and the availability of marine living resources. Dhandapani and Alfred (1998) consider India to be a very strongly conservation oriented country and therefore suggest establishing *in situ* conservation programmes for species present in the coastal areas of India. Over the years a number of authors (Silas and Fernando, 1985; Lal Mohan, 1991; Kumaran, 2002; Pandey *et al.*, 2010; D'Souza and Patankar, 2011; Vivekanandan and Rajagopalan, 2011; Vivekanandan and Jeyabaskaran, 2012; Kumarran, 2012; Sivakumar, 2013) in India have highlighted the need to study in detail and formulate conservation strategies for the marine mammals of India. A dedicated institute on marine and riverine mammalian research in the country can be of great significance in the coming years, similar to the situation in countries such as the USA, Australia, South Africa and various European nations. The need to vary conservation strategies according to the requirements of different species is also very important. For example a conservation strategy for dugongs will be very different from that for other coastal species. Again, Irrawaddy dolphins need special attention with respect to protection and conservation, compared with other dolphins such as *Tursiops* spp. and *Stenella* sp., because their primary habitat in India is just the Chilika Lagoon at present. Some recommendations have been put forward by Kumaran (2002), Vivekanandan and Jeyabaskaran (2012) and Sivakumar (2013) which can be implemented over the years as the next step in marine mammalian research in India. Public awareness is one of the major ways to conserve marine mammals, especially for coastal species like dugong, bottlenose dolphins, etc. More research is needed on developing pingers and excluder devices and their effective use by fishermen to minimize by-catch of marine mammals, as recommended by Yousuf *et al.* (2008). Research on understanding the stranding of marine mammals along the Indian coastline is required as there have been numerous reports of stranding over the years. Raghunathan *et al.* (2013) have posited that some of the reasons for stranding are related to undersea earthquakes, errors in navigation by marine mammals, or while following fast-moving dolphins and prey, tidal currents that change continuously, and sonic waves from ships and submarines that might interfere with marine mammals' navigational abilities. Vivekanandan and Rajagopalan (2011) suggested establishing a marine mammal stranding network to keep track of regular stranding events and to create a database. Educating fishermen will not only help in reducing accidental catches, but also their sighting reports will help in formulation of conservation strategies and knowledge of areas of aggregation in the EEZ of India. Reviewing existing legislation and implementing new legislation, along with a complete action plan for marine mammals in the coming years, will enhance conservation strategies and will trigger more research with modern tools and techniques in India.

REFERENCES

Abhilash, K.S., Anoop, B., Yousuf, K.S.S.M., Jeyabaskaran, R., Vivekanandan, E., 2011. Occurrence of Killer whale Orcinus orca in Andaman waters. In: Jones, S. (Ed.), Centenary Colloquium on 'Challenges in Marine Mammal Conservation & Research in the Indian Ocean', 26-27 August 2011, Kochi.

Afsal, V.V., Rajagopalan, M., 2007. Strandings of whales along Gulf of Mannar and Palk Bay. Marine Fisheries Information Service, Technical and Extension Series 191, 25–26.

Afsal, V.V., Yousuf, K.S.S.M., Anoop, B., Anoop, A.K., Kannan, P., Rajagopalan, M., Vivekanandan, E., 2008. A note on cetacean distribution in the Indian EEZ and contiguous seas during 2003-07. J. Cetacean Res. Manage 10, 209–216.

Andrews, R.D., Pitman, R.L., Ballance, L.T., 2008. Satellite tracking reveals distinct movement patterns for Type B and Type C killer whales in the southern Ross Sea. Antarctica. Polar Biol. (2008) 31, 1461–1468.

Anon, 2010. First census of marine life 2010 Highlights of a decade of Discovery, p. 64.

Anoop, A.K., Rohit, P., Dineshbabu, A.P., Nayak, T.H., Kemparaju, S., 2004. Record of stranded whales along Karnataka coast. Marine Fisheries Information Service, Technical and Extension Series 182, 16.

Archer, F., Gerrodette, T., Dizon, A., Abella, K., Southern, S., 2001. Unobserved kill of nursing dolphin calves in a tuna purse-seine fishery. Marine Mammal Science 17, 540–554.

Baby, K.G., 1996. On a whale stranded at Anchangadi, Trichur district, Kerala. Marine Fisheries Information Service. Technical and Extension Series 142, 17.

Balan, V., 1976. A note on a juvenile Indian porpoise *Neomerisphocaenoides* (Cuvier) caught off Calicut. Indian J. Fisheries 23, 263–264.

Balasubramanium, T.S., 2000. On a sei whale *Balaenoptera borealis* stranded at Vellapatti along the Gulf of Mannar coast. Marine Fisheries Information Service, Technical and Extension Series 163, 13–14.

Baldwin, R., Minton, G., 2000. Whales and Dolphins of the Arabian Peninsula. Paper presented to The Second Arab International Conference and Exhibition on Environmental Biotechnology (Coastal Habitats), April 2000. Abstracts: p. 39.

Bannister, J.L., 2002. Baleen Whales, Mysticetes. In: Perrin, W.F., Würsig, B., Thewissen, J.G.M. (Eds.), Encyclopaedia of Marine Mammals. Academic Press, San Diego, CA, pp. 62–72.

Barlow, J., Taylor, B., 1998. Preliminary abundance of sperm whales in the north eastern temperate Pacific estimated from a combined visual and acoustic survey. Reports of the International Whaling Commission 20, p18, SC/50/CAWS.

Bengston, J.L., Laws, R.M., 1985. Trends in crabeater seal age at maturity: an Insight into Antarctic marine interactions. In: Siegfried, W.R., Condy, P.R., Laws, R.M. (Eds.), Antarctic nutrient cycles, food, webs. Springer-Verlag, Berlin, pp. 669–675.

Bensam, P., Menon, N.G., 1996. Conservation of Marine Mammals. In: Menon, N.G., Pillai, C.S.G. (Eds.), Marine Biodiversity: Conservation and Management. Central Marine Fisheries Research Institute, Cochin, pp. 133–145.

Best, P.B., Peddemors, V.M., Cockcroft, V.G., Rice, N., 2001. Mortalities of right whales and related anthropogenic factors in South African waters, 1963-1998. J. Cetacean Res. Manage 2 (2), 171–176.

Bilgmann, K., Möller, L.M., Harcourt, R.G., Gibbs, S.E., Beheregaray, L.B., 2007. Genetic differentiation in bottlenose dolphins from South Australia: association with local oceanography and coastal geography. Marine Ecological Progress Series 341, 265–276.

Boyd, I.L., 2002. Antarctic Marine Mammals. In: Perrin, W.F., Würsig, B., Thewissen, J.G.M. (Eds.), Encyclopaedia of Marine Mammals. Academic Press, San Diego, CA, pp. 30–36.

Branch, T.A., 2006. Abundance estimates for Antarctic minke whales from three completed circumpolar sets of surveys. Paper SC/58/IA18 presented to the Scientific Committee of the International Whaling Commission, St Kitts, 2006. p. 28.

Bryden, M., Marsh, H., Shaughnessy, P., 1998. Dugongs, Whales, Dolphins and Seals: A Guide to the Sea Mammals of Autralasia. Allen and Unwin, Australia, p176.

Burns, J.J., 2002. Arctic Marine Mammals. In: Perrin, W.F., Würsig, B., Thewissen, J.G.M. (Eds.), Encyclopaedia of Marine Mammals. Academic Press, San Diego, CA, pp. 39–45.

Calambokidis, J., Schelnder, L., 1998. Gray Whale Photographic Identification in 1997. Final Report to National Marine Mammal Laboratory, Seattle, WA, p. 21.

Calambokidis, J., Falcone, E.A., Quinn, T.J., 2008. "SPLASH: structure of populations, levels of abundance and status of humpback whales in the North Pacific," Final Report for Contract AB133F-03-RP-00078, U.S. Department of Commerce, Washington, DC, USA.

Capella, L.F., Falk, P., 2001. Mortality and anthropogenic harassment of humpback whales along the Pacific coast of Colombia. Memoirs of the Queensland Museum 47 (2), 547–553.

Carwardine, M., 1995. Whales, dolphins and porpoises. Dorling Kindersley Limited, London, p. 256.

Chandrakumar, N.P., 1998. Mass entanglement of dolphins in a shore seine near Balaramapuram, Srikakulam District, Andhra Pradesh. Marine Fisheries Information Service. Technical and Extension Series 155, 19.

Cisneros-Montemayor, A.M., Sumaila, U.R., Kaschner, K., Pauly, D., 2010. The global potential for whale watching. Marine Policy 34 (6), 1273–1278.

Clark, C.W., Ellison, W.T., 1989. Numbers and distributions of bowhead whales, *Balaenamysticetus*, based on the 1986 acoustic study off Pt. Barrow. Alaska. Reports of the International Whaling Commission 39, 297–303.

Clark, C.W., Fristrup, K.M., 1997. Whales 95: A combined visual and acoustic survey of blue and fin whales off Southern California. Rep. Int. Whal. Comm 47, 583–600.

CMFRI, 2006. Marine Fisheries Census 2005, Part 1. Kochi: Central Marine Fisheries Research Institute, p. 104.

CMFRI, 2007. Studies on Marine Mammals of Indian EEZ and the Contiguous Seas. New Delhi: Central Marine Fisheries Research Institute, Final Report submitted to Ministry of Earth Sciences, p. 212.

Constantine, R., 2001. Increased avoidance of swimmers by wild bottlenose dolphins (*Tursiops truncatus*) due to long term exposure to swim-with-dolphin tourism. Mar. Mamm. Sci. 17, 689–702.

Dalla Rosa, L., Secchi, E.R., Kinas, P.G., Santos, M.C.O., Martins, M.B., Zerbini, A.N., Bethlem, C.B.P., 2001. Photo identification of humpback whales, *Megaptera novaeangliae*, off the Antarctic Peninsula from 1997/98 to 1999/2000. Memoirs of the Queensland Museum 47 (2), 555–561.

Das, H.S., 1996. The vanishing mermaids of Andaman and Nicobar Islands. Sirenews: Newsletter of the IUCN/SSC Sirenia Specialist group 26, 5–6.

Dawbin, W.A., 1966. The seasonal migratory cycle of humpback whales. In: Norris, K. (Ed.), Whales, dolphins and porpoises. University of California Press, Berkley, pp. 145–170.

Dhandapani, P., 1992. Status of Irrawady River Dolphin *Orcaella brevirostris* in Chilika Lake. J. Marine Biol. Assoc. 34, 90–93.

Dhandapani, P., Alfred, J.R.B., 1998. Conservation of marine mammals in the EEZ of India. Indian J. Marine Sci. 27, 506–508.

D'souza, E., Patankar, V., 2009. First underwater sighting and preliminary behavioural observations of Dugongs (*Dugong dugon*) in the wild from Indian waters. Andaman Islands. J. Threatened Taxa 1 (1), 49–53.

D'souza, E., Patankar, V., 2011. Ecological studies on the Dugong dugon of the Andaman and Nicobar Islands: A step towards species conservation. Nature Conservation Foundation, Mysore, p. 19.

D'Souza, E., Patankar, V., Arthur. R., Alcoverro T., Kelkar, N., 2011. Long-term occupancy trends in a data-poor dugong population in the Andaman and Nicobar Archipelago. PLoS ONE 8(10): e76181. doi:10.1371/journal.pone.0076181.

George, J.C., Zeh, J., Suydam, R., Clark, C., 2004. Abundance and population trend (1978-2001) of western arctic bowhead whales surveyed near Barrow. Alaska. Marine Mammal Science 20 (4), 755–773.

Gordon, J.C.D., Gillespie, D., Potter, J., Frantzis, A., Simmonds, M.P., Swift, R., 1998. The effects of seismic surveys on marine mammals. In: Tasker, M.L. Weir, C. (Eds.), Proceedings of the seismic, marine mammals, workshop., London, pp. 6.1–6.13.

Haase, P.A., Schneider, K., 2001. Birth demographics of bottlenose dolphins, *Tursiops truncatus*, in Doubtful Sound, Fiordland. New Zealand preliminary findings. New Zealand J. Marine Freshwater Res. 35, 675–680.

Hale, P.T., Barreto, A.S., Ross, G.J.B., 2000. Comparative morphology and distribution of the aduncus and truncates forms of bottlenose dolphin *Tursiops* in the Indian and Western Pacific Oceans. Aquatic Mammals 26, 101–110.

Hedley, S., Bravington, M., Gales, N., Kelly, N., Peel, D., 2007. Aerial Survey for Minke Whales off Eastern Antarctica. p. 47.

Heide-Jørgensen, M.P., Acquarone, M., 2002. Size and trends of the bowhead whale, beluga and narwhal stocks wintering off West Greenland. NAMMCO Scientific Publication 4, 191–210.

Hemmi, S., 2011. The dolphin drive hunt: Appropriate management? Observation from the emergency extension of the hunting season of the dolphin drive hunt in Taiji. Tsukuba, Japan: Elsa Nature Conservancy.

Hoyt, E., 2002. Whale Watching. In: Perrin, W.F., Würsig, B., Thewissen, J.G.M. (Eds.), Encyclopaedia of Marine Mammals. Academic Press, San Diego, CA, pp. 1305–1310.

IWC, 1980. Chairman's Report of the thirty-first Annual Meeting. Rep. Int. Whal. Commn. 30, 25–41.

Jayaprakash, A.A., Nammalwar, S., Pillai, S.K., Elayath, M.N.K., 1995. Incidental by-catch of dolphins at Fisheries Harbour, Cochin, with a note on their conservation and management in India. J. Marine Biol. Assoc. India 37, 126–133.

Jayasankar, P., Anoop, B., Afsal, V.V., Rajagopalan, M., 2006. Species and sex of two baleen whales identified from their skin tissues using molecular approach. Marine Fisheries Information Service, Technical and Extension Series 190, 16–17.

Jayasankar, P., Anoop, B., Rajagopalan, M., 2008. PCR-based sex determination of cetaceans and dugong from the Indian Seas. Current Science 94, 1513–1516.

Jayasankar, P., Anoop, B., Vivekanandan, E., Rajagopalan, M., Yousuf, K.M.M., Reynold, P., Krishnakumar, P.K., Kumaran, P.L., Afsal, V.V., Krishnan, A.A., 2008a. Molecular Identification of delphinids and finless porpoises (Cetacea) from Arabian Sea and Bay of Bengal. Zootaxa 1853, 57–67.

Jayasankar, P., Patel, A., Khan, M., Das, P., Panda, S., 2011. Mitochondrial DNA diversity and PCR-based sex determination of Irrawaddy dolphin (*Orcaella brevirostris*) from Chilika Lagoon. India. Molecular Biology Reports 38, 1661–1668.

Jefferson, T.A., Stacey, P.I., Baird, R.W., 2008. Marine Mammals of the World: A Comprehensive Guide to their Identification. Academic Press, p. 592.

Jensen, S., Silber, G.K., 2003. Large whale ship strike database. NOAA Technical Memorandum NMFS-OPR-25, U.S. Department of Commerce, Washington, DC, USA.

Kannan, P., Rajagopalan, M., 2013. Sightings of marine mammals in Bay of Bengal, Andaman and Nicobar Islands waters. In: Venkataraman, K., Sivaperuman, C., Raghunathan, C. (Eds.), Ecology and Conservation of Tropical Marine Faunal Communities. Springer-Verlag, Berlin and Heidelberg, pp. 323–330.

Kasamatsu, F., 2000. Species diversity of the whale community in the Antarctic. Marine Ecology Progress Series 200, 297–301.

Kasamatsu, F., Joyce, G.G., 1995. Current status of odontocetes in the Antarctic waters. Antarct Sci. 7, 365–379.

Kasamatsu, F., Matsuoka, K., Hakamada, T., 2000. Interspecific relationships in density among the whale community in the Antarctic. Polar Biol. 23, 466–473.

Knowlton, R., Kraus, S.D., 2001. Mortality and serious injury of northern right whales (*Eubalaenaglacialis*) in the western North Atlantic Ocean. J. Cetacean Res. Manage. 2, 1–15.

Kumaran, P.L., 1989. Systematics and organ weights of the dolphins and a whale collected from Porto Novo, southeast coast of India. MSc dissertation, CAS in Marine Biology. Annamalai University, India, p. 52.

Kumaran, P.L., 2002. Marine mammal research in India: a review and critique of the methods. Current Science 83 (10), 1210–1220.

Kumarran, R.P., 2012. Cetaceans and cetacean research in India. J. Cetacean Res. Manage. 12, 159–172.

Lagerquist, B.A., Mate, B.R., Ortega-Ortiz, J.G., Winsor, M., Urb an-Ramirez, J., 2008. Migratory movements and surfacing rates of humpback whales (*Megaptera novaeangliae*) satellite tagged at Socorro Island. Mexico. Marine Mammal Science 24, 815–830.

Laist, D.W., 1997. Impacts of marine debris: entanglement of marine life in marine debris including a comprehensive list of species with entanglement and ingestion records. In: Coe, J.M., Rogers, D.B. (Eds.), Marine Debris: Sources, Impacts and Solutions. Springer, New York, pp. 99–140.

Laist, D.W., Knowlton, A.R., Mead, J.G., Collet, A.S., Podesta, M., 2001. Collisions between ships and whales. Marine Mammal Science 17 (1), 35–75.

Lal Mohan, R.S., 1991. Research needs for the better management of dolphins and dugongs of Indian Coast. Proceedings of the National Symposium on Research and Development in Marine Fisheries Sessions. CMFRI Bulletin 44, 662–666.

Lal Mohan, R.S., 1992. Observations on the whales Balaenopteraedeni. B. musculus and Megapteranovaeangliae washed ashore along the Indian coast with a note on their osteology. J. Marine Biol. Assoc. India 34, 253–255.

Lal Mohan, R.S., 1994. Review of gillnet fisheries and cetacean bycatches in the northeastern Indian Ocean. Reports of the International Whaling Commission. Special Issue 15, 329–343.

Laws, R.M., 1985. The ecology of the Southern Ocean. Am. Sci. 73, 26–40.

Leaper, R., Gillespie, D., Papastavrou, V., 2000. Results of passive acoustic surveys for odontocetes from the British Antarctic Survey research vessel James Clark Ross in the Southern Ocean. 300W to 700W. J. Cetacean Res. Manage 2 (3), 187–196.

Leatherwood, S., Reeves, R.R., 1982. Bottlenose dolphin (*Tursiops truncatus*) and other toothed cetaceans. In: Chapman, J.A., Feldhamer, G.A. (Eds.), Wild mammals of North America: biology, management and economics. John Hopkins University Press, Baltimore, pp. 369–414.

LeDuc, R.G., Perrin, W.F., Dizon, A.E., 1999. Phylogenetic relationship among the delphinids cetaceans based on full cytochrome b sequences. Mar. Mamm. Sci. 15, 619–648.

Lusseau, D., 2005. The residency pattern of bottlenose dolphins (*Tursiops* spp.) in Milford Sound, New Zealand, is related to boat traffic. Marine Ecology Progress Series 295, 265–272.

Marsh, H., 1999. Reproduction in sirenians. In : Boyd, I.L., C., Lockyer, H.D., Marsh, Chapter 6 Reproduction in Marine Mammals. In : Reynolds, J.E., J.R., Twiss, (Eds). Marine Mammals. Smithsonian Institute Press, Washington DC. pp. 243–256.

Mate, B.R., Gisiner, R., Mobley, J., 1998. Local and migratory movements of Hawaiian humpback whales tracked by satellite telemetry. Canadian Journal of Zoology 76, 863–868.

McCauley, R.D., 1994. Seismic surveys. In: Swan, J.M., Neff, J.M., Young, P.C. (Eds.), Environmental implications of offshore oil, gas development in Australia – findings of an independent scientific review. Australian Petroleum Exploration Association, Sydney, pp. 19–121.

McCreary, J.P., Kundu, P.K., Molinari, R.L., 1993. A numerical investigation of dynamics, thermodynamics and mixed-layer processes in the Indian Ocean. Prog. Oceanog. 31 (3), 181–244.

Merriman, M.G., 2007. The occurrence and behavioural ecology of bottlenose dolphins (*Tursiops truncatus*) in the Marlborough Sounds, New Zealand. MSc Thesis. Massey University, Auckland, New Zealand.

Mullin, K., Rosel, P., Hohn, A., Garrison, L., 2007. Bottlenose dolphin stock structure research plan for the central northern Gulf of Mexico. NOAA Technical Memorandum NMFS-SEFSC-563. p. 27.

Neilson, J.L., Gabriele, C.M., Jensen, A.S., Jackson, K., Straley, J.M., 2012. Summary of Reported Whale-Vessel Collisions in Alaskan Waters. J. Marine Biology 2012, 1–18.

Norris, T.F., McDonald, M., Barlow, J., 1999. Acoustic detections of singing humpback whales (*Megaptera novaeangliae*) in the eastern North Pacific during their northbound migration. J. Acoust. Soc. Am. 106, 506–514.

O'Connor, S.O., Campbell, R., Cortez, H., Knowles, T., 2009. Whale Watching Worldwide: Tourism Numbers, Expenditures and Expanding Economic Benefits. A Special Report from the International Fund for Animal Welfare. IFA Wand Economists at Large, Yarmouth, MA, USA.

Pandey, C.N., Tatu, K.S., Anand, Y.A., 2010. Status of dugong (*Dugong dugon*) in India. GEER Foundation, Gandhinagar, 146 p.

Parsons, E.C.M., 1998. Observations of Indo-Pacific humpbacked dolphins, *Sousa chinensis*, from Goa. Western India. Marine Mammals Science 14, 166–170.

Perrin, W.F., Donovan, G.P., Barlow, J., 1994. Gillnets and cetaceans. *Reports of International Whaling Commission*. Special Issue, 15, p. 629.

Punt, E., Wade, P.R., 2010. Population status of the eastern North Pacific stock of gray whales in 2009. NOAA Technical Memorandum NMFS-AFSC-207, U. S. Department of Commerce, Washington, DC.

Raghunathan, C., Kumar, S.S., Kannan, S.D., Mondal, T., Sreeraj, C.R., Raghuraman, R., Venkataraman, K., 2013. Mass stranding of pilot whale *Globicephala macrorhynchus* Gray. 1846 in North Andaman coast. Current Science 104 (1), 37–41.

Rajagopalan, M., Menon, N.G., 2003. Marine turtles and mammals. In: Jayaprakash, A.A., Mohan Joseph, M. (Eds.), Status of exploited marine fishery resources of India. Central Marine Fisheries and Research Institute, Kochi.

Rao, P.V., Livingston, P., Atmaram, M., 1989. A report on the whales sighted off Mandapam on the Palk Bay side on 5th July, 1988. Marine Fisheries Information Service, Technical and Extension Series 95, 10.

Read, A.J., Drinker, P., Northridge, S., 2006. Bycatch of marine mammals in U.S. and global fisheries. Biol. Conserv. 20, 163–169.

Rice, D.W., 1998. Marine mammals of the world: systematics and distribution. Special publication No. 4, Society for Marine Mammalogy. Allen Press, Lawrence, KS.

Richardson, W.J., Greene, Jr., C.R., Malme, C.I., Thomson, D.H., 1995. Marine mammals and noise. Academic Press, San Diego, p. 576.

Rughet, D.J., Shelden, K.W., Sims, C.L., Mahoney, B.A., Smith, B.K., Litzley, L.K., Hobbs, R.C., 2005. Aerial surveys of belugas in Cook Inlet, Alaska, June 2001, 2002, 2003, and 2004. U.S. Dep. Commer., NOAA Tech. Memo. NMFS-AFSC-149. p. 71.

Sellas, A.B., Wells, R.S., Rosel, P.E., 2005. Mitochondrial and nuclear DNA analyses reveal fine scale geographic structure in bottlenose dolphins (*Tursiops truncatus*) in the Gulf of Mexico. Conservation Genetics 6 (5), 715–728.

Shantha, G., Rajaguru, A., Natarajan, R., 1987. Incidental catches of dolphins (Delphinidae: Cetacea) along Porto Novo, southwest coast of India. Proceedings of the National Symposium on Research and Development in Marine Fisheries, CMFRI, Mandapam Camp (India), 6 Sep. 1987. Abstract No. 118.

Silas, E.G., Fernando, A.B., 1985. The Dugong in India – is it going the way of the Dodo. Proceedings of Symposium on Endangered Marine Mammals and Marine Parks. J. Mar. Biol. Assoc. India 2, 268–271.

Silas, E.G., Pillay, C.K., 1960. The stranding of two false Killer whales (*Pseudorca crassidens* Owen) at Pozhikara, north of Cape Comorin. J. Mar. Biol. Assoc. India 2, 268–271.

Simmonds, M.P., 2012. Cetaceans and Marine Debris: The Great Unknown. J. Mar. Biol. 2012, 1–8.

Sinha, R.K., 2004. The Irrawaddy dolphin *Orcaella brevirostris* of Chilika Lagoon. India. J. Bombay Nat. Hist. Soc. 101, 244–451.

Sivakumar, K., 2013. Status of Dugong dugon in India: Strategies for species recovery. In: Venkataraman, K., Sivaperuman, C., Raghunathan, C. (Eds.), Ecology and Conservation of Tropical Marine Faunal Communities. Springer-Verlag, Berlin and Heidelberg, pp. 419–432.

Sivaprakasam, T.E., 1980. On the unusual occurrence of the common dolphin *Delphinus delphis* Linnaeus in longline catches at Port Blair. Andaman. J. Bombay Nat. Hist. Soc. 77, 320–321.

Smetacek, V., 2008. Are declining Antarctic krill stocks a result of global warming or of the decimation of the whales? In: Daurte, C.M. (Ed.), Impacts of global warming on polar ecosystems. Fundación BBVA, pp. 45–83.

Smith, T.D., Allen, J., Clapham, P.J., Hammond, P.S., Katona, S., Larsen, F., Lein, J., Mattila, D., Palsbol, P.J., Sigurjonsson, J., Stevick, P.T., Oien, N., 1999. An Ocean-Basin-Wide mark-recapture study of the north Atlantic Humpback whale (*Megaptera novaeangliae*). Marine Mammal Science 15 (1), 1–32.

Stensland, E., Berggren, P., 2007. Behavioural changes in female Indo-Pacific bottlenose dolphins in response to boat-based tourism. Marine Ecological Progress Series 332, 225–234.

Sutaria, D., Jefferson, T.K., 2004. Records of Indo-Pacific humpback dolphins (*Sousa chinensis*, Osbeck, 1965) along the coasts of India and Sri Lanka: an overview. Aquatic Mammals 30, 125–136.

Swartz, S.L., Cole, T., Mcdonald, M.A., Hildebrand, J.A., Oleson, E.M., Martinez, A., Clapham, P.J., Barlow, J., Jones, M.L., 2003. Acoustic and Visual Survey of Humpback Whale (*Megaptera novaeangliae*) distribution in the Eastern and Southeastern Caribbean Sea. Caribbean J. Sci. 39 (20), 195–208.

Thiagarajan, R.S., Pillai, K., Balasubramanium, T.S., Chellam, A., 2000. Accidental catch of three Risso's dolphin at Beemapally, near Vizhinjam. Marine Fisheries Information Service. Technical and Extension Series 163, 10.

Tonnessen, J.N., Johnsen, A.O., 1982. The history of modern whaling. C. Hurst & Co Publishers, London.

Treacy, S.D., 1996. Aerial surveys of endangered whales in the Beaufort Sea, Fall 1995. OCS Study MMS 96-0006. Anchorage, Alaska: U.S. Department of the Interior, Minerals Management Service, Alaska OCS Region (unpublished). p. 128.

Vail, C.S., Risch, D., 2006. Driven by demand: Dolphin drive hunts in Japan and the involvement of the aquarium industry. Whale and Dolphin Conservation Society (WDCS), Wiltshire, UK.

Venkataramana, P., Achayya, P., 1998. On the capture of a bottlenose dolphin off Kakinada. Marine Fisheries Information Service. Technical and Extension Series 155, 20.

Vivekanandan, E., Jeyabaskaran, R., 2012. Marine mammal species of India. Central Marine Fisheries Research Institute, Kochi, p. 228.

Vivekanandan, E., Rajagopalan, M., 2011. Challenges in Marine Mammal Conservation & Research in the Indian Ocean. Central Marine Fisheries Research Institute, Kochi.

Wells, R.S., Scott, M.D., 2002. Bottlenose dolphins – *T. truncates* and *T. aduncuns*. In: Perrin, W.F., Würsig, B., Thewissen, J.G.M. (Eds.), Encyclopaedia of Marine Mammals. Academic Press, San Diego, CA, pp. 122–128.

Wiley, D.N., Asmutis, R.A., Pitchford, T.D., Gannon, D.P., 1995. Stranding and mortality of humpback whales, *Megaptera novaeangliae*, in the mid-Atlantic and southeast United States, 1985-1992. Fishery Bulletin 93, 196–205.

Wilson, B., Hammond, P.S., Thompson, P.M., 1999. Estimating size and assessing trends in a coastal bottlenose dolphin population. Ecological Applications 9 (1), 288–300.

Wise, L., Silva, A., Ferreira, M., Silva, M.A., Sequeira, M., 2001. Interactions between small cetaceans and the purse-seine fishery in western Portuguese waters. Scientia Marina 71, 405–412.

Würsig, B., Jefferson, T.A., 1990. Methods of photo-identification for small cetaceans. Reports of the International Whaling Commission (Special Issue) 12, 43–55.

Würsig, B., Würsig, M., 1977. Photographic determination of group size, composition, and stability of coastal porpoises (*Tursiops truncatus*). Science 198, 755–756.

Yousuf, K.S.S.M., Anoop, A.K., Anoop, B., Afsal, V.V., Vivekanandan, E., Kumarran, R.P., Rajagopalan, M., Krishnakumar, P.K., Jayasankar, P., 2008. Observations on incidental catch of cetaceans in three landing centres along the Indian coast. Biodiversity Records JMBA 2 UK Published online, p. 1-6.

Zeh, J.E., Clark, C.W., George, J.C., Withrow, D., Carroll, G.M., Koski, W.R., 1993. Current population size and dynamics. In: Burns, J.J., Montague, J.J., Cowles, C.J. (Eds.), The Bowhead Whale, Special Publication No. 2, The Society for Marine Mammalogy. Allen Press, Lawrence, KA, pp. 409–489.

Zerbini, N., Waite, J.M., Laake, J.L., Wade, P.R., 2006. Abundance, trends and distribution of baleen whales off western Alaska and the central Aleutian Islands. Deep-Sea Research 53 (11), 1772–1790.

Chapter 19

Coastal and Marine Biodiversity of India

K. Venkataraman* and C. Raghunathan[†]

*Zoological Survey of India, Kolkata, West Bengal, India; [†]Andaman and Nicobar Regional
Centre, Zoological Survey of India, Port Blair, Andaman and Nicobar Islands, India

INTRODUCTION

Oceans and major seas cover 70.8 percent or 362 million km^2 of the earth, with a global coastline of 1.6 million km. Coastal and marine ecosystems occur in 123 countries around the world. Marine ecosystems are strongly connected through a network of surface and deep-water currents and they are among the most productive ecosystems in the world. Coastal and marine ecosystems include sand dune areas where freshwater and seawater mix, nearshore coastal areas, and open ocean marine areas. Marine systems extend from low water mark, i.e., 50 m depth, to the high seas, and coastal systems from less than 50 m depth to the coastline. The Indian Ocean accounts for 29 percent of the global ocean area, 13 percent of the marine organic carbon synthesis, 10 percent of the capture fisheries, 90 percent of the culture fisheries, 30 percent of coral reefs, 10 percent of mangroves and has 246 estuaries draining a hinterland greater than 2000 km^2, besides coastal lagoons and backwaters. Being landlocked in the north, and with the largest portion of it lying in the tropics, the Indian Ocean is found to be a region of high biodiversity, with one of the countries in the region, India, rated as one of the mega-biodiversity centres of the world. In the current context of international trade and intellectual property regimes, it is important for all of the Indian Ocean countries to understand their marine biodiversity. India is one among 12 mega-biodiversity countries and 25 hotspots of the richest and highly endangered eco-regions of the world. In terms of marine environment, India has a coastline of about 8000 km. The exclusive economic zone (EEZ) of the country has an area of 2.02 million km^2 comprising 0.86 million km^2 on the west coast, 0.56 million km^2 on the east coast and 0.6 million km^2 around the Andaman and Nicobar islands. Adjoining the continental regions and the offshore islands are a very wide range of coastal ecosystems such as estuaries,

Marine Faunal Diversity in India. DOI: 10.1016/B978-0-12-801948-1.00019-7

lagoons, mangroves, backwaters, salt marshes, rocky coasts, sandy stretches and coral reefs, which are characterized by unique biotic and abiotic properties and processes. A network of 14 major, 44 medium and numerous minor rivers together with their tributaries cover practically the entire country except for the western arid region of the Rajasthan Desert. The total length of the rivers is estimated at over 40,000 km.

The dissimilarities between the Indian west and east coasts are remarkable. The west coast is generally exposed, with heavy surf and rocky shores and headlands, whereas the east coast is generally shelving with beaches, lagoons, deltas and marshes. The west coast is a region of intense upwelling associated with the southwest monsoon (May to September), whereas the east coast experiences only a weak upwelling associated with the northeast monsoon (October to January), resulting in marked differences in hydrographic regimes, productivity patterns and qualitative and quantitative composition of fisheries. All islands on the east coast are continental islands whereas the major island formations in the west coast are oceanic atolls.

Among the Asian countries, India is perhaps the only one that has a long record of inventories of coastal and marine biodiversity dating back at least two centuries. However, these are so diverse in space, time and taxon that it is almost impossible to review all records and reports. The synthesis of what is known of coastal and marine biodiversity in India attempted in this overview relies mainly on systematic accounts, records and reports of two major institutions concerned with surveys and inventories of fauna and flora—the Zoological Survey of India and the Botanical Survey of India—as well as other research organizations such as the Central Marine Fisheries Research Institute and the National Institute of Oceanography.

MARINE ALGAE

Marine macro-algae (seaweeds) from Indian coasts have been fairly well surveyed over several decades. Oza and Zaidi (2001) listed 844 species (including forma and varieties) a decade ago. However, a recent report reveals a total of 936 species of marine algae from different areas of India (Rao, 2010). The most abundant among them are rhodophytes (434 species in 136 genera), followed by chlorophytes (216 species in 43 genera), phaeophytes (191 species in 37 genera) and xanthophytes (3 species in 1 genus). Among these, the maximum number of species has been recorded from Tamil Nadu (302), followed by Gujarat (202), Maharashtra (159), Lakshadweep (89), Andhra Pradesh (79) and Goa (75). The scanty records in other maritime states, especially the Andaman and Nicobar Islands, may not necessarily mean a paucity of algal species but may rather reflect a lack of intensive surveys. However, recent studies conducted by the Botanical Survey of India recorded 206 species of seaweeds from the Andaman and Nicobar Islands (Palanisamy, 2012).

FIGURE 19.1   **Marine macro-algae (seaweeds)** (a) *Gracillaria* sp.; (b) *Turbinaria* sp. (Please see color plate at the back of the book.)

The seaweeds are harvested mainly for use as raw materials for the production of agar, alginates and seaweed liquid fertilizer. The estimated total standing crop of seaweeds in intertidal and shallow waters of the Indian coast is 91,345 tonnes wet weight and 75,373 tonnes in deep water, which consists of 6000 tonnes of agar-yielding seaweeds (Roy and Gosh, 2009). The red algae (*Gelidiella acerosa, Gracilaria edulis, G. crassa, G. foliifera* and *G. verrucosa*) are used for manufacture of agar, and the brown algae (*Sargassum* spp.*, Turbinaria* spp. and *Cystoseira trinodis*) for alginates and seaweed liquid fertilizers (Figure 19.1). The bulk of the harvest is from the natural seaweed beds of the Gulf of Mannar Islands. The data collected by the Central Marine Fisheries Research Institute (CMFRI) on seaweed landings in Tamil Nadu from 1978 to 2000 reveal that the quantity (dry wt) exploited in a year during this period was in the range 102–541 t for *Gelidiella acerosa*, 108–982 t for *Grcilaria edulis,* 2–96 t for *G. crassa,* 3–110 t for *G. foliifera* and 129–830 t for *G. verrucosa* (Silas and Kalimuthu, 1997; Ramalingam, 2000). Recently, *Euchema cottonii* has been introduced in the Gulf of Mannar for commercial farming. Its effect on native species, not known so far, remains a matter of great concern.

SEAGRASSES

Global composition of seagrass ranges from 0.1 to 0.2 percent of the aquatic flora. Seagrasses provide a habitat for *Dugong dugon,* the only herbivorous marine mammal that exists in the sea (Figure 19.2). They also provide exceptional habitats for a wide variety of marine organisms, both plant and animal; these include meiofauna and flora, benthic flora and fauna, epiphytic organisms, plankton and fish, not to mention microbial and parasitic organisms. Seagrass meadows account for 15 percent of the ocean's total carbon storage. The ocean currently absorbs 25 percent of global carbon emissions. It is estimated that seagrasses per square metre are capable of binding about 1 kg of carbon every

FIGURE 19.2 **Seagrass and dugong** (a) Seagrass meadow; (b) *Dugong dugon* grazing on seagrass. (Please see color plate at the back of the book.)

year. Sixty species of seagrass are described from the world's oceans. Fourteen species of seagrass under 6 genera are known from Indian seas (Table 19.1). They are often found in association with coral reef areas. Eleven species are known from the Palk Bay, which include *C. serrulata, H. ovalis* sub sp. *ovalis, K. pinifolia* and *S. isoetifolium.* Thirteen species occur in the Gulf of Mannar Biosphere Reserve, with *Halophila, Halodule, Enhalus* and *Cymodocea* being common among them. *Thalassia* and *Syringodium* are dominant in coral reef areas and coral rubbles, whereas the others are distributed in muddy and fine sandy soils. Along the west coast, only *Halophila* and *Halodule* species are cosmopolitan in distribution, and *Cymodocea* sp. and *Syringodium isoetifolium* occur as very small patches at the southernmost end of Thiruvananthapuram. Nine species, among which *Thalassia hemprichii and Cymodocea rotundata* are dominant, occur in the Andaman and Nicobar Islands. From the Lakshadweep Islands, seven species are known, among which *Thalassia hemprichii* is dominant.

MANGROVES

Mangroves occur in 112 countries, primarily in the tropical regions of the world, with an area of 189,399 km². One hundred species of true mangroves have been described so far from the world. The Indian mangroves cover about 4827 km², with about 57 percent of them along the east coast, 23 percent along the west coast and the remaining 20 percent in Andaman and Nicobar Islands (Figure 19.3). The mangrove formations are of three types: deltaic, backwater-estuarine and insular. The deltaic mangroves occur mainly along the east coast, the backwater-estuarine type along the west coast and the insular in the Andaman and Nicobar Islands. A total of 69 mangrove species belonging to 42 genera and 22 families are known from India (Table 19.2). While several of them are cosmopolitan in distribution, five of them —*Aegialitis rotundifolia, Heritiera fomes, H. kanikensis, Rhizophora annamalayana* and *R. stylosa*—are

TABLE 19.1 Distribution of Sea Grasses in Five Regions of the Indian Coast

Sl. No	Species	Coromandel / East coast	Palk Bay and Gulf of Mannar	West coast	A and N Islands	Lakshadweep
1.	Enhalus acoroide (L.f.) Royle	–	+	–	+	+
2.	Halophilia beccarii (Asch).	+	+	+	–	–
3.	Halophilia decipiens Osten.f.	–	+	–	–	–
4.	Halophilia ovalis (R.Br.) Hook.f. subsp. ovalis	+	+	+	+	–
5.	Halophilia ovalis (Hook). f.subsp. ramamurthiana	+	–	–	–	–
6.	Halophilia ovata (Gaud)	+	+	+	+	+
7.	Halophilia stipulacea (Forsk) Asch.	–	+	–	–	–
8.	Thalassia hemprichii (Ehrenb.) Asch.	–	+	–	+	+
9.	Cymodocea rotundata (Ehren & Hempr Ex Asch)	–	+	+	+	+
10.	Cymodocea serrulata (R. Br.) Asch. & Magnus	–	+	+	+	+
11.	Holodule pinifolia (Miki) Hartog	+	+	+	+	–
12.	Halodule uninervis (Forsk) Asch.	+	+	+	+	+
13.	Holodule wrightii (Asch).	+	+	–	–	–
14.	Syringodium isoetifolium (Asch.) Dandy	–	+	–	+	+
	Total	7	13	7	9	7

Source: Status of sea grasses of India, Kannan et al., 1999.

FIGURE 19.3   **Mangroves in North Andaman.**

TABLE 19.2 Mangrove Diversity in India

Sl. No.	State	Area (km²)	No. of species
1.	Andhra Pradesh	353	31
2.	Goa	17	not known
3.	Gujarat and Daman & Diu	1047	12
4.	Karnataka	3	29
5.	Kerala	5	27
6.	Maharashtra	186	26
7.	Orissa	221	60
8.	Tamil Nadu & Pondicherry	39	24
9.	West Bengal	2152	57
10.	Andaman and Nicobar	615	44

Source: Bhatt *et al.* (2013).

restricted to the east coast, and one, *Lumnitzera littorea*, is present only in the Andaman and Nicobar Islands (Kathiresan, 1999). Mangroves serve as an abode for several faunal communities, and in India mangroves harbor 2359 species (Table 19.3). Twenty species of mangroves fall under the category of rare, endemic and restricted (Table 19.4).

DIATOMS

Diatoms are the dominant component of phytoplankton in all the Indian estuaries and the coastal waters from which detailed inventories of floristic composition and seasonal changes are available. Among the estuaries of the east coast, phytoplankton composition has been studied in detail only from Hooghly, Rushikulya, Godavari, Couum, Ennore, Adyar and Vellar. A total of 102 species of diatoms belonging to 17 families are known from the east coast, with the

TABLE 19.3 Mangrove-associated Fauna in India

Sl. No.	Faunal element	No. of species
1.	Prawns	55
2.	Crabs	138
3.	Molluscs	308
4.	Insects	711
5.	Fishes	546
6.	Amphibians	13
7.	Reptiles	85
8.	Birds	433
9.	Mammals	70

TABLE 19.4 Rare, Endemic and Restricted Category Mangroves in India

Sl. No.	Species	Category
1.	Acanthus ebracteatus	Restricted to ANI
2.	Aegialitis rotundifolia	Confined to WB, Orissa and AP
3.	Aglaia cuculata	Restricted to WB and Orissa
4.	Brownlowia tersa	Restricted to WB, Orissa and AP
5.	Heritiera fomes	Restricted to WB and Orissa
6.	Heritiera kanikensis	Endemic to Bhitarkanika
7.	Lumnitzera littorea	Restricted to ANI
8.	Merope angulata	Confined to WB and Orissa
9.	Nypa fruticans	Restricted to WB and ANI
10.	Phoenix paludosa	Restricted to ANI and AP
11.	Rhizophora annamalayana	Endemic to Pitchavaram
12.	Rhizophora stylosa	Confined to Orissa
13.	Scyphiphora hydrophyllacea	Restricted to ANI and AP
14.	Sonneratia apetala	Rare in several areas
15.	Sonneratia griffithii	Restricted to WB, Orissa and ANI
16.	Tylophora tenuis	Restricted to WB and Orissa
17.	Urochondra setulosa	Endemic to Gujarat
18.	Thespesia pipulneoides	Restricted to WB and Orissa
19.	Xylocarpus makongensis	Restricted to WB, Orissa, ANI
20.	Xylocarpus molluccensis	Restricted to ANI

ANI, Andaman and Nicobar Islands; AP, Andhra Pradesh; WB, West Bengal.

largest diversity pertaining to Naviculaceae (21 spp.) and Chaetoceraceae and Coscinodiscaceae (11 species each). Several other families like Biddulphiaceae, Lithodesmiaceae, Nitzchiaceae, Thalassionemataceae and Thalassiosiraceae are represented by fewer species. (Figure 19.4).

The diatom diversity along the west coast is higher, with 148 species under 22 families. Naviculaceae is dominant with 22 species, followed by Biddulphia-ceae (16 spp.), Lithodesmiaceae (15 spp.) and Thalassiosiraceae (12 spp.). Five families—Hemidiscaceae, Stellarimaceae, Stephanodisceae, Streptothecaceae and Heliopeltaceae—with one to three species are known so far only from the west coast (Figure 19.5).

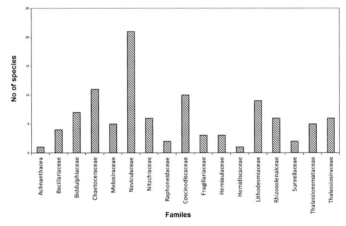

FIGURE 19.4   Species diversity of diatoms along the east coast.

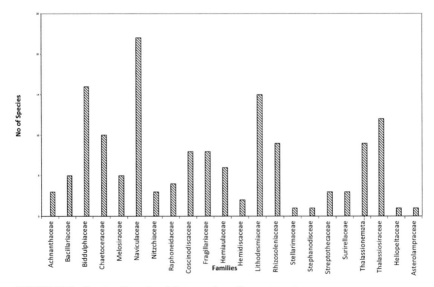

FIGURE 19.5   Species diversity of diatoms along the west coast.

The families Bacillariaceae, Biddulphiaceae, Chaetoceracae, Naviculaceae, Thalassiosiraceae, Thalassionemataceae and Rhizosoleniaceae are the most cosmopolitan in distribution. Of the few groups of marine organisms, planktonic algae appear to have been more completely catalogued (Venkataraman, 1939; Sournia *et al.*, 1991). Their compilation suggests that the number of pennate diatoms in the world oceans could range from 500 to 784, and that of centric diatoms from 865 to 999. Compared with these, no more than 25 percent of diatoms species are recorded in Indian waters.

DINOFLAGELLATES

The dinoflagellate species diversity in the east coast estuaries is small (15 species in 7 families) compared with the west coast estuaries (76 species from 10 families). The family Dinophyceae is dominant with 18 species, followed by Peridiniaceae and Ceratiaceae with 13 and 10 species, respectively.

Unlike the diatoms, the estimated number of dinoflagellate species in the marine environment varies from one thousand to two thousand. Compared with these, the current inventory of dinoflagellates in Indian waters appears too small. Such inventories, however, do not distinguish between truly tropical species and others that are cosmopolitan.

PROTOZOA

The known number of protozoan species from Indian seas is 2577, equivalent to about 8 percent of the total world protozoan fauna. Among them, 52 percent are free-living, and the remaining are parasitic species. Out of seven protozoan phyla, only one, Labyrinthomorpha, has not yet been reported from India.

Foraminifera

Foraminiferans are eukaryotic unicellular organisms with the general characteristics of protists. Their exoskeleton is commonly made of calcium carbonate while the rest have agglutinated shells made up of sediments or shells of dead organisms. Due to their diversity, which is a function of their ecological adaptation, each environment is characterized by ecological assemblages. Their small size, sensitivity to small change in the environment, and ability to preserve these changes in their hard part, give them an immense applicability in the field of palaeoclimatic reconstruction and environmental monitoring (Rana *et al.*, 2007). It has been estimated that the total number of foraminiferans might be approximately four thousand living species. Forminiferan is one of the relatively well-studied groups, with the earliest descriptions of new species dating back to the 18th Century (Fichtel and Moll, 1798). The most important phase of documentation of foraminiferan fauna began with the *Challenger* Expedition (1873–76), giving rise to detailed descriptions of deep and shallow water Foraminifera (Brady, 1884; Chapman, 1895; Hofker, 1927-1951). Contemporary studies began with the International Geophysical year in 1958 and the International Indian Ocean Expedition (1962–65). The major part of the work on this group has been done along the east coast of India. These are by Bhatia and Bhalla (1959) (14 benthic species from Puri Beach), Sarojini (1958) (Waltair Coast), Subba Rao and Vedantam (1968) (distribution of 32 species on the continental shelf off Visakhapatnam at depths of 20–200 m), Bhalla (1970) (16 species from beach sand of Visakhapatnam), Bhatt (1969) (15 planktonic species off Visakhapatnam), Bhalla (1970) (15 species from Madras Marina Beach), Gnanamuthu (1943) (47 littoral benthic species from Krusadai Island, Gulf of Mannar) and Ameer Hamsa (1974) (description of four new records from Palk Bay). Comparatively less work has been done on the west coast of India and the Arabian Sea. Mention may be made of the work of Antony (1968) (description and distribution of 164 species from Kerala coast), Siebold (1974) (12 species of benthic foraminifera from Kochi backwaters), Chapman (1895) (description of 277 species from bottom samples near Lakshadweep Islands), Chatterjee and Gururaja (1967) (unidentified species from 16–20 m depth off Mangalore coast), Chaudhury and Biswas (1954) (12 species from Juhu Beach, Bombay) and Rao (1970, 1971a,b) (a series of papers describing 84 species from shallow waters of Gulf of Cambay). The study of Frerichs (1967) on the distribution and ecology of benthonic and planktonic forms in the sediments of the Andaman Sea appears to be the only one from the Andaman and Nicobar Islands.

Tintinnid

The order Tintinnida comprises more than one thousand species of marine ciliates that form an important component of the microzooplankton. A total of 32 species belonging to 12 genera are known from Indian waters. The degree of abundance of tintinnid populations seems to coincide with diatom and dinoflagellate blooms; however, the persistence of such "swarms" appears to be controlled by the larger zooplankton grazers, such as copepods, chaetognaths, bivalve and gastropod veligers. More studies on diversity, biology and other ecological aspects of the tintinnids are considered necessary.

PORIFERA

The phylum Porifera, commonly known as sponges (Figure 19.6), are the most primitive of the multicellular animals and have existed on earth for more than 700–800 million years. It is interesting to note that sponges had managed to conceal their true animal nature for several centuries amid the evolutionary changes. Some workers even considered them to be houses of nonliving matter secreted by worms. Robert Grant in 1826 was the first to recognize and prove that sponges are animals. Sponges occur from the intertidal region to the deepest part of the ocean at 2000 to 8840 m. Approximately fifteen thousand species of sponge

FIGURE 19.6   Sponges (a) *Paratetilla bacca;* (b) *Xestospongia testudinaria.* (Please see color plate at the back of the book.)

are known across the world, most of them found in the marine environment whereas only about one percent of the species inhabit freshwater. However, there are a total of 8424 valid species identified to date and still several more waiting for identification in several museum collections, as well as the unexplored seas. In fact, they are more diverse than scleractin corals in coral reef ecosystems around the world. Sponges play a major role in the coral ecosystem, such as in facilitating regeneration of broken reefs and binding live coral to the reef frame. Sponges are broadly classified into three classes: Demospongiae, Calcarea and Hexactinellidae, of which the Demospongiae is the most diverse class containing more than 85 percent of the sponges identified to date. Most of the studies of marine sponges in India were carried out on the Gulf of Mannar, Kerala coast, Lakshadweep, Gulf of Kachchh and Gulf of Cambay. So far, 486 species under 3 classes, 17 orders, 65 families and 169 genera have been described in India (Thomas, 1998). The sponge fauna of India is dominated by species of Desmospongia, followed by those of Hyalospongiae and Calciospongiae. Also 34 species of coral-boring sponges (20 from Gulf of Mannar and Palk Bay, 5 from Andaman and Nicobar Islands and 18 from Lakshadweep reefs) have been recorded. The Gulf of Mannar and Palk Bay region has the highest diversity (319 species), followed by Lakshadweep (82 species) and Gulf of Kachchh (25 species). In order to assess the status and diversity of sponges, the Zoological Survey of India has conducted extensive underwater surveys employing SCUBA diving along the coral reef areas of Andaman and Nicobar Archipelago from 2008 to 2013 at the area between the intertidal region and 40 metres. As a result of surveys a total of 114 species of sponge were identified in the coral reef ecosystem of these islands. Of these, 26 species are new records to Indian waters. So far, only 400 Calcarean sponges have been identified from world oceans. Of these, in Indian waters, only 9 species of Calcarean sponges were reported; among them, 6 species from Gujarat and Maharashtra coasts, 2 species from Andaman and one species from Gulf of Mannar were identified. All 9 species of calcarean sponges are protected under Schedule III category of the Wildlife (Protection) Act, 1972.

CNIDARIA

The global estimates of cnidarian diversity vary between nine thousand and twelve thousand species. In India 212 species of Hydrozoa, 37 species of Scyphozoa, 5 species of Cubozoa and about one thousand species of Anthozoa have been reported until now. Since no group of Cnidarians has received adequate attention from Indian taxonomists, the above figures cannot be taken as final. Except for the pioneering works of Annandale (1907, 1915, 1916), Leloup (1934) and Menon (1931a) other studies are few and scattered. Comprehensive accounts are available only for siphonophores by Daniel (1985), scyphomedusae by Chakrapany (1984) and scleratinian corals by Pillai (1991).

Hydrozoa

The first description of hydrozoans (Figure 19.7) in India was by Annandale (1915) from Chilika Lagoon and subsequently by Menon (1931), reporting 35 species under 28 genera, and of Mammen (1963, 1965) who has inventoried 116 species belonging to 13 families. Among these forms, species of the orders Milliporina, Stylasterina and Trachylina have received only scant attention so far.

Siphonophora

Siphanophora (Figure 19.8) are abundant in the Indian seas and constitute an important part of the marine plankton. The siphanophores from the Indian Ocean have been studied by several workers: Browne (1926) from the Seychelles, Mauritius and Chagos Archipelago; Sundara Raj (1927a), Leloup (1934) and Daniel and Daniel (1963) from the Madras Coast; Patriti (1970) from off the southeast coast of Africa and Madagascar; Totton (1954) from the SE coast of Africa, the SE, NW and South Indian Ocean, Gulf of Aden, Aquaba and Red Sea; Alvarino (1974) from the tropico-equatorial region; Rengarajan (1974)

FIGURE 19.7 **Hydrozoa** (a) *Eudendrium* sp.; (b) *Macrorhynchia philippina*. (Please see color plate at the back of the book.)

FIGURE 19.8   **Syphonaphora** *Physalia physalis* (Linnaeus, 1758). (Please see color plate at the back of the book.)

from the west coast of India and Daniel (1966, 1974) from the west and east coasts of India and those collected by RV *Vityaz* along 90–110° E longitude down 35° S latitude. A comprehensive account of Siphanophora of India (1985) shows 116 valid species, while one variety and 3 species are yet to be confirmed.

Anthozoa

Scleractinia

Studies on taxonomy of coral reefs (Figure 19.9) started in India as early as 1847 with Rink in the Nicobar Islands and was pursued in 1989 by Thurston in the Gulf of Mannar region. Brook (1893) recognized 8 species of *Acropora* from Rameswaram, out of which *A. multicaulia, A. thurstoni* and *A. indica* were described as new. Subsequent contributions to inventory of coral species were made by Alcock (1893, 1898), Gardiner (1903–1906), Matthai (1924), Gravely

FIGURE 19.9   **Corals** (a) *Acropora humilis* (Dana, 1846); (b) *Psammocora digitata* MED & H, 1851. (Please see color plate at the back of the book.)

(1927a) and Sewell (1935). Contemporary studies on corals are those of Pillai (1967) and Venkataraman *et al.* (2003) which list a total of 218 species under 60 genera and 15 families. The consecutive surveys made between 2008 and 2013 by the Zoological Survey of India resulted in the occurrence of 519 scleractinian corals including two new species from the Andaman and Nicobar Islands. Among the four major reef areas of India, the Andaman and Nicobar Islands are rich in coral species diversity whereas those of the Gulf of Kachchh are poorer. The Lakshadweep Islands have more species than those of the Gulf of Mannar. Among the deepwater (ahermatypic) corals, so far 720 species belonging to 110 genera and 12 families have been reported from the world, of which 227 species belonging to 71 genera and 12 families have been reported from the Indian Ocean region (Cairns and Kitahara, 2012; Venkataraman and Satynarayana, 2012). However, meager attention has been paid so far to inventory of the deepwater corals, and as a result only 44 species are known until now from Indian seas (Venkataraman *et al.,* 2003)

Gorgonians

Gorgonians are marine coelenterates of the class Anthozoa, which include sea fan, sea whips, corals, sea anemones and other related species. The gorgonians popularly called sea fans and sea whips are marine sessile coelenterates with colonial skeleton and living polyps. A total about 135 species of Gorgonians are reported from Indian seas. Venkataraman *et al.* (2004) reported 27 species of gorgonians belonging to 8 families and 19 genera from India. Among them, 12 species of gorgonians belonging to 4 families and 9 genera have been reported from the northeast coast of India (Thomas *et al.*, 1995). However in the Andaman and Nicobar Islands, only 10 species under 4 families and 9 genera have been recorded (Venkataraman *et al.,* 2004). Recently, Venkataraman *et al.* (in press) reported 51 species belonging to 25 genera, 8 families and 3 suborders. Among them, 44 species belonging to 24 genera and 7 families are new to India (Figure 19.10).

FIGURE 19.10 Gorgonians (a) *Nicella flabellata* (Whitelegge, 1897); (b) *Echinogorgia flora* (Nutting, 1910). (Please see color plate at the back of the book.)

Scyphozoa: Scyphomedusae

The earliest records of Scyphozoa in Indian seas were made by Browne (1905, 1906, 1916) from Lakshadweep, Maldives, Sri Lanka and Okhamandal Coast of Kattiawar, followed by Chakrapany (1984), Annandale (1916), Menon (1930, 1931b), Panikkar (1944) and Nair (1945, 1954). In the Indian seas several cruises of the RIMS *Investigator* and coastal surveys by the Officers of the Zoological Survey of India have yielded a collection of 24 species, which form the Indian National Collections in the Zoological Survey of India, Kolkata (Chakrapany, 1915). In addition, several cruises of the RV *Chota Investigator* along the Chennai coast from 1972 to 1983 revealed the occurrence of 19 species of which 11 were already known from the Indian seas. Thus, out of the 200 species of Scyphomedusae known from the world's oceans, 34 are known from the Indian seas (Chakrapany, 1915) (Figure 19.11).

Actinaria

Sea anemones are brightly colored, classified under the phylum Cnidaria, inhabit coastal waters throughout the world but are particularly abundant in tropical oceans (Figure 19.12). They are distributed from intertidal zone to deep

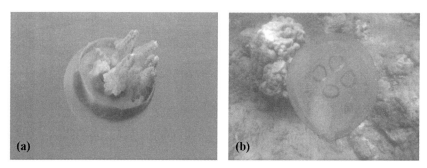

FIGURE 19.11   Scyphomedusae (a) *Aurelia aurita* (Linnaeus 1758); (b) *Crambionella masti-gophora* (Maas 1903). (Please see color plate at the back of the book.)

FIGURE 19.12   Sea anemones (a) *Entacmaea quadricolor* (Rüppell & Leuckart, 1828); (b) *Heteractis magnifica* (Quoy & Gaimard,1833). (Please see color plate at the back of the book.)

oceans and live attached to rocks, sea floor, shells, and some forms burrow in the mud or sand. There are over one thousand species of sea anemone reported worldwide. They are usually about 2.5–10 cm across, but a few grow up to 1.8 m across. As evinced by scant literature, though India has 7600 km of coastline, studies on sea anemones in Indian waters are very limited. The earliest studies were made by Annandale (1907, 1915), Panikkar (1936, 1937a–c, 1939) and Parulekar (1967, 1968, 1969a,b, 1971). However, these studies have discontinuously been made and are better known through new species and new genera, rather than the magnitude and diversity of the fauna itself (Parulekar, 1990). Parulekar (1990) enumerated 40 species of sea anemone belonging to 33 genera under 17 families from India, of which 13 species were reported for the first time. Out of 40 species, 24 are marine, 13 estuarine and 3 are common to both habitats. The actiniarian sea anemone fauna of India is so far known from few places: West Bengal (Port Canning), Orissa (Chilika Lake), Tamil Nadu (Adyar backwaters and Gulf of Mannar), Kerala (Cochin backwaters and Ashtamudi), Gujarat (Gulf of Kachchh), Maharashtra (Mumbai, Malvan), Goa, northern Karnataka and the Andaman and Nicobar Islands. Probably since the description of two new species—*Edwardsia jonessi* (Seshaiya and Cutress, 1969) from Porto Novo and *Paracondylactis sagarensis* (Battacharya, 1979)—in India, no species has been added to the sea anemone list of India. The studies on sea anemones of the Andaman and Nicobar Islands are *terra incognita* except for a few reports. The Indian sea anemones listed by Parulekar (1990) include *Anthoplerua panikkarii, Bunodactis nicobarica* and *Parabunodactis inflexibilis* from the Andaman and Nicobar Islands. Madhu and Madhu (2007) reported the occurrence of 10 species of sea anemone at 14 sites from these islands; Raghunathan *et al.* (in press) reported 19 species of sea anemones from these islands and compiled 54 species of sea anemones reported from India.

Ctenophora

Only 12 species of Ctenophores, among the 100–150 species known from the world's oceans, occur in Indian seas (Figure 19.13). This inventory is derived from sporadic studies carried out several decades ago by Annandale and Kemp

FIGURE 19.13   **Ctenophore** *Ceonoplana* sp. (Please see color plate at the back of the book.)

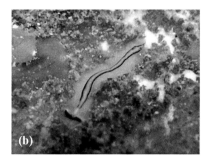

FIGURE 19.14   **Polyclads** (a) *Pseudobiceros hymanae*; (b) *Prosthiostomum trilineatum*. (Please see color plate at the back of the book.)

(1915), Varadarajan (1934) and Devanesan and Varadarajan (1939). Since then there have been no studies of Ctenophores in India.

Polyclads

Polycladida, a highly diverse group of free-living flatworms, belongs to a taxonomic subgroup named Turbellaria. In India, 37 flatworm species of the order Polycladida are known (Figure 19.14). Out of these, 22 species are recorded from Lakshadweep, 12 from the Andaman and Nicobar Islands, and 3 are common to both areas. The first account of these animals from India came in 1902 by F.F. Laidlaw in which he gave information of about 15 species from Laccadive (Lakshadweep) archipelago. After this the only accounts are given by Sreeraj and Raghunathan (2011, 2013) from the Andaman and Nicobar Islands and Apte and Pitale (2011) from Lakshadweep. Recently, studies conducted by the Zoological Survey of India revealed 38 species of polyclads from the Andaman and Nicobar Islands (Venkataraman *et al.*, 2013 in press)

ANNELIDA

Archiannelida

Pioneering studies on archiannelids of India were made by Aiyar and Alikunhi (1944) and Alikunhi (1946, 1948a) along the Madras coast, from which 2 species of *Polygordius,* 2 species of *Protodrilus* and 4 species of *Saccocirrus* were described as new to science. Rao and Ganapati (1968) recorded 15 species of archiannelids from the beach sands along Waltair coast. Thus, compared with the vast stretch of Indian coast, the investigations hitherto carried out on Archiannelida are quite limited, and any further intensive surveys of the fauna in other areas are quite likely to yield interesting results. The world records of Archiannelida hitherto made fall under 5 families, 18 genera and over 90 species, of which about 20 species are reported from Indian coasts.

Polychaeta

In the phylum Annelida, the Polychaeta have received considerable attention since 1909. Polychaetes form an important component in the marine food chain, especially for demersal fish. The worldwide number of polychaete species is estimated as about eight thousand. Survey of this group actually started with Southern's (1921) work on "Polychaeta of Chilika Lake" followed by the littoral fauna of Krusadai Island in the Gulf of Mannar by Gravely (1927b) (nearly 36 species under 11 families) and by Fauvel (1930) (119 species under 22 families). Perusal of literature shows that most of the records pertaining to this group are either from the Madras coast or the Gulf of Mannar (Aiyar, 1924, 1931; Subramaniam, 1938; Aiyar and Alikunhi, 1940; Gravely, 1942; Alikunhi, 1941, 1942, 1946, 1947, 1948b; Krishnan, 1946; George, 1905; Ganapati and Radhakrishna, 1958; Ghosh, 1963; Banse, 1959; Krishnamoorthi, 1963; Tampi and Rangarajan, 1963). The Central Marine Fisheries Research Institute has listed 200 species under 46 families in the catalogue of types and reference collections. From the collections of the Zoological Survey of India and the Indian Museum, Fauvel (1932) described 300 species under 30 families and in his later monograph (Fauvel, 1953) raised this to 450 species. Hartman (1947), while dealing with polychaetes of the Indian Ocean, recorded 244 species of which 116 are considered new to the region. The catalogue of the polychaetous annelids from India lists 883 species (Figure 19.15).

Oligochaeta

Marine Oligochaete fauna is poorly known in India, and most of the species are recorded from littoral zones of small freshwater bodies like ponds, tanks, pools, ditches, etc., all over the country. The Enchytraeidae (pot-worms) occur in terrestrial, littoral and marine habitats, being abundant in acidic soils with high organic matter. As compared with the world fauna, only 3% of enchytraeid species have so far been reported from this region, mainly from Orissa.

FIGURE 19.15 **Polychaetes** (a) *Neries* sp.; (b) *Bispira brunnea*. (Please see color plate at the back of the book.)

Sipuncula

The pioneering work on the Indian Sipuncula dates back to Shipley (1903), followed by a rather scattered series of taxonomic contributions made by Gravely (1927c), Prashad (1936), Johnson (1964, 1969, 1971), Haldar (1975, 1976, 197 7, 1978, 1985a,b), Cutler (1977) and Cutler and Cutler (1979). Of the 320 species (Salinas, 1993) known from the world's oceans, 35 species under 10 genera and 5 families occur on the Indian coasts (Figure 19.16). So far as the distributional pattern of the sipunculan fauna is concerned the major areas of species concentration are the Andaman and Nicobar Islands, Lakshadweep Islands, Gulf of Mannar and Gulf of Kachchh.

Echiura

The phylum Echiura comprises 129 species under 32 genera and 5 families (Stephen and Edmonds, 1972). Studies on Indian echiuran fauna began only in the early 20th Century when Annandale and Kemp (1915) described two new species of the genus *Anelassorhynchus* from Chilika Lagoon. Subsequent studies (Prashad, 1919, 1935; Menon *et al.,* 1964; Dattagupta, 1967, 1976; Dattagupta and Menon, 1963, 1964) enriched knowledge on Indian echiuroids so much that the current inventory of 33 species under 11 genera is fairly rich in comparison with what is known (43 species under 14 genera) from the Indian Ocean. Maximum abundance of echiurans is in Gulf of Kachchh, Gulf of Kambath, Lakshadweep, Andaman and Nicobar Islands and Gulf of Mannar. Mud-dwelling forms are few in numbers and are found in Kerala, West Bengal and Orissa.

Chaetognatha

Chaetognaths (arrow worms or glass worms) rank second in terms of abundance after copepods in marine zooplankton and are cosmopolitan in distribution. They are mostly marine, but a few species are estuarine. Among the 120 species known from the world's oceans, about 32 are reported from the Indian

FIGURE 19.16   **Peanut worm *Siponculus* sp.** (Please see color plate at the back of the book.)

seas. Chaetognaths have been extensively studied in Indian waters and from various coastal and oceanic sites: Bombay Harbour (Lele and Gae, 1936); Chennai coast (Subramaniam, 1937); Kurusadai Island (Varadarajan and Chacko, 1943) Trivandrum coast (Pillai, 1944; Menon, 1945); Malabar coast (George, 1952); Mandapam area (Prasad, 1956; Sudarsan, 1961); Lawson's Bay, Waltair (Rao, 1958, 1966; Rao and Ganapati, 1958; Rao and Kelly, 1962); Ennore estuary (Srinivasan, 1977; 1980), Andaman Sea (Nair *et al.,* 1981) and coastal and offshore waters (Nair, 1967, 1971, 1973, 1974, 1975; Nair and Rao, 1973; Nair and Selvakumar, 1979; Silas and Srinivasan, 1968, 1969, 1970; Srinivasan, 1972, 1979, 1996). In contrast with these numerous studies, chaetognaths of the deeper waters of the seas around India and those of the Andaman and Nicobar Islands and central and northern parts of the Bay of Bengal are not well known.

Tardigrada

Tardigrades occur as meiofauna in the sandy beaches up to 2–3 m from the sea's edge. Among the three orders of the phylum Tardigrada, the Heterotardigrada are found in marine, freshwater and high altitude mountains. So far 214 species are reported from the world under 5 families and 20 genera. However, in India only 10 species under 2 families and 3 genera have been reported as meiofauna of marine regions (Venktaraman, 1998).

ARTHROPODA: CRUSTACEA

Global estimate of Crustacean species diversity is one hundred and fifty thousand, of which forty thousand have been described so far. Of the 2934 species of Crustacea that have been reported from India (Venktaraman, 1998) so far, marine species (94.85%) contribute most to this diversity. In India as many as 139 species of stomatopods (4 families and 26 genera), 26 species of lobsters (4 families, 11 genera), 162 species of hermit crabs (3 families, 40 genera), 705 brachyuran crabs (28 families, 270 genera), 84 species of shrimps and prawns (7 families, 19 genera) and 159 species of Caridea (15 families, 56 genera) have been recorded so far. Other than these, 540 species of copepods, 104 species of cirripeds and 120 species of ostracods have also been recorded.

Copepoda

Copepods are the most widely studied group among the marine zooplankton. There are approximately 210 described families, 2280 genera and over 14,000 species in the world. Important contributions to systematics of copepods from Indian waters are those of Sewell (1929), Krishnaswamy (1950, 1952, 1953) and Pillai (1967). Largely as a result of these studies as well as several others since then (e.g. Madhu Pratap, 1979), it is now known that there are over 540 copepod species in Indian waters (Figure 19.17). Among these, the most dominant group is Calanoida, with the Cyclopoida and Harpacticoida being less important. Major studies on Cyclopoids are again those of Krishnaswamy

(1950, 1952, 1953). Only very few papers dealing with marine Harpacticoida of India and neighboring seas have been published so far (Swell, 1924, 1940). Studies on sand-dwelling forms are still fewer: only those of Krishnaswamy (1950, 1952, 1953, 1956, 1957) provide an account of 17 sand-dwelling harpacticoids under 5 families together with discussion on their adaptation as well as their ecology. A total of 106 species belonging to 23 families are known from the east coast estuaries (Figure 19.18). Among them the calanoids are the dominant, distributed in 16 families, followed by harpacticoids (5 families and

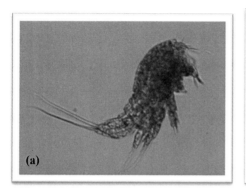
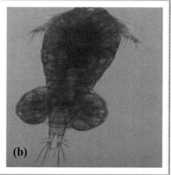

FIGURE 19.17   Copepods (a) *Microsetella norvegica*; (b) *Euterpina acuitiferans.*

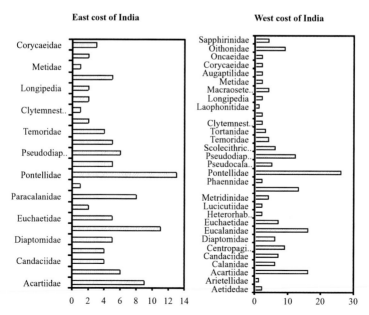

FIGURE 19.18   **Diversity and distribution of copepod families on the east and west coast of India.**

cyclopoids (2 families). The diversity in the west coast estuaries is relatively higher, with 179 species in 31 families. Calanoids are dominant with 20 families. Though the number of families of harpacticoids and cyclopoids are the same (6 families), the latter is more diverse, with 22 species compared with 7 species of harpacticoids.

Ostracoda

The Ostracoda are one of the most successful aquatic crustacean groups with approximately eight thousand living species. The six extant and extinct orders are ubiquitous and diverse, with over fifty thousand named species and genera and more awaiting study. Except for the studies of Poulsen (1965) very little is known of Indian ostracods. Only 60+ species of ostracods are known from the Indian coast, of which 38 species are known from the east coast and 28 from the west coast (George *et al.,* 1975; George and Nair, 1980).

Branchiura

Our knowledge of this group from the Indian region is rather scanty. It was not until 1951, when Ramakrishna contributed to our knowledge of the Indian species of arguulids found parasitic on fishes, that the group received adequate attention. He described five species of the genus *Argulus*, of which three were described as new to science.

Cirripedes

Information on the diversity of Cirripedia of the Indian coast is far from complete, with only 36 species having been recorded so far. Even this record (Weltner, 1894; Borradaile, 1903; Annandale, 1907, 1909, 1910, 1913, 1914, 1924; Nilsson, 1938; Panikkar and Aiyar, 1937; Daniel, 1952, 1953, 1956, 1958, 1962, 1963, 1971, 1974, 1975) is rather sketchy and has low geographical coverage.

Malacostraca: Mysidacea

Mysids are shrimp-like animals with a shield-shaped carapace of fused first three segments. A total number of 780 species are well known worldwide under 120 genera (Mauchline, 1980). Mysidaceans, with a total number of about 75 species, are known so far only from the works of Tattersall (1922) and Pillai (1968, 1973).

Cumacea

A total of 1300 species of marine cumaceans are reported till now (Jaume and Boxshall, 2008). Cumacean species are also little known except for the studies of Calman (1904), Kemp (1916) and Kurian (1954, 1965). Earlier, 23 species of Bodotriidae, 3 species of Disstylidae, 4 species of Nannastaeidae and 1 species of Camylaspididae are known from the Indian region. Recent studies reveal that a total of 71 species under 8 genera and 3 families are reported as Indian cumaceans (Chatterjee and Pesic, 2010; Petrescu and Chatterjee, 2011).

Tanaidacea

Tanaidacea is an order of crustaceans which includes benthic macro-invertebrate species. A report of 2008 implied that a total of 900 marine tanaidacean species are available worldwide (Jaume and Boxshall, 2008). Tanaidacea is categorized into four suborders: Anthracocaridomorpha, Apseudomorpha, Neotanaidomorpha and Tanaidomorpha; the first is a fossil species, but the others include a total of 1052 species (Apseudomorpha with 457 species under 93 genera, Neotanaidomorpha with 45 species under 4 genera and Tanaidomorpha with 550 species under 120 genera) (Blazewicz-Paszkowycz *et al.*, 2012). Our knowledge of Tanaidacea is rather poor from the Indian region. Chilton (1923) contributed a paper dealing with a species of the group from the Chilika Lake.

Isopoda

Very little is known about the marine isopods when compared with terrestrial isopods of India. Chopra (1923a) contributed a monumental monograph on the Bopyrid isopods of Indian Macrura wherein 33 species pertaining to 13 genera were described from the Andaman Islands, Ganges delta, Madras and other areas. Chopra (1923b) contributed another paper on the Bopyrid isopods of Indian Macrura. The collection included 12 species pertaining to 7 genera collected mostly from the Andaman and Nicobar Islands, Ganges delta, Gulf of Mannar and Bombay. The contributions on the marine woodborers from 1963 to 1968 by various Indian authors revealed six species of the genus *Sphaeeroma* and nine species of *Limnoria* from Indian waters.

Amphipoda

Studies on the amphipods of the Indian and neighboring waters received the attention of zoologists only as late as 1885 when Giles published a paper on the occurrence of two species of amphipods from Bengal. His subsequent works raised the number to 27. Gravely (1927a) and Sundara Raj (1927b) reported 16 species of amphipods from Krusadai Island, Gulf of Mannar and the neighboring waters. Bernard (1935) reported amphipods from the collection made from Travancore, Cochin and Bengal coasts by the Zoological Survey of India. This added to the record of three species of amphipods off the coast of Mahabalipuram by Giles (1888, 1890) and a brief note about the occurrence of three species of amphipods at Adyar in Madras. In the last half of the previous century, Sivaprakasam (1966, 1967, 1968, 1970) in a series of contributions enriched our knowledge of the amphipods of the east coast of India and listed 61 species. Nayar (1959, 1966) dealt with the amphipods of the Madras coast and Gulf of Mannar. In his monographs on the Gammaridean amphipods of the Gulf of Mannar he described 78 species, of 26 families. Surya Rao (1975) enumerated a detailed account of the intertidal Gammarid amphipods from the Indian coasts and listed 132 species pertaining to 54 genera.

Euphausiacea

The earliest account of Indian euphasids is from the work of Wood-Mason and Alcock (1891) and Alcock and Anderson (1893, 1894). Tattersall (1911) gave an account of them from the Indian Ocean. Among the Indian coasts, 23 species of euphasids from the Laccadives and Maldives and adjoining regions and two species from the southwest coast of India have been recorded so far.

Stomatopoda

Kemp (1913) published a monograph on Indo-Pacific stomatopods comprising 139 species and varieties known till then. Kemp and Chopra published papers on the stamatopods form collection of the John Murray Expedition (1933–34) made by Sewell. Tiwari and Biswas (1952) published a paper based on material accumulated since Chopra's work. After a gap of two decades Ghosh (1975, 1976) and Tiwari and Ghosh (1975) have contributed a series of papers highlighting the present knowledge of Stomatopoda in Indian waters. The study of Stomatopoda of India is, however, far from complete.

Decapoda: Macrura

Decapoda as a whole have received good attention from scientific workers compared with other groups. The earliest to contribute was de Man (1908) who, in a series of papers, referred to the decapod collection from the brackish water ponds of Lower Bengal. The contributions of Kemp to the study of Indian Crustacea are among the most noteworthy of the group. His contributions on decapod crustaceans of the Indian Museum published in 24 parts in the *Records of the Indian Museum* contains systematic accounts of various marine and brackish water forms belonging to the families Hippolytidae, Carangonidae, Disciadidae, Palaemonidae, Pasiphasidae, Stylodactylidae, Rhynchocinetidae, Pacdalidae and Anchistodidae, in which species from most varied habitats have been reported. Alcock (1901, 1906) contributed a comprehensive catalogue on the penaeid prawns of India. Since then, several Indian researchers have contributed to inventories of this group. Although a large number of species of prawns and lobsters are known to occur in and along the Indian coast, work on this group of species is very limited. Worldwide,17 families, 67 genera and 383 species have been recorded as commercially important. A total of 55 species of commercial shrimps and prawns have been recorded in India. The east coast of India contributes about 24.5% and the west coast 75.3% of the country's shrimp production.

Brachyura

The earliest works on the crabs of Indian Seas were those of Milne Edwards (1834–1837), Henderson (1893) and de Man (1887, 1895, 1896, 1898, 1899, 1900, 1901). The first comprehensive study of the crabs of the west coast was that of Borradaile (1900, 1902, 1903). Alcock (referred elsewhere) in his contributions entitled "Materials for a carcinological fauna of India" published detailed accounts of marine and brackish water crabs. Kemp (1915) dealt with 38 species under six families collected from the Chilika Lake and in 1923 accounted for crabs collected from the mouth of Hooghly River. Chopra (1930, 1931, 1933, 1934, 1939) in a series of contributions entitled "Further Notes on Crustacea Decapoda in the Indian Museum" published in seven parts dealt with Hymenosomatid, Dromiacea, Oxystomata, Oxyrhyncha, Brachyrhyncha and Potamonid crabs. These series were in continuation of Kemp's series entitled "Notes on Crustacea Decapoda in the Indian Museum." Many other Indian authors added to the earlier works, raising the total carcinological fauna to over 250 species. There are about 254 species of crabs belonging to 120 genera under 24 families recorded along the west coast of India. Among these, the names of 100 species have been revised. Twenty-two families and 37 subfamilies represent brachyuran crabs (Figure 19.19). Family Leucosiidae contains the highest number of species, with 20, followed by subfamily Thalamitae of family Portunidae (19 species). Family Xanthidae alone is represented by 10 subfamilies of which the subfamily Zosiminae is represented by 14 species.

Anomura

In 2009, a total of 1069 species of anomura were documented by De Grave *et al.*, while a recent report presented a checklist of 1106 species worldwide

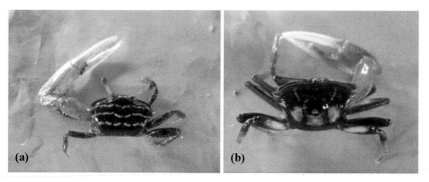

FIGURE 19.19 **Brachyuran crabs** (a) *Uca crassipes*; (b) *Uca coarctata*. (Please see color plate at the back of the book.)

(McLaughlin *et al.*, 2010). Sarojini and Nagabhushanam (1972) gave a detailed account on the Porcellanids from the Waltair coast. Reddy and Ramakrishna (1972) listed 20 species pertaining to the families Paguridae and Coenobitidae. Study of the Anomuran crabs is far from complete, and more studies are needed on this group.

MOLLUSCA

The number of species of molluscs recorded from various parts of the world is in the range eighty thousand to one hundred and fifty thousand. The history of mala-cological study in India is immense and interesting. Studies on Indian molluscs were initiated by the Asiatic Society of Bengal (1784) and the Indian Museum, Kolkata (1814). Benson was perhaps the first author to publish a scientific paper on Mollusca. Between the years 1830 and 1865 he published a total of about 90 papers dealing with the land and freshwater molluscs of the Indian subcontinent. The beginning of the 20th century is the most productive and significant period in the history of Indian malacology, with the Zoological Survey of India, Central Marine Fisheries Research Institute and several maritime universities contributing immensely to the knowledge of the molluscan fauna. In India, to date, 5100 species of Mollusca have been recorded from freshwater (22 families, 53 genera 183 species), land (26 families, 140 genera and 1487 species) and marine habitats (242 families 591 genera, 3400 species) (Subba Rao, 1991, 1998). From the available data, it is possible to identify certain areas having rich molluscan diversity. The Andaman and Nicobar Islands have a rich molluscan diversity, including over one thousand marine species (Subba Rao, 2000); the Gulf of Mannar and Lakshadweep have 428 and 424 species, respectively (Venkataraman *et al.*, 2004). Eight species of oyster, two species of mussel, 17 species of clam, six species of pearl oyster, four species of giant clam, one species of window-pane oyster and other gastropods such as sacred chank, *Trochus, Turbo*, as well as 15 species of cephalopods are exploited commercially from the Indian seas (Figure 19.20).

Opisthobranchia

The Opithobranchs or sea slugs are shell-less molluscs (Figure 19.21). About four thousand species of opisthobranchs are reported worldwide. The earliest works on opithobranchs were initiated during the 1880s (Alder and Hancock, 1864). The first report on opithobranchs was made by Eliot (1906), which deals with 42 species belonging to 10 families. A total of 260 species of opisthobranchs have been reported to date form Indian waters, of which 200 species are recorded from the Andaman and Nicobar Islands (Ramakrishna *et al.*, 2010).

Bryozoans

Although regarded traditionally as a minor phylum, this group contains as many as twenty thousand described species, actually occupying an intermediate

FIGURE 19.20   **Molluscs** (a) Chambered nautilus *Nautilus pompilius* (Linnaeus, 1758); (b) lesser spider conch *Lambis scorpius indomaris* (Abbott, 1961). (Please see color plate at the back of the book.)

FIGURE 19.21   **Sea slugs** (a) *Phyllidiella zeylanica*; (b) *Hypselodoris bullock*. (Please see color plate at the back of the book.)

position in the hierarchy of animal phyla in respect of species representation. Of these, approximately four thousand species are living. At least two hundred valid species occur in India (Satynarayana Rao, 1998). The bryozoa are grouped under three classes: Phylactolaemata (freshwater species), Stnolaemata and Gynolaemata. A total of 126 families are recognized: 100 from Gymnolaemata (15 from the order Ctenostomata and 85 from Cheilostomata), 21 from Stenolaemata and five from Phylactolaemata. In India, as in other parts of the world, only a few species of bryozoans inhabit freshwater lakes and rivers (Phylactolaemata) and most others are marine or estuarine. It is however to be noted that vast stretches of the long Indian coastline still remain unexplored and the biology and ecology of several species still remain uninvestigated.

Entoprocta

The Entoprocta are predominantly marine, having about 60 species known from the world, with the exception of one genus in freshwater. Reports of Entoprocta from India are scanty except for the brackish and marine water species reported by Annandale (1908, 1916) and Harmer (1915). The diversity in Entoprocta is limited and restricted basically to the following three families: Loxosomatidae, Pedicellinidae and Urnatellidae. Family Loxosomatidae is commonly represented in India by two genera: *Loxosoma* and *Loxocalyx*. The family Pedicellinidae is represented by the genera *Pedicellina, Myosoma, Chitaspis, Loxosomatoides, Pedicellinopsis, Barentsia, Gonypodaria* and *Arthropodaris*. The family Urnatellidae occurs in freshwater only.

ECHINODERMATA

Plancus and Gualtire made the first report on Indian echinoderms from Goa in 1743, and the next was by Collier (1830) on the *bêche-de-mer*. Subsequently, the accounts of Müller (1849), Lütken (1865, 1872) and Marktanner-Turneretscher (1887) included a few new species from the Bay of Bengal. Most of what we know of the echinoderm fauna is from examination of the collections from expeditions such as *Investigator, Challenger, Valdivia* and *John Murray*. India has 765 species (Crinoidea: 13 families, 43 genera 95 species; Astereroidea: 20 families, 81 genera and 180 species; Ophiuroidea: 15 families, 67 genera 150 species; Echioidea: 28 families, 79 genera 150 species; Holothuroidea: 14 families, 62 genera, 160 species) recorded to date, and about 257 species are known from the Andaman and Nicobar Islands (Sastry, 1998; James, 1986). A recent investigation on echinoderms stated that 12 species of echinoderms, including 5 species of crinoids, 6 species of ophiuroids and 1 species of astereroids, were new records to India, reported from the Andaman and Nicobar Islands (Sadhukhan and Raghunathan, 2011, 2012a–c), which updates the total Indian echinoderm database to 777 species (Figure 19.22). Lakshadweep has 77 species and the Gulf of Mannar 112. Economically, only Holothuroidea are exploited on a commercial scale for export. Twelve species of Holothurians belonging to the genera *Actinopyga, Bohadschia, Holothuria, Stichopus* and *Thelenota* are known to be of commercial importance in

FIGURE 19.22   Echinoderms (a) *Mespilia globulus*; (b) *Comanthina nobilis*. (Please see color plate at the back of the book.)

India. However, only three species—*Bohadschia marmorata, Holothuria scabra* and *H. spinifera*—are being exploited to a large extent in the Gulf of Mannar. All holothurians are now included under Schedule 1 of the Wildlife Protection Act, 1972.

HEMICHORDATA

Phylum Hemichordata is divided into three classes: Enteropneusta, Pterobranchia and Planctosphaeroidea. Of the four families known from the world only three are recorded from India. So far, 238 species have been recorded from the world, of which 12 are known from India (Dhandapani, 1998a; Zhang, 2011). Genera such as *Ptychodera, Glossobalanus, Glandiceps* are collected from the Gulf of Mannar, Gulf of Kachchh, Andaman Islands, Lakshadweep and Maldive Seas, and Tamil Nadu coast up to Cape Comerin, and the *Saccoglossus* is recorded from the high saline marshy areas of Sunderbans in West Bengal. Once commonly available the enteropneust worm *Ptychodera flauva* is at present restricted to Krusadai island in the Gulf of Mannar.

PROTOCHORDATA

This phylum includes two subphyla: Cephalochordata and Urochordata. Worldwide the diversity of cephalochordates includes two families, two genera and 24 species, and in India six species are reported under two families and two genera. The subphylum Urochordata is divided into class Ascidiacea (sea squirts) that are sessile or benthos attached to substratum on the coral reef, class Thaliacea (= salps) and class Larvacea that are planktonic. About two thousand species of ascidians are reported from all over the world, of which 47 are reported in India (9 families, 21 genera). Out of 57 species of Thaliacea reported from the world, 48 species (4 families and 19 genera) occur in India, and out of 25 species of Larvacea reported from the world 18 (2 families 14 genera) are reported from India (Dhandapani, 1998b; Renganathan, 1986).

Urochordata

Tunicates or sea squirts are well known as Urochordates and represent the taxon Protochordata. The most important feature of this group is presence of notochord during the early developmental stage. The animals lose their myomeric segmentation during the adult life stage. These marine animals are usually sessile and well recognized filter feeders. The structures of these animals is just like a sac. To date 2150 species have been recorded worldwide, including 281 species from India. There are about 25 species identified by the Zoological Survey of India in the Andaman and Nicobar Islands (Figure 19.23). The close proximity with the chordates gives enormous scope to researchers to work with this group of animals to elucidate their life processes.

FIGURE 19.23   **Urochordates** (a) *Didemnum molle*; (b) *Clavinella moluccensis.*

FISH

The history of ichthyology in India is colossal and interesting. Brief histories of early Indian ichthyology may be found in Day (1875–1878, 1889a,b) and Whitehead and Talwar (1976). Among the books published on Indian fish, Francis Day's (1875–1878) treatise *The Fishes of India* is of greatest importance. The publications on *Commercial Sea Fishes of India* by Talwar and Kacker (1984) and *Fishes of the Laccadive Archipelago* by Jones and Kumaran (1980) are noteworthy contributions to our knowledge of fish faunal resources of India, besides many research publications by other scientists. Fish comprise about half the total number of vertebrates. The estimated number of living fish species worldwide may be close to twenty-eight thousand. Day (1889a,b) described 1418 species of fish under 342 genera from British India. Talwar (1991) described 2546 species of fish belonging to 969 genera, 254 families and 40 orders. The distribution of marine fishes is rather wide, and some genera are common to the Indo-Pacific and the Atlantic regions. Fifty-seven percent of marine fish genera found in India are common to the Indian Ocean and to the Atlantic and Mediterranean.

The exact number of species associated with coral reefs of India is still to be found; however, the number of fish species in coastal and marine ecosystems of India is over 2546, of which Chondrichthyes (cartilaginous fish) account for 154 and Actinopterigii (bony fishes) for more than 2275 (Figure 19.24). The Lakshadweep Islands have a total of 603 species of fish (Jones and Kumaran, 1980). Recent studies increased the database of Indian fish to 2629 (Ramakrishna *et al.*, 2010). Over one thousand species are found in the Andaman and Nicobar Islands and about 538+ in the Gulf of Mannar Biosphere Reserve. The categories of fish occurring in coral reef ecosystems of India include groups such as the damselfish (76+ species), butterfly fish (40+ species), parrot fish (24+ species), sea bass, groupers and fairy basslets (57+ species), cardinal fish (45+ species), jacks and kingfish (46+ species), wrasse (64+ species), combtooth blennies (58+ species), gobies (110+ species), surgeonfish, tangs and unicornfish (40+ species) (Ramakrishna *et al.*, 2010). Another 20 percent is composed of cryptic

FIGURE 19.24   (a) Clownfish *Amphiprion ocellaris* (Cuvier, 1830); (b) Lionfish *Pterois antennata* (Bloch,1787). (Please see color plate at the back of the book.)

and nocturnal species that are confined primarily to caverns and reef crevices during daylight periods.

REPTILES

A total of 32 species of marine reptiles are reported from Indian seas, including 26 species of sea snake belonging to the family Hydrophiidae, five species of sea turtle and one species of saltwater crocodile *Crocodylus porosus* (Figure 19.25). All the sea snakes, four species of turtles and a crocodile are also known from the Andaman and Nicobar Islands. Studies on sea turtles occurring in the coastal waters of India and their nesting grounds were neglected until Smith (1931) focused our attention on these giants among the sea reptiles. Seven species of sea turtle are found in the world's warm oceans, of which five are reported in India. Of these, the Leatherback sea turtle, *Dermochelys coriacea* is the sole representative of the family Dermochelyidae and is a rare species. The remaining four species—the Green turtle (*Chelonia* mydas), the Olive Ridley (*Lepidochelys olivacea*), the Hawksbill (*Eretmochelys imbricata*), and the Loggerhead (*Caretta caretta*)—are contained in a single family, Cheloniidae.

FIGURE 19.25   (a) Saltwater crocodile *Crocodylus porosus* (Schneider, 1801); (b) Leatherback sea turtle *Dermochelys coriacea* (Vandelli, 1761).

SEABIRDS

The marine ecosystem offers a variable feeding and breeding ground for a number of birds. Although not exhibiting spectacular diversity, a number of seabirds are found regularly in marine and estuarine ecosystems. There are some special species, which are exclusively dependent on the coral reef ecosystem, while a few are generalists without much dependence on it. Some of the pelagic seabirds reported, notably boobies (Sulidae), shearwaters (Procellariidae) and terns (Sternidae), rarely nest on the Andaman and Nicobar Islands. Smaller numbers of waders and other seabirds are also found on or near coral reefs. These include sandpipers, oystercatchers, turnstones and plovers. Egrets and herons are also widespread, often feeding across the reef flat at low tide. Pelicans and flamingos have been recorded in the Gulf of Kachchh. Birds of prey including ospreys and sea eagles are likewise occasional visitors to the marine region. In the Gulf of Kachchh Marine Park area 123 species of waterfowl and 85 species of terrestrial birds have been reported in 2002. Waterfowl with moderately good population that have been found in Kachchh are the lesser flamingo, Kentish plover, ruff, crab plover, black tailed godwit and avocet. From the Gulf of Mannar Marine National Park in 1985–88 a total of 187 species of birds were recorded, of which 84 were aquatic and the remaining terrestrial. At Manali and Hare Islands 23 species of migratory birds were found to over-summer every year. Waders uncommon to India, such as knot *Calidris canuta,* eastern knot *Calidris tenuirostris,* curlew *Numenius auquata,* whimbrel *Numenius phaeopus* and bar tailed godwit *Limosa lapponica,* were recorded as regular winter visitors to this area (Balachandran, 1995) (Figure 19.26).

MARINE MAMMALS

Marine mammals are classified under three major orders: Cetacea (whales, dolphins and porpoises), Sirenia (manatees and dugong) and Carnivora (sea otters, polar bears and pinnipeds). Order Cetacea consists of two suborders: Mysticeti (baleen whales) and Odontoceti (toothed cetaceans). Mysticeti comprises four

FIGURE 19.26   (a) Black-naped tern *Sterna sumatrana* (Raffles, 1822); (b) Whimbrel *Numenius phaeopus* (Linnaeus, 1758).

families of 14 species, while Odontoceti comprises 10 families of 73 species. In total, 130 marine mammal species have been recognized in the world's oceans (Jefferson *et al.*, 2008). The Indian seas support 24 species of marine mammals, which include Delphinidae, Physeteridae, Kogiidae, Ziphiidae, Phocoenidae and Platanistidae (Kumaran, 2002; Vivekanandan and Jeyabaskaran, 2012). Of the 25 species of cetaceans, five are baleen whales, and the rest are Odontoceti, which includes Delphinidae, Physeteridae, Kogiidae, Ziphiidae, Phocoenidae and Platanistidae (Kumaran, 2002). However, most of these are oceanic forms, and occasionally a few individuals may be stranded on the shore. Stranding and sighting records report that the Indian seas are a habitat for 20 species of cetaceans and one species of sirenian (Vivekanandan and Jeyabaskaran, 2012). The sea cow, *Dugong dugon,* occurs in nearshore waters of the Gulf of Mannar, Gulf of Kachchh and Andaman and Nicobar Islands. All reported marine mammal species are protected under the Wildlife (Protection) Act, 1972.

The information on species distribution and abundance of marine mammals in marine regions of India remains scanty. Absence of capacity of marine mammalogists for surveys and research impedes progress in research to gain knowledge of species level distribution, abundance, biology and ecological characteristics. Recently Vivekanandan and Jeyabaskaran (2012) published the results of marine mammal surveys conducted in the Southern Ocean on distribution of cetaceans in Antarctic waters.

MARINE BIODIVERSITY OF INDIA AND THE GLOBE

The current inventory of coastal and marine biodiversity of India is summarized in Table 19.5. It indicates that a total of 17,802 species of faunal and floral communities have been reported from the seas around India. The data revealed that India contributes 6.75 percent of global marine biodiversity. Further intensive

TABLE 19.5 Marine Biodiversity of India in Comparison with Global Marine Biodiversity

Sl. No.	Taxon/group	Species in India	Species in world[1]
1.	**Bacteria**	–	4800
2.	**Sea weeds:** Cyanophyta	936	1000
3.	Chlorophyta		2500
4.	Phaeophyta		1600
5.	Rodhophyta		6200
6.	**Seagrass**	14	60
7.	**Mangroves**	69	100
8.	Other Protoctista	–	23,000
9.	**Bacillariopyta** (Diatoms)	200	5000

(Continued)

TABLE 19.5 Marine Biodiversity of India in Comparison with Global Marine Biodiversity *(cont.)*

Sl. No.	Taxon/group	Species in India	Species in world[1]
10.	Dinophyceae: (Dinoflagellates)	90	–
11.	Euglenophyta	–	250
12.	Chrysophyceae	–	500
13.	Dinomastigota	–	4000
14.	Radiolaria	–	550
15.	Porista	532	–
16.	Foraminifera	500	10,000
17.	**Porifera**	**512**	**5500**
18.	**Cnidaria**	**1042**	**9924**
19.	Actinaria	54	1000
20.	Hydrozoa	212	–
21.	Scyphozoa	37	220
22.	Cubozoa	5	–
23.	Anthozoa	600	–
24.	Siphonophora	89	–
25.	Ctenophora	12	166
26.	**Platyhelminthes** (Polyclads)	38	15,000
27.	**Arthropoda: Crustacea** Copepoda	540	2280
28.	Ostracoda	60	–
29.	Branchiura	5	–
30.	Cirrepedia	36	–
31.	**Malacostraca:** Mysidacea	75	780
32.	Cumacea	71	1300
33.	Tanidacea	1	900
34.	Isopoda	33	–
35.	Amphipoda	132	–
36.	Euphausiacea	23	–
37.	Stomataopoda	139	–
38.	**Decapoda**: Macrura	55	–
39.	Brachyura	250	–
40.	Anomura	20	1069
41.	Nemertina	5	1230
42.	Gnathostomulida	4	97
43.	Rhombozoa	–	82
44.	Orthonectida	–	24

TABLE 19.5 Marine Biodiversity of India in Comparison with Global Marine Biodiversity *(cont.)*

Sl. No.	Taxon/group	Species in India	Species in world[1]
45.	**Gastotricha**	38	400
46.	**Rotifera**	3+	50
47.	**Ciliata**	21	−
48.	**Kinorhyncha**	10	186
49.	**Loricifera**	−	18
50.	**Acanthocephala**	229	600
51.	Cycliphora	−	1
52.	**Entoprocta**	10	170
53.	**Nematoda**	45+	12,000
54.	**Nematomorpha**	−	5
55.	**Ectoprocta**	10	5700
56.	**Phornida**	1	10
57.	**Brachiopoda**	2	550
58.	Mollusca	3400	52,525
59.	Opisthobranchia	260	4000
60.	Bryozoa	200	4000
61.	Priapulida	−	8
62.	Siponcula	38	144
63.	Echiura	43	176
64.	**Annelida:** Polychaeta	883	12,000
65.	Achianellida	20	90
66.	**Tardigrada**	10	212
67.	Chelicerata	−	2267
68.	**Crustacea**	2394	44,950
69.	**Pogonophora**	−	148
70.	**Echinodermata**	777	7000
71.	**Chaetognatha**	32	121
72.	**Protochordata**	119	−
73.	**Hemichordata**	12	106
74.	**Urochordata**	281	4900
75.	**Cephalochordata**	−	32
76.	Pisces	2629	16,475
77.	Reptiles	37	−
78.	Birds	187	−
79.	Mammalia	24	110
80.	Fungi	−	500
	Total	**17,802**	**263,657**

− No reliable data available.
[1]*Bouchet (2006); Groombridge and Jenkins (2000).*

explorations and taxonomical studies may reveal several more species in the Indian context.

MARINE PROTECTED FAUNA

To safeguard nature and the animals of the wild, 683 areas (including National Parks, Wildlife Sanctuaries, Community Reserves and Conservation Reserves) have been designated protected areas to date in India, while 17 Biosphere Reserves were also designated to protect the entire ecosystem of that place. According to the series of updated notifications issued under the Wildlife (Protection) Act 1972 by the Ministry of Environment and Forests, Government of India, so far 885 species (Table 19.6) of marine fauna belong to eight phyla— Porifera, Coelenterata, Arthropoda, Mollusca, Echinodermata, Pisces, Reptilia and Mammalia—are well protected under the legislation of different categories (Schedules I, II, III, IV) as their natural population is depleting.

TABLE 19.6 Protected Marine Faunal Species in India

Faunal group	Number of protected species	Legislation
Porifera		
Calcareous sponges (all species)	10	Schedule III
Coelenterata		
Reef-building corals (all scleractinians)	519	Schedule I
Black corals (all antipatharians)	8	Schedule I
Organ pipe coral (*Tubipora misica*)	1	Schedule I
Fire coral (all millepora species)	5	Schedule I
Sea fan (all gorgonians)	86	Schedule I
Arthropoda		
Robber Crab (Crustacea)	1	Schedule I
Horseshoe crab (Merostomata)	2	Schedule IV
Mollusca		
Gastropoda	20	Schedule I, IV
Bivalvia	4	Schedule I, IV
Echinodermata		
Sea cucumber (all holothurians)	163	Schedule I
Fishes		
Elasmobranchs (sharks and rays)	10	Schedule I
Seahorse (all syngnathidians)	23	Schedule I
Giant grouper	1	Schedule I

TABLE 19.6 Protected Marine Faunal Species in India *(cont.)*

Faunal group	Number of protected species	Legislation
Reptiles		
Marine turtles	5	Schedule I
Saltwater crocodile	1	Schedule I
Mammals		
Marine mammals	26	Schedule I, II
Total	**885**	

CONCLUSION

The world's biodiversity is estimated to be one and three-quarter million species excluding microbial species (Heywood and Watson, 1996); another estimate is of a range from five to one hundred and twenty million species (Reaka-Kudla, 1997). Approximately three hundred thousand marine species are known, compared with more than one and a half million known terrestrial species. This indicates only a lack of knowledge of marine species rather than greater terrestrial diversity (Alder, 2003). It is thus evident that most of the world's marine species are unknown, and may remain so given a 500 year projection to describe all species based on a current knowledge of 1.36 million animal species and possibility that ten times this number actually exist (Heywood and Watson, 1996). In the context of Indian seas, it is also evident that even many of the better known taxonomic groups remain poorly studied. We may of course never know the full extent of biodiversity in any of the world's oceans, and the rate at which we increase our understanding (Keesing and Irvine, 2005) is likely to be lowest in Indian seas. The impacts of climate change will alter coastal marine ecosystems, affecting the range of species and their ecology at a rate faster than we are able to record them (Keesing and Irvine, 2005). It is evident that comprehensive taxonomic coverage of the marine biota of the any region remains a monumental task, beyond the capacity of existing local taxonomic expertise and subject to existing taxonomic impediments. It would be quite appropriate if cautious placement of systematic studies is made rather than haphazard and opportunistic description of new species as and when they are discovered, as remarked by Griffiths (2004).

REFERENCES

Aiyar, R.G., Alikunhi, K.H., 1940. Rec. Indian. Mus., Calcutta 42, 89–107.
Aiyar, R.G., Alikunhi, K.H., 1944. Proc. Nat. Inst. Sci. 10 (1), 113–140.
Aiyar, R.G., 1931. Proc 18th Indian. Sci. Congr, p. 244.
Aiyar, R.G., 1924. Proc 11th Indian Sci. Congr. p. 112.

Alcock, A., 1893. J Asiat Soc Bengal 62 (2), 138–149.

Alcock, A. (1898). Mem. Indian Mus. pp. 1–29.

Alcock, A., 1901. A descriptive catalogue of the Indian deep-sea Crustacea, Decapoda Macrura and Anomura in the Indian Museum being a revised account of the deep-sea species collected by the Royal Indian Marine Survey Ship, "Investigator". Indian Museum, Calcutta, pp. 1–286.

Alcock, A., 1906. Catalogue of the Indian Decapod Crustacea in the collection of Indian Museum Part III. Macura Fasciculus I. The prawns of the Peneus group. Indian Museum, Calcutta, 1-55

Alcock, A., Anderson, R.S., 1894. J. Asiat. Soc. Bengal. 43, 141–185.

Alcock, A.W., Anderson, R.S., 1893. Ann. Mag Nat Hist. 3 (7), 1–27, 278–292.

Alder, E., 2003. A world of neighbours: UNEP's Regional Seas Programme (UNEP) 2003 [www.unep.ch/seas/Library/-neighbours.pdf]

Alder, J., Hancock, A., 1864. Trans. Zool. Soc. London 5, 117–147.

Alikunhi, K.H., 1941. Proc. Indian Sci. Congr. 27, 152, Proc Indian. Acad. Sci, B, 13(3): 193–238.

Alikunhi, K.H., 1942. Proc. Indian. Sci. Congr. 28, 173, 29(1943); 149–150.

Alikunhi, K.H., 1946. Curr. Sci. 15, 140.

Alikunhi, K.H., 1947. Proc. Nat. Inst. Sci. India 13 (3), 105–127.

Alikunhi, K.H., 1948a. J. Royal Asiat. Soc. Bengal 14 (1), 17–25.

Alikunhi, K.H., 1948b. Proc. Nat. Inst. Sci. India. 14 (8), 373–383.

Alvarino, A., 1974. J. Mar. Biol. Assoc. India 14 (2), 713–722.

Ameer Hamsa, K.M.S., 1974. J. Mar. Biol. Assoc. India 14, 418–423.

Annandale, N., 1907. Proc. Asiat. Soc. Bengal NS III (2), 79–81.

Annandale, N., 1908. Rec. Indian Mus. 2, 24–32.

Annandale, N., 1909. Notes on freshwater sponges. X. Report on a small collection from Travancore. Rec. Indian Mus. 3, 101–104.

Annandale, N., 1913. Rec. Indian Mus. 9, 227–236.

Annandale, N., 1914. Rec. Indian Mus. 10, 273–280.

Annandale, N., 1915. Mem Indian Mus. 5, 65–114.

Annandale, N., 1916. Mem. Asiat. Soc. Bengal 5, 18–24.

Annandale, N., 1924. Rec. Indian Mus. 8, 61–68.

Annandale, N., Kemp, S., 1915. Mem. Indian Mus. 5, 55–63.

Antony, A., 1968. Bull. Dept. Mar. Biol. Oceanogr. 4, 11–154.

Apte, D., Pitale, R.D., 2011. J. Bombay Nat. Hist. Soc. 108 (2), 109–113.

Balachandran, S., 1995. J. Bombay Nat. Hist. Soc. 92 (3), 303–313.

Banse, K., 1959. J. Mar. Biol. Assoc. India 1 (2), 165–177.

Bernard, K.H., 1935. Rec. Indian. Mus. 37, 279–319.

Bhalla, S.N., 1970. Contr. Cushman. Fdn. Foramin. Res. 21 (4), 156–163.

Bhatia, B., Bhalla, S.N.J., 1959. J. Paleont. Soc. India 4, 78.

Bhatt, D.K., 1969. Contr. Cushman Fdn. Foramin. Res 20, 30.

Bhatt, J.R., Ramakrishna, Sanjappa, M., Remadevi, O.K., Nilaratna, B.P., Venkataraman, K., 2013. Mangroves of India – their biology and uses. Zoological Survey of India, Kolkata, p. 640.

Blazewicz-Paszkowycz, M., Bamber, R., Anderson, G., 2012. PLoS ONE 7 (4), e33068.

Borradaile, L.A., 1900. Proc. Zool. Soc. Lond., 568–596.

Borradaile, L.A., 1902. The fauna and geography of the Maldives and Laccadive archipelagoes, the account of the work carried on and of the collections made by an expedition during the years 1899 and 1900 by J. Stanley Gardiner. I, Marine crustaceans. On Varieties 1 (2), 191–208.

Borradaile, L.A., 1903. Marine Crustaceans. IV. Some remarks on the classification of the crabs. In: J. St. Gardiner, The Fauna and Geography of the Maldive and Laccadive Archipelagoes 1(4): 424–429.

Bouchet, P., 2006. The Magnitude of Marine Biodiversity. In: Duarte, C.M. (Ed.), The Exploration of Marine Biodiversity, Scientific and Technological Challenges. Foundation BBVA, (2006).

Brady, H.B., 1884. Rep. Sci. Res. Voyage Challenger Zool. 9, 21.

Brook, G., 1893. The Genus Madrepora. Catalogue of the Madreporarian Corals in the British Museum (Nat. Hist.), 1–212.

Browne, E.T., 1905. Rep. Govt. Ceylon Pearl Oyster 4, 131–166.

Browne, E.T., 1906. Trans Linn Soc London (Zool) 10, 163–187.

Browne, E.T., 1916. Trans Linn Soc London (Zool) 17, 169–210.

Browne, E.T., 1926. Trans Linn Soc (Zool) London 2, 55–86.

Cairns, S.D., Kitahara, M.V., 2012. Zoo. Keys. 227, 1–47.

Calman, W.T., 1904. Rep. Ceylon Pearl Oysters Fish. Gulf of Mannar 2, 159–180.

Chakrapany, S., 1984. Studies on Marine Invertebrates. Scyphomedusae of the Indian and adjoining seas. PhD Thesis submitted to the University of Madras. p. 206.

Chapman, F., 1895. Proc. Zool. Soc., 4–55.

Chatterjee, B.P., Gururaja, M.N., 1967. Bull. Nat. Inst. Sci. India 38, 393.

Chatterjee, T., Pesic, V., 2010. Cahiers de Biol. Mar. 51, 289–299.

Chaudhury, A., Biswas, B., 1954. Recent perforate Foraminifera from Juhu beach, Bombay. Micropal. 8 (4), 30–32.

Chilton, C., 1923. Mem. Indian. Mus. 5, 877–895.

Chopra, B., 1923a. Bopyrid isopods parasitic on Indian Decapoda Macura. Rec. Indian Mus. Calcutta, 25:411–550.

Chopra, B., 1923b. Bopyrid isopods on Indian Decapod Macrura. Rec. Indian Mus. Calcutta, 39a: 1-51.

Chopra, B., 1931. Rec. Indian Mus. 33, 303–324.

Chopra, B., 1933. Rec. Indian Mus. 35 (1), 25–32.

Chopra, B., 1934. Rec. Indian Mus. 37 (4), 463–514.

Chopra, B., 1939. Rec. Indian Mus. 32, 413–429.

Collier, C., 1830. Edinburgh New Phil. J. 8, 46–52.

Cutler, E.B., 1977. Steenstrupia 4 (12), 151–155.

Cutler, E.B., Cutler, N.J., 1979. Bull. Mus. Nat. Hist. Paris 4 (1), 941–990.

Daniel, A., 1952. Ann. Mag. Nat. Hist. 12 (5), 400–403.

Daniel, A., 1953. J. Zool. Soc. India 5 (2), 235–238.

Daniel, A., 1956. Bull. Madras Govt. Mus. 6 (2), 1–39.

Daniel, A., 1958. Ann. Mag. Nat. Hist. 13 (1), 305–308.

Daniel, A., 1962. Ann. Mag. Nat. Hist. 13 (5), 193–197.

Daniel, A., 1963. Ann Mag. Nat. Hist. 13 (6), 641–645.

Daniel, A., 1971. J. Mar. Biol. Assoc. India 13 (1), 82–85.

Daniel, A., 1974. Proc. Indian Nat. Sci. Acad. 38 (3&4), 179–189.

Daniel, A., 1975. J. Mar. Biol. Assoc. India 16 (2), 182–210.

Daniel, R., 1966. Ann. Mag. Nat. Hist. 9 (13), 689–692.

Daniel, R., 1974. Mem. Zool. Surv. India 15 (2), 865–868.

Daniel, R., 1985. *Zool. Surv. India* p. 440.

Daniel, R., Daniel, A., 1963. J. Mar. Biol. Assoc. India 5 (2), 185–220.

Dattagupta, A.K., 1976. Proc. Int. Symp. Biol. Sipuncula and Echiura Kotov 2, 111–118.

Dattagupta, A.K., 1967. Bull. Nat. Inst. Sci. 34, 365–370.

Dattagupta, A.K., Menon, P.K.B., 1963. Ann. Mag. Nat. Hist. 6, 57–63.

Dattagupta, A.K., Menon, P.K.B., 1964. Ann. Mag. Nat. Hist. 7, 57–63.

Day, F.T. (1875–1878). In: The fishes of India: being a natural history of the fishes known to inhabit the seas and fresh waters of India, Burma and Ceylon. 778 p. Reprinted William Dawson and Sons Ltd., London.

Day, F.T., 1889. The fauna of British India, including Ceylon and Burma. Fishes 1. Taylor & Francis, London, p. 548.

Day, F.T., 1889. The fauna of British India, including Ceylon, Burma, Fishes, Vol., II, Taylor, Francis Ltd., London, p. 509.

de Man, J.G., 1887. Zool. Jahrb. (Syst). 2, 639–689.

de Man, J.G., 1895. Zool. Jahrb. (Syst). 64 (2), 157–289.

de Man, J.G., 1896. Zool. Jahrb. (Syst). 65 (2), 134–296.

de Man, J.G., 1898. Zool. Jahrb. (Syst). 67 (2), 67–233.

de Man, J.G., 1899. Zool. Jahrb. (Syst). 68 (2), 1–104.

de Man, J.G., 1900. Zool. Jahrb. (Syst). 68 (3), 11–119.

de Man, J.G., 1901. Zool. Jahrb. (Syst). 69 (3), 279–486.

de Man, J.G., 1908. Rec. Indian Mus. 2, 211–231.

Devanesan, D.W., Varadarajan, S., 1939. Curr. Sci. 8 (4), 157–159.

Dhandapani, P., 1998a. Hemichordata. Faunal Diversity in India. Zoological Survey of India, Kolkata, pp. 406–409.

Dhandapani, P., 1998b. Protochordata. Faunal Diversity in India. Zoological Survey of India, Kolkata, pp. 412–415.

Eliot, C., 1906. On the nudibranchs of South India and Ceylon with special reference to the drawings of Kelaart and the collection belonging to Alder and Hancock preserved in the Hancock Museum at Newcastle-on-Tyne. Proc. Zool. Soc. London pt 111, 636–691, and 997–1008.

Fauvel, P., 1932. Bull. Madras Govt. Mus. I (2), 1–72.

Fauvel, P., 1930. Mem. Indian Mus. XII (1), 1–262.

Fauvel, P., 1953. Fauna of India including Pakistan, Ceylon, Burma and Malaya. Annelida, Polychaeta. The Indian Press, Allahabad, p. 507.

Fichtel, L. Von and Von Moll, J.P.C. (1798). Testacea microscopica, aliaqua minuta ex generibus Argpmauta et Mautilus and naturam pieta et descripta, 7: 123.

Frerichs, W.E., 1967. Diss. Abstr. 28B, 940–941.

Ganapati, P.N., Radhakrishna, Y., 1958. Andhra Univ. Mem. Oceanogr. 2, 210–237.

Gardiner, J.S. (1903–1906). In: J.S. Gardiner (Ed.) The fauna and geography of the Maldive and Laccadive Archipelagoes. Cambridge 2(1): 1679.

George, A.I., Nair, V.R., 1980. Mahasagar Bull. Nat. Insat. Oceanogr. 13 (1), 29–44.

George, A.I., 1905. Proc. Indian. Acad. Sci. Sec. B32, 215–221.

George, A.I., Purushan, K.S., Madhu Pratap, M., 1975. Indian J. Mar. Sci. 4, 201–202.

George, P.C., 1952. Proc. Nat. Inst. Sci. India 18, 657–689.

Ghosh, A., 1963. J. Mar. Biol. Assoc. India 5, 239–245.

Ghosh, H.C., 1975. Crustaceana. Rec. Zool. Surv. India 28 (1), 33–36.

Ghosh, H.C., 1976. Rec. Zool. Surv. India 71, 51–55, (1976).

Giles, G.M., 1885. J. Asiat. Soc. Bengal 54, 69–71.

Giles, G.M., 1887. J. Asiat. Soc. Bengal 56, 212–229.

Giles, G.M., 1888. J. Asiat. Soc. Bengal 57, 220–255.

Giles, G.M., 1890. J. Asiat. Soc. Bengal 59, 63.

Gnanamuthu, C.P., 1943. Bull. Madras Govt. Mus. New Ser. Nat. Hist. Section 1 (2), 21.

Gravely, F.H., 1942. Shells and other animal remains found on the Madras Beach II. Snails etc. (Mollusca : Gastropoda). Bull. Madras Govt. Mus. N S (N H) 5 (2), 104.

Gravely, F.H., 1927a. Bull. Madras Govt. Mus. New Series 1 (1), 41–51.

Gravely, F.H., 1927b. Bull. Madras Govt. Mus. New Series 1 (1), 123–124.

Gravely, F.H., 1927c. Bull. Madras Govt. Mus. New Series 1 (2), 123–128.

Groombridge, B., Jenkins, M.D. (Eds.), 2000. Global biodiversity: Earth's living resources in the 21st Century. World Conservation Press, Cambridge.

Haldar, B.P. (1975). In: Proceeding Int. Symp. Biol. Sipuncula and Echiura Kotor, 1: 51–92.

Haldar, B.P., 1976. Newsl. Zool. Surv. India 3 (3), 120–123.

Haldar, B.P., 1977. Rec Zool. Surv. India 70 (1), 12–13.

Haldar, B.P., 1978. Bull. Zool Surv. India 1 (1), 37–42.

Haldar, B.P. (1985a). In: State of the Art Report: Estuarine Biology, Workshop, Berhampore (Orissa), 9., p. 13.

Haldar, B.P. (1985b). In: Second National Seminar on Marine Intertidal Ecology, Waltair, India. Abstract 26 (1985b).

Harmer, S., 1915. Siboga Expedition Monographs, 28a. 565 pp.

Hartman, O., 1947. J. Mar. Biol. Assoc. India 16 (2), 609–644.

Henderson, J.R., 1893. Trans. Linn. Soc. London 5 (2), 325–458.

Heywood, V.H., Watson, R.T. (Eds.), 1996. Global biodiversity assessment. Cambridge University Press, New York.

Hofker, J., 1927–1951. The foraminifera of the SIBOGA Expedition. Part 1 (1927); Part II (1930); Part III (1951). Brill, Leiden.

James, D.B., 1986. In: James, P.S.B.R. (Ed.), Recent Advances in Marine Biology. Today and Tomorrow's Printers and Publishers, New Delhi, pp. 569–591.

Jaume, D., Boxshall, G.A., 2008. Hydrobiologia 595, 225–230.

Jefferson, T.A., Webber, M.A., Pitman, R.L., 2008. Marine mammals of the world: A comprehensive guide to their identification. Academic Press, San Diego, 592 p.

Johnson, P., 1964. Ann. Mag. Nat. Hist. 7 (13), 331–335.

Johnson, P., 1969. J. Bombay Nat. Hist. Soc. 66 (1), 43–46.

Johnson, P., 1971. J. Bombay Nat. Hist. Soc. 68 (3), 596–608.

Jones, S., Kumaran, M., 1980. Fishes of the Laccadive archipelago. Nature Conservation and Aquatic Sciences Service, Trivandrum, Kerala, 760 p.

Kannan, L., Thangaradjou, T., Anantharaman, P., 1999. Seeweed Res. Utiln. 21 (1&2), 25–33.

Kathiresan, K., 1999. Mangrove atlas and status of species of India. Project report. Ministry of Environment and Forests, Government of India, New Delhi, 235 p.

Keesing, J., Irvine, T., 2005. Coastal Biodiversity of Indian Ocean: the known, the unknown and the unknowable. Indian J. Mar. Sci. 34 (1), 11–26.

Kemp, S., 1913. Mem. Indian Mus. 4, 10–17.

Kemp, S., 1915. Mem. Indian Mus. 5, 199–325.

Kemp, S., 1916. Rec. Indian Mus 12 (8), 386–405.

Krishnamoorthi, B., 1963. J. Mar. Biol. Assoc. India 5, 97–102.

Krishnan, G. (1946). Studies on the Polychaetes from Madras. PhD Thesis, University of Madras.

Krishnaswamy, S., 1950. Rec. Indian Mus. 48, 117–120.

Krishnaswamy, S., 1952. Rec. Indian Mus. 50, 324.

Krishnaswamy, S., 1953. J. Madras Univ. 23B (1&2), 61–75.

Krishnaswamy, S., 1956. Rec. Indian Mus. 54, 29–32.

Krishnaswamy, S., 1957. Studies on the Copepoda of Madras. PhD Thesis, University of Madras.

Kumaran, P.L., 2002. Curr. Sci. 83 (10), 1210–1220.

Kurian, C.V., 1965. J. Mar. Biol. Assoc. India 2, 630–633.

Kurian, C.V., 1954. Rec. Indian Mus. 52 (2–4), 275–311.

Laidlaw, F.F., 1902. Fauna and Geology of the Maldive and Laccadive Archipelagoes 1: 282–312.

Lele, S.H., Gae, P.B., 1936. J. Univ. Bombay 4, 105–113.

Leloup, E., 1934. Bull. Mus. Hist. Nat. Belg. Bruxelles 10 (9), 1–5.

Lütken, C., 1865. Vidensk Meddr dansk naturh. Foren., 123–169, 1864.

Lütken, C., 1872. Overs k danske Videns. Selsk. Forh. 77, 75–178.

Madhu Pratap, M., 1979. Indian J. Mar. Sci. 8 (1), 1–8.

Madhu, R., Madhu, K., 2007. J. Mar. Biol. Ass. India 49 (2), 118–126.

Mammen, T.A., 1963. J. Mar. Biol. Assoc. India 5 (1), 27–61.

Mammen, T.A., 1965. J. Mar. Biol. Assoc. India 5 (1), 1–57.

Marktanner-Turneretscher, G., 1887. Annln. Naturh. Mus. Wein. 2, 291–316.

Matthai, G., 1924. Mem. Indian Mus. 8, 1–59.

Mauchline, J., 1980. Adv. Mar. Biol. 18, 1–369.

McLaughlin, P.A., Komai, T., Lemaitre, R., Rahayu, D.L., 2010. Raf. Bull. Zool. Suppl. No. 23, 5–107.

Menon, M.A.S., 1930. Bull Madras Govt. Mus. NS (NH) 3 (1), 1–28.

Menon, K.S., 1931a. Rec. Indian Mus. 33, 489–516.

Menon, M.A.S., 1931b. Bull Madras Govt. Mus. (NH) III 2, 32.

Menon, M.A.S., 1945. Proc. Indian Acad. Sci. 22, 31–62.

Menon, P.K.B., Dattagupta, A.K., Johnson, P., 1964. Ann. Mag. Nat. Hist. 7, 49–57.

Milne-Edwards, H., 1834–1837. Histoire Naturelle des Crustacés. Parts I–II. Librairie encyclopédique de Roret, Paris, 532 pp.

Müller, J., 1849. Über die Larven un die Metamorphose der Holothurien. Müller's Archiv, Berlin, 1849: 364–399.

Nair, K.K., 1945. Proc. Indian Sci. Congr. 32nd Sess. 3, 97.

Nair, K.K., 1954. Bull. Cent. Res. Inst. Univ. Travancore 2 (1), 47–75.

Nair, R.V., 1967. Proc. Symp. Indian Ocean Bull. NISI 38, 747–752.

Nair, R.V., 1971. J. Mar. Biol. Assoc. India 13 (2), 226–233.

Nair, R.V., 1973. IOB C Hand Book 5, 87–96.

Nair, R.V., 1974. J. Mar. Biol. Assoc. India 16 (3), 721–730.

Nair, R.V., 1975. Mahasagar 8 (1&2), 81–86.

Nair, R.V., Selvakumar, R.A., 1979. Magasagar Bull. Nat. Inst. Oceanogr. 12, 17–25.

Nair, R.V., Achuthankutty, C.T., Nair, S.S.R., Madhupratap, M., 1981. Indian J. Mar. Sci 10 (3), 270–273.

Nair, R.V., Rao, T.S.S., 1973. In: Zeitschel, B. (Ed.), The biology of Indian Ocean. Chapman and Hall, London, p. 549.

Nayar, K.N., 1959. Bull. Madras Govt. Museum (NH) VI (3), 59.

Nayar, K.N., 1966. Proc. Symp. Crustacea MBAI 1, 133–168.

Nilsson Cantell, C.A., 1938. Mem. Indian Mus. 13, 1–81.

Oza, R.M., Zaidi, S.H., 2001. A revised checklist of Indian Marine algae. CSMCRI, Bhavnagar, 296 p.

Palanisamy, M., 2012. Seaweeds of South Andaman: Chidiyatapu, North Bay and Viper Island. In: Proceedings of the International Day for Biological Diversity, Marine Biodiversity. 22 May, Uttar Pradesh State Biodiversity Board, Lucknow. pp. 49–58.

Panikkar, N.K., 1936. Proc. Zool. Soc. London 106, 39–52.

Panikkar, N.K., 1936. Proc. Zool. Soc. London 106, 39–52.

Panikkar, N.K., 1937a. Zool. Jahb. Abt. Anat. 3, 62–71.

Panikkar, N.K., 1937b. Proc. Indian Acad. Sci. 5 (2), 33–41.

Panikkar, N.K., 1937c. Rec. Indian Mus. XXXIX:, P.IV.

Panikkar, N.K., 1939. Proc. Zool. Soc. London 108, 4–7.

Panikkar, N.K., 1944. Curr. Sci. 13, 238–239.

Panikkar, N.K., Aiyar, R.G., 1937. Proc. Indian Acad. Sci. 6 (5), 284–336.

Parulekar, A., 1967. J. Bombay Nat. Hist. Soc. 64 (3), 524–529.

Parulekar, A., 1968. J. Bombay Nat. Hist. Soc. 65 (1), 138–147.

Parulekar, A., 1969a. J. Bombay Nat. Hist. Soc. 66 (1), 57–62.

Parulekar, A., 1969b. J. Bombay Nat. Hist. Soc. 66 (3), 590–595.

Parulekar, A., 1971. J. Bombay Nat. Hist. Soc. 68 (1), 291–295.

Parulekar, A. (1990). In: Marine Bio fouling, Power, Plants. (Eds.) K.V.K. Niltil and V.P. Venegopalan. pp. 218–228.

Patriti, G., 1970. Marsielle Fac. Ser. Suppl. 10, 285–303.

Petrescu, I., Chatterjee, T., 2011. Zootaxa 2966, 51–57.

Pillai, C.S.G., 1991. Bull Cent Mar Fish Res Inst 7, 23–30, (1967).

Pillai, N.K., 1944. Proc. Indian Sci. Congress 31, 99.

Pillai, N.K., 1968. J. Zool. Soc. India 20, 6–24.

Pillai, N.K., 1973. Mysidacea of the Indian Ocean. International Indian Ocean Expedition. Handbook of Zooplankton collection 6, 1–126.

Pillai, P.P., 1967. J. Mar. Biol. Assoc. India 13, 162–172.

Poulsen, E.M., 1965. Dana Report 65, 1–484.

Prasad, P.R., 1956. Indian J. Fish III, 1–42.

Prashad, B., 1919. Rec. Indian Mus. 16, 399–402.

Prashad, B., 1935. Rec. Indian Mus. 37, 39–43.

Prashad, B., 1936. Rec. Indian Mus. 38, 231–238.

Raghunathan, C., Smitanjali Choudhury, Raghuraman, R. and Venkataraman, K. (in press). Actinarian sea anemones of India with special reference to Andaman and Nicobar Islands. Zoological Survey of India, Kolkata.

Ramakrishna, Sreeraj, C.R., Raghunathan, C., Sivaperuman, C., Yogesh Kumar, J.S., Raghuraman, R., Immanuel, T., Rajan, P.T., 2010. Guide to Opisthobranchs of Andaman and Nicobar Islands. Zoological Survey of India, Occasional Paper p. 196.

Ramalingam, J.R. 2000. Golden jubilee Celebrations Souvenir 2000, Mandapam R C of CMFRI, Mandapam Camp. pp. 81–83.

Rana, S.S., Nigam, R., Panchang, 2007. Indian J. Mar. Sci. 36 (4), 355–360.

Rao, K.K., 1970. J. Bombay Nat. Hist. Soc. 67, 259.

Rao, K.M., 1971a. Proc. Indian Acad. Sci. 73, 155.

Rao, K.M., 1971b. J. Bombay Nat. Hist. Soc. 68 (1), 9–19.

Rao, M.U., 2010. Nat. Symp. Mar. Plants Parangipettai pp. 5–6.

Rao, T.S.S., 1958. Andhra Univ. Mem. Oceanogr. 2, 137–146.

Rao, T.S.S., 1966. J. Bombay Nat. Hist. Soc. 62, 544–548.

Rao, T.S.S., Ganapati, P.N., 1958. Andhra Univ. Mem. Oceanogr. 2, 147–163.

Rao, G.C., Ganapati, P.N., 1968. Proc. Indian. Acad. Sci. Sec. 67B (1), 24–29.

Rao, T.S.S., Kelly, S., 1962. J. Zool. Soc. India 14, 219–225.

Reaka-Kudla, M.L., 1997. The global biodiversity of coral reefs: A comparison with rain forests. In: Reaka-Kudla, M.L., Wilson, D.E., Wilson, E.O. (Eds.), Biodiversity II: Understanding and protecting our biological resources. Joseph Henry Press, Washington, DC, pp. 36–50.

Reddy, K.N., Ramakrishna, G., 1972. Rec. Zool. Surv. India 66 (1–4), 19–30.

Renganathan, T.K. 1986. Studies on the Ascidians of South India. PhD Thesis, Madurai Kamaraj University, Madurai. p. 249.

Rengarajan, K., 1974. J. Mar. Biol. Ass. India 16, 280–286.

Roy, A. and Ghosh, A. 2009. Ocean. Published by Sheuli Chatterjee, Sea Explorer's Institute, Kolkata, p. 256.

Sadhukhan, K., Raghunathan, C., 2011. World J. Zool. 6 (4), 334–338.

Sadhukhan, K., Raghunathan, C., 2012a. Int. J. Plant, Ani. Env. Sci. 2 (1), 183–189.

Sadhukhan, K., Raghunathan, C., 2012b. Int. J. Biol. Phar. All. Sci. 1 (1), 44–55.

Sadhukhan, K., Raghunathan, C., 2012c. Int. J. Sci. Nat. 3 (1), 167–169.

Salinas, J.I.S., 1993. J. Nat. Hist. 27 (3), 535–555.

Sarojini, R., Nagabhushanam, R., 1972. Rec. Zool. Surv. India 66, 249–272.

Sastry, D.R.K., 1998. Echinodermata. Faunal Diversity in India. Zoological Survey of India, Kolkata, pp. 398–403.

Satyanarayana Rao, K., 1998. Bryozoa. Faunal Diversity in India. Zoological Survey of India, Kolkata, pp. 371–377.

Sewell, R.B.S., 1924. Fauna of Chilka Lake. Crustacea Copepoda. Mem. Indian Mus. 5, 771–851.

Sewell, R.B.S., 1929. Mem. Asiat. Soc. Bengal 9, 133–205.

Sewell, R.B.S., 1935. Mem. Asiat. Soc. Bengal 9, 461–540.

Sewell, R.B.S., 1940. Copepoda Harpacticoida. Scient. Rep. John Murray Exped. 7, 117–382.

Shipley, A.E., 1903. Echiuroidea. Gardiner, J.S. (Ed.), Fauna and Geography of the Maldive and Laccadive Archipelagos, vol 1, Cambridge University Press, Cambridge, pp. 131–140.

Siebold, I., 1974. Distribution: Archaias angulatus is a typical Caribbean species. Rev. Esp. Micropal. 7, 175–213.

Silas, E.G., Kalimuthu, S., 1997. Commercial exploitation of seaweeds in India. Bull. Cent. Mar. Fish. Res. Inst. 41, 55–59.

Silas, E.G., Srinivasan, M., 1968. On the little known Chaetognatha Sagitta bombayensis Lele and Gae (1936) from Indian waters. J. Mar. Biol. Assoc. India 9 (1), 84–95.

Silas, E.G., Srinivasan, M., 1969. A new species of *Eukrohnia* from the Indian seas, with notes on three other species of Chaetognatha. J. Mar. Biol. Assoc. India 10 (2), 1–33.

Silas, E.G., Srinivasan, M., 1970. Chaetognatha of the Indian Ocean with a key for their identification. Proc. Indian Acad. Sci. 71, 177–192.

Sivaprakasam, T.E., 1966. J Mar. Biol. Ass. India 12 (1&2), 81–92.

Sivaprakasam, T.E., 1967. Notes on some amphipods from the south east coast of India. J. Mar. Biol. Assoc. India, 372–383.

Sivaprakasam, T.E., 1968. Leucothoid amphipoda from the Madras coast, India. J. Mar. Biol. Assoc. India 14, 34–51.

Sivaprakasam, T.E., 1970. Amphipods of the genus Lembos Bate from the south-east Coast of India. J. Mar. Biol. Assoc. India 16, 81–92.

Smith, M.A., 1931. The Fauna of British India. Reptilia and Amphibia, vol 1: Loricata, Testudines. Taylor and Francis, London, xxviii+p. 185.

Sournia, A., Dinet, M.J.C., Richard, M.J., 1991. Marine phytoplankton: How many species in the world ocean? Plankton Res 13 (5), 1093–1099.

Southern, R., 1921. Polychaeta of the Chilka Lake and also of fresh and brackish waters in other parts of India. Mem. Indian Mus. 5, 563–659.

Sreeraj, C.R., Raghunathan, C., 2011. New records of pseudocerotid polyclads from Andaman and Nicobar Islands. J. Mar. Biol. Assoc. UK 4 e73, 1–15.

Sreeraj, C.R., Raghunathan, C., 2013. Pseudocerotid polyclads (Platyhelminthes, Turbellaria, Polycladida) from Andaman and Nicobar Islands, India. Proc. Int. Acad. Ecol. Environ. Sci. 3, 36–41.

Srinivasan, M., 1979. Taxonomy and Ecology of Chaetognatha of the west coast of India. Zool. Surv. India, Techn. Monogr. 3, 1–47.

Srinivasan, M., 1972. Two new records of meso- and bathy-planktonic chaetognaths from the Indian Seas. J. Mar. Biol. Assoc. India 13 (1), 130–133.

Srinivasan, M., 1976. Distribution of Chaetopaths with special reference to *Sagitta decipiens* as an indicator of upwelling along the west coast of India. J. Mar. Biol. Assoc. India 16 (3), 126–143.

Srinivasan, M., 1977. J. Mar. Biol. Assoc. India 16 (3), 836–838.

Srinivasan, M., 1980. Bull. Zool. Surv. India 3, 55–61.

Srinivasan, M., 1996. Results of FORV Sagar Sampada. NIO, Goa, pp 139–148..

Stephen, A.C., Edmonds, S.J., 1972. The Phyla Sipuncula and Echiura. Trustees of the British Museum (Natural History), London.

Subba Rao, N.V., 1998. Mollusca. Faunal Diversity in India. Zoological Survey of India, Kolkata, pp. 104–117.

Subba Rao, N.V., 1991. Mollusca. Animal Resources of India. Zoological Survey of India, Kolkata, pp. 125–147.

Subba Rao, M., Vedantam, D., 1968. Bull Nat. Inst. Sci. India 38, 49–501.

Subramaniam, M.K., 1937. Distribution of the genus Sagitta during the several months of the year in the Indian seas. Curr. Sci. 6, 284–288.

Subramaniam, M.K., 1938. Proc. Indian Acad. Sci. Sec. B7, 270–276.

Sundara Raj, B., 1927a. Suborder Caprellidea (Laemodipoda). The littoral fauna of Krusadai Island in the Gulf of Mannar. Bull. Madras Govt. Mus. 1 (1), 125–128.

Sundara Raj, B., 1927b. Littoral fauna of Krusadai Island in the Gulf of Mannar. Siphonophora. Bull. Madras Govt. Mus. 1, 21–23.

Surya Rao, K.V., 1975. Intertidal amphipods from the Indian Coast. Proc. Indian Nat. Sci. Acad. 38, 190–205.

Talwar, P.K., 1991. Pisces. Animal Resources of India: Protozoa to Mammalia, vol 1, Zoological Survey of India, Kolkata, pp. 577–630.

Talwar, P.K., Kacker, R.K., 1984. Commercial sea fishes of India, Handbook. Zoological Survey of India, Kolkata, 997 p.

Tampi, P.R.S., Rangarajan, K., 1963. On the occurrence of *Arenicola brasiliensis nonato* (Family: Arenicolidae Polychaeta) in Indian waters. J. Mar. Biol. Assoc. India 5 (1), 108–112.

Tattersall, W.M., 1922. Indian Mysidacea. Rec. Indian Mus. 24, 445–504.

Thomas, P.A., 1998. Porifera. In: Alfred, J.R.B., Das, A.K., Sanyal, A.K. (Eds.), Faunal Diversity of India. Zoological Survey of India, Kolkata, pp. 27–36.

Thomas, P.A., George, M.A., Lazarus, S., 1995. J. Mar. Biol. Assoc. India 37, 134–142.

Tiwari, K.K., Biswas, S., 1952. On two new species of the genus *Squilla Fabri.*, with notes on other stomatopods in the collection of the Zoological Survey of India. Rec. Indian Mus. 49 (3 & 4), 349–363.

Tiwari, K.K., Ghosh, H.C., 1975. Redescription of *Squilla bengalensis* Tiwari and Biswas (Crustacea: Stomatopoda). Proc. Zool. Soc. Calcutta 26, 33–37.

Totton, A.K., 1954. Disc. Rep. Cambridge 27, 161.

Varadarajan, S., 1934. *Pentaceros hedemanni* in the sea off Krusadai Island, Gulf of Mannar. Curr. Sci. 8, 3–6.

Varadarajan, S., Chacko, P.I., 1943. On the arrow-worms of Krusadai. Proc. Nat. Inst. Sci. India 9, 245–248.

Venkataraman, G., 1939. A Systematic account of some South Indian diatoms. Proc. Indian Acad. Sci. 10(B), 293–368.

Venkataraman, K., 1998. In: Alfred, J.R.B., Sanyal, A.K., Das, A.K. (Eds.), Faunal Diversity in India. Zoological Survey of India, Kolkata, pp. 391–395.

Venkataraman, K., Satynarayana, Ch., 2012. Corals identification manual. Zoological Survey of India, Kolkata, 1–136.

Venkataraman, K., Jeyabaskaran, R., Raghuram, K.P., Alfred, J.R.B., 2004. Bibliography and checklist of corals and coral reef associated organisms of India. Rec. Zool. Surv. India, Occ. Paper 226, 1–648.

Venkataraman, K., Satynarayanan, Ch., Alfred, J.R.B., Wolstenholme, J., 2003. Handbook on Hard Corals of India. Zoological Survey of India, Kolkata, p. 266.

Vivekanandan, E., Jeyabaskaran, R., 2012. Marine mammal species of India. Central Marine Fisheries Research Institute, Kochi, p. 228.

Weltner, W, 1894. Zwei neue Cirripediea aus dem Indischen Ocean *S.B. Ges Natur Fr Berlin* (1894).

Whitehead, P.J.P., Talwar, P.K., 1976. Francis Day (1829–1889) and his collections of Indian Fishes. Bull. Br. Mus. Nat. Hist. (Sr.) 5, 1–189.

Wood Mason, J., Alcock, A., 1891. Ann. Mag. Nat. Hist. 7, 1–19, 186–202, 258–272.

Zhang, Z.-Q., 2011. Zootaxa 3148, 7–12.

Part II

Ecology and Conservation

Chapter 20

DNA Barcoding of Marine Venomous and Poisonous Fish of Families Scorpaenidae and Tetraodontidae from Andaman Waters

V. Sachithanandam*, P.M. Mohan* and N. Muruganandam[†]

*Department of Ocean Studies and Marine Biology, Pondicherry University, Port Blair, Andaman Islands, India; [†]Regional Medical Research Centre (ICMR), Port Blair, Andaman Islands, India

INTRODUCTION

An estimate of the earth's biodiversity shows that humans are one among 10 to 15 million species inhabiting the blue planet (Hammond, 1992). In traditional taxonomy, species are described on the basis of morphological characters, and taxonomic keys are developed. Taxonomic identification keys originated with Aristotle and were systematically organized by Linnaeus. The conventional taxonomic approach tends to be very tedious and is a matter of subjectivity on the part of the taxonomist who chooses those morphological characters believed to delineate species (Coyne and Orr, 2004). In the last 250 years, the world's taxonomists have described and made inventory only for 25 percent of species delineated through the traditional taxonomy protocols (Packer *et al.*, 2009). Currently, shortage of trained personnel limits the development of taxonomic knowledge. The above factors have created a taxonomic impediment to biodiversity studies (Hebert *et al.*, 2003).

Oceans cover more than 70 percent of the Earth. Bouchet (2006) noted that 229,602 marine species had been described but that the total species living on Earth may exceed more than 10 million as estimated by Grassle and Maciolek (1992).

Worldwide, about thirty-two thousand species of fin fish have been catalogued, in six classes, 62 orders and 540 families (Eschmeyer, 2010), accounting for more than half of all vertebrate species. Major contributions to Indian ichthyology have been made by the pioneering work of Day (1878); later, Talwar

Marine Faunal Diversity in India. DOI: 10.1016/B978-0-12-801948-1.00020-3

351

(1990) reported 2546 fish species belongings to 969 genera, 254 families and 40 orders. Fifty-seven percent of marine fish genera are common to the Indo-Pacific and Atlantic and Mediterranean regions (Venkataraman and Wafar, 2005).

Fish diversity in the seas of the Andaman and Nicobar Islands also is of special interest in terms of marine zoogeography because they lie at the confluence of the Andaman Sea with the western Pacific and the Indian Ocean (Rajan, 2010). A total of 1485 species of fish under 603 genera belonging to 177 families is represented from these islands, of which 400 species have commercial significance as food. Among the fish species, 1089 (73.38%) are recorded from coral reef environment, 277 from mangroves, 152 from seagrass meadow, 23 from freshwater streams and 101 from offshore environment, while 158 species are commonly observed in mangrove, seagrass, coral reefs and offshore ecosystems (Rao *et al.*, 2000; Rajan, 2001, 2010; Rao, 2003; Rajan and Sreeraj *et al.*, 2012). The catalogued endemic biodiversity of the islands comprises 16 species of fish, 31 species of reptile and 8 species of amphibian (Ramakrishna *et al.*, 2010).

The exploitation of commercially important marine fish species from the Andaman and Nicobar Islands has been marked by unregulated trading practices, lack of scientific data, poor fish conservation management and overfishing, and excessive by-catch, which have led to decline in fish stocks. Further, as regards fish species that are not commercially important, efforts to study, identify and conserve them have been limited.

According to Venkataraman and Wafar (2005) who summarized the Indian coastal and marine biodiversity reported at that time, minor phyla, non-commercially important fishes, remote parts of the islands and minor estuaries remain untouched for taxonomic studies. Lack of trained taxonomists is another reason why environmental biodiversity work on marine fauna from the islands' waters is still at a very early stage. As a contribution to overcoming this problem, the present work was designed for molecular taxonomy of marine venomous and poisonous species through mitochondrial DNA Cytochrome C Oxidase subunit I (COI) gene sequences.

DNA barcoding is a method, based on short standardized gene sequences of DNA (Hebert et al., 2003), used for species identification. It has been used *inter alia* for identification of fish species (Hebert *et al.*, 2004; Ward *et al.*, 2005; Hubert *et al.*, 2008; Persis *et al.*, 2009; Steinke *et al.*, 2009a; Lakra *et al.*, 2010; Sachithanandam *et al.*, 2011a, 2012). Molecular taxonomy through COI gene sequence analysis can be used to identify fish species from a small portion of muscle from whole specimens or any larval stage of the species. Nowadays, the DNA barcoding tool is revealing cryptic species and enabling new species assignments (Hebert *et al.*, 2004; Hubert *et al.*, 2012).

DNA barcoding is being applied in the field of fish conservation (Holmes *et al.*, 2009; Persis *et al.*, 2009; Steinke *et al.*, 2009a), fisheries managements aspects (Rasmussen *et al.*, 2009) and food safety analysis, with mislabeling being revealed through COI sequences (Wong and Hanner, 2008). In 2005 FISH-BOL

(Fish Barcode of Life Initiative) was set up to establish a library of COI gene sequences of all fish species, to enable global taxonomic identification (Ward *et al.,* 2009 and Eschmeyer, 2010).

Of the 1561 species of the order Scorpaeniformes catalogued worldwide on the basis of morphology, 500 (32%) have been barcoded (Becker *et al.,* 2011). The order Tetraodontiformes comprises 431 morphologically reported species. COI gene sequences studies of this order suggest that 193 species (45%) have been analyzed.

FISH-BOL's primary work is led by a research team with responsibility for collection of samples, traditional identification using morphological keys, and COI gene sequencing of species in their geographic regions. Current progress of DNA barcoding in the Arctic and Antarctic regions is 74 percent and 50 percent, respectively. Other regions from tropical to subtropical show good progress, with 20 percent coverage from Australia, Mesoamerica continental and Oceania. However, species-rich regions of Asia, South America and Africa show low progress (Becker *et al.,* 2011). In India, 11,023 fish species have been morphologically reported (Nelson, 2006; Mecklenburg *et al.,* 2010). Reports suggest that only 1918 of these (17.4%) have been barcoded (Becker *et al.,* 2011). Therefore this study focused on the collection of non-commercial fishes of the families Scorpaenidae and Tetraodontidae from Andaman waters and their COI gene sequencing.

The present study was carried out on marine venomous and poisonous fishes collected from Andaman mangrove and intertidal areas. Specimens were identified according to morphological keys as described by Randall and Eschmeyer (2001). The morphologically identified fish specimens were analyzed for molecular taxonomy, which is first-time documentation from this waters. The studied family of Scorpaenidae, genera *Pterois,* are morphologically ambiguous in nature, which has led to misidentification in different regions (Randall and Eschmeyer, 2001). Similarly, family Tetraodontidae (Puffer Fish), genus *Arothron*, morphologically resemble sister species, which has also confused species identification and confirmation in the field as well as in the laboratory.

METHODS

Study Area

The Andaman and Nicobar Islands (92–94° East; 06–14° North) are an archipelago with 572 islands/islets, stretching over 700 km from north to south, in the Bay of Bengal. These islands are near to east Asian countries and more than 1400 km from mainland India. Andaman is a volcanic rock land mass surrounded by various endemic flora and fauna (Tikader *et al.,* 1986; Rao *et al.,* 2000; Rajan, 2001; Rao, 2003). The fish samples belonging to the families Scorpaenidae and Tetraodontidae were collected from mangrove regions and tidal pools off Port Blair, the capital (Figure 20.1).

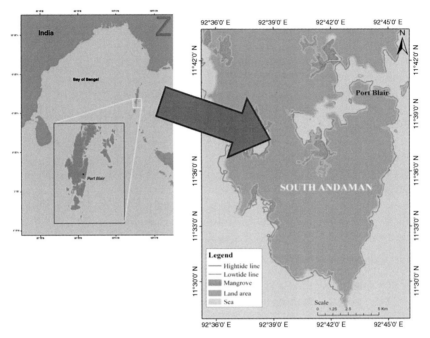

FIGURE 20.1   Study area.

Morphological Identification

Traditional taxonomic analyses were carried out using the morphological keys set out below (Rao *et al*., 2000; Randall and Eschmeyer, 2001; Rao, 2003).

Diagnostic Features of Scorpaenidae

Genus: *Pterois*

Dorsal spines XIII, 9–11; anal spines, 3; anal soft rays, 6–7. Pectoral rays branched, broad and feather like structure. Body slightly robust, snout, pre-orbital and pre-opercula had margin with dermal filaments. Color: Head and body reddish brown with numerous dark brown cross bars and numerous narrow pale inter-spines. Dorsal, anal and caudal fins have black spots. Ventral side fins with small pearly spots and black markings. Reddish to tan or grey in color, with numerous thin dark bars on body and head; tentacle above eye may be faintly banded. Adults have a band of small spines along the cheek and small spots in the median fins. Standard Length (SL) maximum is 35.0 cm.

Ambiguous Nature *Pterois miles* and *Pterois volitans* morphologically resemble each other. *Pterois miles* is known as *Pterois volitans* in Sumatra (Randall and Eschmeyer, 2001).

Genus: *Scorpaenodes*

Dorsal spines XII, 9; anal spines III, 5. Pectoral rays, 18–19. Nasal spines present; a few spines present in a row under eye. Color pattern: dark brown with dark and light mottling; fins with white and brown spots in rows. A black spot nearly as large as eye on inner surface of pectoral fins near base of first 5 rays; no black mark inside mouth at front of upper jaw. Ascending process of premaxilla narrow and its maximum width is 1.8–2.2 in orbit diameter. A series of papillae or nodules noticed across inter-orbital space between supra-ocular spines; nasal spine single. SL maximum is 21.0 cm. Usually found immobile at the bottom camouflaged among rocks and coral.

Diagnostic features of Tetraodontidae (Puffers)

Genus: *Arothron*

Dorsal profile 10–12 rays; anal rays 10–11; pectoral fins rays 16–19. Body heavy and broad; small spinules present on head and body. Caudal fin rounded and posterior caudal peduncle positions spines absent.

1. Caudal fin margins black and a dark blotch at base of pectoral fin. Body brown above side. No other marks or spots are present on the body. → *A. immaculatus.*
2. Cheek and snot, sides back, caudal peduncle and fin with white spots present on body. Ventrolateral side curved stripes present. → *A. hispidus.*
3. Back, sides, caudal peduncle and fin with white spots and belly with longitudinal stripes ascending to cheeks and snout forming oblique lines. → *A.reticularis.*

DNA Barcoding

Total DNA was extracted from 0.25 g of tissue by using lysis buffer and followed by standard proteinase-K / phenol-chloroform-isoamyl alcohol-ethanol precipitation method (Sambrook et al., 1987) and fish DNA modified isolation protocol followed as described by Sachithanandam *et al.*, (2012). The concentration of DNA was estimated using UV spectrophotometer method at 260/280 nm.

Subsequently the DNA was diluted to final concentration of 100 ng/μL for further use. The 650–655 bp section of the mitochondrial (mt) DNA genome from the COI gene was amplified using a published universal degenerated primer set (Ward *et al.*, 2005) synthesized by Sigma Aldrich Chemicals India Pvt. Ltd.

The polymerase chain reaction (PCR) was carried out in 25 μL consisting of 100 ng/μL of DNA and PCR master mix described in Table 20.1. PCR was carried out in an Applied Bio systems AB-2720 Thermal cycler, and PCR thermal condition was followed (Sachithanandam et al., 2011b and 2012). PCR products were resolved in 1 percent agarose containing 0.5 μg/mL of ethidium bromide and viewed under UV Transilluminator and documented.

TABLE 20.1 Master Mix Preparation

Reagent	Volume (μL)
10X buffer	5.0
dNTP (5 mM)	1.50
COI gene forward Primer F1 (0.5 μM) 5′TCAACCAACCACAAAGACATTGGCAC3′	1.0
COI gene reverse primer R1 (0.5 μM) 5′TAGACTTCTGGGTGGCCAAAGAATCA3′	1.0
Taq polymerase (3 U)	1.0
MgCl$_2$	0.25
RNase free water	10.25
Total volume	20
5 μL DNA was mixed with 20 μL reaction mix.	

Nucleotide sequencing was performed using the Sanger *et al.* (1977) method. Sequencing was performed using a BigDye Terminator Cycle Sequencing kit, following manufacturer's instructions (Applied Biosystems, Foster City, CA, USA). The sequencing was done both in the forward and reverse directions.

Sequence Analysis

The DNA sequences were analyzed both in forward and reverse directions for each individual fish and assembled using the SeqMan II version 5.03 (DNA-STAR) and ChromaxSeq Version 3.1. The sequence analysis was done along with reference sequences of various species belonging to the family Scorpaenidae and Tetraodontidae, retrieved from the National Center for Biotechnology Information (NCBI) GenBank and Barcode of Life Data System (BOLD). Nucleic acid sequence multiple and pairwise alignment was done using the CLUSTALW tool, and phylogenetic molecular evolutionary analyses were conducted using MEGA (Molecular Evolutionary Genetics Analysis) version 4.1. A neighbor-joining (NJ) tree was constructed using MEGA 4.1. Bootstrap values for the NJ tree were estimated using searches with 1000 pseudo replicates (Tamura *et al.*, 2007; Saitou and Nei, 1987; Kimura, 1980). The aligned sequences were also included for nucleotide search carried out in the NCBI Basic Local Alignment Search Tool (BLAST) 2.2.26+ (Zhang *et al.*, 2000) and in the BOLD system (Ratnasingham and Hebert, 2007) to determine genetic identity and further strengthen our results.

RESULTS AND DISCUSSION

DNA sequences for six studied species belonging to the families Scorpaenidae and Tetraodontidae were submitted to GenBank (NCBI PubMed).

Family Scorpaenidae COI Gene Sequence Analysis

Three species of the Scorpaenidae family genera *Pterois* and *Scorpaendoes* were DNA barcoded on the basis of samples from Andaman waters: *Scorpaendoes gibbosa*, *Scorpaendoes guamesis* and the lionfish *Pterois miles*. The COI gene sequences obtained were compared with data in the NCBI BLAST and BOLD databases to confirm species identification and for observation of haplogroup similarity.

The BLAST nucleotide analysis results for the two species of *Scorpaendoes* studied showed no record available for *S. guamesis* in the BOLD database (Figure 20.2) and that *S. gibbosa* had genetic identity of 99.84 percent

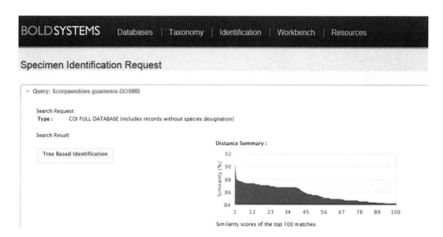

Phylum	Class	Order	Family	Genus	Species	Specimen Similarity (%)
Chordata	Actinopterygii	Scorpaeniformes	Synanceiidae	Synanceia	horrida	90.28
Chordata	Actinopterygii	Scorpaeniformes	Synanceiidae	Inimicus	sp.B	87.96
Chordata	Actinopterygii	Scorpaeniformes	Tetrarogidae	Ocosia	cf.zaspilota	87.81
Chordata	Actinopterygii	Scorpaeniformes	Synanceiidae	Inimicus	sinensis	87.65
Chordata	Actinopterygii	Scorpaeniformes	Synanceiidae	Inimicus	sinensis	87.65
Chordata	Actinopterygii	Scorpaeniformes	Synanceiidae	Synanceia	verrucosa	87.5
Chordata	Actinopterygii	Scorpaeniformes	Synanceiidae	Synanceia	verrucosa	87.42
Chordata	Actinopterygii	Scorpaeniformes	Synanceiidae	Synanceia	verrucosa	87.35
Chordata	Actinopterygii	Scorpaeniformes	Synanceiidae	Synanceia	verrucosa	87.35
Chordata	Actinopterygii	Scorpaeniformes	Synanceiidae	Synanceia	horrida	87.35
Chordata	Actinopterygii	Scorpaeniformes	Synanceiidae	Synanceia	verrucosa	87.35
Chordata	Actinopterygii	Scorpaeniformes	Tetrarogidae	Liocranium	praepositum	87.35
Chordata	Actinopterygii	Scorpaeniformes	Scorpaenidae	Scorpaenopsis		87.35
Chordata	Actinopterygii	Scorpaeniformes	Tetrarogidae	Ocosia	sp.1	87.19
Chordata	Actinopterygii	Scorpaeniformes	Tetrarogidae	Liocranium	praepositum	87.19
Chordata	Actinopterygii	Scorpaeniformes	Synanceiidae	Inimicus	caledonicus	87.19
Chordata	Actinopterygii	Scorpaeniformes	Tetrarogidae	Liocranium	praepositum	87.04
Chordata	Actinopterygii	Scorpaeniformes	Tetrarogidae	Liocranium	praepositum	87.04
Chordata	Actinopterygii	Scorpaeniformes	Synanceiidae	Inimicus	cf.caledonicus	87.04
Chordata	Actinopterygii	Scorpaeniformes	Tetrarogidae	Liocranium	pleurostigma	87.04
Chordata	Actinopterygii	Scorpaeniformes	Synanceiidae	Inimicus	caledonicus	87.04
Chordata	Actinopterygii	Scorpaeniformes	Synanceiidae	Inimicus	caledonicus	87.04
Chordata	Actinopterygii	Scorpaeniformes	Tetrarogidae	Liocranium	praepositum	86.88
Chordata	Actinopterygii	Scorpaeniformes	Synanceiidae	Inimicus	sp.b	86.87
Chordata	Actinopterygii	Scorpaeniformes	Synanceiidae	Inimicus	sinensis	86.87

FIGURE 20.2   **BLAST nucleotide analysis results for *Scorpaendoes guamesis* species from Andaman waters.**

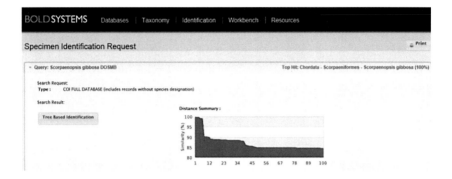

Phylum	Class	Order	Family	Genus	Species	Specimen Similarity (%)
Chordata	Actinopterygii	Scorpaeniformes	Scorpaenidae	*Scorpaenopsis*	gibbosa	100
Chordata	Actinopterygii	Scorpaeniformes	Scorpaenidae	*Scorpaenopsis*	gibbosa	99.84
Chordata	Actinopterygii	Scorpaeniformes	Scorpaenidae	*Scorpaenopsis*	gibbosa	99.84
Chordata	Actinopterygii	Scorpaeniformes	Scorpaenidae	*Scorpaenopsis*	oxycephala	99.84
Chordata	Actinopterygii	Scorpaeniformes	Scorpaenidae	*Scorpaenopsis*	gibbosa	99.22
Chordata	Actinopterygii	Scorpaeniformes	Scorpaenidae	*Scorpaenopsis*	oxycephala	99.22
Chordata	Actinopterygii	Scorpaeniformes	Scorpaenidae	*Scorpaenopsis*	oxycephala	99.05
Chordata	Actinopterygii	Scorpaeniformes	Scorpaenidae	*Scorpaenopsis*	cirrosa	90.5
Chordata	Actinopterygii	Scorpaeniformes	Scorpaenidae	*Scorpaenopsis*	vittapinna	90.34
Chordata	Actinopterygii	Scorpaeniformes	Scorpaenidae	*Scorpaenopsis*	vittapinna	90.34
Chordata	Actinopterygii	Scorpaeniformes	Scorpaenidae	*Scorpaenopsis*	vittapinna	90.19
Chordata	Actinopterygii	Scorpaeniformes	Scorpaenidae	*Scorpaenopsis*	cirrosa	90.03
Chordata	Actinopterygii	Scorpaeniformes	Scorpaenidae	*Scorpaenopsis*	possi	89.1
Chordata	Actinopterygii	Scorpaeniformes	Scorpaenidae	*Scorpaenopsis*	s.p.	89.1
Chordata	Actinopterygii	Scorpaeniformes	Scorpaenidae	*Scorpaena*	scrofa	88.94
Chordata	Actinopterygii	Scorpaeniformes	Scorpaenidae	*Scorpaenopsis*	venosa	88.94
Chordata	Actinopterygii	Scorpaeniformes	Scorpaenidae	*Scorpaenopsis*	venosa	88.94
Chordata	Actinopterygii	Scorpaeniformes	Scorpaenidae	*Scorpaenopsis*	venosa	88.94
Chordata	Actinopterygii	Scorpaeniformes	Scorpaenidae	*Scorpaenopsis*	venosa	88.94
Chordata	Actinopterygii	Scorpaeniformes	Scorpaenidae	*Scorpaenopsis*	possi	88.94
Chordata	Actinopterygii	Scorpaeniformes	Scorpaenidae	*Scorpaenopsis*	possi	88.79
Chordata	Actinopterygii	Scorpaeniformes	Scorpaenidae	*Scorpaenopsis*	possi	88.79
Chordata	Actinopterygii	Scorpaeniformes	Scorpaenidae	*Scorpaenopsis*	cf.possi	88.79
Chordata	Actinopterygii	Scorpaeniformes	Scorpaenidae	*Scorpaenopsis*	possi	88.75
Chordata	Actinopterygii	Scorpaeniformes	Scorpaenidae	*Scorpaenopsis*	possi	88.71
Chordata	Actinopterygii	Scorpaeniformes	Scorpaenidae	*Scorpaenopsis*	venosa	88.63
Chordata	Actinopterygii	Scorpaeniformes	Scorpaenidae	*Scorpaenopsis*	venosa	88.63
Chordata	Actinopterygii	Scorpaeniformes	Scorpaenidae	*Scorpaenopsis*	venosa	88.63
Chordata	Actinopterygii	Scorpaeniformes	Scorpaenidae	*Scorpaenopsis*	venosa	88.63
Chordata	Actinopterygii	Scorpaeniformes	Scorpaenidae	*Scorpaenopsis*	venosa	88.63
Chordata	Actinopterygii	Scorpaeniformes	Scorpaenidae	*Scorpaenopsis*	venosa	88.63
Chordata	Actinopterygii	Scorpaeniformes	Scorpaenidae	*Scorpaenopsis*	possi	88.63

FIGURE 20.3 **BLAST nucleotide analysis results for *Scorpaendoes gibbosa* species from Andaman waters.**

(Figure 20.3) with reference sequences of the same species from other regions. Further, an NJ tree was constructed with the studied species' COI gene sequences and 42 intra- and interspecific reference COI gene sequences retrieved from the NCBI and BOLD databases. Subsequently, pairwise genetic analysis was carried out using MEGA software (Tamura *et al.*, 2007). The pairwise genetic distances (K2P) of the Andaman studied species of *Scorpaendoes* with intra- and interspecific reference sequences are shown in Table 20.2. From this analysis it was found that the two species *S. gibbosa* and *S. guamesis* had 0.141 percent genetic distance. This suggests that intrageneric genetic distances are very clearly differentiated through COI gene sequence analysis. Genetic

TABLE 20.2 Pairwise Genetic Distances of *Scorpaendoes Gibbosa* and *Scorpaendoes Guamesis* with Intra- and Interspecific Reference Sequences

Sl. No.	GenBank No. and species	K2P (%)	Country
1	– *Scorpaendoes gibbosa* DOSMB	0.000	Andaman Islands
2	JQ350359 *Scorpaenopsis gibbosa*	0.020	Madagascar: Antananarivo
3	JQ350357 *Scorpaenopsis gibbosa*	0.021	Madagascar: Antananarivo
4	JQ350358 *Scorpaenopsis gibbosa*	0.024	Madagascar: Antananarivo
5	JQ350356 *Scorpaenopsis gibbosa*	0.021	Madagascar: Antananarivo
6	HM424135 *Scorpaena isthmensis*	0.096	Brazil
7	JQ841963 *Scorpaena grandicornis*	0.106	USA: Florida
8	JQ841790 *Scorpaena albifimbria*	0.095	Belize: Stann Creek District
9	GU805078 *Pterois miles*	0.102	South Africa "29.867 S 31.048 E"
10	FJ584042 *Pterois volitans*	0.101	Viet Nam: Thanh Pho-Ho Chi Minh
11	– *Scorpaendoes guamesis* DOSMB	0.000	Andaman Islands
12	HM376359 *Scorpaeniformes* sp.	0.048	Australia 21.738 S 149.603 E
13	HQ956462 *Scorpaeniformes* sp.	0.060	Australia 9.08567 S 143.94 E
14	– *Scorpaendoes gibbosa* DOSMB[1]	0.067	Andaman Islands
15	HQ956458 *Scorpaeniformes* sp.	0.050	Australia 10.6023 S 141.602 E
16	GU673139 *Scorpaeniformes* sp.	0.066	Australia 24.356 S 152.671 E
17	GU673141 *Scorpaeniformes* sp.	0.070	Australia 24.356 S 152.671 E

DOSMB, Department of Ocean Studies and Marine Biology sample.
[1]*Morphologically identified.*

distances have been similarly differentiated in other fish (Ward *et al.*, 2005; Dooh *et al.*, 2006; Clare *et al.*, 2007; Ward *et al.*, 2008; Hubert *et al.*, 2008; Persis *et al.*, 2009; Steinke *et al.*, 2009a,b; Oliveira *et al.*, 2009; Rasmussen *et al.*, 2009; Lakra *et al.*, 2010; Zhang and Hanner, 2011; Hubert *et al.*, 2012; Lorz *et al.*, 2012) and other animals (Hebert *et al.*, 2004; Hajibabaei *et al.*, 2006a,b; Elias-Gutierrez *et al.*, 2008; Aliabadian *et al.*, 2009; Radulovici *et al.*, 2010).

The *Pterois miles* COI gene BLAST nucleotide analysis results (Figure 20.4) revealed that the Andaman species had 99 percent genetic similarity with reference sequences of the same species from other parts of the world—India, Sri Lanka, Indonesia, Madagascar and South Africa—and 92 percent genetic identity with the species *P. lunulata* and *P. russelii* from the South China Sea.

Further, the NJ tree (Figure 20.5) and pairwise genetic distances were calculated using MEGA software (Tamura *et al.*, 2007). Table 20.3 shows the pairwise genetic distances of the studied species (3 specimens) with reference sequences for the same species from other parts of the world (mean K2P = 0.004 percent) and for sister species of the same genus (mean K2P = 0.05 percent). The Andaman species of *P. miles* exhibited a close genetic relationship with

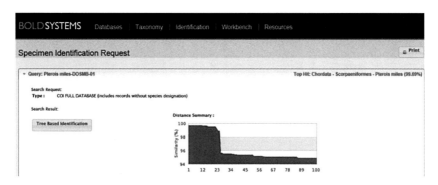

Phylum	Class	Order	Family	Genus	Species	Specimen Similarity (%)
Chordata	Actinopterygii	Scorpaeniformes	Scorpaenidae	*Pterois*	*miles*	99.69
Chordata	Actinopterygii	Scorpaeniformes	Scorpaenidae	*Pterois*	*miles*	99.69
Chordata	Actinopterygii	Scorpaeniformes	Scorpaenidae	*Pterois*	*miles*	99.69
Chordata	Actinopterygii	Scorpaeniformes	Scorpaenidae	*Pterois*	*miles*	99.69
Chordata	Actinopterygii	Scorpaeniformes	Scorpaenidae	*Pterois*	*miles*	99.69
Chordata	Actinopterygii	Scorpaeniformes	Scorpaenidae	*Pterois*	*miles*	99.69
Chordata	Actinopterygii	Scorpaeniformes	Scorpaenidae	*Pterois*	*miles*	99.69
Chordata	Actinopterygii	Scorpaeniformes	Scorpaenidae	*Pterois*	*miles*	99.69
Chordata	Actinopterygii	Scorpaeniformes	Scorpaenidae	*Pterois*	*miles*	99.69
Chordata	Actinopterygii	Scorpaeniformes	Scorpaenidae	*Pterois*	*miles*	99.69
Chordata	Actinopterygii	Scorpaeniformes	Scorpaenidae	*Pterois*	*miles*	99.69
Chordata	Actinopterygii	Scorpaeniformes	Scorpaenidae	*Pterois*	*miles*	99.68
Chordata	Actinopterygii	Scorpaeniformes	Scorpaenidae	*Pterois*	*miles*	99.68
Chordata	Actinopterygii	Scorpaeniformes	Scorpaenidae	*Pterois*	*miles*	99.67
Chordata	Actinopterygii	Scorpaeniformes	Scorpaenidae	*Pterois*	*miles*	99.67
Chordata	Actinopterygii	Scorpaeniformes	Scorpaenidae	*Pterois*	*miles*	99.54
Chordata	Actinopterygii	Scorpaeniformes	Scorpaenidae	*Pterois*	*miles*	99.54
Chordata	Actinopterygii	Scorpaeniformes	Scorpaenidae	*Pterois*	*miles*	99.54
Chordata	Actinopterygii	Scorpaeniformes	Scorpaenidae	*Pterois*	*miles*	99.54
Chordata	Actinopterygii	Scorpaeniformes	Scorpaenidae	*Pterois*	*miles*	99.54
Chordata	Actinopterygii	Scorpaeniformes	Scorpaenidae	*Pterois*	*miles*	99.54
Chordata	Actinopterygii	Scorpaeniformes	Scorpaenidae	*Pterois*	*miles*	99.53
Chordata	Actinopterygii	Scorpaeniformes	Scorpaenidae	*Pterois*	*miles*	99.18
Chordata	Actinopterygii	Scorpaeniformes	Scorpaenidae	*Pterois*	*miles*	98.92
Chordata	Actinopterygii	Scorpaeniformes	Scorpaenidae	*Pterois*	*miles*	98.92
Chordata	Actinopterygii	Scorpaeniformes	Scorpaenidae	*Pterois*	*lunulata*	95.68
Chordata	Actinopterygii	Scorpaeniformes	Scorpaenidae	*Pterois*	*russelii*	95.68
Chordata	Actinopterygii	Scorpaeniformes	Scorpaenidae	*Pterois*	*lunulata*	95.52
Chordata	Actinopterygii	Scorpaeniformes	Scorpaenidae	*Pterois*	*lunulata*	95.52
Chordata	Actinopterygii	Scorpaeniformes	Scorpaenidae	*Pterois*	*russelii*	95.52
Chordata	Actinopterygii	Scorpaeniformes	Scorpaenidae	*Pterois*	*russelii*	95.52
Chordata	Actinopterygii	Scorpaeniformes	Scorpaenidae	*Pterois*	*russelii*	95.52
Chordata	Actinopterygii	Scorpaeniformes	Scorpaenidae	*Pterois*	*russelii*	95.52

FIGURE 20.4 BLAST nucleotide analysis results for *Pterois miles* species from Andaman waters.

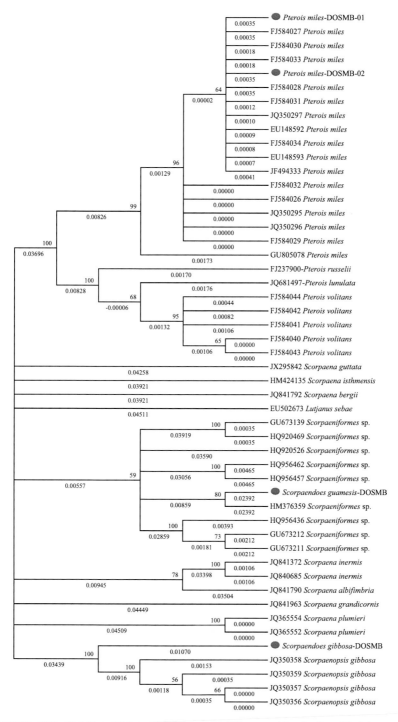

FIGURE 20.5 **Neighbor-joining phylogram tree of Scorpaenidae family with intra- and interspecific and intergeneric reference sequences.** Shows 50 percent collapsed bootstrap values and branch length values.

TABLE 20.3 Pairwise Genetic Distances of *Pterios miles* with Intra- and Interspecific Reference Sequences

Sl. No.	GenBank No. and species	K2P (%)	Country
1	– *Pterois miles* DOSMB01	0.000	Andaman Islands
2	EU148592 *Pterois miles*	0.002	India
3	EU148593 *Pterois miles*	0.002	India
4	JQ350296 *Pterois miles*	0.002	Madagascar; 13.48 S 48.23 E
5	FJ584034 *Pterois miles*	0.002	Indonesia: Jakarta Raya
6	FJ584032 *Pterois miles*	0.002	Indonesia: Jakarta Raya
7	FJ584030 *Pterois miles*	0.002	Sri Lanka: Western
8	FJ584028 *Pterois miles*	0.002	Indonesia: Jakarta Raya
9	FJ584026 *Pterois miles*	0.002	Canada
10	FJ584033 *Pterois miles*	0.002	Indonesia: Jakarta Raya
11	FJ584031 *Pterois miles*	0.002	Sri Lanka: Western
12	FJ584029 *Pterois miles*	0.002	Sri Lanka: Western
13	FJ584027 *Pterois miles*	0.002	Sri Lanka: Western
14	JF494333 *Pterois miles*	0.002	South Africa: Vetch's Pier
15	JF494332 *Pterois miles*	0.002	South Africa: KwaZulu-Natal
16	GU805078 *Pterois miles*	0.007	South Africa: Vetch's Pier
17	JQ350295 *Pterois miles*	0.007	Madagascar: Antananarivo
18	JQ681497 *Pterois lunulata*	0.046	South China Sea
19	FJ237900 *Pterois russelii*	0.044	China: South China Sea
20	FJ237899 *Pterois russelii*	0.044	China: South China Sea
21	FJ584042 *Pterois volitans*	0.051	Vietnam: Thanh Pho – Ho Chi Minh
22	FJ584040 *Pterois volitans*	0.051	Vietnam: Thanh Pho – Ho Chi Minh
23	FJ584043 *Pterois volitans*	0.051	Vietnam: Thanh Pho – Ho Chi Minh
24	FJ584041 *Pterois volitans*	0.053	Vietnam: Thanh Pho – Ho Chi Minh

DOSMB, Department of Ocean Studies and Marine Biology sample.

reference sequences of the same species from India, Madagascar, Indonesia, South Africa and Canada. The genetic relationships with sister *Pterois* species were marked by K2P values in the range 0.044 percent (*P. russeli*) to 0.053 percent (*P. volitans*). The genetic distances between the studied *Scorpaendoes* species from Andaman waters are shown in Table 20.4.

TABLE 20.4 Pairwise Genetic Distances of Studied *Scorpaenidae* Species

Sl. No.	Species relation	K2P (%)
1	*Scorpaendoes guamesis* DOSMB vs. *Pterois miles* DOSMB-01	0.232
2	*Scorpaendoes guamesis* DOSMB vs. *Scorpaendoes gibbosa* DOSMB	0.067
3	*Scorpaendoes gibbosa* DOSMB vs. *Pterois miles* DOSMB-01	0.232

Regarding the NJ tree for the Scorpaenidae family, the COI gene sequences for the studied species and the reference sequences for the same species from different regions form an unambiguous cluster (Figure 20.5). However, the latter group (from South African, Canadian, Madagascar and Sri Lankan waters) formed a subcluster in the tree. These results suggest the detection of phylogeographic signals through this analysis. Similar results are mentioned in earlier work in fish (Ward *et al.*, 2009) and are in line with the proposals of Hajibabaei *et al.* (2007). Further, the morphologically similar species *Pterois miles* and *Pterois volitans* (Randall and Eschymers, 2001) are clearly distinguished through this analysis at interspecies genetic distances of 0.051 percent. The same level of K2P values has been observed in other marine fish (Ward *et al.*, 2005; Hubert *et al.*, 2008; Ward *et al.*, 2008; Persis *et al.*, 2009; Kartavtsev *et al.*, 2009; Steinke *et al.*, 2009a; Rasmussen *et al.*, 2009; Zemlak *et al.*, 2009; Turanov *et al.*, 2012) as well as other animals such as bumblebee species (0.033–0.044%) (Carolan *et al.*, 2012).

Family Tetraodontidae (Puffer) COI Gene Sequence Analysis

The genus *Arothron* consists of three species—*A. hispidus, A. immaculatus* and *A. reticularis*—individuals of which were morphologically identified and barcoded. The genetic distances were calculated through MEGA software with 29 reference sequences at intra- and interspecific level. To further strengthen our results, standard integrated tools of species identification analysis were used in the BOLD system. Species genetic identity or similarity values were recorded (Figures 20.6 to 20.8).

BLAST results suggested that the studied species *A. hispidus* had 99 percent genetic similarly with a haplogroup of same-species COI gene sequences from India, Sri Lanka, Japan, Mozambique and the Indian Ocean to the west of India.

The NJ phylogram and K2P analysis for *A. hispidus* were carried out in MEGA4.1 software (Tamura *et al.*, 2007). The pairwise genetic distances analysis (Table 20.5) suggested that the Andaman species has a close genetic relationship with reference sequences of the same species from India, Sri Lanka,

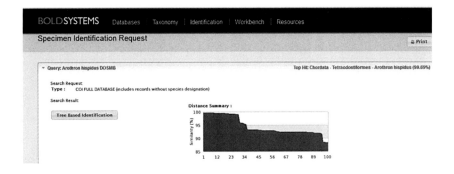

<table>
<tr><td>BOLD SYSTEMS</td><td>Databases</td><td>Taxonomy</td><td>Identification</td><td>Workbench</td><td>Resources</td></tr>
</table>

Specimen Identification Request

▾ Query: Arothron hispidus DOSMB

Top Hit: Chordata - Tetraodontiformes - Arothron hispidus (99.69%)

Search Request:
Type :   COI FULL DATABASE (includes records without species designation)

Search Result:

Tree Based Identification

Distance Summary :

Phylum	Class	Order	Family	Genus	Species	Specimen Similarity (%)
Chordata	Actinopterygii	Tetraodontiformes	Tetraodontidae	Arothron	hispidus	99.69
Chordata	Actinopterygii	Tetraodontiformes	Tetraodontidae	Arothron	hispidus	99.69
Chordata	Actinopterygii	Tetraodontiformes	Tetraodontidae	Arothron	hispidus	99.69
Chordata	Actinopterygii	Tetraodontiformes	Tetraodontidae	Arothron		99.69
Chordata	Actinopterygii	Tetraodontiformes	Tetraodontidae	Arothron	hispidus	99.69
Chordata	Actinopterygii	Tetraodontiformes	Tetraodontidae	Arothron	hispidus	99.69
Chordata	Actinopterygii	Tetraodontiformes	Tetraodontidae	Arothron	hispidus	99.69
Chordata	Actinopterygii	Tetraodontiformes	Tetraodontidae	Arothron	hispidus	99.69
Chordata	Actinopterygii	Tetraodontiformes	Tetraodontidae	Arothron	hispidus	99.68
Chordata	Actinopterygii	Tetraodontiformes	Tetraodontidae	Arothron	hispidus	99.53
Chordata	Actinopterygii	Tetraodontiformes	Tetraodontidae	Arothron	hispidus	99.53
Chordata	Actinopterygii	Tetraodontiformes	Tetraodontidae	Arothron	hispidus	99.53
Chordata	Actinopterygii	Tetraodontiformes	Tetraodontidae	Arothron	hispidus	99.53
Chordata	Actinopterygii	Tetraodontiformes	Tetraodontidae	Arothron	hispidus	99.53
Cnidaria	Anthozoa	Alcyonacea	Alcyoniidae	Sinularia		99.53
Chordata	Actinopterygii	Tetraodontiformes	Tetraodontidae	Arothron	hispidus	99.53
Chordata	Actinopterygii	Tetraodontiformes	Tetraodontidae	Arothron	hispidus	99.53
Chordata	Actinopterygii	Tetraodontiformes	Tetraodontidae	Arothron	hispidus	99.51
Chordata	Actinopterygii	Tetraodontiformes	Tetraodontidae	Arothron	hispidus	99.44
Chordata	Actinopterygii	Tetraodontiformes	Tetraodontidae	Arothron	hispidus	99.38
Chordata	Actinopterygii	Tetraodontiformes	Tetraodontidae	Arothron	hispidus	99.38
Chordata	Actinopterygii	Tetraodontiformes	Tetraodontidae	Arothron	hispidus	99.22
Chordata	Actinopterygii	Tetraodontiformes	Tetraodontidae	Arothron	hispidus	99.22
Chordata	Actinopterygii	Tetraodontiformes	Tetraodontidae	Arothron		99.22
Chordata	Actinopterygii	Tetraodontiformes	Tetraodontidae	Arothron	hispidus	99.22
Chordata	Actinopterygii	Tetraodontiformes	Tetraodontidae	Arothron	hispidus	99.22
Chordata	Actinopterygii	Tetraodontiformes	Tetraodontidae	Arothron	hispidus	99.19
Chordata	Actinopterygii	Tetraodontiformes	Tetraodontidae	Arothron	hispidus	99.19
Chordata	Actinopterygii	Tetraodontiformes	Tetraodontidae	Arothron	hispidus	99.14
Chordata	Actinopterygii	Tetraodontiformes	Tetraodontidae	Arothron	meleagris	95.97

FIGURE 20.6 BLAST nucleotide analysis results for *Arothron hispidus* species from Andaman waters.

Japan and Mozambique. Another reference sequence from Mozambique, designated as Tetraodontiformes sp., showed a close relationship. This sequence may be from *A. hispidus*, given the close genetic distance with the studied species as well as reference sequences of *A. hispidus* from other regions. The NJ tree showed that the studied species *A. hispidus* formed a cluster with the same-species haplogroup (Figure 20.9).

The COI gene sequences obtained for *A. immaculatus* were compared with data in the NCBI BLAST and BOLD databases. The BOLD results revealed that sequences for the same species from different regions had 97 percent genetic similarity with the Andaman studied species. However, *A. manilesis* had 97 percent genetic identity with the studied species COI gene sequences. This species might be misidentified, or same genetic thresholds need detailed study in future.

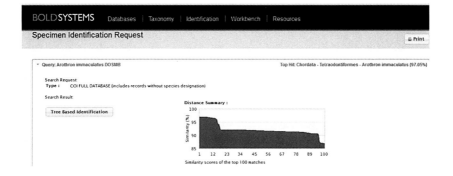

FIGURE 20.7   BLAST nucleotide analysis results for *Arothron immaculatus* species from Andaman waters.

The pairwise genetic distances analysis (Kimura, 1980) was carried out for *A. immaculatus* in MEGA4 software. The K2P results for intraspecific genetic distances (Table 20.6) revealed close genetic distances with reference sequences of the same species from India and South Africa. In addition, three reference sequences designated as Tetraodontiformes sp. had similarly close genetic distance, and these may be considered to be from *A. immaculatus*. Studied specimen interspecific genetic distances are shown in Table 20.7.

For the species *Arothron reticularis* from Andaman waters, analyses were carried out as described above for other species. The BLAST analysis results (Figure 20.8) showed 94 percent genetic identity with sequences of *A. hispidus*. The BOLD species identification results indicate that no *A. reticularis* COI gene sequence is available in either the BOLD system or the NCBI database. Becker

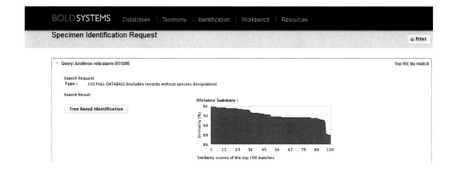

FIGURE 20.8 BLAST nucleotide analysis results for *Arothron reticularis* species from Andaman waters.

TABLE 20.5 Pairwise Genetic Distances of *Arothron hispidus* with Intraspecific Reference Sequences

Sl. No.	GenBank No. and species	K2P (%)	Country
1	– *Arothron hispidus* DOSMB	0.000	Andaman Islands
2	EU148579 *Arothron hispidus*	0.003	India
3	FJ582875 *Arothron hispidus*	0.003	Sri Lanka: Western
4	JF952687 *Arothron hispidus*	0.003	Japan
5	HQ972642 Tetraodontiformes sp.	0.005	Mozambique
6	HQ561518 *Arothron hispidus*	0.005	Mozambique: Pomene
7	EU148578 *Arothron hispidus*	0.008	India

DOSMB, Department of Ocean Studies and Marine Biology sample.

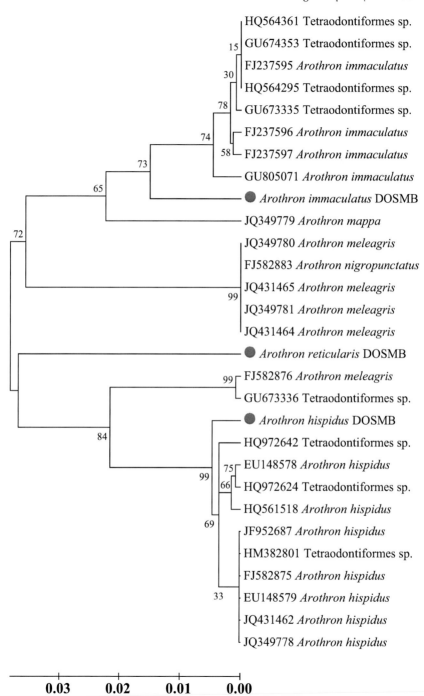

FIGURE 20.9   Neighbor-joining phylogram tree of Tetraodontidae family with intra- and interspecific and intergeneric reference sequences. Shows 50 percent collapsed bootstrap values and branch length values.

TABLE 20.6 Pairwise Genetic Distances of *Arothron immaculatus* with Intraspecific Reference Sequences

Sl. No.	GenBank No. and species	K2P (%)	Country
1	– *Arothron immaculatus* DOSMB	0.000	Andaman Islands
2	FJ237595 *Arothron immaculatus*	0.028	India 9.16 N 79.08 E
3	FJ237596 *Arothron immaculatus*	0.031	India 9.16 N 79.08 E
4	HQ564361 Tetraodontiformes sp.	0.028	Indonesia
5	GU673335 Tetraodontiforme sp.	0.029	BOLD
6	GU674353 Tetraodontiformes sp.	0.028	BOLD
7	FJ237597 *Arothron immaculatus*	0.031	India 9.16 N 79.08 E
8	GU805071 *Arothron immaculatus*	0.036	South Africa

DOSMB, Department of Ocean Studies and Marine Biology sample.

et al. (2011) summarized fish species' identification of genetic similarity as being in the range 99–95 percent identity. A similar value of 98 percent for marine fishes' genetic identity is described by Ward *et al.* (2008). The above-mentioned finding clearly supports the first documentation of *A. reticularis* COI gene sequence from this study.

Calculation of interspecific genetic distances with sister species of *A. reticularis* produced the following K2P values (%): *A. immaculatus* 0.063; *A. hispidus* 0.088; reference sequences of *A. mappa* 0.060 and *A. meleagris* 0.088 from Andaman, Madagascar and Reunion waters. Similar results have been observed in other marine fish (Clare *et al.*, 2007; Persis *et al.*, 2009; Pramual *et al.*, 2011; Sachithanandam *et al.*, 2011a, 2012; Hubert *et al.*, 2012; Lorz *et al.*, 2012).

TABLE 20.7 Pairwise Genetic Distances of *Arothron immaculatus* with Interspecific Reference Sequences

Sl. No.	GenBank No. and species	K2P (%)	Country
1	– *Arothron immaculatus* DOSMB	0.000	Andaman Islands
2	– *Arothron reticularis* DOSMB	0.063	Andaman Islands
3	– *Arothron hispidus* DOSMB	0.088	Andaman Islands
4	JQ349779 *Arothron mappa*	0.060	Madagascar
5	JQ349781 *Arothron meleagris*	0.088	Reunion
6	HQ972624 Tetraodontiformes sp.	0.097	Mozambique
7	HM382801 Tetraodontiformes sp.	0.091	Tanzania

DOSMB, Department of Ocean Studies and Marine Biology sample.

CONCLUSION

The present work once again confirms the utility and effectiveness of COI gene sequences for species identification in marine fish. The study showed unambiguous identification of species from the families Scorpaenidae and Tetraodontidae through COI gene sequencing. From the Scorpaenidae family: for the studied species *Pterois miles* we found intraspecific mean genetic distances of 0.004 percent and interspecific K2P values of 0.044 to 0.051 percent with the sister species *P. russelii, P. lunulata* and *P. volitans*; in the genus *Scorpaenopsis*, for the studied species of *S. gibbosa* and *S. guamesis* we found mean intraspecific K2P values of 0.004 percent and interspecific genetic distances of 0.096 to 0.106 percent. From the family Tetraodontidae, genus *Arothron*, we found for the studied species: *A. immaculatus* had mean K2P of 0.027 percent with intraspecific reference sequences; *A. hispidus* had mean intraspecific genetic distances of 0.003 percent; and *A. reticularis* had interspecific genetic distances between 0.063 and 0.088 percent. Similar results have been found through COI gene sequencing for Australian marine fishes (Ward *et al.*, 2005), Amphipoda (Lorz *et al.*, 2012), bumblebee species (Carolan *et al.*, 2012), Indo-Pacific coral reefs fishes (Hubert *et al.*, 2012) and Andaman groupers (Sachithanandam *et al.*, 2012). Further, the study showed that COI gene sequencing and analysis revealed phylogeographical information and regional closeness (Hajibabaei *et al.*, 2007). Moreover, COI gene sequencing enabled species differentiation beyond doubt. Detailed future study with increased sample numbers will improve our ability to use the technique as the standard for region-wide unambiguous identification of species.

ACKNOWLEDGEMENTS

The authors thank the Centre for Marine Living Resources and Ecology (CMLRE), Ministry of Earth Sciences (MoES), Kochi for financial support of this study. The authors are also obliged to Dr P. Vijayachari, Director, Regional Medical Research Centre (RMRC), Indian Council of Medical Research, Port Blair for providing laboratory facilities. Dr C. Raghunathan and Shri. P.T. Rajan, Zoological Survey of India, Port Blair are also acknowledged for their timely help. One of the author (VS) also thanks the University Grand Commission (UGC) for providing a Research Fellowship in Science to Meritorious Student (RFSMS).

REFERENCES

Aliabadian, M., Kaboli, M., Nijman, V., Vences, M., 2009. Molecular identification of birds: performance of distance-based DNA barcoding in three genes to delimit parapatric species. PLoS One. 4, e4119.

Becker, S., Hanner, R., Steinke, D., 2011. Five years of FISH-BOL: brief status report. Mitochondrial DNA 22 (S.1), 3–9.

Bouchet, P., 2006. The magnitude of marine biodiversity. In: Duarte, C.M. (Ed.), The Exploration of Marine Biodiversity: Scientific and Technological Challenges Fundacion BBVA, Bilbao, Spain, pp. 31–64.

Carolan, J.C., Murray, T.E., Fitzpatrick, U., Crossley, J., Schmidt, H., 2012. Colour Patterns Do Not Diagnose Species: Quantitative Evaluation of a DNA Barcoded Cryptic Bumblebee Complex. PLoS One 7 (1), e29251.

Clare, E.L., Kerr, K.C., Von- Konigslow, T.E., Wilson, J.J., Hebert, P.D.N., 2007. Diagnosing Mitochondrial DNA Diversity: Applications of a Sentinel Gene Approach. J. Molecular Evol. 66, 362–367.

Coyne, J.A., Orr, H.A., 2004. Speciation. Sinauer Associates, Sunderland, MA, USA, p. 545.

Day, F., 1878. The Fishes of India; Being a Natural History of the Fishes known to inhabit the Seas and Fresh Waters of India, Burma, and Ceylon. Parts 1-4. William Dawson and Sons Ltd., London. p. 778.

Dooh, R.T., Adamowicz, S.J., Hebert, P.D.N., 2006. Comparative phylogeography of two North American 'glacial relict' crustaceans. Molecular Ecol. 15 (14), 4459–4475.

Elias-Gutierrez, M., Jeronimo, F.M., Iavanova, N.V., Valdes-Moreno, M., Hebert, P.D.N., 2008. DNA barcodes for Cladocera and Copepoda from Mexico and Guatemala, highlights and new discoveries. Zootaxa 1839, 1–42.

Eschmeyer, W.N., 2010. Catalog of fishes. Available at: http://research.calacademy.org/research/ichthyology/catalog/speciesByFamily.asp.

Grassle, J.F., Maciolek, N., 1992. Deep Sea species richness: Regional and local diversity estimates from quantitative bottom samples. American Nat. Soc. 39, 313–334.

Hajibabaei, M., Janzen, D.H., Burns, J.M. Hallwachs, W., Hebert, P.D.N., 2006a. DNA barcodes distinguish species of tropical Lepidoptera. Proceedings of the National Academy of Sciences, USA, V.103, pp. 968–971.

Hajibabaei, M., Smith, M.A., Janzen, D.H., Rodriguez, J.J., Whitefield, J.B., Hebert, P.D.N., 2006b. A minimalist barcode can identify a specimen whose DNA is degraded. Molecular Ecol. Notes 6, 959–964.

Hajibabaei, M., Singer, G.A.C., Hebert, P.D.N., Hickey, D.A., 2007. DNA Barcoding: how it complements taxonomy, molecular phylogenetics and population genetics. Trends in Genetics 23 (4), 167–172.

Hammond, P., 1992. Global biodiversity: Status of earth living resources. World Conservation Monitoring Centre, USA, pp.17–39.

Hebert, P.D.N., Cywinska, A., Ball, S.L., DeWaard, J.R., 2003. Biological identifications through DNA barcodes. Proceedings of the Royal Society of London, Series B 270, 313–321.

Hebert, P.D.N., Penton, E.H., Burns, J.D.H., Hallwachs, W., 2004. Ten species in one: DNA barcoding reveals cryptic species in the neotropical skipper butterfly *Astraptes fulgerator*. Proceeding of National Academic Science USA V101 (41), 14812–14817.

Holmes, B.H., Steinke, D., Ward, R.D., 2009. Identification of shark and ray fins using DNA barcoding. Fisheries Res. 95, 280–288.

Hubert, N., Hanner, R., Holm, E., Mandrak, N.E., Taylor, E., 2008. Identifying Canadian Freshwater Fishes through DNA Barcodes. PLoS One 3 (6), e2490.

Hubert, N., Meyer, C.P., Bruggemann, H.J., Guerin, F., Komeno, R.J.L., 2012. Cryptic Diversity in Indo-Pacific Coral-Reef Fishes Revealed by DNA-Barcoding Provides New Support to the Centre-of-Overlap Hypothesis. PLoS One 7 (3), e28987.

Kartavtsev, Y.P., Sharina, S.N., Goto, T., Rutenko, O.A., Zemnukhov, V.V., Semenchenko, A.A., Pitruk, D.L., Hanzawa, N., 2009. Molecular phylogenetics of pricklebacks and other percoid fishes from the Sea of Japan. Aquatic Biology 8, 95–103.

Kimura, A., 1980. A simple method for estimating evolutionary rate of base substitutions through comparative studies of nucleotide sequences. J. Molecular Evol. 15, 111–120.

Lakra, W.S., Verma, M.S., Goswami, M., Lal, K.K., Mohindra, V., Punia, P., Gopalakrishnan, A., Singh, K.V., Ward, R.D., Hebert, P.D.N., 2010. DNA barcoding Indian marine fishes. Molecular Ecol. Res. 11, 60–71.

Lorz, A.N., Linse, K., Smith, P.J., Steinke, D., 2012. First Molecular Evidence for Underestimated Biodiversity of Rhachotropis (Crustacea, Amphipoda), with Description of a New Species. PLoS One 7 (3), e32365.

Mecklenburg, C.W., Moller, P.R., Steinke, D., 2010. Biodiversity of arctic marine fishes: Taxonomy and zoogeography. Marine Biodiv. 41 (1), 109–140.

Nelson, J.S., 2006. Fishes of the world, 4th Edition John Wiley and Sons, Inc, New York, p. 601.

Oliveira, C., Pereira, L.H.G., Henriques, J.M., Foresti, F., 2009. DNA Barcode of freshwater fishes from upper Parana Basin, Brazil. 3rd International Barcode of Life Conference, Mexico.

Packer, L., Gibbs, J., Sheffield, C., Hanner, R., 2009. DNA barcoding and the mediocrity of morphology. Molecular Ecol. Res. 9, 42–50.

Persis, M., Reddy, A.C.S., Rao, L.M., Khedkar, G.D., Ravinder, K., Nasruddin, K., 2009. COI (Cytochrome oxidase – I) sequence based studies of Carangid fishes from Kakinada coast. India. Mol. Biol. Rep. 36, 1733–1740.

Pramual, P., Wongpakam, K., Adler, P.H., 2011. Cryptic biodiversity and phylogenetic relationships revealed by DNA barcoding of Oriental black flies in the subgenus Gomphostilbia (Diptera: Simuliidae). Genome 54, 1–9.

Radulovici, A.E., Archambault, P., Dufresne, F., 2010. Review DNA Barcodes for Marine Biodiversity: Moving Fast Forward. Diversity 2, 450–472.

Rajan, P.T., 2001. A Field Guide to Grouper and Snapper Fishes of Andaman and Nicobar Islands. Zoological Survey of India, Kolkata, p. 106.

Rajan, P.T., 2010. Diversity of butterfly fishes (Chaetodontidae) of Andaman and Nicobar Islands: indicators in coral reef habitat monitoring and managements. pp. 337–342. In: Recent Trends in Biodiversity of Andaman, Nicobar Islands, (Eds.) Ramakrishna, Ragunanathan, C. and Sivaperuman, C. Zoological Survey of India, Kolkata.

Rajan, P.T. and Sreeraj, C.R., 2012. Structure of Reef fish communities of Seven Islands of Andaman an Nicobar Islands, India. pp. 127–146. In: Ecology of Faunal communities on the Andaman, Nicobar Islands, (Eds.) K. Venkataraman, C. Raghunathan and C. Sivaperuman. Springer-Verlag, Berlin and Heidelberg.

Ramakrishna, Raghunathan, C. and Sivaperuman, C., 2010. Biodiversity of Andaman and Nicobar Islands - an overview. pp. 1–42. In: Recent Trends in Biodiversity of Andaman, Nicobar Islands, (Eds.) Ramakrishna, Ragunanathan, C. and Sivaperuman, C. Zoological Survey of India, Kolkata.

Randall, J.E., Eschmeyer, W.N., 2001. Revision of the Indo-Pacific scorpionfish genus Scopaenopsis, with descriptions of eight new species. Indo-Pacific Fishes (34), 79.

Rao, D.V., 2003. Guide to reef fishes of Andaman and Nicobar Islands. Zoological Survey of India, Kolkata, p. 555.

Rao, D.V., Devi, K., Rajan, P.T., 2000. An Account of Ichthyofauna of Andaman and Nicobar Islands, Bay of Bengal. Rec. Zool. Surv. India. Occ. Paper No. 178, 434.

Rasmussen, R.S., Morrissey, M.T., Hebert, P.D.N., 2009. DNA Barcoding of Commercially important Salmon and Trout Species in North America. 3rd International Barcode of Life Conference, Mexico City, p. 84.

Ratnasingham, S., and Hebert, P.D.N., 2007. BOLD: The Barcode of Life Data System (www.barcodinglife.org) Molecular Ecol. Notes, 7(3), 355–364.

Sachithanandam, V., Mohan, P.M., Dhivya, P., Muruganandam, N., Baskaran, R., Chaaithanya, I.K., Vijayachari, P., 2011a. DNA barcoding, phylogenetic relationships and speciation of Genus: Plectropomus in Andaman coast. J. Res. Biol. 3, 179–183.

Sachithanandam, V., Mohan, P.M., Dhivya, P., Muruganandam, N., Chaaithanya, I.K., Vijayachari, P., 2011b. DNA Barcoding of Spotted Coral Grouper *Plectrophomus maculatus* (Serranidae) from Andaman Islands. India. J. Res. Anal. Eval. 2, 1–4.

Sachithanandam, V., Mohan, P.M., Muruganandam, N., Chaaithanya, I.K., Dhivya, P., Baskaran, R., 2012. DNA barcoding, phylogenetic study of *Epinephelus* spp. from Andaman coastal region, India. Indian J. Geo-Marine Sci. 41, 203–211.

Saitou, N., Nei, M., 1987. The Neighbour-Joining method – a new method for reconstructing phylogenetic trees. Molecular Biology and Evolution 4, 406–425.

Sambrook, J., Fritsch, E.F., Maniatus, T., 1987. Molecular Cloning: A Laboratory Manual. Cold Spring Harbor Laboratory Press, New York, p. 1659.

Sanger, F., Nicklen, S., Coulson, A.R., 1977. DNA Sequencing with chain terminating inhibitors. Proceedings of the National Academy of Science, USA, 74(12): 5463–5467.

Steinke, D., Zemlak, T.S., Hebert, P.D.N., 2009a. Barcoding Nemo: DNA-Based identifications for the ornamental fish Trade. PLoS One 4, e6300.0.

Steinke, D., Zemlak, T.S., Boutillier, J.A., Hebert, P.D.N., 2009b. DNA barcoding of Pacific Canada's fishes. Marine Biology 156, 2641–2647.

Talwar, P.K., 1990. Fishes of the Andaman and Nicobar Islands: A Synoptic analysis. J. Andaman Sci. Assoc. 6 (2), 71–102.

Tamura, K., Dudley, J., Nei, M., Kumar, S., 2007. MEGA4: Molecular Evolutionary Genetics Analysis (MEGA) Software Version 4. 0. Mol. Biol. Evol. 24 (8), 1596–1599.

Tikader, B.K., Daniel, A., Subbarao, N.V., 1986. Sea Shore Animals of Andaman and Nicobar Islands. Zoological Survey of India, Calcutta, p. 239.

Turanov, S.V., Kartavtseva, I.F., Zemnukhov, V.V., 2012. Molecular phylogenetic study of several eelpout fishes (Perciformes, Zoarcoidei) from far eastern seas on the basis of the nucleotide sequence of the mitochondrial cytochrome oxidase 1 gene (Co-1). Genetika 48, 235–252.

Venkataraman, K., Wafar, M., 2005. Coastal and marine biodiversity of India. Ind. J. Mar. Sci. 34 (1), 57–75.

Ward, R.D., Zemlak, T.S., Innes, B.H., Last, P.R., Hebert, P.D.N., 2005. DNA barcoding Australia's fish species. Philosophical Trans. Royal Soc., Series Biol. 360, 847–1857.

Ward, R.D., Costa, F.O., Holmes, B.H., Steinke, D., 2008. DNA barcoding of shared fish species from the North Atlantic and Australasia: minimal divergence for most taxa, but Zeus faber and Lepidopus caudatus each probably constitute two species. Aquatic Biol. 3, 71–78.

Ward, R.D., Hanner, R., Hebert, P.D.N., 2009. The campaign to DNA barcode all fishes, FISH-BOL. J. Fish Biol. 74 (2), 329–356.

Whitehead, D. and Talwar, P.K., 1976. Clupeoid Fishes of the world (Suborder: clupeoidei): Engraulididae. American Museum of Natural History, New York, Part 4. Fishes, 121: 118–119.

Wong, E.H.K., Hanner, R., 2008. DNA barcoding detects market substitution in North American seafood. Food Research International 41, 828–837.

Zemlak, T.S., Ward, R.D., Connell, A.D., Holmes, B.H., Hebert, P.D.N., 2009. Barcoding Vertebrates. DNA barcoding reveals overlooked marine fishes. Molecular Ecol. Res. 9 (Suppl 1), 237–242.

Zhang, J.B., Hanner, R., 2011. DNA barcoding is a useful tool for the identification of marine fishes from Japan. Biochemical Systematics and Ecology 39, 31–42.

Zhang, Z., Schwartz, S., Wagner, L., Miller, W., 2000. A greedy algorithm for aligning DNA sequences. J. Computational Biol. 7 (1–2), 203–214.

Chapter 21

Molecular Taxonomy of Serranidae, Subfamily Epinephelinae, Genus *Plectropomus* (Oken, 1817) of Andaman Waters by DNA Barcoding Using COI Gene Sequence

V. Sachithanandam*, P.M. Mohan*, N. Muruganandam[†],
I.K. Chaaithanya[†] and R. Baskaran*
*Department of Ocean Studies and Marine Biology, Pondicherry University, Port Blair,
Andaman Islands, India; [†]Regional Medical Research Centre, Indian Council of Medical Research,
Port Blair, Andaman Islands, India

INTRODUCTION

The phylum Pisces is a highly diverse group among the vertebrates, exhibiting wide phylogenetic variation (Nelson, 2006). The subfamily Epinephelinae of family Serranidae contains genera commonly known as groupers, rock cods, hinds and sea basses. The subfamily comprises about 159 species of marine fish (Randall and Heemstra, 1991; Heemstra and Randall, 1993).

Groupers (Parrish, 1987) have high morphological diversity and spectacular color variation, and are abundant on reefs throughout shallow, tropical seas (Heemstra, 1991). These features have attracted the enthusiasm and inquiry of divers, scientists, and coral reef conservationists (Sadovy, 1999; Ottolenghi *et al.*, 2004). In the Indo-Pacific region, there are 110 species of grouper (Randall and Heemstra, 1991). India has a rich natural heritage and is one of the most biodiverse countries. Out of 31,100 fish species recorded worldwide, 2438 are known from the Indian subcontinent (Froese and Pauly, 2000).

The fish fauna of the Andaman and Nicobar Islands (ANI) comprises about twelve hundred diverse species (Rajan, 2001; Rao, 2003). The estimated potential

Marine Faunal Diversity in India. DOI: 10.1016/B978-0-12-801948-1.00021-5

marine fishery resource in the Andaman Exclusive Economic Zone (EEZ) region is 148,000 tonnes, some 10 percent of that in the entire Indian EEZ (James *et al.*, 1996; Rajan, 2001; Rao, 2003). About 43 species of 11 genera in the subfamily Epinephelinae have been recorded in ANI. The diversity of groupers of ANI is about 39 percent of that of the Indo-Pacific region (Rajan, 2001; Rao *et al.*, 2000). The economy of ANI depends on the exploitation of their marine resources (Sachithanandam *et al.*, 2012). About three-quarters of the species of grouper from these islands are commercially exploitable (Rajan, 2001; Rao, 2003; Mustafa, 2011). During 1999–2000, 24,876 tonnes of groupers from Indian seas were landed (CMFRI, 2001). A total of 577 tonnes of groupers were exported from the ANI in 2010 (Mustafa, 2011).

However, a difficulty faced by exporters of grouper catch to foreign countries is the problem of morphological identification of species. The morphological ambiguity of this family leads to misidentification of fishes by commercial vendors, resulting in the wrong species being dispatched to market (Govindaraju and Jayasankar, 2004). During quality check, if any piece of fish is identified as being of the wrong species, the whole consignment is considered to be wrongly supplied (MPEDA, 2000) and is rejected by the buyer. So, a reliable, fast tool for the identification of species will be a major asset.

DNA barcoding is one such efficient method for species-level identification. It uses an array of species-specific molecular tags derived from the 59 region of the mitochondrial DNA Cytochrome C Oxidase I (COI) gene sequence (Hebert *et al.*, 2003a; Ward *et al.*, 2005; Hebert *et al.*, 2005a,b). Phylogenetic systems, in combination with conservation genetics, provide a critical framework for understanding diversity (Feral, 2002) and predict vulnerability to exploitation of tropical reef fishes (Simon *et al.*, 1999). The amount of study of groupers of the Andaman coral reef environment has been very limited (Rao *et al.*, 2000; Rajan, 2001; Rao, 2003). A more precise taxonomy is required for the differentiation of species of this morphologically ambiguous group (Mustafa, 2011; Sachithanandam *et al.*, 2012).

Traditionally, *Plectropomus* spp. groupers are identified on the basis of visible morphometric and anatomical characters (Heemstra and Randall, 1993). Taxonomic ambiguities exist due to morphological similarities, leading to misidentification (Sachithanandam *et al.*, 2012). Fish Base (www.fishbase.org) citations contain many synonyms which indicate ambiguities in morphological identification of grouper species. The genus *Plectropomus* consists of seven species which are available in Indo-Pacific waters, as reported by Randall and Hoese (1986) and Randall and Heemstra (1991). However, only three species— *Plectropomus laevis, Plectropomus leopardus* and *Plectropomus maculatus*— have been found in Andaman waters (Rao *et al.*, 2000; Rajan, 2001; Rao, 2003; Mustafa, 2011).

The present study was carried out in the Andaman coastal region, which is representative of areas of commercial fishing of groupers. Morphological and molecular taxonomic (DNA barcoding) analysis was carried out to elucidate genetic relationships among species of the genus *Plectropomus*. DNA barcoding of *Plectropomus* species from Andaman waters had not previously been carried out. The COI gene sequences obtained were compared with equivalent sequences retrieved from the relevant database of the National Center for Biotechnology Information (NCBI), and genetic relationships among them were analyzed.

METHODS

Study Area

The Andaman and Nicobar Islands (92–94° East; 06–14° North) are an archipelago with 572 islands/islets, stretching over 700 km from north to south, in the Bay of Bengal. These islands are near to east Asian countries and more than 1400 km from mainland India. Andaman is a volcanic rock land mass surrounded by various endemic flora and fauna (Tikader *et al.*, 1986; Rao *et al.*, 2000; Rajan, 2001; Rao, 2003). The fish samples were collected from landing centers in and around Port Blair: Junglighat, Guptapara, Wandoor, Panighat and Dignabad (Figure 21.1).

Morphological Identification

Classical Taxonomy of Plectropomus *genus*

Collected specimens (Figure 21.2) were identified according to traditional morphology-based taxonomic keys (Randall and Heemstra, 1991; Heemstra and Randall, 1993; Rajan, 2001; Rao, 2003).

Diagnostic Features of Plectropomus *genus*

Body robust, elongate, the depth less than head length and contained 2.9 to 3.9 times in standard length; body width contained 1.6 to 2.1 times in its depth. Head length contained 2.8 to 3.2 times in standard length; snout distinctly longer than eye diameter, snout length 2.8 to 3.6 times in head length; pre-orbital depth contained 5.6 to 10 times in head length; inter-orbital area concave or flat, the dorsal head profile convex. Pre-opercle broadly rounded, with 3 large spines, ventrally directed spines (hidden by skin) along lower half and lower developed gill rakers 4 to 10.

Dorsal fin with VII or VIII slender spines and 10 to 12 rays, the fin membranes distinctly incised between the spines, the third or fourth spine usually longest, longest dorsal-fin ray 2.2 to 3.2 times in head length; pectoral-fin rays 16 to 18; pectoral-fin and pelvic-fin length 2.1 to 2.4 times in head length; caudal-fin length 1.5 to 1.8 times in head length.

Caudal fin emarginate, the caudal concavity 5 to 12 times in head length; no scales on inter-orbital area; blue spots round to oblong; head and body pale, with large saddle-like dark brown or black bars and a few small blue spots,

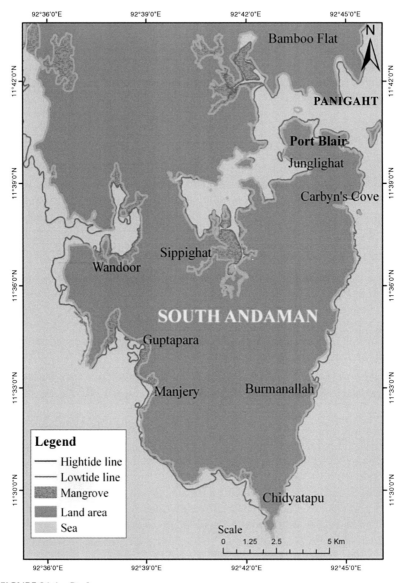

FIGURE 21.1   Study area.

the fins yellow; or head and body brownish with numerous small blue spots and with or without faint dark bars → *P. laevis*.

Color pattern of fish specimens are blue spots round to oblong; head and body covered (except ventrally) with minute round blue spots, which are about the size of the nostrils, the distance between the spots more than twice their diameter; median fins also covered with blue spots → *P. leopardus*.

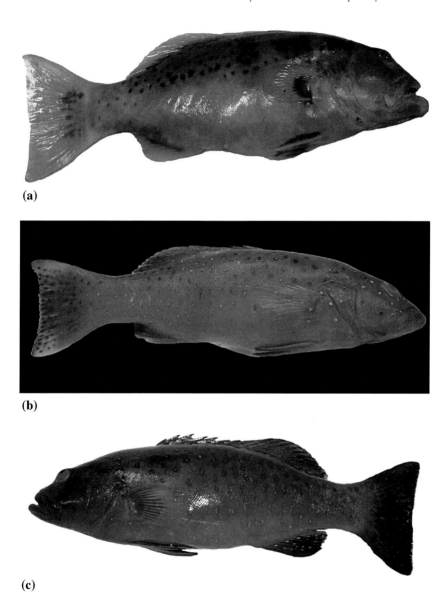

(a)

(b)

(c)

FIGURE 21.2 *Plectropomus* **species from South Andaman.** (a) *P. laevis* (total length 30 cm); (b) *P. leopardus* (total length 32 cm); *P. maculatus* (total length 30 cm). (Please see color plate at the back of the book.)

Most blue spots on head and body more than twice the size of nostrils; some spots on head and body elongate (except juveniles); pelvic fins without blue spots; some spots on body of adults horizontally elongate; gill raker at angle of first gill arch longer than longest gill filament; pelvic-fin length 1.7 to 2.1 times in head length; nostrils subequal → *P. maculatus*.

Molecular Taxonomy Analysis

Total DNA was extracted from 0.25 g of tissue by using lysis buffer and followed by standard proteinase-K / phenol-chloroform-isoamyl alcohol-ethanol precipitation method (Sambrook et al., 1989) and fish DNA modified isolation protocol followed as described by Sachithanandam et al., (2012). The concentration of DNA was estimated using UV spectrophotometer method at 260/280 nm.

Subsequently the DNA was diluted to final concentration of 100 ng/μL for further use. The 650–655 bp section of the mitochondrial (mt) DNA genome from the COI gene was amplified using a published universal degenerated primer set (Ward et al., 2005) synthesized by Sigma Aldrich Chemicals India Pvt. Ltd.

The polymerase chain reaction (PCR) was carried out in 25 μL consisting of 100 ng/μL of DNA and PCR master mix described in Table 21.1.

Thermal cycle condition

PCR was carried out in Applied Bio systems AB-2720. The initial denaturation was performed at 95°C for 5 min, followed by denaturation at 94°C for 30 s, annealing at 56°C for 30 s and extension at 72°C for 60 s for 40 cycles, followed by final extension at 72°C for 10 min.

Agarose Gel Electrophoresis

The PCR products obtained after amplification of COI genes were resolved and analyzed by electrophoresis using 2 percent agarose gel (Sigma) prepared in 1X TAE buffer (Tris Acetate EDTA) containing EtBr. DNA marker (100 bp DNA ladder – 1 μg/L, Sigma) was used for estimation of size of the PCR products,

TABLE 21.1 Master Mix Preparation

Reagent	Volume (μL)
10X buffer	5.0
dNTP (5 mM)	1.50
COI gene forward Primer F1 (0.5 μM) 5'TCAACCAACCACAAAGACATTGGCAC3'	1.0
COI gene reverse primer R1 (0.5 μM) 5'TAGACTTCTGGGTGGCCAAAGAATCA3'	1.0
Taq polymerase (3 U)	1.0
MgCl$_2$	0.25
RNase free water	10.25
Total volume	20
5 μL DNA was mixed with 20 μL reaction mix.	

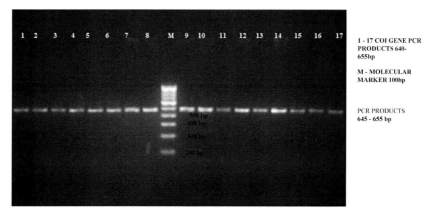

FIGURE 21.3   Gel electrophoresis.

which were visualized under UV-Transilluminator (548 nm), and photographs of the gels (Figure 21.3) were recorded using Gel documentation system (Bio-Rad).

Gel Extraction and Purification of PCR Products

All PCR products were excised from the gel and purified using QIAquick gel extraction kit (Qiagen, Germany). Three volumes (300 μL) of QG buffer were added to 1 volume (100 mg gel slice) of gel containing excised DNA fragment. The incubation was carried out at 56°C for 15–20 min or until the gel was completely dissolved in QG buffer. Isopropanol (100 μL) was added in a proportion equal to gel volume (100 mg) and mixed properly. The mixture was applied to the QIAquick column placed on a collection tube (provided in the kit) and centrifuged at 10,000 rpm for 1 min at room temperature. After removal of impurities, the pure DNA was eluted in the elution buffer or water. The purified DNA was stored at −20°C until utilized.

Nucleotide Sequencing

Nucleotide sequencing was performed using the Sanger *et al.* (1977) method. Sequencing was performed using a BigDye Terminator Cycle Sequencing kit, following manufacturer's instructions (Applied Biosystems, Foster City, CA, USA). The sequencing was done both in the forward and reverse directions.

Sequence Analysis

The DNA sequences were analyzed both in forward and reverse directions for each individual fish and assembled using the SeqMan II version 5.03 (DNA-STAR) and ChromaxSeq Version 3.1. The sequence analysis was done along with reference sequences of various species belonging to the family Serranidae, retrieved from the National Center for Biotechnology Information (NCBI) GenBank and Barcode of Life Data System (BOLD). Nucleic acid sequence multiple and pairwise alignment was done using the CLUSTALW tool. Phylogenetic molecular evolutionary analyses were conducted by using MEGA (Molecular Evolutionary Genetics Analysis) version 4.1 software. Neighbor-joining (NJ) trees were constructed using this software. Bootstrap values for the NJ trees were estimated using searches with 1000 pseudo replicates (Tamura et al., 2007; Saitou and Nei, 1987; Kimura, 1980). The aligned sequences were also included for nucleotide search carried out in the NCBI Basic Local Alignment Search Tool (BLAST) 2.2.26+ (Zhang et al., 2000) and in the BOLD system (Ratnasingham and Hebert, 2007) to determine genetic identity and further strengthen our results.

RESULTS AND DISCUSSION

Morphological and Meristic Findings

Values of morphological and meristic parameters for the specimens from Andaman waters are shown in Tables 21.2 and 21.3.

Genetic Distance Analysis

The studied fish specimens' assembled COI gene sequences were subject to further analysis: NJ trees were constructed and pairwise genetic divergences were calculated using MEGA 4.1 beta version software (Tamura et al., 2007).

TABLE 21.2 Morphological Parameter Mean Values[1] for Specimens of *Plectropomus* Genus from Andaman Coast

Parameter	*P. leopardus*	*P. maculatus*	*P. laevis*
TL	26	28	32
SL	23.5	23	29
DS	8	8	8
DR	11	12	12
AS	3	3	3
AR	7	8	8
PFR	15	15	17
PFL	2	2	2.2
LL	95	93	90
LLS	105	110	110

AR, anal ray; AS, anal spines; DR, dorsal ray; DS, dorsal spines; LL, lateral line scales; LLS, lateral line scale series; PFL, pectoral fin length; PFR, pectoral fin ray; SL, standard length; TL, total length.
[1]Seventy-five individuals from each species.

TABLE 21.3 Meristic Parameter Mean Values[1] for Specimens of *Plectropomus* Genus from Andaman Coast

Parameter	*P. laevis*	*P. leopardus*	*P. maculatus*
TL	32	26	29.5
SL	29	21.5	25.5
BD	7	7.5	9
CPD	4.2	3	4
HL	9	6.6	9
PRDFL	9.5	7.7	9
HD	5.4	7	6
PRVFL	9.1	6.3	7
VDOL	7.1	7	10
ADFEL	14	5	14
DFBL	15	10.5	11.5
VOAEFL	14	11	13
SpDAEFL	16.4	11.2	14.2
DEVOFL	14.8	12	9.5
VEAOFL	8	6.5	11.5
DVCFL	3.8	4	4
DEDCFL	5.5	4.5	5.5
AEVCFL	4.9	5	4.5
DEVCL	5.8	5.5	6.5
AEDCFL	6.8	6.5	7.5
ED	1.4	1	1
SNL	2.8	1.9	1.9

ADFEL, distance b/w anal and dorsal fin ends; AEDCFL, distances b/w anal fin end and dorsal caudal fin origin; AEVCFL, distances b/w anal fin end and ventral caudal fin origin; BD, body depth; CPD, caudal peduncle depth; DEDCFL, distance b/w dorsal fin end and dorsal caudal fin origin; DEVCL, distance b/w dorsal fin end and ventral caudal fin origin; DEVOFL, distance b/w dorsal fin end and ventral fin origin; DFBL, dorsal fin base length; DVCFL, distance b/w dorsal and ventral caudal fin origin; ED, eye diameter; HD, head depth; HL, head length; PRDFL, pre-dorsal fin length; PRVFL, pre-ventral fin length; SL, snout length; SNL, snout length; SpDAEFL, distance b/w first spine of dorsal fin and end of anal fin; TL, total length; VDOL, distance b/w ventral and dorsal fins origin; VEAOFL, distance b/w ventral fin end and anal fin origin; VOAEFL, distance b/w ventral fin origin and end of anal fin.
[1] *Seventy-five individuals from each species.*

Plectropomus laevis *(Lacepède, 1801), Black Saddled Coral Grouper*

The morphological and meristic characters suggest that *Plectropomus laevis* is highly similar to its sibling species *P. maculatus* (Talbot, 1959; Morgans, 1982; Randall and Hoese, 1986; Myers, 1989; Randall *et al.*, 1990; and Randall and Heemstra, 1991).

The COI gene sequences obtained were compared with data in the NCBI BLAST database. BLAST nucleotide analysis results showed 99 percent genetic identity with two reference sequences of the same species (JQ350229 and DQ107908) from Australia and 96 percent identity with unclassified Perciformes species (HQ564565 and JN313020) from Mauritius and Indonesia.

Further, an NJ tree (Figure 21.4) was constructed with the studied species' COI gene sequences and 61 reference sequences from the NCBI GenBank, and pairwise genetic distances (K2P) were estimated (Table 21.4) using MEGA 4.1 software (Tamura *et al.*, 2007).

The pairwise genetic distance results showed that 18 intra- and interspecific reference sequences were genetically close to those of the studied species of *P. laevis*, with K2P values in the range 0.001 to 0.067 percent. The four specimens of *P. laevis* from Andaman showed intraspecific mean K2P value of 0.002 percent with each other. The closest intraspecific genetic relationship (K2P = 0.001%) was observed with *P. laevis* from Australia and with Perciformes species from Mauritius and Indonesia. These Perciformes species, hitherto not more precisely classified, may be considered to be *P. laevis*. The next level of genetic distance was with Perciformes species from Indonesia (K2P = 0.028%), showing that these species also may be *P. laevis*. In summary the barcoding results for the studied species of *P. laevis* showed intraspecific mean K2P value of 0.001 percent, with 99 percent genetic similarity, with other COI sequences from the same species. These results confirm the genetic level identification of Andaman waters *P. laevis* as reported by Becker *et al.* (2011).

The interspecies genetic distance analysis carried out for 12 reference sequences resulted in K2P values in the range 0.052 to 0.066 percent (Table 21.4). Similar values for genetic distance have been found in other studies on Australian fish (Ward *et al.*, 2005), Indian carangids from Andhra coastal region (Persis *et al.*, 2009) and Indian marine fish subject to barcoding (Lakra *et al.*, 2010). The K2P values found in the present study of *Plectropomus* species suggest genetic closeness of *P. laevis*, *P. leopardus* and *P. maculatus*. Further, the morphological ambiguity of *P. maculatus* as reported by Talbot (1959), Morgans (1982), Randall and Hoese (1986), Myers, (1989), Randall *et al.* (1990), Randall and Heemstra (1991) and Leis (1986) was clearly discriminated using COI gene sequence analysis.

Plectropomus leopardus *(Lacepède, 1802) Leopard Coral Grouper*

The morphological features study supported the ambiguity between *Plectropomus leopardus* and *P. maculatus* reported by Fourmanoir and Laboute (1976), Randall and Hoese (1986) and Randall and Heemstra (1991).

BLAST nucleotide analysis showed that the studied species had 99 percent genetic identity with reference sequences of the same species (DQ107921, JN021314, JF952815, JF750763 and EU595233) from Australia, USA, Japan and China. Ninety-eight percent genetic identity was observed with unclassified Perciformes species HQ564562 from Mauritius.

Further, an NJ tree (Figure 21.5) was constructed with the studied species' COI gene sequences and 61 reference sequences from the NCBI GenBank, and

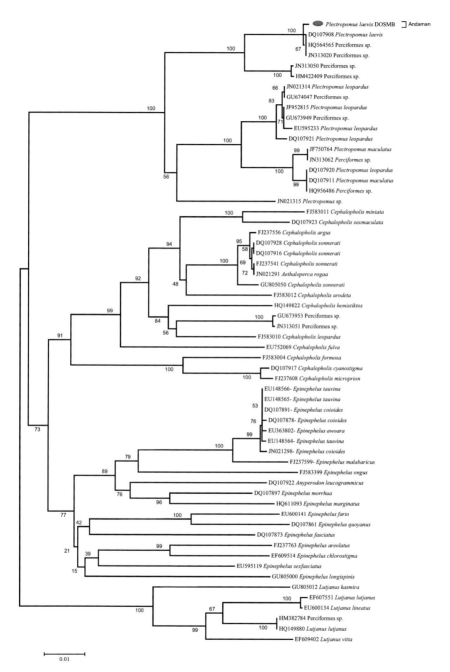

FIGURE 21.4   **Neighbor-joining tree and evolutionary relationships of** *Plectropomus laevis* **with 61 COI sequences.** Taxa inferred using bootstrap 1000 replicates. Branches corresponding to partitions reproduced in fewer than 50% bootstrap replicates are collapsed.

TABLE 21.4 Pairwise Genetic Distances of *Plectropomus laevis* with Intra- and Interspecific Reference Sequences

Sl. No.	GenBank No. and Species	K2P (%)	Country
1	– *Plectropomus laevis* DOSMB	0.000	Andaman Islands
2	DQ107908 *Plectropomus laevis*	0.001	Australia: Queensland, Bowen "20.00 S 148.00 E"
3	HQ564565 Perciformes sp.	0.001	Mauritius "16.6505 S 59.6669 E"
4	JN313020 Perciformes sp.	0.001	Indonesia "8.75 S 115.15 E"
5	JN313050 Perciformes sp.	0.028	Indonesia "8.75 S 115.15 E"
6	HM422409 Perciformes sp.	0.028	Indonesia "8.75 S 115.167 E"
7	JF952815 *Plectropomus leopardus*	0.052	Japan: Okinawa, Ishigaki "24.00 N 124.00 E"
8	GU673949 Perciformes sp.	0.052	Indonesia "8.75 S 115.167 E"
9	JN021314 *Plectropomus leopardus*	0.053	USA
10	DQ107921 *Plectropomus leopardus*	0.053	Australia: Queensland
11	GU674047 Perciformes sp.	0.053	Indonesia "8.75 S 115.167 E"
12	EU595233 *Plectropomus leopardus*	0.054	China: South China Sea "16.64 N 113.34 E"
13	JF750764 *Plectropomus maculatus*	0.065	China "24.4 N 118.1 E"
14	JN313062 Perciformes sp.	0.065	Indonesia "8.75 S 115.15 E"
15	DQ107911 *Plectropomus maculatus*	0.066	Australia: Western Australia, North of Cape Lambert "20.1167 S 117.10 E"
16	DQ107920 *Plectropomus leopardus*	0.066	Australia: Western Australia, East of Barrow Island
17	HQ956486 Perciformes sp.	0.066	Australia "10.9103 S 141.658 E"
18	JN021315 *Plectropomus* sp.	0.067	USA

DOSMB, Department of Ocean Studies and Marine Biology sample.

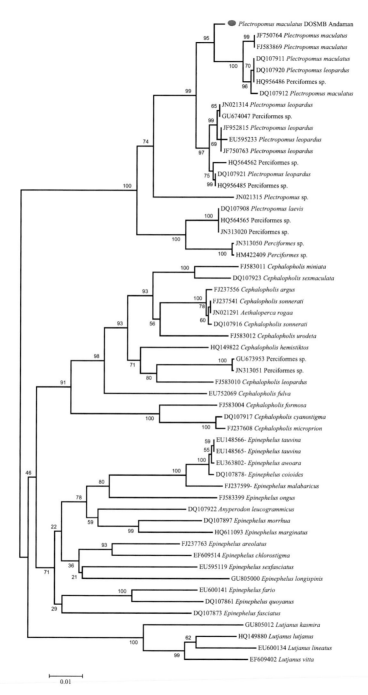

FIGURE 21.5   Neighbor-joining tree and evolutionary relationships of *Plectropomus leopardus* with 61 COI sequences. Taxa inferred using bootstrap 1000 replicates. Branches corresponding to partitions reproduced in fewer than 50% bootstrap replicates are collapsed.

pairwise genetic distances (K2P) were estimated (Table 21.5) using MEGA 4.1 software (Tamura *et al.*, 2007).

Only nine sequences were found to have close genetic distance to the studied species, in the range of 0.001 to 0.010 percent. The closest genetic distance was for the unclassified Perciformes species (HQ564562, K2P = 0.001%) from Mauritius. This unclassified Perciformes species from Mauritius may therefore be considered to be *P. leopardus*. Similarly, the Perciformes species HQ956485 from Australia had a K2P value of 0.003 percent with the studied *P. leopardus* and reference sequence DQ107921 and thus may also be considered to be *P. leopardus*. The present study concluded that the studied Andaman species of *P. leopardus* exhibited mean intraspecific genetic distance of 0.006 per cent and genetic identity of 99 percent from BLAST nucleotide analysis, matching the results of earlier work (Becker *et al.*, 2011).

However, the next cluster represented in the NJ tree consists of *P. maculatus* reference sequences with one reference sequence of *P. leopardus*. This *P. leopardus* had close genetic distance (K2P = 0.030%) with *P. maculatus* DQ107920 from Australia. This *P. leopardus* species might be considered to be *P. maculatus* (K2P values: DQ107911 Australia 0.030%; JF750764 China 0.030%; FJ583869 Philippines 0.030%) because similar genetic distance was found for the Andaman studied species of *P. maculatus*. This is consistent with the reports of Randall and Hoese (1986), and Randall and Heemstra (1991), who mentioned that this species is often misidentified in various regions. Further, the effectiveness of COI gene sequencing and evaluation of genetic distances for identification of marine fish (Ward *et al.*, 2005; Persis *et al.*, 2009; Hubert *et al.*, 2008; Steinke *et al.*, 2009a,b; Sachithanandam *et al.*, 2011, 2012; Hubert *et al.*, 2012) was supported by this study.

Plectropomus maculatus *(Bloch, 1790), Spotted Coral Grouper*

Morphological characterization studies have suggested that *Plectropomus maculatus* species is indistinct from the sibling species *P. pessuliferus* (Randall and Heemstra, 1991; Heemstra and Randall, 1993). The species is geographically distributed over western Pacific and Indo-Malay countries: Thailand, Singapore, Philippines, Indonesia, Papua New Guinea, Solomon Islands, and Australian waters (Western Australia to southern Queensland). No report of this species in Andaman waters was available.

BLAST nucleotide analysis showed that the studied *P. maculatus* species had 99 percent genetic identity with reference sequences of the same species (DQ107911, DQ107912, JF750764, FJ583869) and one of *P. leopardus* (DQ107920), from Australia, China and the Philippines.

Further, an NJ tree (Figure 21.6) was constructed with the studied species' COI gene sequences and 56 reference sequences from the NCBI GenBank, and pairwise genetic distances (K2P) were estimated (Table 21.6) using MEGA 4.1 software (Tamura *et al.*, 2007). The NJ tree showed an intraspecies cluster, and sibling or morphologically ambiguous species of *P. leopardus* formed another cluster with an interspecific relationship at 99 percent of bootstrap value.

TABLE 21.5 Pairwise Genetic Distances of *Plectropomus leopardus* with Intra- and Interspecific Reference Sequences

Sl. No.	GenBank No. and Species	K2P (%)	Country
1	– *Plectropomus leopardus* DOSMB	0.000	Andaman and Nicobar Islands
2	HQ564562 Perciformes sp.	0.001	Mauritius "16.6505 S 59.6669 E"
3	DQ107921 *Plectropomus leopardus*	0.003	Australia: Queensland
4	HQ956485 Perciformes sp.	0.003	Australia "10.4123 S 142.69 E"
5	GU674047 Perciformes sp.	0.007	Indonesia "8.75 S 115.167 E"
6	JN021314 *Plectropomus leopardus*	0.007	USA
7	JF952815 *Plectropomus leopardus*	0.008	Japan: Okinawa, Ishigaki "24.00 N 124.00 E"
8	JF750763 *Plectropomus leopardus*	0.008	China "24.4 N 118.1 E"
9	EU595233 *Plectropomus leopardus*	0.010	China: South China Sea "16.64 N 113.34 E"
10	DQ107912 *Plectropomus maculatus*	0.030	Australia: Western Australia, North of Cape Lambert "20.2167 S 117.083 E"
11	FJ583869 *Plectropomus maculatus*	0.030	Philippines: Manila "14.246 N 120.479 E"
12	DQ107920 *Plectropomus leopardus*	0.030	Australia: Western Australia, East of Barrow Island
13	JF750764 *Plectropomus maculatus*	0.030	China "24.4 N 118.1 E"
14	DQ107911 *Plectropomus maculatus*	0.030	Australia: Western Australia, North of Cape Lambert "20.1167 S 117.10 E"
15	HQ956486 Perciformes sp.	0.030	Australia "10.9103 S 141.658 E"
16	JN021315 *Plectropomus* sp.	0.049	USA
17	DQ107908 *Plectropomus laevis*	0.052	Australia: Queensland, Bowen "20.00 S 148.00 E"
18	HQ564565 Perciformes sp.	0.052	Mauritius "16.6505 S 59.6669 E"
19	JN313020 Perciformes sp.	0.052	Indonesia "8.75 S 115.15 E"
20	JN313050 Perciformes sp.	0.058	Indonesia "8.75 S 115.15 E"
21	HM422409 Perciformes sp.	0.058	Indonesia "8.75 S 115.167 E"

DOSMB, Department of Ocean Studies and Marine Biology sample.

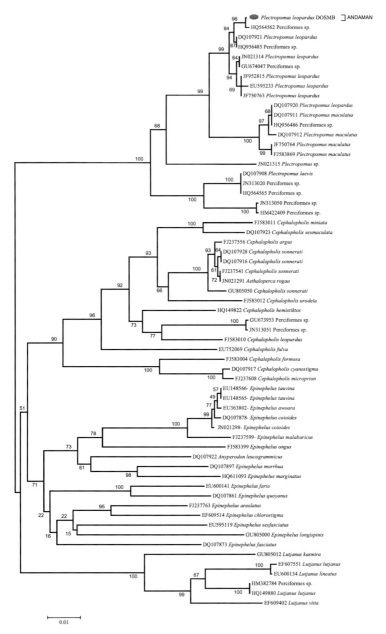

FIGURE 21.6   Neighbor-joining tree and evolutionary relationships of *Plectropomus maculatus* with 56 COI sequences. Taxa inferred with the sum of branch length = 0.99076496 are shown.

TABLE 21.6 Pairwise Genetic Distances of *Plectropomus maculatus* with Intra- and Interspecific Reference Sequences

Sl. No.	GenBank No. and Species	K2P (%)	Country
1	– *Plectropomus maculatus* DOSMB	0.000	Andaman and Nicobar Islands
2	HQ564562 Perciformes *sp.*	0.013	Mauritius "16.6505 S 59.6669 E"
3	DQ107911 *Plectropomus maculatus*	0.014	Australia: Western Australia, North of Cape Lambert "20.1167 S 117.10 E"
4	DQ107912 *Plectropomus maculatus*	0.014	Australia: Western Australia, North of Cape Lambert
5	DQ107920 *Plectropomus leopardus*	0.014	Australia: Western Australia, East of Barrow Island
6	HQ956486 Perciformes sp.	0.014	Australia "10.9103 S 141.658 E"
7	DQ107921 *Plectropomus leopardus*	0.014	Australia: Queensland
8	HQ956485 Perciformes sp.	0.014	Australia
9	GU674047 Perciformes sp.	0.016	Indonesia "8.75 S 115.167 E"
10	JF750764 *Plectropomus maculatus*	0.016	China "24.4 N 118.1 E"
11	FJ583869 *Plectropomus maculatus*	0.016	Philippines: Manila
12	JN021314 *Plectropomus leopardus*	0.016	USA
13	JF952815 *Plectropomus leopardus*	0.017	Japan: Okinawa, Ishigaki "24.00 N 124.00 E"
14	JF750763 *Plectropomus leopardus*	0.017	China "24.4 N 118.1 E"
15	EU595233 *Plectropomus leopardus*	0.019	China: South China Sea "16.64 N 113.34 E"
16	JN021315 *Plectropomus* sp.	0.047	USA
17	DQ107908 *Plectropomus laevis*	0.053	Australia: Queensland, Bowen
18	HQ564565 Perciformes sp.	0.053	Mauritius "16.6505 S 59.6669 E"
19	JN313020 Perciformes sp.	0.053	Indonesia "8.75 S 115.15 E"
20	JN313050 Perciformes sp.	0.058	Indonesia "8.75 S 115.15 E"
21	HM422409 Perciformes sp.	0.058	Indonesia "8.75 S 115.167 E"

DOSMB, Department of Ocean Studies and Marine Biology sample.

The genetic distance results showed a close relationship of the Andaman species with eight reference sequences, at a mean K2P value of 0.014 percent. The closest genetic relationship (K2P = 0.013%) of the Andaman species was with the reference sequence of unclassified Perciformes sp. HQ956486 from Australia, suggesting that this species may be considered to be *P. maculatus*. Since, *P. leopardus* DQ107920 showed a similar genetic distance from the studied *P. maculatus* and other closely related reference sequences, it may be concluded that DQ107920 may be *P. maculatus* misidentified or misnamed on account of morphological ambiguity.

The clear genetic distance between *P. maculatus* and *P. leopardus* found in this study supports the use of COI gene sequence analysis as a reliable method for species identification in the *Plectropomus* genus (Ward *et al.*, 2005; Hubert *et al.*, 2008; Carolan *et al.*, 2012).

Discussion

Species conservation is often hampered by inaccurate species identification arising from the limitations of conventional taxonomic methods. The present study used a molecular taxonomic approach, through COI gene sequencing, to identify three species of the *Plectropomus* species from Andaman waters.

For the presently studied Andaman species, the constructed neighbor-joining (NJ) tree suggested that *P. laevis* had a distant relationship with the sibling species *P. maculatus* (K2P = 0.103%). A closer genetic relationship of *P. laevis* was with *P. leopardus* (K2P = 0.093%). However, *P. leopardus* and *P. maculatus* had a close relationship (K2P = 0.024%), and in the relevant NJ tree these two species formed a subgroup of a major group. Similar values for genetic distance have been found in other studies on Australian fish (Ward *et al.*, 2005), Indian carangids from Andhra coastal region (Persis *et al.*, 2009) and Indian marine fish subject to barcoding (Lakra *et al.*, 2010).

P. leopardus showed a close genetic relationship (K2P = 0.028%) with ambiguous or sister species *P. maculatas* but was clearly differentiated through COI gene sequence analysis. Through COI gene sequence analysis: the same value of 0.028 percent was found for interspecies genetic divergence in Indian carangids (Persis *et al.*, 2009); Pacific Canadian fish genera discrimination threshold was 3.75 percent (Steinke *et al.*, 2009a); Acanthuridae genus *Acanthurus* was 0.063 percent (Hubert *et al.*, 2012); interspecific genetic distances were 0.033 to 0.044 percent for bumblebee species (Carolan *et al.*, 2012); and for eelpout fish from far-eastern seas, mean K2P within genus was 0.37 percent (Turanov *et al.*, 2012). However, earlier morphological taxonomy studies in Andaman waters suggested that *P. maculatus* is indistinct from the sibling species *P. pessuliferus* (Rajan, 2001; Rao, 2003). A COI sequence study on *P. pessuliferus* may sort out this morphological indistinctness.

The present study found intraspecific mean pairwise genetic distances of 0.006 percent (range 0.001 to 0.008 percent) and interspecific mean K2P values of 0.045 percent, which were slightly low values when compared with previous studies of

Australian marine fish (mean K2P = 0.39%) (Ward *et al.*, 2005), DNA barcoding of carangid fish from the Andhra coast (0.2%), pricklebacks and other percoid fishes from the Sea of Japan (intraspecies $0.11 \pm 0.04\%$, intragenus $1.87 \pm 0.68\%$) (Kartavtsev *et al.*, 2009), ornamental fish (intraspecific K2P = 0.26%, congeneric K2P = 10.31%) (Steinke *et al.*, 2009b) and Amphipoda *Rhachotropis* (interclade 0.143 to 0.370% with an overall average divergence of 0.284%) (Lorz *et al.*, 2012).

DNA barcoding using COI gene sequence based analysis effectively establishes species identification, and this is supported by the present study. Clustering of most congeneric and confamilial species is evident (Ward *et al.*, 2005; Hajibabaei *et al.*, 2007). The present study found 100 percent clustering of studied species with reference sequences of the same species in NJ trees. Similarly, barcoding studies in Australian marine fishes and Canadian Pacific coral ornamental fishes achieved 100% discrimination of species (Ward *et al.*, 2005; Steinke *et al.*, 2009a,b), supporting the importance of barcoding of all marine fish for identification through molecular taxonomy.

CONCLUSION

The present study has demonstrated the utility of COI gene sequence analysis of species divergences in marine fish, especially in *Plectropomus* genus. Intraspecific and interspecific genetic distances were revealed that allow unambiguous discrimination of species.

ACKNOWLEDGEMENTS

The authors thank the Centre for Marine Living Resources and Ecology (CMLRE), Ministry of Earth Sciences (MoES), Kochi for financial support of this study. The authors also are obliged to Dr P. Vijayachari, Director, and Regional Medical Research Centre (RMRC), Port Blair for providing laboratory facilities. One of the authors (VS) also thanks the University Grand Commission (UGC) for providing a Research Fellowship in Science to Meritorious Student (RFSMS).

REFERENCES

Becker, S., Hanner, R., Steinke, D., 2011. Five years of FISH-BOL: brief status report. Mitochondrial DNA 22, 3–9.

Carolan, J.C., Murray, T.E., Fitzpatrick, U., Crossley, J., Schmidt, H., 2012. Colour Patterns Do Not Diagnose Species: Quantitative Evaluation of a DNA Barcoded Cryptic Bumblebee Complex. PLoS One 7, e29251.

CMFRI., 2001. Annual Report, 2000. Central Marine Fisheries Research Institute, Kochi, p. 163.

Feral, J.P., 2002. How useful are the genetic markers in attempts to understand and manage marine biodiversity? *J.* Experimental Marine Biol. and Ecol. 268, 121–145.

Fourmanoir, P., and Laboute, P., 1976. Poissons de Nouvelle Calédonie et des Nouvelles Hébrides. Les Éditions du Pacifique, Papeete, Tahiti. p. 376.

Froese, R., and Pauly, D., 2000. Fish Base 2000 Concepts, design and data sources. ICLARM Contribution, 1594. ICLARM: Los Baños. p. 344.

Govindaraju, G.S., Jayasankar, P., 2004. Taxonomic Relationship among Seven species of Groupers (Genus *Epinephelus*; Family Serranidae) as revealed by RAPD Fingerprinting. Marine Biotech. 6, 229–237.

Hajibabaei, M., Singer, G.A.C., Hebert, P.D.N., Hickey, D.A., 2007. DNA Barcoding: how it complements taxonomy, molecular phylogenetics and population genetics. Trends in Genetics 23, 167–172.

Hebert, P.D.N., Cywinska, A., Ball, S.L., DeWaard, J.R., 2003. Biological identifications through DNA barcodes. Proc. Roy. Soc. Lond., Ser. B 270, 313–321.

Hebert, P.D.N., Gregory, T.R., 2005a. The promise of DNA barcoding for taxonomy. Syst. Biol. 54, 852–859.

Hebert, P.D.N., Barrett, R.D.H., 2005b. Identifying spiders through DNA barcodes. Can. J. Zool. 83, 481–491.

Heemstra, P.C., 1991. A taxonomic revision of the eastern Atlantic groupers (Pisces: Serranidae). Bol. Mus. Mun. Funchal 43, 5–71.

Heemstra, P.C., and Randall, J.E., 1993. Groupers of the world (Family Serranidae, Subfamily Epinephelinae). An annotated and illustrated catalogue of the grouper, rockcod, hind, coral grouper and lyretail species known to date. Food and Agriculture Organization (FAO) Species Catalogue, FAO Fisheries Synopsis no. 125, Vol 16. Rome. FAO, p. 522.

Hubert, N., Hanner, R., Holm, E., Mandrak, N.E., Taylor, E., 2008. Identifying Canadian Freshwater Fishes through DNA Barcodes. PLoS One 3, 2490.

Hubert, N., Meyer, C.P., Bruggemann, H.J., Guerin, F., Komeno, R.J.L., 2012. Cryptic Diversity in Indo-Pacific Coral-Reef Fishes Revealed by DNA-Barcoding Provides New Support to the Centre-of-Overlap Hypothesis. PLoS One 7 (3), e28987.

James. P.S.B.R., Murty, V.S., and Nammalwar, P., 1996. Groupers and snappers of India: biology and exploitation *[Meros y pargos de la India: biologia y explotacion]*. In: Arreguin-Sanchez, F., Munro, J.L., Balgos, M.C., Pauly, D., (eds.) Biology, fisheries and culture of tropical groupers and snappers. ICLARM Conf. Proc. 48: 449.

Kartavtsev, Y.P., Sharina, S.N., Goto, T., Rutenko, O.A., Zemnukhov, V.V., Semenchenko, A.A., Pitruk, D.L., Hanzawa, N., 2009. Molecular phylogenetics of pricklebacks and other percoid fishes from the Sea of Japan. Aquatic Biol. 8, 95–103.

Kimura, A., 1980. A simple method for estimating evolutionary rate of base substitutions through comparative studies of nucleotide sequences. J. Mol. Evol. 15, 111–120.

Lakra, W.S., Verma, M.S., Goswami, M., Lal, K.K., Mohindra, V., Punia, P., Gopalakrishnan, A., Singh, K.V., Ward, R.D., Hebert, P.D.N., 2010. DNA barcoding Indian marine fishes. Mol. Ecol. and Res. 11, 60–71.

Leis, J.M., 1986. Larval development in four species of Indo-Pacific coral trout *Plectropomus* (Pisces: Serranidae: Epinephelinae) with an analysis of the relationships of the genus. Bull. Marine Sci. 38, 525–552.

Lorz, A.N., Linse, K., Smith, P.J., Steinke, D., 2012. First Molecular Evidence for Underestimated Biodiversity of Rhachotropis (Crustacea, Amphipoda), with Description of a New Species. PLoS One 7, e32365.

Marine Products Export and Development Authority (MPEDA)., 2000. Statistics of Marine Products Exports. MPEDA, Kochi, India, pp. 59–61.

Morgans, J.F.C., 1982. Serranid fishes of Tanzania and Kenya. Ichthyology. Bull. J. L. B. Smith Inst. Ichthyology 46, 1–44.

Mustafa, A.M., 2011. Perch Fishery in Andaman Islands. Directorate of Fisheries, Andaman and Nicobar Islands, p. 25.

Myers, R.F., 1989. Micronesian Reef Fishes. A Practical Guide to the Identification of the Coral Reef Fishes of the Tropical Central and Western Pacific. Coral Graphics, Guam, p. 298.

Nelson, J.S., 2006. Fishes of the World, fourth ed. John Wiley & Sons, Hoboken, NJ.

Ottolenghi, F., Silvestri, C., Giordano, P., Lovatelli, A., 2004. The fattening of eels, groupers, tunas and yellowtails. Captured-based Aquaculture. FAO, Rome, p. 308.

Parrish, J.D., 1987. The trophic biology of snappers and groupers, pp. 405–463. In: Polovina, J.J., Ralston, S. (Eds.), Tropical Snappers and Groupers: Biology and Fisheries Management. Westview Press, Boulder, CO, p. 659.

Persis, M., Reddy, A.C.S., Rao, L.M., Khedkar, G.D., Ravinder, K., Nasruddin, K., 2009. COI (cytochrome oxidase – I) sequence based studies of Carangid fishes from Kakinada coast. India. Molecular Biology Report 36, 1733–1740.

Rajan, P.T., 2001. A Field Guide to Grouper and Snapper Fishes of Andaman and Nicobar Islands. Zoological Survey of India, Kolkata, p. 106.

Randall, J.E., Hoese, D.F., 1986. Revision of the groupers of the Indo-Pacific genus *Plectropomus* (Perciformes: Serranidae). Indo-Pacific Fishes 13, 1–31.

Randall, J.E., Allen, G.R., Steene, R.C., 1990. Fishes of the Great Barrier Reef and Coral Sea. Crawford House Press, Bathurst, Australia, p. 507.

Randall, J.E., Heemstra, P.C., 1991. Revision of Indo-pacific Groupers (Perciformes: Serranidae: Epinephelinae), with descriptions of five new species. Bernice Pauahi Bishop Museum, Honolulu, p. 332.

Rao, D.V., 2003. Guide to reef fishes of Andaman and Nicobar Islands. Zoological. Survey of India, Kolkata, p. 555.

Rao, D.V., Devi, K. and Rajan, P.T., 2000. An account of ichthyofauna of Andaman and Nicobar Islands, Bay of Bengal. *Rec. Zool. India, Occ. Paper* No. 178, p. 434.

Ratnasingham, S., and Hebert, P.D.N., 2007. BOLD: The Barcode of Life Data System (www.barcodinglife.org). Molecular Ecology Notes, 7(3), 355–364.

Sachithanandam, V., Mohan, P.M., Dhivya, P., Muruganandam, N., Baskaran, R., Chaaithanya, I.K., Vijayachari, P., 2011. DNA barcoding, phylogenetic relationships and speciation of Genus: *Plectropomus* in Andaman coast. J. Res. Biol. 3, 179–183.

Sachithanandam, V., Mohan, P.M., Muruganandam, N., Chaaithanya, I.K., Dhivya, P., Baskaran, R., 2012. DNA barcoding, phylogenetic study of *Epinephelus* Spp. from Andaman Coastal Region. Indian J. Marine Sci. 41, 203–211.

Sadovy, Y., 1999. The case of the disappearing grouper: *Epinephelus striatus*, the Nassau grouper, in the Caribbean and western Atlantic. Proceedings of the Gulf and Caribbean Fisheries Institute 45, 5–22.

Saitou, N., Nei, M., 1987. The Neighbour-Joining method – a new method for reconstructing phylogenetic trees. Mol. Biol. Evol. 4, 406–425.

Sambrook, J., Fritsch, E.F., Maniatus, T., 1989. Molecular Cloning: A Laboratory Manual, 2nd edition. Cold Spring Harbor Laboratory Press, NY, p. 1959.

Sanger, F., Nicklen, S., Coulson, A.R., 1977. DNA Sequencing with chain terminating inhibitors. Proceedings of the National Academy of Science, USA 74, 5463–5467.

Simon, J., Reynods, J.D., Polunin, N.V.C., 1999. Predicting the vulnerability of tropical reef fishes to exploitation with phylogenies and life histories. Conserv. Biol. 13, 1466–1475.

Steinke, D., Zemlak, T.S., Hebert, P.D.N., 2009a. Barcoding Nemo: DNA-based identifications for the ornamental fish trade. PLoS One 4, e6300.

Steinke, D., Zemlak, T.S., Boutillier, J.A., Hebert, P.D.N., 2009b. DNA barcoding of Pacific Canada's fishes. Marine Biology 156, 2641–2647.

Talbot, F.H., 1959. On *Plectropomus maculatus* (Bloch) and *Plectropomus marmoratus* (n. sp.) from East Africa (Pisces, Serranidae). Ann. Mag. Nat. Hist. 1, 748–752.

Tamura, K., Dudley, J., Nei, M., Kumar, S., 2007. MEGA4: Molecular Evolutionary Genetics Analysis (MEGA) Software Version 4.0. Mol. Biol. Evol. 24, 1596–1599.

Tikader, B.K., Daniel, A., Subbarao, N.V., 1986. Sea Shore Animals of Andaman and Nicobar Islands. Zoological Survey of India, Calcutta, India, p. 239.

Turanov, S.V., Kartavtseva, I.F., Zemnukhov, V.V., 2012. Molecular phylogenetic study of several eelpout fishes (Perciformes, Zoarcoidei) from far eastern seas on the basis of the nucleotide sequence of the mitochondrial cytochrome coxidase 1 gene (Co-1). Genetika 48, 235–252.

Ward, R.D., Zemlak, T.S., Innes, B.H., Last, P.R., Hebert, P.D.N., 2005. DNA barcoding Australia's fish species. Philosophical Transactions of the Royal Society Series Biology 360, 1847–1857.

Zhang, Z., Schwartz, S., Wagner, L., Miller, W., 2000. A greedy algorithm for aligning DNA sequences. J. Computational Biol. 7 (1–2), 203–214.

Chapter 22

Diversity of Antagonistic *Streptomyces* Species in Mangrove Sediments of Andaman Island, India

R. Baskaran,* P.M. Mohan,* R. Vijayakumar[†] and V. Sachithanandam*
*Department of Ocean Studies and Marine Biology, Pondicherry University, Port Blair, Andaman and Nicobar Islands, India; [†]Department of Microbiology, Bharathidasan University Constituent College for Women, Orathanadu, Tamil Nadu, India

INTRODUCTION

The actinobacteria can produce a wide array of enzymes for exploiting nutrients and have been the source of a broad range of bioactive metabolites of industrial and medical importance: for example, compounds with antibiotic activity against fungal and bacterial competitors. *Streptomyces* is the largest genus of actinobacteria (Kampfer, 2006). They produce over two-thirds of the clinically useful antibiotics of natural origin, e.g., neomycin, and chloramphenicol (Kieser *et al.*, 2000). Currently, the incidence of life-threatening infections caused by various microorganisms is rapidly increasing. Many species of bacteria are resistant to one or more of the hundred antibiotics currently in use. In order to overcome this, microorganisms have been screened for their biologically active compounds from unexploited environments (Kokare *et al.*, 2004). In this aspect, members of the actinobacteria of marine environments are poorly understood, and only a few reports are available pertaining to actinobacteria from mangrove environments of east coast regions of south India (Sivakumar, 2001; Vijayakumar *et al.*, 2011; Baskaran *et al.*, 2011).

Marine soils of India, especially those of the Andaman and Nicobar Islands, are known for their rich microbial biodiversity, including actinobacteria. Surprisingly, they have not been extensively explored for novel secondary metabolites. Keeping this in mind and recognizing the significance of marine actinobacteria, especially *Streptomyces* species, as a source of novel bioactive compounds, an effort was made in the present study to understand the diversity

Marine Faunal Diversity in India. DOI: 10.1016/B978-0-444-63258-6.00022-6
395

of marine actinobacteria and to screen the antagonistic activity of streptomycetes occurring in the mangrove sediments of South Andaman Island.

METHODS

Soil Sample Collection

The soil samples were collected from seven different locations of South Andaman: namely Guptapara (A), Manjery (B), Carbyn's Cove (C), Burmanallah (D), Chidyatappu (E), Wandoor (F) and Sippighat (G) (latitude $11°32'00''$N to $11°35'3''$N; longitude $92°38'00''$E to $92°41'00''$E) during February 2009. The sediment samples were collected from 0.5 cm to 15 cm depth in sterile plastic containers, transferred immediately to the laboratory and stored for further study.

Physico-Chemical Parameters of Sediments

The temperature, pH, and salinity of sediments were determined in accordance with Jackson (1973). The major elements and trace elements of the sediments were analyzed by X-ray fluorescence spectrometer (XRF) (Bruker S4-Pioneer) method (Subramanian and Angleja, 1976). Values of Spearman's correlation coefficient between total streptomycetes population and physico-chemical parameters were calculated by statistical software SPSS 16.

Isolation of Actinobacteria

The sediment samples were air dried and ground aseptically with sterile mortar and pestle. The powdered samples were mixed thoroughly and passed through a 2 mm sieve to remove gravel and debris. The samples were kept at 70°C for 15 minutes in a separate glass container for pre-treatment (Seong *et al.*, 2001). Then, 10-fold serial dilutions of the sediment samples were made using sterile 50 percent sea water. Three different culture media—Kuster's agar (KU), Starch Casein Agar (SCA) and Actinomycetes Isolation Agar (AIA)—were prepared using 50 percent sea water, and antibiotics—80 μg/mL of cycloheximide and 75 μg/mL of nalidixic acid (Himedia, Mumbai)—were added to prevent other bacterial and fungal growth (Baskaran *et al.*, 2011). An aliquot of 0.1 mL of the diluted sample was spread over the culture media in triplicate.

After 7 days of incubation at 28 ± 2°C, the actinobacterial colonies were counted and recorded. Each morphologically distinct actinobacterial colony was purified on yeast and malt extract agar medium (ISP2) by streak plate technique (Shirling and Gottlieb, 1966) and maintained in ISP2 agar slants in refrigerated condition for further investigation. These isolates were identified on the basis of their colony morphology, aerial mycelia, substrate mycelia and microscopic examination.

Screening for Antagonistic Activity of Streptomycetes

The ability of the streptomycetes isolates to produce antimicrobial compounds was screened by cross streak method (Baskaran *et al.*, 2011). After observing a good ribbon-like growth of the *Streptomyces* on Modified Nutrient Agar (MNA), the bacterial pathogens *Klebsiella pneumoniae* (MTCC-3040), *Salmonella infantis* (MTCC-1167), *Staphylococcus aureus* (MTCC-3160), *Lactococcus lactis* (MTCC-440), *Escherichia coli* (ICMR), *Vibrio cholerae* (ICMR), *Shigella flexneri* (ICMR), *Pseudomonas* sp., *Proteus* sp., *Citrobacter diserus* and *Bacillus* sp. (MTCC-3133) and fungal pathogens *Aspergillus niger* (CAS), *Aspergillus flavus* (CAS), *Aspergillus fumigatus* (CAS), *Penicillium* sp. (CAS), *Fusarium* sp. (CAS) and *Candida magnoliae* (MTCC-3602) were streaked at right angles to the original streak of streptomycetes and incubated at 37°C for bacteria and 27°C for fungi. The inhibition zone was measured after 24 h and 48 h. Based on the presence and absence of inhibition zones, streptomycetes displaying potential to produce antimicrobial compounds were selected for further investigation.

The selected *Streptomyces* spp. were inoculated in a conical flask containing 100 mL of production media and incubated at 28 ± 2°C on a rotary shaker at 200 rpm for 216 h. The fermented broth was centrifuged at 10,000 rpm at 4°C for 20 min. Supernatant was filtered using Whatman No.1 (0.45 μm) filter (Millipore), the filtrate was collected and used to determine the antimicrobial activity against pathogenic microbes by using the well diffusion method. Triplicate wells (6 mm diameter) were prepared in seeded agar plates, and to each well was added 100 μL of the culture filtrate of streptomycetes. These plates were incubated at 37°C for 24 h (for bacteria) and 28 ± 2°C for 48 h (for fungi). After incubation, the diameter of each inhibition zone was measured (Baskaran *et al.*, 2011).

Taxonomic Identification

All 86 isolates were subject to preliminary identification by colony morphology with respect to color, aerial mycelium, size and nature of colony, reverse side color and pigmentation, and then the isolates were observed under a light microscope for identification using the standard systematic keys (Burkholder *et al.*, 1954). Based on secondary metabolic activity, only two strains were selected (DOSMB-A107 and DOSMB-D105) for further identification by:

- spore chain morphology under high power magnification in a phase contrast microscope at 450× (Pridham *et al.*, 1958);
- cultural characteristics on various culture media, namely yeast malt extract agar (ISP2), oat meal agar (ISP3), inorganic salt starch casein agar (ISP4), glycerol asparagine agar (ISP5), tyrosine agar (ISP7), Kuster's agar (KU) and actinomycetes isolation agar (AIA); and
- cell wall characteristics.

In addition, various biochemical tests were performed to identify the individual *Streptomyces* species (Shirling and Gottlieb, 1966): citrate utilization test, hydrogen sulfide production test, nitrate reduction test, urease test, catalase test, oxidase test, starch hydrolysis, casein hydrolysis, gelatine hydrolysis, and lipid hydrolysis.

Physiological tests, such as effect of pH, temperature, inhibitory compounds, antibiotic sensitivity, amino acid utilization and utilization of carbon sources, were carried out (Flowers and Williams, 1977; Shimizu *et al.* 2000).

Analysis of Cell Wall Amino Acids

One millilitre of 6N HCl was added to 50 mg freeze-dried streptomycetes in a vial with a screw cap and heated for 4 h at 100°C (Becker *et al.*, 1965). Two millilitre of distilled water was added to the vial and then dried off. This step was repeated several times to remove HCl. The final dried material was dissolved in 0.2 mL of water. About 5 μL of the sample and 1 μL of standard diaminopimelic acid and glycine were separately spotted onto a TLC silica gel plate. The plate was developed in the solvent (methanol : distilled water : 6N HCl : pyridine = 80:26:4:10) for 3 h, air dried in a chemical hood for 2 h and sprayed with 0.1 percent ninhydrin, in acetone followed by heating at 120°C for 10 min on a hot plate.

Analysis of Whole Cell Sugars

One millilitre of 1N H_2SO_4 was added to 50 mg freeze-dried streptomycetes in a vial with a screw cap and heated at 100°C for 2 h (Lechevalier and Lechevalier, 1970). The mixture was centrifuged at 3000 rpm for 10 min. The pH of the supernatant was adjusted to 5 with saturated barium hydroxide, followed by centrifugation at 6000 rpm for 10 min. The supernatant was filtered. About 5 μL of the fluid and 3 μL of standard 1 percent sugar solution (galactose, glucose, arabinose and ribose) and the filtrate were spotted separately onto a TLC silica gel plate. The plate was developed in the solvent (acetonitrile : water = 92.5:7.5 v/v) for 20 min and dried in a chemical hood for 2 h. Then it was sprayed with aniline phthalate and heated at 100°C for 4 min on a hot plate (Myertons *et al.*, 1988).

RESULTS AND DISCUSSION

Isolation of *Streptomyces*

A total of 188 streptomycete colonies, including white, brown, grey and pink colored, were isolated from seven different locations of mangrove environment. Among them, the maximum 41 (21.80%) colonies were isolated from Guptapara, followed by Manjery 33 (17.55%), Carbyn's Cove 26 (13.82 %), Burmanallah 28 (14.89%), Chidyatappu 24 (12.76%), Wandoor 19 (10.10%)

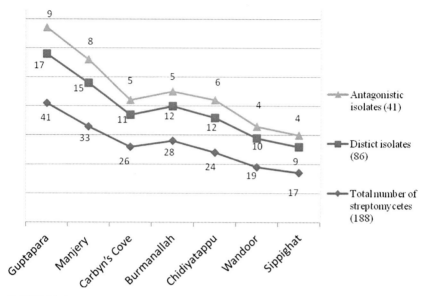

FIGURE 22.1   **Total isolates, distinct isolates and antagonistic *Streptomyces* spp.**

and Sippighat 17 (9.04%) (Figure 22.1). Based on the preliminary morphological properties, such as colony morphology and aerial and substrate mycelium color (Figure 22.2), it was characterized that, out of 188 colonies, 86 were morphologically distinguished. All the 86 isolates belonged to the Streptomycetaceae family and genus *Streptomyces* and were grouped by station as follows: Guptapara 17, Manjery 15, Burmanallah and Chidyatappu 12 each, Carbyn's Cove 11, Wandoor 10 and Sippighat 9 (Table 22.1). The percentage frequency of morphologically discriminated isolates by station was 19.76, 17.44, 13.95, 13.95, 12.79, 11.62 and 10.46 from Guptapara, Manjery, Chidyatappu, Burmanallah, Carbyn's Cove, Wandoor and Sippighat, respectively (Figure 22.1).

The present study recorded more isolates than were recorded from Muthupet mangrove, Point Calimere seashore and Vedaranyam saltpan, in the Palk Strait region of the Bay of Bengal, by Vijayakumar *et al.* (2007). Actinobacteria, especially streptomycetes, had been reported from marine sub-habitats such as marine sediments (Goodfellow and Haynes, 1984; Jensen *et al.*, 1991; Kokare *et al.*, 2004; Baskaran *et al.*, 2011) and marine soils (Haung *et al.*, 1991; Vijayakumar *et al.*, 2011).

The physico-chemical properties of the mangrove sediments such as potassium (59,500 ppm), phosphorus (1560 ppm), silicon (629,800 ppm), zirconium (1920 ppm), rubidium (67 ppm), niobium (13 ppm), cesium (5 ppm), lanthanum (28 ppm), barium (297 ppm), cerium (79 ppm) and thorium (12 ppm) were high in Guptapara compared with other sampling stations, whereas temperature (29°C), pH (6.6) and salinity (5 psu) were moderate in all locations

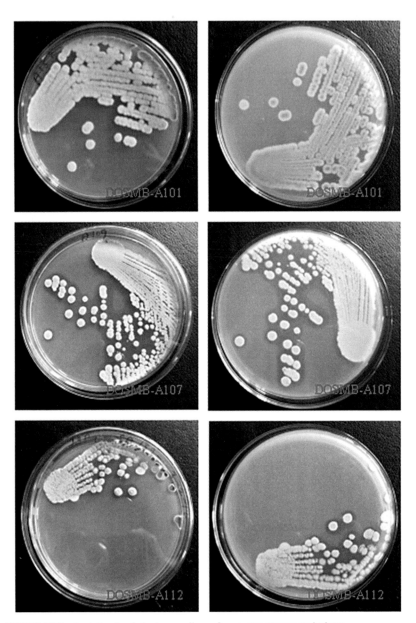

FIGURE 22.2 Aerial and substrate mycelium of some streptomycete isolates.

(Table 22.2). The above conditions may explain why the actinobacterial population was highest at Guptapara. It was concluded that the actinobacterial population was directly influenced by the physico-chemical parameters of the mangrove sediments. This conclusion was supported by the positive correlation between the physico-chemical properties of soil and total *Streptomyces*

TABLE 22.1 Morphologically Distinguished Streptomycetes Isolated from Mangrove Sediments of Andaman Island

Sl. No.	Isolate code	Color of aerial mycelium	Color of substrate mycelium and size of the colony
Guptapara			
1	DOSMB.A101	Grey	4.5mm, pink
2	DOSMB.A102	Light grey	3mm, elevated colony with flat margin
3	DOSMB.A103	Grey	6mm, button shape, pale brown
4	DOSMB.A104	Creamy white	9mm, pale yellow
5	DOSMB.A105	Grey	5mm, centrally raised with flat margin
6	DOSMB.A106	Dark grey	3mm, dark brown
7	DOSMB.A107	Grey	9mm, whitish grey
8	DOSMB.A108	Red	2.5mm, button shape
9	DOSMB.A109	Green	6mm, flat margin
10	DOSMB.A110	Light yellow	3mm, light red color
11	DOSMB.A111	White	2mm, pale brown
12	DOSMB.A112	Grey	7.5mm, whitish grey
13	DOSMB.A113	Grey	4 mm, brown
14	DOSMB.A114	White	4.2mm, brown
15	DOSMB.A115	Grey	5.6mm, round flat margin
16	DOSMB.A116	Dull white	6mm, brown
17	DOSMB.A117	Bluish white	2.5mm bark brown
Manjery			
18	DOSMB.B101	Dark grey	5mm, button shape
19	DOSMB.B102	Grey	6mm, flat colonies
20	DOSMB.B103	Grey	5mm, elevated colony with dull white
21	DOSMB.B104	Grey	3.2mm, light brown
22	DOSMB.B105	Grey	12mm, colored with white
23	DOSMB.B106	White	3mm, flat with pearls like arrangement pale yellow diffusible pigment
24	DOSMB.B107	Light grey	2.5mm, button shape green diffusible pigment
25	DOSMB.B108	Grey	3mm, brown
26	DOSMB.B109	Dark grey	4mm, flat, centrally grey with white color surrounding

(Continued)

TABLE 22.1 Morphologically Distinguished Streptomycetes Isolated from Mangrove Sediments of Andaman Island *(cont.)*

Sl. No.	Isolate code	Color of aerial mycelium	Color of substrate mycelium and size of the colony
27	DOSMB.B110	Grey	5mm, centrally raised with flat margin
28	DOSMB.B111	White	6mm, light orange,
29	DOSMB.B112	Yellow	5mm, creamy yellow
30	DOSMB.B113	Dark grey	5mm, yellow
31	DOSMB.B114	Grey	8mm, light yellow
32	DOSMB.B115	Grey	3mm, dark brown
Carbyn's Cove			
33	DOSMB.C101	Dark grey	6mm, elevated colony
34	DOSMB.C102	Grey	3mm, pale brown
35	DOSMB.C103	Yellow	4mm, dark yellow
36	DOSMB.C104	Grey	3mm, dark brown
37	DOSMB.C105	Bluish white	4mm, scattered margin
38	DOSMB.C106	Grey	4.5mm, dark brown
39	DOSMB.C107	White	7mm, pale brown
40	DOSMB.C108	White	4.2mm, pale yellow
41	DOSMB.C109	Grey	3mm, button shape with irregular margin
42	DOSMB.C110	Grey	4mm, scattered margin
43	DOSMB.C111	Grey	2.5mm, dark brown
Burmanallah			
44	DOSMB.D101	White	7mm, dark pink
45	DOSMB.D102	Dark grey	5mm, whitish grey
46	DOSMB.D103	Grey	5mm, yellow
47	DOSMB.D104	Dark grey	4mm, fully elevated with white margin
48	DOSMB.D105	Light grey	8mm, dark yellow
49	DOSMB.D106	Grey	3mm, dark brown
50	DOSMB.D107	Grey	4.2mm pale yellow
51	DOSMB.D108	Grey	4.5mm, pale yellow
52	DOSMB.D109	Light yellow	6mm, pale brown
53	DOSMB.D110	White	4mm, orange color
54	DOSMB.D111	White	6.2mm, green color
55	DOSMB.D112	Grey	4mm, yellow

TABLE 22.1 Morphologically Distinguished Streptomycetes Isolated from Mangrove Sediments of Andaman Island *(cont.)*

Sl. No.	Isolate code	Color of aerial mycelium	Color of substrate mycelium and size of the colony
Chidiyatapu			
56	DOSMB.E101	White	7mm, pink
57	DOSMB.E102	Dark grey	3mm, dark brown
58	DOSMB.E103	Pale yellow	6mm, dark yellow
59	DOSMB.E104	Grey	5mm, dark brown
60	DOSMB.E105	Dark grey	4mm, pale grey
61	DOSMB.E106	Pink	4.2mm, dark pink
62	DOSMB.E107	Grey	4.8mm, pale brown
63	DOSMB.E108	White	4mm, orange color
64	DOSMB.E109	Dark grey	3.5mm, white spore margin brownish yellow
65	DOSMB.E110	Red	2.5mm, button shape
66	DOSMB.E111	Green	2.5mm, pale green
67	DOSMB.E112	White	4.2mm, brown
Wandoor			
68	DOSMB.F101	Bluish white	4.5mm, scattered margin
69	DOSMB F102	Grey	5mm, dark brown
70	DOSMB.F103	Dark grey	5.2mm, dark brown
71	DOSMB.F104	White	8mm, brown
72	DOSMB.F105	White	4mm, yellow
73	DOSMB.F106	Grey	3mm, button shape irregular margin
74	DOSMB.F107	Grey	2.5mm dark brown
75	DOSMB.F108	Grey	4.8mm, pale yellow
76	DOSMB.F109	Dark grey	6.2mm, pink
77	DOSMB.F110	Grey	6mm, Grey
Sippighat			
78	DOSMB.G101	Grey	5.5mm, yellow
79	DOSMB.G102	Dark grey	5mm, brown
80	DOSMB.G103	Grey	4.2mm, brown
81	DOSMB.G104	Dark grey	3mm, creamy
82	DOSMB.G105	Light yellow	3.5mm, light red color
83	DOSMB.G106	Dark grey	4.5mm, fully elevated white margin
84	DOSMB.G107	Grey	7mm, light yellow
85	DOSMB.G108	Dark grey	5mm, button shape
86	DOSMB.G109	Grey	3.5mm, brown

TABLE 22.2 Physico-Chemical Parameters of Mangrove Sediments

Physico-chemical parameter	A	B	C	D	E	F	G
Temperature (°C)	29	29	32	28	28	29	30
pH	6.6	6.8	7.6	6.6	7.4	6.5	7.2
Salinity (ppt)	5	5	5	6	4	4	4
Al (ppm)	133,800	123,200	165,300	141,200	117,000	171,300	97,750
Br (ppm)	170	160	310	370	400	290	290
Ca (ppm)	7710	7400	26,920	82,450	91,240	19,750	14,400
Cl (ppm)	19,900	32,200	42,600	42,700	46,090	25,500	69,330
Cr (ppm)	490	510	900	1680	4800	1490	300
Cu (ppm)	190	160	290	320	230	300	170
Fe (ppm)	77,280	71,370	143,400	172,300	166,900	184,000	56,080
K (ppm)	59,500	51,550	36,820	19,120	10,800	32,900	58,220
Mg (ppm)	13,370	15,290	23,850	33,440	99,510	37,490	14,450
Mn (ppm)	370	390	790	1600	1570	860	430
Na (ppm)	19,100	24,630	31,170	43,590	40,630	26,970	34,190
Ni (ppm)	150	170	340	400	2000	1600	0
P (ppm)	1560	1190	1310	1450	1260	1360	860
S (ppm)	19300	36230	32310	32310	15500	31620	62000
Si (ppm)	629,800	619,700	475,900	408,200	391,000	445,900	577,100
Sr (ppm)	330	290	300	640	520	320	340

TABLE 22.2 Physico-Chemical Parameters of Mangrove Sediments (*cont.*)

Physico-chemical parameter	A	B	C	D	E	F	G
Ti (ppm)	14,700	13,600	15,890	17,610	10,100	17,450	11,900
Zn (ppm)	190	200	380	310	210	370	160
Zr (ppm)	1920	1580	940	220	210	340	1830
Sc (ppm)	11	11	29	39	31	36	8
V (ppm)	95	86	183	252	179	240	61
Co (ppm)	9	11	28	40	69	41	6
Ga (ppm)	9	9	15	14	12	14	7
As (ppm)	7	9	14	12	8	8	9
Rb (ppm)	67	62	55	28	18	43	62
Y (ppm)	25	25	28	27	24	25	21
Nb (ppm)	13	13	9	6	4	7	13
Mo (ppm)	4	3	6	4	2	4	7
Sn (ppm)	5	6	6	6	4	6	6
Sb (ppm)	5	5	4	5	4	6	4
Cs (ppm)	5	4	3	0	0	1	3
Ba (ppm)	297	235	119	29	12	148	276
La (ppm)	28	19	24	8	11	20	16
Ce (ppm)	79	68	46	1	5	32	61
Pb (ppm)	25	21	27	22	16	20	23
Th (ppm)	12	11	9	3	3	6	10
U (ppm)	4	3	3	2	1	3	7

A, Guptapara; B, Manjery, C, Carbyn's Cove; D, Burmanallah; E, Chidyatappu; F, Wandoor; G, Sippighat .

population (TSP). Positive correlations were found between TSP and salinity, Al, K, P, Si, Zr, Rb, Y, Nb, Sb, Cs, Ba, La, Ce, Th and Ti. Whereas, negative correlations were found between TSP and temperature, pH, Br, Ca, Cl, Cr, Cu, Fe, Mg, Mn, Na, Ni, S, Sr, Zn, Sc, V, Co, Ga, As, Mo, Sn and U (Table 22.3). Positive correlations between salinity, pH, organic content of marine sediments and actinobacterial population have been reported by various researchers (Jiang and Xu, 1990; Jensen et al., 1991; Sivakumar, 2001). However, variations in temperature, pH and dissolved phosphate were found to be insignificant, though the variations in total nitrogen and organic matter were significant, in the population studied in Alexandria by Ghanem et al. (2000).

Antagonistic Activity of *Streptomyces* spp.

In the present study, among the 86 isolates of streptomycetes, only 41 (47.67%) were identified as antimicrobially active (Figure 22.3). Guptapara showed the maximum number (9) of antagonistic isolates among the sampling stations (Figure 22.1). Among 41 isolates, 15 (36.58%) showed both antibacterial and antifungal activity, 16 (39%) showed only antibacterial activity and 10 (24.39%) showed only antifungal activity. On the whole, among the 41 antagonistic streptomycete isolates, 31 (75.6%) showed antibacterial activity and 25 (60.97%) showed antifungal activity. Of the 31 antibacterial isolates, 30 (96.77%) showed activity against Gram-negative bacteria, 30 (96.77%) showed activity against Gram-positive bacteria, and 29 (93.54%) showed activity against both Gram-positive and Gram-negative bacteria (Tables 22.4 and 22.5).

Screening of their antimicrobial activity showed that the streptomycetes of South Andaman Island sediments displayed higher antibacterial than antifungal activity. In a similar way, 10 potential marine actinobacteria from 34 isolates have been reported from the Tamil Nadu coastal area (Pugazhvendan et al., 2010). Among the 68 marine actinobacteria of the Palk Strait regions of the Bay of Bengal reported by Vijayakumar et al. (2011), only 25 (36.8%) isolates demonstrated antimicrobial activity against the various pathogens tested: 22 (88%) with antibacterial activity, 16 (64%) with antifungal activity and 13 (52%) with both antibacterial and antifungal properties. Fifteen (68%) of the 22 isolates with antibacterial activity inhibited the growth of Gram-positive bacteria, 16 (72.7%) had activity against Gram-negative bacteria and nine (40.9%) showed activity against both Gram-positive and Gram-negative bacteria. Dhanasekaran et al., (2005) reported that, out of 107 actinobacteria from different coastal areas of Tamil Nadu, only 22 (20.65%) isolates possessing antifungal activities were found. Compared with the findings of the above-mentioned studies, the present study reports that the highest percentage of potentially bioactive actinobacterial isolates was found in the coastal regions of the Andaman Islands, India.

Out of 41 antagonistic *Streptomyces* spp., the isolates with high activity (DOSMB-A102, DOSMB-A107, DOSMB-B104, DOSMB-B109, DOSMB-C107,

TABLE 22.3 Correlations between Physico-Chemical Parameters and Total Streptomycete Population (TSP)

Sl. No.	Variables	Spearman's r
1	TSP vs. Temperature (°C)	−0.157
2	TSP vs. pH	−0.325
3	TSP vs. Salinity	0.581
4	TSP vs. Aluminium	0.019
5	TSP vs. Bromine	−0.607
6	TSP vs. Calcium	−0.188
7	TSP vs. Chlorine	−0.652
8	TSP vs. Chromium	−0.222
9	TSP vs. Copper	−0.276
10	TSP vs. Iron	−0.324
11	TSP vs. Potassium	0.303
12	TSP vs. Magnesium	−0.279
13	TSP vs. Manganese	−0.255
14	TSP vs. Sodium	−0.511
15	TSP vs. Nickel	−0.381
16	TSP vs. Phosphorus	0.647
17	TSP vs. Sulfur	−0.539
18	TSP vs. Silicon	0.496
19	TSP vs. Strontium	−0.065
20	TSP vs. Titanium	0.106
21	TSP vs. Zinc	−0.253
22	TSP vs. Zirconium	0.395
23	TSP vs. Scandium	−0.320
24	TSP vs. Vanadium	−0.263
25	TSP vs. Cobalt	−0.321
26	TSP vs. Gallium	−0.180
27	TSP vs. Arsenic	−0.149
28	TSP vs. Rubidium	0.336
29	TSP vs. Yttrium	0.380
30	TSP vs. Niobium	0.366
31	TSP vs. Molybdenum	−0.377
32	TSP vs. Tin	−0.217
33	TSP vs. Antimony	0.153
34	TSP vs. Cesium	0.566

(Continued)

TABLE 22.3 Correlations between Physico-Chemical Parameters and Total Streptomycete Population (TSP) *(cont.)*

Sl. No.	Variables	Spearman's *r*
35	TSP vs. Barium	0.239
36	TSP vs. Lanthanum	0.448
37	TSP vs. Cerium	0.418
38	TSP vs. Lead	0.313
39	TSP vs. Thorium	0.427
40	TSP vs. Uranium	−0.222

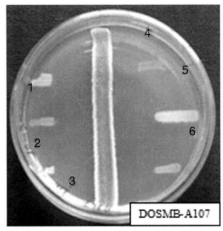

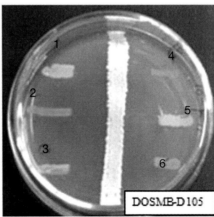

FIGURE 22.3   Antagonistic activity of strains DOSMB-A107 and DOSMB-D105 by cross streak method. From top left anticlockwise: 1, *K. pneumoniae;* 2, *S. infantis;* 3, *S. aureus;* 4, *L. lactis;* 5, *E. coli;* 6, *V. cholerae.*

TABLE 22.4 Preliminary Screening of Streptomycetes for Antibacterial Activity

Sl. No.	Isolate	Test bacteria										
		B1	B2	B3	B4	B5	B6	B7	B8	B9	B10	B11
Guptapara												
1	DOSMB.A101	++	++	++	++	++	++	-	-	+	+	+
2	DOSMB.A102	++	+++	++	++	+++	+++	-	++	++	++	++
3	DOSMB.A103	++	-	++	-	-	-	-	++	++	-	+++
4	DOSMB.A104	+	-	++	-	-	-	-	++	++	-	++
5	DOSMB.A107	+++	+++	+++	+++	+++	+++	++	+++	+++	+++	+++
6	DOSMB.A110	+++	-	++	++	+++	++	-	++	++	-	-
7	DOSMB.A117	++	-	++	-	-	-	-	++	++	-	-
Manjery												
8	DOSMB.B104	+++	++	++	++	++	++	+	+++	++	++	+++
9	DOSMB.B109	++	++	+++	++	++	++	++	+++	++	++	++
10	DOSMB.B113	++	-	-	-	-	-	-	-	+	-	+
Carbyn's Cove												
11	DOSMB.C102	+++	-	++	++	+++	+	-	++	++	+++	++
12	DOSMB.C105	++	-	+++	++	-	++	-	+++	-	+++	-
13	DOSMB.C107	+++	++	++	+++	++	+++	++	+++	++	+++	+++
14	DOSMB.C109	-	-	-	-	-	-	-	-	-	-	-
Burmanallah												
15	DOSMB.D102	++	-	++	-	-	-	-	++	++	-	-
16	DOSMB.D105	+++	++	+++	+++	+++	+++	++	+++	+++	+++	+++
17	DOSMB.D106	++	++	-	++	++	+++	-	++	-	++	-
18	DOSMB.D110	++	++	++	++	++	+++	+	++	++	-	++
19	DOSMB.D111	++	++	-	++	++	-	+	+++	-	++	-

(Continued)

TABLE 22.4 Preliminary Screening of Streptomycetes for Antibacterial Activity (cont.)

Sl. No.	Isolate	Test bacteria										
		B1	B2	B3	B4	B5	B6	B7	B8	B9	B10	B11
Chidyatappu												
20	DOSMB.E101	++	-	++	-	++	-	-	++	++	-	++
21	DOSMB.E106	++	++	-	++	++	-	-	++	++	++	-
22	DOSMB.E110	++	++	++	++	++	++	+	++	++	++	++
23	DOSMB.E112	-	-	-	-	-	-	-	++	++	-	-
Wandoor												
24	DOSMB.F103	++	++	++	++	++	++	+	++	++	++	++
25	DOSMB.F104	++	-	++	-	-	-	-	++	++	-	++
26	DOSMB.F108	++	-	-	++	-	-	-	++	++	-	++
27	DOSMB.F110	-	-	++	++	++	-	-	++	++	-	++
Sippighat												
28	DOSMB.G101	++	-	++	++	-	+	-	+	+	-	+
29	DOSMB.G105	-	++	+	++	+	-	-	++	+	-	++
30	DOSMB.G107	++	++	+++	++	++	++	-	++	+	++	++
31	DOSMB.G109	++	++	+	+	++	++	-	+	++	-	+

B1, *K. pneumoniae*; B2, *S. infantis*; B3, *S. aureus*; B4, *L. lactis*; B5, *E. coli*; B6, *V. cholerae*; B7, *S. flexneri*; B8, *Pseudomonas* sp.; B9, *Proteus* sp.; B10, *C. Diserus*; B11, *Bacillus* sp.
+, <10 mm; ++, 10–20 mm; +++, >20 mm.

TABLE 22.5 Preliminary Screening of Streptomycetes for Antifungal Activity

		Pathogenic fungus					
Sl. No.	Isolate	*F1*	*F2*	*F3*	*F4*	*F5*	*F6*
Guptapara							
1	DOSMB.A101	++	++	-	-	-	-
2	DOSMB.A102	++	++	+	++	++	-
3	DOSMB.A107	+++	++	++	++	+++	++
4	DOSMB.A108	++	++	-	++	-	-
5	DOSMB.A113	+++	-	-	+	-	-
Manjery							
6	DOSMB.B102	+++	-	-	-	-	-
7	DOSMB.B103	++	++	-	-	-	-
8	DOSMB.B104	++	++	++	+		++
9	DOSMB.B105	++	++	++	++	++	-
10	DOSMB.B106	-	-	-	++	-	++
11	DOSMB.B109	++	+	++	+	-	+
Carbyn's Cove							
12	DOSMB.C101	++	++	-	-	-	-
13	DOSMB.C102	++	-	-	-	-	-
14	DOSMB.C107	+++	-	++	++	-	++
Burmanallah							
15	DOSMB.D105	+++	++	++	++	++	++
16	DOSMB.D110	++	++	-	++	-	-
17	DOSMB.D111	++	-	++	++	++	-
Chidyatappu							
18	DOSMB.E102	++	-	++	++	-	-
19	DOSMB.E110	++	-	++	++	+	++
20	DOSMB.E111	-	++	++	++	-	++
21	DOSMB.E112	++	-	++	-	++	++
Wandoor							
22	ANI.F103	++	-	++	++	-	+
23	ANI.F104	++	++	++	-	++	+
Sippighat							
24	DOSMB.G101	++	+	++	+	+	++
25	DOSMB.G107	++	+	++	++	+	++

F1, *A. niger*; F2, *A. flavus*; F3, *A. fumigatus*, F4, *Penicillium* sp.; F5, *Fusarium* sp.; F6, *C. magnoliae*.
+, <10 mm; ++, 10– 20 mm; ++ + , >20 mm.

DOSMB-D105, DOSMB-E110, DOSMB-F103 and DOSMB-G107) in the primary screening were selected for secondary metabolic activity. From the nine selected isolates, DOSMB-A107 and DOSMB-D105 showed prominent antimicrobial activity against all pathogens tested (Table 22.6). Maximum antibacterial activity of the isolate DOSMB-A107 was observed against *S. infantis*, followed by *V. cholerae*, *K. pneumoniae*, *E. coli*, *Bacillus* sp., *L. lactis*, *S. flexneri*, *Proteus* sp., *S. aureus* and *C. diserus*. The maximum antifungal activity of this isolate was recorded against *A. niger*, followed by *Penicillium* sp., *A. flavus*, *A. fumigatus*, *C. magnoliae* and *Fusarium* sp. The antibacterial activity of the isolate DOSMB-D105 was maximum against *Pseudomonas* sp., followed by *Proteus* sp., *S. aureus*, *Bacillus* sp., *K. pneumoniae*, *E. coli*, *C. diserus*, *S. flexneri*, *V. cholerae* and *S. infantis*. Maximum antifungal activity of this isolate was observed against *A. niger* followed by, *A. flavus*, *C. magnoliae*, *A. fumigatus*, *Penicillium* sp. and *Fusarium* sp.

Previous studies found that *Streptomyces* spp. had great antifungal and antibacterial activity against *Candida albicans*, *C. tropicalis*, *Aspergillus fumigatus*, *Fusarium solani*, *Pseudomonas* sp. and *Salmonella* sp. (Fukuda *et al.*, 1990; Pisano *et al.*, 1992). *Streptomyces* spp. also showed maximum antifungal activity against *Fusarium* sp. and *Penicillium* sp. along with certain bacteria (Yon *et el.*, 1995; Ouhdouch *et al.*, 1996). Generally, the screening of the antibiotic production ability of actinobacteria is a key process which can be influenced by the methods adopted for screening, the load of the producers and the test pathogens, and the media used. From the present investigation, it is concluded that streptomycetes isolated from mangrove sediments of Andaman Island are a tremendous potential source of novel antibiotics with great pharmaceutical interest.

Identification of the Isolates

The members of *Streptomyces* are characterized by certain diverse but stable and distinct morphological characters, which are used for species differentiation. Colony formation, vegetative and aerial mycelium, structure of sporophores and spores are the most important features for the identification of *Streptomyces* (Waksman, 1961; Kuster, 1963).

Color of aerial mycelium is considered to be an important character for the grouping and identification of actinobacteria (Pridham and Tresner, 1974). In the present study, grey isolates were predominant (Table 22.1). Such dominance of grey series has already been reported in different soils (Kim *et al.*, 1999; Ndonde and Semu, 2000). Two potential strains (DOSMB-D105 and DOSMB-A107) were examined under phase contrast microscope, and it was found that both *Streptomyces* were spiral in nature (Figure 22.4). Similar findings have been reported by previous researchers (Ohnishi *et al.*, 2008; Moncheva *et al.*, 2002).

The two selected *Streptomyces* spp. were grown on various media. The strain DOSMB-D105 produced yellow mycelia and grey melanin pigment whereas

TABLE 22.6 Secondary Metabolic Activity of Potent Streptomycetes[1]

Test organism	DOSMB. A102	DOSMB. A107	DOSMB. B104	DOSMB. B109	DOSMB. C107	DOSMB. D105	DOSMB. E110	DOSMB. F103	DOSMB. G107
B1	17.66 ± 0.19	24.66 ± 0.9	9.33 ± 0.19	10.66 ± 0.19	22.66 ± 0.19	25 ±0.33	8.33 ± 0.19	13.33 ± 0.19	16.66 ± 0.19
B2	15.33 ± 0.19	25.33 ± 0.38	11.66 + 0.19	13 ± 00	20.66 ± 0.19	20.33 ± 0.19	18.33 ± 0.38	17.33 ± 0.19	10.66 ± 0.19
B3	17.33 ± 0.19	22 ± 00	14.33 ± 0.19	15.66 ± 0.19	22.66 ± 0.19	30.33 ± 0.33	14.66 ± 0.38	18.33 ± 0.19	14.33 ± .19
B4	14.66 ± .19	22.66 ± 0.50	14.66 ± 0.19	11 ± 0.33	17.66 ± .019	29.33 ± 0.38	12 ± 0.00	11.33 ± 0.19	12.33 ± 0.38
B5	18.66 ± .19	23.66 ± .19	18.66 ± 0.19	22 ± 033	9 ± 0	24.66 ± 0.50	16.66 ± 0.19	20.66 ± 0.19	18.33 ± 0.38
B6	16.66 ± .0.19	25 ± 0.33	16 ± .0	22.33 ± 0.19	7.33 ± 0.19	22 ± 00	9.33 ± 0.19	18.66 ± 0.19	20.33 ± 0.38
B7	7.66 ± 0.19	22.33 ± 0.38	8.0 ± 0	8.66 ± 0.19	5.66 ± 0.19	23.66 ± 0.19	7.66 ± 0.19	10.66 ± 0.19	7.66 ± 0.19
B8	13.66 ± 0.19	14.33 ± .38	12.66 ± 0.19	17.33 ± 0.19	16 ± 0	40 ± 0.33	11.33 ± 0.19	11.66 ± 0.19	15.33 ± 0.19
B9	12 ± 00	22.33 ± .19	10.33 ± 0.19	16 ± 0.19	12 ± 0	35.0 ± 0	11.66 ± 0.19	9.33 ± 0.19	13.33 ± 0.19
B10	8.66 ± 0.19	20.33 ± 0.19	11.66 ± 0.19	18.33 ± 0.19	20.66 ± 0.19	24.66 ± 0.19	15.66 ± 0.19	10.33 ± 0.19	12.33 ± 0.19
B11	10 ± 00	23.33 ± 0.19	10 ± 0	15.66 ± 0.19	18.66 ± 0.19	30 ± 0.00	13.33 ± 0.19	13.66 ± 0.19	16.66 ± 0.19
F1	10.33 ± 0.19	20.66 ± 0.38	11.66 ± 0.19	12.66 ± 0.19	13.66 ± 0.19	37.33 ± 0.19	12.66 ± 0.38	10.66 ± 0.19	8.33 ± 0.19
F2	11.33 ± 0.19	15.33 ± 0.19	13.66 ± 0.19	10.66 ± 0.19	14.66 ± 0.19	18.33 ± 0.19	11.66 ± 0.19	0 ± 0	11.33 ± 0.19
F3	0 ± 0	13.66 ± 0.19	12.33 ± 0.19	13 ± 0.33	12.33 ± 0.19	14.33 ± 0.38	13.33 ± 0.19	13.33 ± 0.19	7.33 ± 0.19
F4	12.33 ± 0.19	16.33 ± 0.19	8.33 ± 0.19	9.33 ± 0.19	11.33 ± 0.19	13.33 ± 0.19	8.33 ± 0.19	14.33 ± 0.19	9.33 ± 0.19
F5	9.33 ± 0.19	11.33 ± 0.19	0 ± 0	0 ± 0	0 ± 0	12.33 ± 0.19	9.33 ± 0.19	0 ± 0	8.66 ± 0.19
F6	10.33 ± 0.19	13.33 ± 0.19	0 ± 0	15.33 ± 0.19	12.66 ± 0.19	15.66 ± 0.19	6.66 ± 0.19	9.33 ± 0.19	8.33 ± 0.19

B1, K. pneumoniae; B2, S. infantis; B3, S. aureus; B4, L. lactis; B5, E. coli; B6, V. cholerae; B7, S. flexneri; B8, Pseudomonas sp.; B9, Proteus sp.; B10, C. Diserus; B11, Bacillus sp.; F1, A. niger; F2, A. flavus; F3, A. fumigatus; F4, Penicillium sp.; F5, Fusarium sp.; F6, C. magnoliae.

[1]Mean values are shown for zone of inhibition (mm) per 100 μL.

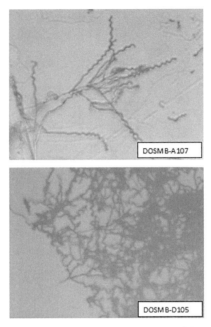

FIGURE 22.4 Aerial spore morphology of strains DOSMB-A107 and DOSMB-D105.

DOSMB-A107 did not produce any pigment on the culture media (Table 22.7). The Kuster's agar medium was better than other media tested, because it was found that *Streptomyces* could easily utilize all nutrients and sporulated in this medium. Previously, it has been reported that the culture growth on different media is one of the investigational tools for the identification of actinobacteria (Gesheva and Gesheva, 1993).

In the present study, the cell wall composition, such as cell wall amino acid and whole cell sugar, of the antagonistic *Streptomyces* spp. DOSMB-D105 and DOSMB-A107 was investigated. Both organisms showed the presence of LL-diaminopimelic acid and glycine and did not contain any diagnostic sugar in their cells. Thus, the present investigation reports that the isolates DOSMB-D105 and DOSMB-A107 have a type-I cell wall. The most useful diagnostic marker is that diaminopimelic acid occupies an anchor position in the tetra-peptide of the wall peptitoglycon, since many actinobacteria contain meso-di-aminopimelic acid. Of the various sugars occurring in cell hydrolysate of acti-nobacteria, four sugars have proved to be of taxonomic relevance: arabinose, galactose, madurose and xylose; members of the family Streptomycetaceae do not contain any of these sugars, and in most cases they are devoid of any other carbohydrates. The cell chemistry of the *Streptomyces* spp. has been reported by earlier researchers (Lechevalier and Lechevalier, 1970; Semedo *et al.*, 2001).

Various biochemical characterizations of the *Streptomyces* spp. were used for their identification (Manfio *et al.*, 2003). Biochemical and physiological characterization of the isolates DOSMB-A107 and DOSMB-D105 from the

TABLE 22.7 Cultural Characteristics of DOSMB-A107and DOSMB-D105 on Different Media

Medium	Morphological feature	DOSMB.A107	DOSMB.D105
ISP2	Aerial mycelium	Grey	Grey
	Substrate mycelium	Brown	Yellow
	Diffusible pigment	Nil	Nil
	Melanin pigment	Nil	Whitish grey
ISP3	Aerial mycelium	Whitish grey	Dark yellow
	Substrate mycelium	Black	Nil
	Diffusible pigment	Nil	Nil
	Melanin pigment	Nil	Whitish grey
ISP4	Aerial mycelium	Grey	Dark yellow
	Substrate mycelium	Brown	Yellow
	Diffusible pigment	Nil	Nil
	Melanin pigment	Nil	Grey
ISP5	Aerial mycelium	Grey	Yellow
	Substrate mycelium	Black	Nil
	Diffusible pigment	Nil	Nil
	Melanin pigment	Nil	Grey
ISP7	Aerial mycelium	Grey	Yellow
	Substrate mycelium	Brown	Nil
	Diffusible pigment	Nil	Positive
	Melanin pigment	Nil	Grey
KU	Aerial mycelium	Grey	yellow
	Substrate mycelium	Black	Yellow
	Diffusible pigment	Nil	Nil
	Melanin pigment	Nil	Grey
AIA	Aerial mycelium	Grey	Yellow
	Substrate mycelium	Brown	Yellow
	Diffusible pigment	Nil	Nil
	Melanin pigment	Nil	Nil

AIA, actinomycetes isolation agar; ISP2, yeast malt extract agar; ISP3, oat meal agar; ISP4, inorganic salt starch casein agar; ISP5, glycerol asparagine agar; ISP7, tyrosine agar; KU, Kuster's agar.

present study are given in Tables 22.8 and 22.9. Previously, it has been reported that nitrate reducing ability, production of H_2S, urease, catalase, oxidase and citrase of *Streptomyces* have also been useful characters for the identification of streptomycetes (Gotoh *et al.*, 1982; Shirling and Gottlieb, 1966). Toleration study of actinobacteria against some inhibitory compounds is the suitable method for examination of *Streptomyces* environmental conditions (Tresner *et al.*, 1968). The different species/strains vary in their response to different

TABLE 22.8 Biochemical Characteristics of Isolates DOSMB-A107and DOSMB-D105

Biochemical test	DOSMB-A107	DOSMB-D105
Citrate utilization	+	+
H$_2$S production	+	−
Nitrate utilization	+	−
Urease	+	+
Catalase	+	−
Oxidase	+	−
Starch hydrolysis	+	+
Gelatin hydrolysis	−	−
Lipid hydrolysis	−	−
Casein hydrolysis	+	+

+, positive; −, negative.

TABLE 22.9 Physiological Characteristics of Isolates DOSMB-A107 and DOSMB-D105

Sl. No.	Test	DOSMB-A107	DOSMB-D105
1.	**Temperature (°C)**		
a.	4	−	−
b.	15	−	−
c.	25	+	+
d.	28	+	+
e.	35	+	+
f.	42	+	+
g.	55	−	−
2.	**pH**		
a.	4	−	−
b.	6	+	+
c.	7	+	+
d.	8	+	+
e.	9	+	+
	10	−	−

TABLE 22.9 Physiological Characteristics of Isolates DOSMB-A107 and DOSMB-D105 *(cont.)*

Sl. No.	Test	DOSMB-A107	DOSMB-D105
3.	**Antibiotic sensitivity**		
a.	Cephalothin (30 mcg)	R	R
b.	Clindamycin (2 mcg)	R	18 mm
c.	Co-Trimoxazole (25 mcg)	R	R
d.	Erythromycin (1 5mcg)	12 mm	R
e.	Gentamycin (10 mcg)	22 mm	21 mm
f.	Oflaxin (1 mcg)	R	R
g.	Penicillin (10 unit)	R	R
h.	Vancomycin (mcg)	25 mm	R
i.	Amikain	34 mm	35 mm
4.	**Inhibitory compound (%w/v)**		
a.	Crystal violet (0.0001)	+	+
b.	Potassium tellurite (0.001)	−	−
c.	Sodium peruside (0.01)	+	+
d.	Sodium chloride		
	1%	−	−
	3%	+	+
	5%	+	+
	7%	+	+
	10%	+	−
5.	**Utilization of amino acids**		
a.	L-tyrosine	+	+
b.	D-tryptophan	+	+
c.	L-proline	−	+
d.	L-methionine	+	+
e.	L-lysine	+	+
f.	L-arginine	+	+

+, positive reaction; −, negative reaction; R, drug resistant.

inhibitory compounds, depending on their genetic makeup. *Streptomyces* spp. DOSMB-D105 and DOSMB-A107 are distinctly different from one another in their ability to tolerate inhibitory compounds. Antibiotic sensitivities of *Streptomyces* spp. have also been reported by previous researchers (Raytapudar and Paul, 2001). Utilization of various carbon sources confirms the identity of *Streptomyces* spp. (Chun *et al.*, 1997). In the present study both the isolates were able to utilize dextrose, cellulose, xylose, arabinose, raffinose, mannitol, fructose, sorbitol, maltose, lactose, rhamnose, adonital and starch. Further, neither organism utilized inosital or L-rhamose..

The present investigation revealed that Andaman Island has in its mangrove sediments numerous *Streptomyces* spp. capable of producing anti-infective agents. It convincingly demonstrates the potential to find novel antibiotics from these mangrove sediments.

ACKNOWLEDGEMENTS

The authors thank the University Grand Commission (UGC) for providing a Research Fellowship in Science to Meritorious Student (RFSMS) to R. Baskaran. The authors are also obliged to Prof. J.A.K. Tareen, Vice chancellor, Pondicherry University for providing necessary facilities at Port Blair Campus and to Central Instrumentation Facility Section (CIF), Pondicherry University campus for electron microscopy experiment.

REFERENCES

Baskaran, R., Vijayakumar, R., Mohan, P.M., 2011. Enrichment method for the isolation of bioactive actinomycetes from mangrove sediments of Andaman Island, India. Malaysian J. Microbiol 7: 22–28.

Becker, B., Lechevalier, M.P., Lechevalier, H.A., 1965. Chemical composition of cell wall preparations from strains of various form genera of aerobic actinomycetes. Appl. Microbiol. 13, 236–243.

Burkholder, P.R., Sun, S.H., Ehrlich, J., Anderson, L., 1954. Criteria of speciation in the genus *Streptomyces*. Ann. New York Acad. Sci. 60, 102–123.

Chun, J., Youn, H.D., Yim, H.I., Lee, H., Kim Yung Chil Hah, M.Y., Kang, S., 1997. *Streptomyces seoulensis* sp. nov. Int. J. Syst. Bacteriol. 47, 240–245.

Dhanasekaran, D., Panneerselvam, A., Thajuddin, N., 2005. Antifungal actinomycetes in marine soils of Tamil Nadu. Geobios. 32, 37–40.

Fukuda, D.S., Mynderse, J.S., Baker, P.J., Berry, D.M., Boeck, L.D., Yao, R.C., Mertz, F.P., Nakatsukasa, W.M., Mabe, J., Ott, J., Counter, F.T., Ensminger, P.W., Allen, N.E., Alborn, W., Hobbs, J.N., 1990. A new antibiotic complex produced by *Streptomyces aculeolatus*. Discovery, taxonomy, fermentation, isolation, characterization and antibacterial evaluation. J. Antibiot. 43, 623–633.

Flowers, T.H., Williams, S.T., 1977. The influence of pH on the growth rate and variability of neutrophilic and acidophilic streptomycetes. Microbes 18, 223–228.

Gesheva, V., Gesheva, R., 1993. Structure of the population of *Streptomyces hygrocopicus* and characteristics of its variants. Actinomycetes 4, 65–71.

Ghanem, N.B., Sabry, S.A., El-Sherif, Z.M., Abu El-Ela, G.A., 2000. Isolation and enumeration of marine actinomycetes from seawater and sediments in Alexandria. J. Gen. Appl. Microbiol. 46, 105–111.

Gotoh, T., Nakahara, K., Lwami, M., Aoki, H., Imanaka, H., 1982. Studies on a new immunoactive peptide, Fk-156 I. Taxonomy of the producing strains. J. Antibiot. 35, 1280–1285.

Goodfellow, M., Haynes, J.A., 1984. Actinomycetes in marine sediments. In: Oritz-Oritz, L., Bojail, C.F., Yakoleff, V. (Eds.), Biological, biochemical and biomedical aspects of actinomycetes. Academic press, New York & London, pp. 453–463.

Haung, W., Fang, J., Su, G., Liu, T., 1991. Marine actinomycetes from seashore of Fujina area and its antibiotic substances. Chinese J. Mar. Drugs 10, 1–6.

Jackson, M.L.(1973). Soil chemical analysis. Prentice Hall of India, New Delhi. p. 498.

Jensen, P.R., Dwight, R., Fenical, W., 1991. Distribution of actinomycetes in near shore tropical marine sediments. Appl. Environ. Microbiol. 57, 1102–1108.

Jiang, C.L., Xu, L.H., 1990. Characteristics of the populations of soil actinomycetes in Yunnan. Actinomycetes 1, 67–74.

Kampfer, P., 2006. The Family Streptomycetaceae, Part I: Taxonomy. In: Dworkin, M. et al., (Ed.), The prokaryotes: a handbook on the biology of bacteria. Springer, Berlin, pp. 538–604.

Kieser, T., Bibb, M.J., Buttner, M.J., Chapter, K.F., Hopwood, D.A., 2000. Practical Streptomyces genetics, 2nd ed. John Innes Foundation, Norwich, England, ISBN 0-7084-0623-8.

Kim, B.S., Sahin, N., Minnikin, D.E., Screwinska, J.Z., Mordarski, M., Goodfellow, M., 1999. Classification of thermophilic streptomycetes, including the description of *Streptomyces thermoalcalitolerans* sp. nov. Int. J. Syst Bacteriol. 49, 7–17.

Kokare, C.R., Mahadik, S.S., Kadam, Chopade, B.A., 2004. Isolation, characterization and antimicrobial activity of marine halophilic *Actinopolypora* species AH 1 from the west coast of India. Curr. Sci. 86, 593–597.

Kuster, E., 1963. Morphological and physicological aspects of the taxonomy of streptomycetes. Microbiol. Espanola 16, 193–202.

Lechevalier, M.P., Lechevalier, H.A., 1970. Chemical composition as a criterion in the classification of aerobic actinomycetes. Int. J. Syst. Bacteriol. 20, 435–443.

Manfio, G.P., Atalan, E., Zakrzewska, C.J., Mordarski, M., Rodriguez, C., Collins, M.D., Goodfellow, M., 2003. Classification of novel soil streptomycetes as *Streptomyces aureus* sp. nov., *S. laceyi* sp. nov. and *S. sangelieri* sp. nov. Antonie Van Leeuwenhoek 83, 245–255.

Moncheva, P., Tishkov, S., Dimitrova, N., Chipeva, V., Nikolova, S.A., Bogatzevska, N., 2002. Characteristics of soil actinomycetes from Antarctica. J. Cul. Collec. 3, 3–14.

Myertons, J.L., Labeda, D.P., Cote, G.L., Lechevalier, M.P., 1988. A thin layer chromatographic method for whole cell sugar analysis of *Micromonospora* species. Actinomycetes 20, 182–192.

Ndonde, M.J.M., Semu, E., 2000. Preliminary characterization of some *Streptomyces* species from four Tanzanian soils and their antimicrobial compounds. Asian J. Microbial. Biotech. Environ. Sci. 7, 121–124.

Ohnishi, Y., Ishikawa, J., Hara, H., Suzuki, H., Ikenoya, M., Ikeda, H., Yamashita, A., Hattori, M., Horinouchi, S., 2008. Genome sequence of the streptomycin-producing microorganism *Streptomyces griseas* IFO 13350. J. Bacteriol., 4050–4060.

Ouhdouch, Y., Jana, M., Imziln, B., Boissaid, A., Finance, C., 1996. Antifungal activities of actinomycetes isolated from Moroccan habitats. Actinomycetes 7, 12–22.

Pisano, M.A., Sommer, M.J., Taras, L., 1992. Bioactivity of chitinolytic actinomycetes of marine origin. Appl. Microbiol. Biotechnol. 36, 553–555.

Pridham, T.G., Hesseltine, C.W., Benedict, R.G., 1958. A guide for the classification of streptomycetes according to selected groups. Appl. Microbiol. 6, 52–79.

Pridham, T.G., Tresner, H.D., 1974. Streptomycetaceae, 8th Edition In Bergey's Manual of Determinative BacteriologyWilliams and Wilkins, Baltimore, p. 747.

Pugazhvendan, S.R., Kumaran, S., Alagappan, K.M., Prasad, G., 2010. Inhibition of fish bacteriology pathogens by antagonistic marine actinomycetes. Eur. J. Appl. Sci. 2, 41–43.

Raytapudar, S., Paul, A.K., 2001. Production of an antifungal antibiotic by *Streptomyces aburavi-ensis* 1DA-28. Microbiol. Res. 155, 315–323.

Semedo, L.T.A.S., Linhares, A.A., Gomes, R.C., Manfio, G.P., Alviano, C.S., Linhares, L.F., Coel-ho, R.R.R., 2001. Isolation and characterization of actinomycetes from Brazilian tropical soils. Microbiol. Res. 155, 291–299.

Seong, C.H., Choi, J.H., Baik, K.S., 2001. An improved selective isolation of rare actinomycetes from forest soil. J. Microbiol. 17, 23–39.

Shirling, E.B., Gottlieb, D., 1966. Methods for characterization of *Streptomyces* species. Int. J. Syst. Bacteriol. 16, 312–340.

Shimizu, M., Nakagawa, Y., Sato, Y., Furumai, T., Igarashi, Y., Onaka, H., Yoshida, R., Kunch, H., 2000. Studies on endophytic actinomycetes (1) *Streptomyces* sp. isolated from *Rhododendron* and its antimicrobial activity. J. Gen. Pl. Pathol. 66, 360–366.

Sivakumar, K.(2001). Actinomycetes of an Indian Mangrove (Pichavaram) environment: An Inven-tory. PhD thesis, Annamalai University, India, p.91.

Subramanian, V., Angleja, B.F., 1976. Methodology of X-ray fluorescence analysis of suspended in estuarine waters: *Mar.* Geol. 22, M1–M6.

Tresner, H.D., Hayes, J.A., Bakus, E.J., 1968. Differential tolerance of Streptomycetes to sodium chloride as a taxonomic aid. Appl. Microbiol. 16, 1134–1136.

Vijayakumar, R., Muthukumar, C., Thajudin, N., Panneerselvam, A., Saravanamuthu, R., 2007. Studies on the diversity of actinomycetes in the Palk Strait region of Bay of Bengal, India. Actinomycetologica 21, 59–65.

Vijayakumar, R., Pannerselvam, K., Muthukumar, C., Thajuddin, N., Panneerselvam, A., Saravan-muthu, R., 2011. Antimicrobial potentiality of a halophilic strain of *Streptomyces* sp. VPTSA 18 isolated from the saltpan environment of Vedaranyam, India. Ann Microbiolol. 11, 345–501.

Waksman, S.A.(1961) The actinomycetes: classification, identification and descriptions of genera and species, Vol. II, Williams and Wilkins, Baltimore. p. 363.

Yon, C., Suh, J.W., Chang, J.H., Lim, Y., Lee, C.H., Lee, Y.S., Lee, Y.W., 1995. Al702, a novel anti-legionella antibiotic produced by *Streptomyces* sp. J. Antibiot. 48, 773–779.

Chapter 23

Impact of Anthropogenic Activity and Natural Calamities on Fringing Reef of North Bay, South Andaman

R. Raghuraman and C. Raghunathan
Andaman and Nicobar Regional Centre, Zoological Survey of India, Port Blair, Andaman and Nicobar Islands, India

INTRODUCTION

The Andaman and Nicobar Islands (ANI), situated in the Bay of Bengal within 6° to 14° N latitude and 92° to 94° E longitude, have remarkable marine biodiversity and more than 30 percent endemicity (Savant, 2009). The coral reef areas of the islands are observed to be the most diverse and extensive in the Indian Ocean. There are about 572 islands in the Andaman and Nicobar group, which includes six National Parks, 94 Sanctuaries and one biosphere.

In India, reefs cover approximately 5790 km^2 and are divided into three major zones: the ANI; the coral reefs of mainland India, distributed along the east and west coasts at restricted places; and the Lakshadweep Islands in the Arabian Sea. All the major reef types are represented in India (Venkataraman *et al.*, 2003). Rink (1847) initiated the study of coral reefs in Indian waters and provided a brief account of coral reef in the Nicobar Islands. Later, Lt. Col. R.B.S. Sewell conducted studies on corals of India (Sewell, 1922, 1925). Taxonomic studies of Indian corals were restricted to the pioneering works of Pillai (1971a,b, 1972), Pillai and Patel (1988), Pillai and Jasmine (1989) during the late 20th century. The Zoological Survey of India carried out studies on coral reefs during the last few decades (Reddiah, 1970a,b, 1977; Venkataraman and Rajan, 1998; Jeyabaskaran, 1999; Turner *et al.* 2001, Venkataraman, *et al.*, 2003). Venkataraman *et al.* (2003) added 42 species to the list of coral of the ANI and 13 to the Lakshadweep Islands. Raghuram and Venkataraman (2005) added two more species from the Gulf of Mannar and Andaman waters. Recently, Raghuraman *et al.* (2012) compiled a consolidated report on corals of the ANI.

Marine Faunal Diversity in India. DOI: 10.1016/B978-0-12-801948-1.00023-9
421

Two anthropogenic factors contributing to coral reef decline are eutrophication (Koop *et al.,* 2001) and damage from tourist activities such as snorkeling and SCUBA diving (Chadwick-Furman, 2002). Tourism plays a major role in uplifting the economy of the ANI. Coral reefs are one of the main attractions for tourists in bay islands. In addition, tourism is potentially one of the most benign and profitable uses of coral reefs, but this beneficence can be degraded by overuse and damage to corals (Woodland and Hooper, 1977). Sedimentation, which may be increased by anthropogenic activities also is known to affect coral community structure and can damage coral colonies (Roger, 1990). The coral reef at North Bay is subjected to anthropogenic stress from tourist overuse, which has depleted the biodiversity of the shallow benthic community (Raghuraman, 2008). In this study, variations in coral live cover and distribution pattern at three stations that are exposed to unregulated tourism activities were examined.

METHODS

Study Area

This study was carried out from August 2009 to March 2012 at North Bay, South Andaman (Table 23.1 and Figure 23.1). The area of the bay is about 1 km². Sea water enters through the open sea on the northeast of the bay. Corals are fringed in the west bank. The northwest side is overlooked by hills (Mount Harriet) which have a lot of water flow during the monsoon, and the other three sides are open water. North Bay is one of the main tourist sites (Krishnan *et al.,* 2011) in the ANI and is open to tourist visits on all days.

Survey Method

Three different stations were selected and were categorized on the basis of tourist accessibility. Station 1 was characterized as an area of low disturbance, located about 200 m from a tourist landing. Station 2 was characterized as a tourist site, where tourists do snorkeling, swimming and sea walking. Station 3, located 300 m from station 2, was characterized as not easily accessible to tourists, because of high wave action and deep water, which prevents snorkeling activities.

TABLE 23.1 Coordinates of the Study Sites

Stations	Coordinates	Characteristics
1	11° 42′18.1″ N; 92° 45′07.2″ E	Less disturbed area
2	11° 39′48.5″ N; 92° 45′12.5″ E	Tourist accessible area
3	11° 42′14.9″ N; 92° 45′06.8″ E	Tourist inaccessible area

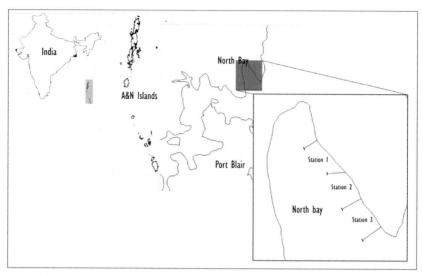

FIGURE 23.1   Study area: North Bay, South Andaman, Andaman and Nicobar Islands.

During the survey, the line intercept transect (LIT) method used to measure the live cover percentage of reef area (English *et al.*, 1997). Three 20 m transects were laid at three different depths at each station. Instead of life forms, in the LITs species were identified *in situ* to illustrate the distribution pattern.

Species diversity was assessed by extensive video and photography of corals using underwater digital cameras. The images were then analyzed for taxonomic identification. Specimens were identified by using standard manuals (Venkataraman *et al.*, 2003; Veron 2000). A total of 36 transects at three stations were surveyed to record the occurrence and density distribution of various marine fauna such as hard corals (= live coral), soft coral, dead coral, sponges, gastropods, bivalves, sea cucumbers, sea stars, sea urchins, crabs, lobsters and sea anemones.

RESULTS AND DISCUSSION

Benthic percent cover among the three stations for 2009 and 2012 was compared and analyzed. Significant reduction in percent cover of coral communities was observed over the period (Table 23.2 and Figure 23.2). Some coral species commonly recorded on the North Bay reef are shown in Figure 23.3.

Reef Status

Station 1

During the initial period of the study (2009), Station 1 was used for practising mariculture. However, coral bleaching occurred during 2010 (Krishnan

TABLE 23.2 Percent Live Cover at Each Station in 2009 and 2012

Station	1		2		3	
	2009	*2012*	*2009*	*2012*	*2009*	*2012*
Live coral	54.9	38.1	34.7	20.3	44.7	34.6
Soft coral	0	1	0	5.35	0	0
Sponge	0	18.5	0.75	5.4	0.48	7.1
Zoanthids	0	0	0	0	0.15	2.8
Others	0	1.25	2.4	1	6.93	0
Dead coral	44.6	33.5	2.38	20	0	15.5
Rubble	0	0	56.7	14.2	0	0
Sand	0.5	7.7	3.13	33.8	47.7	40
Total	100	100	100	100	100	100

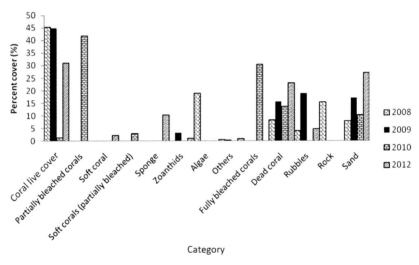

FIGURE 23.2   Percent live cover of North Bay reef, by year 2008 to 2012. 2008 and 2010 data are from Raghuraman (2008) and Krishnan *et al.* (2011), respectively.

et al., 2011), hampering activities, and mariculture was subsequently stopped completely during the latter half of the year, which made Station 1 an undisturbed area. The effect of lack of disturbance was apparent during the study, as decrease in live coral cover was lower here than in the tourist accessible area. Live coral cover at Station 1 decreased during the period of study by close to 17 percent. During a previous survey, the reef was chiefly covered by boulder and dead corals which were covered with turf algae. But in the last survey of the present study, about 19 percent of the reef bottom was found to be covered

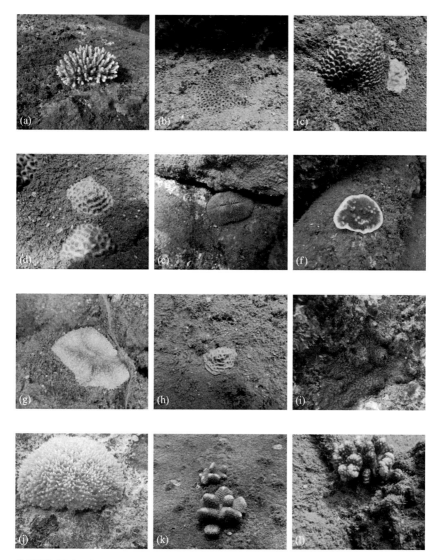

FIGURE 23.3 Coral species commonly recorded on the North Bay reef. (a) *Acropora cerealis* (Dana, 1846); (b) *Favia matthaii* (Vaughan, 1918); (c) *Favia truncatus* (Veron, 2002); (d) *Favites spinosa* (Klunzinger, 1879); (e) *Fungia paumotensis* (Stuchbury, 1833); (f) *Leptoseris striata* (Fenner and Veron, 2002); (g) *Lobophyllia corymbosa* (Forskal, 1775); (h) *Pavona bipartita* (Nemenzo, 1980); (i) *Pavona varians* (Verrill, 1864); (j) *Physogyra lichtensteini* (Milne Edwards and Haime, 1851); (k) *Pocillopora eydouxi* (Milne Edwards and Haime, 1860); (l) *Pocillopora verrucosa* (Ellis and Solander, 1786). (Please see color plate at the back of the book.)

by sponge, soft coral and reef associated fauna (sea cucumber, sea star and gastropods). Significant increase in percent live cover of these reef detrimental communities—sponge and soft coral—is a sign of reef destruction. Smothering of dead corals by sand prevents coral settlement.

Station 2

The decrease in live coral cover of up to 14 percent at Station 2 is probably due to easy access by tourists. The reef flat extends up to 70 m seaward at this station. Maximum coral damage was observed, due to coral walking, swimming, snorkeling and souvenir collection. Only boulder corals dominated the live coral cover at Station 2. Subsequently, sand increased by about 30 per cent, and dead coral cover also increased considerably, by 18 percent. Reef walking and high sediment load prevents the growth and settlement of coral. Abrupt decreasing of rubble indicates souvenir collection of tourists; many pieces of coral rubble can be seen on the beach and at the tourist rest area (Jeyabaskaran *et al.*, 2007). In addition, the newly inaugurated sea walking bridge also destroys the reef considerably. During the survey, no new coral recruits or branching corals were noticed. The physical damage to the reef is high, as a result of factors related to shipping activities, such as increased oil concentration in the water and boat anchoring (Muthiga and McClanahan, 1997). Below 8 m there was no coral reef recorded during the 2012 survey; the entire reef was smothered by high sediment from the reef flat region. Only few Pectinidae corals were recorded at the reef edge.

Station 3

Station 3, though an undisturbed area, has shown a decrease in live coral cover of about 10 percent from 2009 to 2012, which could be attributable to the 2010 bleaching. There was a substantial increase in sponge cover from 0.15 percent in 2009 to 7.10 percent in 2012. Though the live cover at Station 3 is less than at Station 1, the degree of depletion of live coral cover is significantly higher at the latter. Earlier the coral at this station had been found to consist mainly of *Echinopora lamellosa* (Raghuraman, 2008); but after the 2010 bleaching all these plate corals were dead, and the main types were boulder *Porites* and *Acropora*. Percent cover of sponges has increased significantly, which may lead to destruction of the small encrusting and boulder corals.

Diversity and Abundance

Diversity and abundance of corals were recorded from all three study sites. The major diversity indices Shannon-Weiner (H'), Species Richness (S), Total Abundance, Simpson Index and Pielou's Evenness are given in Table 23.3. Shannon-Weiner (H') index showed an increasing trend from 2009 to 2012 at all three stations. Station 1 showed highest diversity index in 2012 (2.43), and

TABLE 23.3 Diversity Indices at Each Station in 2009 and 2012

Station	1		2		3	
	2009	*2012*	*2009*	*2012*	*2009*	*2012*
Shannon-Wiener Diversity Index, H′	1.61	2.44	1.09	1.97	1.54	1.64
Species Richness, S	13	14	10	11	9	7
Total Abundance	59	25	51	24	40	15
Simpson's Index, D	0.35	0.11	0.57	0.22	0.30	0.24
1 − D	0.65	0.89	0.43	0.78	0.70	0.76
1/D	2.89	9.06	1.77	4.65	3.29	4.09
Evenness	0.63	0.92	0.47	0.82	0.70	0.84

the lowest was found at Station 2 in 2009 (1.08). However, species richness showed a decreasing trend at Station 3, from 9 to 7, in the vicinity of the tourist accessible area. The abrupt decrease in Total Abundance (from 40 in 2009 to 15 in 2012) at Station 3 may have been due to the impact of tourism. The other two stations, however, showed a gradual decrease. The Pielou's Evenness index suggested an uneven distribution of species throughout the study period. Diversity studies revealed the extent of change in coral cover and community structure over the years.

The gradual change in benthic cover of North Bay reef is depicted in Figure 23.2. Four years of data have been compared. The 2008 and 2010 data are from Raghuraman (2008) and Krishnan *et al.* (2011), respectively. All the indices of reef health presented here—live coral cover and diversity indices—indicate a sick reef. This situation demands careful consideration from policy makers and planners. Global warming, coral bleaching and overfishing are all responsible for changing reef biodiversity and reducing the quality of reefs over large areas; to protect the biodiversity, we must understand the processes that maintain diversity at this scale.

The results of this study may be helpful in encouraging the imposition of limits on tourism, in order to protect the remaining corals at North Bay. A major goal of management strategies is the protection of habitat over large scale areas. The results of the present study also point to a need to shift our focus from individual taxa to broader habitat-based management strategies (Sandhya *et al.*, 2008), and they highlight the need for national management of reef resources.

REFERENCES

Chadwick-Furman, N.E., 2002. Effects of SCUBA diving on coral reef invertebrates in the US Virgin Islands: Implication for the management of diving tourism. In: Proceedings of the 6th International Conference on Coelenterate biology, pp. 91–100.

English, S., Wilkinson, C., Baker, V., 1997. Survey Manual for Tropical Marine Resources. Australian Institute of Marine Science, Townsville, p. 390.

Jeyabaskaran, R., 1999. Report on Rapid assessment of coral reefs of Andaman & Nicobar Islands. GOI/UNDP/GEF Project on Management of Coral Reef Ecosystem of Andaman & Nicobar Islands. Zoological Survey of India, Port Blair, p. 110.

Jeyabaskaran, R., Venkataraman, K., Alfred, J.R.B., 2007. Implications for Conservation of Coral Reefs in the Andaman and Nicobar Islands, India. In: The Third International Tropical Marine Ecosystem Management Symposium (ITMEMS3), October 16-20, 2006, Cozumel, Mexico.

Koop, K., Booth, D., Broadbent, A., Brodie, J., Bucher, D., Capone, D., Coll, J., Dennison, W., Erdmann, M., Harrison, P., Hoegh-Guldberg, O., Hutchings, P., Jones, G.B., Larkum, A.W.D., O'Neil, J., Steven, A., Tentori, E., Ward, S., Williamson, J., Yellowlees, D., 2001. ENCORE: The effect of nutrient enrichment on coral reefs, synthesis of results of conclusions. Mar. Poll. Bull. 42, 91–120.

Krishnan, P., Dam Roy, S., Grinson George, Srivastava, R.C., Anand, A., Murugesan, S., Kaliyamoorthy, M., Vikas, N., Soundararajan, R., 2011. Elevated sea surface temperature during May 2010 induces mass bleaching of corals in the Andaman. Curr. Sci. 100 (1), 111–117.

Muthiga, N.A., McClanahan, T.R., 1997. The effect of visitors use on the hard coral communities of the Kisite Marine Park, Kenya. Proceedings of the 8th International Coral Reef Symposium, Panama City, 24-29 June 1996. Vol 2. Smithsonian Tropical Research Institute, Panama, pp. 1879–1882.

Pillai, C.S.G., 1971a. Composition of the coral fauna of the southeast coast India and the Laccadives. Symp. Zool. Soc. London 28, 301–327.

Pillai, C.S.G., 1971b. The distribution of shallow water stony corals at Minicoy Atoll in the Indian Ocean with a checklist of species. Atoll. Res. Bull. 141, 21–33.

Pillai, C.S.G., 1972. Stony corals of the seas around India. In: Mukundan, C., Gopinadha Pillai, C.S. (Eds.), Proceedings of the First International Symposium on Corals and Coral Reefs. Marine Biological Association of India, Mandapam Camp, India, pp. 191–216.

Pillai, C.S.G., Patel, M.I., 1988. Scleractinian corals from the Gulf of Kachchh. J. Mar. Biol. Assoc. India 30 (1&2), 54–74.

Pillai, C.S.G., Jasmine, S., 1989. The coral fauna of Lakshadweep. Bull. Central Mar. Fish. Res. Inst. 43, 179–199.

Raghuraman, R., 2008. Distribution of Corals in North Bay, Andamans. Unpublished M.Sc., Thesis, Pondicherry University.

Raghuram, Venkataraman, K., 2005. New records of *Porites annae* Crossland and *Porites cylindrical* Dana from Gulf of Mannar and Andaman Waters. Rec. Zool. Surv. India 105 (Part1–2), 133–138.

Raghuraman, R., Sreeraj, C.R., Titus Immanuel, Raghunathan, C., 2010. Intensive study on the Scleractinian coral diversity of Pongibalu, South Andaman. J. Environ. & Sociobiol. 7 (1), 29–36.

Raghuraman, R., Sreeraj, C.R., Raghunathan, C., Venkataraman, K., 2012. Scleractinian Coral diversity in Andaman and Nicobar Islands in Comparison with other Indian Reefs. International day for Biological Diversity Marine Biodiversity. Uttar Pradesh State Biodiversity Board, pp. 75–92.

Reddiah, K., 1970a. The topography of Appa Island and its fringing reef in the Gulf of Mannar. Indian Science Congress Association Proceedings. 57 (4), 405–406.

Reddiah, K., 1970b. The formation of secondary rock on a reef flat and its effect on reef organisms. In: Proceedings of the International Symposium on the Biology of the Sipuncula and Echiura, Kotor, Montenegro, 18-25 June 1970. Vol 1. Institute for Biological Research "Siniša Stankovic'", Belgrade, pp. 18–25.

Reddiah, K., 1977. The coral reefs of Andaman and Nicobar Islands. Rec. Zool. Sur. India 72, 315–324.

Rink, H.J., 1847. Die Nikobarischen Inseln. Eine geographische Skizze, mit spezieller Berücksichtigung der Geographie. H.G. Klein, Copenhagen.

Rogers, C.S., 1990. Responses of coral reefs and reef organisms to sedimentation. Mar. Eco. Prog. Ser. 62, 185–202.

Sandhya, S., Rani, M.G., Kasinathan, C., 2008. Biodiversity Assessment of Fringing Reef in Palk Bay. India. Fishery Tech. 45 (2), 163–170.

Savant, P.V., 2009. Biodiversity conservation through ecotourism: A new approach. In: 125 years of forestry (1883-2008) Andaman and Nicobar Islands. Department of Environment and Forests, Andaman and Nicobar Administration, p. 3.

Sewell, R.B.S., 1922. A survey season in the Nicobar Islands on the R.I.M.S. "Investigator" October 1921 to March 1922. J. Bombay Nat. Hist. Soc. 28, 970–989.

Sewell, R.B.S., 1925. The geography of the Andaman Sea Basin. Mem. Siat. Soc. Bengal 9 (10), 1–26.

Turner J.R., Vousden, D., Klaus, R., Satyanarayana, Ch., Fenner, D., Venkataraman, K., Rajan, P.T., Subba Rao N.V., 2001. GOI/UNDP GEF Coral reef Ecosystems of the Andaman Islands.

Venkataraman, K., Rajan. P.T., 1998 Coral Reefs of Mahatma Gandhi Marine National Park and crown of thorn phenomenon. In: Gangwar, B. Chandra, K. (9Eds.), Symp. Proc. Islands Ecosystem & Sustainable Development. Andaman Scientific Association and Department of Science and Technology, Andaman and Nicobar Administration, Port Blair, pp. 124 –132.

Venkataraman, K., Satyanarayana, Ch., Alfred, J.R.B., Wolstenholme, 2003. Handbook on Hard Corals of India. Zoological Survey of India, Kolkata.

Veron, J.E.N., 2000. Corals of the World. In: Stafford-Smith, M. (Ed.), Australian Institute of Marine Science, Townsville. Vol 1, p. 463; Vol 2, p. 429; Vol 3, p. 490.

Woodland, D.J., Hooper, J.N.A., 1977. The effects of human trampling on coral reefs. Biol Conserv. 11, 1–4.

Chapter 24

Lucrative Business Opportunities with Shrimp Brood Stocks

V.S. Gowri and P. Nammalwar
Institute for Ocean Management, Anna University, Chennai, Tamil Nadu, India

INTRODUCTION

Many species of shrimp inhabit brackish and marine waters of the world. Most of them are unsuitable for human consumption. The shrimps, whether caught by fishermen or farm raised, are "Penaeids" and belong to the Penaeidae family of Decapod crustaceans. Shrimp farms generally can produce two to three crops per year. Shrimp farming is a lucrative business, and export-oriented shrimp aquaculture has been promoted by aid agencies, financial organizations and governments. In 2000 the leading shrimp-producing countries were Thailand, China, Indonesia, India, Vietnam, Ecuador, the Philippines, Bangladesh, Mexico and Brazil.

Indian farmed shrimp production increased from about 30,000 tons in 1990 to around 115,000 tons during 2002–03. However, in India, the average shrimp productivity was found to be 1.017 tons/ha/yr (FAO, 2007). The demand for shrimp post larvae (PL) increased subsequently. The successful farming of tiger shrimp (*Penaeus monodon*) in India relies mainly on some three hundred hatcheries whose capacity to produce 12,000 million PL annually has provided an assured supply of seed (FAO, 2007).

The area under shrimp aquaculture was about 154,600 ha out of 1,150,000 ha of land available for shrimp culture, which indicates that India has significant potential for aquaculture development (Table 24.1 and Figure 24.1).

However, in India, wild-caught brood stock (gravid females) is preferred over farm-raised brood stock as the source of shrimp seed (Figure 24.2). The sustainable supply of wild-caught brood stock is mainly hampered by limited stocks, seasonal availability and pathogen infection. As the Indian Government plans to double shrimp production, there will be an increasing demand for shrimp brood stocks from the wild by shrimp farmers in the near future.

Marine Faunal Diversity in India. DOI: 10.1016/B978-0-12-801948-1.00024-0

TABLE 24.1 Shrimp Productivity in India

Coastal state	Aquaculture area available (ha)	Area under culture (ha)	Productivity (ton/ha/yr)
West Bengal	405,000	49,925	0.60
Gujarat	376,000	1013	1.49
Andhra Pradesh	150,000	69,640	0.76
Maharashtra	80,000	615	1.60
Kerala	65,000	14,029	0.46
Tamil Nadu and Puducherry	56,800	3214	1.91
Orissa	31,600	12,116	1.02
Goa	18,500	963	0.73
Karnataka	8000	3085	0.59
Total	1,150,000	154,600	1.017 (avg)

SHRIMP FARMING IN INDIA

Life of Shrimps

Penaeid shrimps are short-lived animals, having a life span of about 1–2 years. The larvae and post larvae migrate to nursery grounds in estuaries and other wetlands. During the juvenile stage, they migrate to offshore waters and attain sexual maturity. Maturity stages of females are classified into five categories: immature, early maturing, late maturing, mature and spent. The ovaries of the shrimps pass through a

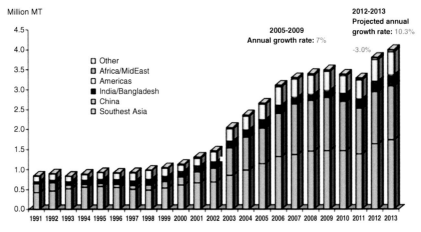

FIGURE 24.1   Shrimp aquaculture production by world region and year. *Source: Valderrama and Anderson (2011).*

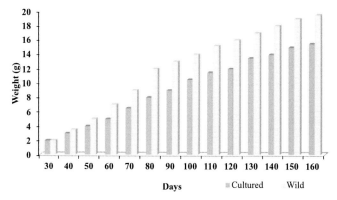

FIGURE 24.2   Growth rate of cultured and wild stock of *Penaeus indicus*. *Source: Emadi et al. (2006).*

series of color changes during the maturation process. Females with green ovaries (green yellow, green white, light green and dark green) are considered as spawning females, while those with other than green ovaries (translucent, white, cream and yellow) are considered as non-spawning (Ayub and Ahmed, 2002). Adult females mature within 6 months and have high fecundity (Esmaeili and Omar, 2003).

Shrimp Spawners

One of the most important aspects of aquaculture and management is the availability of spawners. A double-peaked pattern of spawning among penaeid shrimps is more common (Garcia, 1985). Most shrimps cultured worldwide are either collected from the wild or are offspring of wild-caught brood stock (Arce *et al.*, 2000). The shrimp trawl generally has a cod end mesh size as small as 8–10mm.

Though the production of farm-raised brood stocks is more economic and more reliable, they must be cultured in conditions close to those of wild brood stocks (Emadi *et al.*, 2006). The major incentives for aquaculturists to hunt for wild spawners are natural feeding, lower stress, better growth and higher genetic diversity. However, the current interest lies in formulating harvest policies to ensure that the spawner populations are not overfished.

Brood Stock Landing Centres, Collection Centres and Holding Techniques

The major brood stock landing centres and collection centres in Tamil Nadu, Andhra Pradesh, Orissa and Andaman Islands are shown in Table 24.2.

The All India Shrimp Hatcheries Association, Visakhapatnam, in association with the Department of Fisheries, the Marine Products Export Development Authority (MPEDA) and boat owners, has established the country's first brood stock collection centre at Visakhapatnam fishing harbor for the hygienic handling and maintenance of brood stocks. To minimize stress, trawler operators at

TABLE 24.2 Major Brood Stock Landing Centres

State	Location
Tamil Nadu	Palayar, Nagapattinam, Rameswaram
	Chennai[1], Karaikal[1], Thiruchendur[1]
Andhra Pradesh	Kakinada, Bhiravapalem, Machilipatnam, Nizampatnam, Krishnapatnam, Visakhapatnam[1]
Orissa	Gopalpur, Puri, Paradip
Andaman Islands	

[1]Brood stock collection centre for hygienic handling and maintenance.

the landing centres have been trained in the importance of hygienic and careful landing of brood stocks during catching, holding and transportation to the landing centres. The brood stock is held and auctioned individually in oxygenated tanks or bags chilled to < 29°C with ice to maintain bio-security. Brood stock should not be kept in overcrowded tanks for prolonged periods prior to transport. The use of high quality feeds enriched with vitamin C and astaxanthin (or paprika) and an acceptable probiotic formula help to reduce stress and bacterial levels. Moreover, the shrimps should not be fed for 12 h, as any feces produced during transport will lead to poor quality and possible infection of healthy brood stocks. High quality seawater should be available for the brooders in packing.

High quality brood stocks are reported to be available in the Andaman Islands, because of the pristine nature of the sea. At present, The Indian Government has recently permitted only three operators to gain licences, with the restriction that they may only use 500 brood stocks per operator per year.

Selection of Spawners and Transport

Spawners are in great demand for local hatcheries. The demand for brooders of *Penaeus monodon* by hatcheries has resulted in a vigorous trade. More than three hundred brooders per day are brought for trading in the peak season. At Visakhapatnam, the total length of male brooders ranged from 190 to 246 mm, and for females it varied between 210 and 330 mm. The price for male *Penaeus monodon* is only in the range Rs.50 to Rs.150/-. The female "empties" (with developing ovaries) are generally sold at between Rs.400 and Rs.800/-. The maximum price reported for a spawner was Rs.56,000/- in 2001–02 and Rs.32,500/- in 2002–03. The decline in the price could be due to the prevalence of disease and a crash of culture operations. However, on average, the price of gravid brooder is about Rs.30,000/- (Paul *et al.*, 2004).

A premium price is obtained for high quality disease-free brood stock. Supplying brood stocks has become a big business, and the middlemen have become highly influential. However, a problem in the selection of suitable spawners is poor knowledge of the maturation process. Some of the key considerations in gross examination include: lack of red coloration; rejection of weak or mori-bund animals; clear gill coloration; absence of gill fouling; absence of black spots (necrosis) on the thelycum; absence of white spots; and knowledge about the different maturity stages of ovaries.

Transportation time for live shrimp brood stocks should be minimized by plan-ning and confirming well in advance, and the transport should be done during the cooler night times. Only the hard shelled should be transported, as the animals that moult during transport will die or may be killed by other shrimps. Rubber tubes should be placed over the rostra of the shrimp to avoid puncturing of plastic bags. Shrimps should be packed in individual plastic bags if possible or at low density (500 g of shrimp per 10 litres of seawater). The transport bags should be filled one-third full with the cleanest seawater (already chilled to the desired temperature), and subject to UV or ozone disinfection. About 0.5°C/h is enough to reduce the physical and metabolic activity of the shrimps. Low temperature should be main-tained by enclosing the bags in polystyrene boxes, and exposure to sunlight should be prevented at all times. Dissolved oxygen should be maintained at > 5 ppm by filling the bags two-thirds full with pure oxygen and refilling during shipping if transport time exceeds 24 hours. A few grams of activated charcoal (1 g/L) should be used in each transportation bag to reduce the build-up of ammonia and nitrite in the bags. EDTA at 10 mg/L can be used to chelate heavy metals and inhibit bacte-rial growth, while Tris HCl buffer can be added (10 mg/L) to stabilize the pH of the water. The bumping or dropping of the boxes should be avoided (Paul *et al.* 2004).

Role of Rajiv Gandhi Centre for Aquaculture (RGCA), Chennai in disease-free brood stocks

The shrimp brood stock collection centres of RGCA, Chennai are shown in Figure 24.3. The selected wild brood stocks, collected from the trawlers are packed in clean seawater (1–2 pieces per Styrofoam box with ice cubes). They are transported within 2–8 hours to the hatchery by road and are quarantined in the hatchery. Strict quarantine and screening of brood stock can facilitate production of good quality seeds free of Monodon Baculo Virus (MBV) and White Spot Syndrome Virus (WSSV). The presence of WSSV and MBV in brood stock has been reported in earlier studies (Itami *et al.*, 1998; Uma *et al.*, 2005; Sethi *et al.*, 2011).

Staining of feces with malachite green to detect MBV, and PCR analysis for all major viral pathogens prior to transport is considered. Shrimp that test posi-tive are discarded. Handling of shrimps during collection, holding and packing is reduced to a minimum.

Commercially Important Shrimp Species

Giant Tiger Shrimp (Penaeus monodon)

This species accounts for about a quarter of the farmed shrimp coming out of Asia. The "tigers" are the largest (maximum length 363 mm) and fastest-growing of the farmed shrimp. They tolerate a wide range of salinities, but shortages of

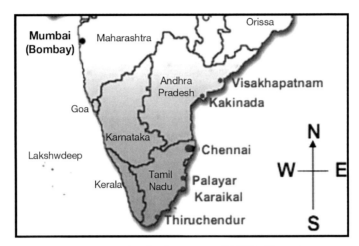

FIGURE 24.3   Shrimp brood stock collection centres of RGCA, Chennai.

wild brood stocks often exist, captive breeding is difficult, and hatchery survivals are low (20–30%). Tigers are very susceptible to two of the most lethal shrimp viruses: yellow head and white spot.

Indian White Shrimp (Penaeus indicus)

This species likes high salinities, high temperatures and high densities and it is readily available from the wild. It tolerates low water quality better than *Penaeus monodon*.

Banana Shrimp (Penaeus merguiensis)

This species is readily available in the wild. It is also a "white" shrimp and tolerates low water quality better than *Penaeus monodon*. It tolerates a wide range of salinities and temperatures. Wild-caught breeders are cheap compared with *Penaeus monodon*. Each female yields between 100,000 and 200,000 eggs per spawn.

Hatcheries in India

Currently many hatcheries depend on sourcing of gravid females. There are about three hundred hatcheries (Table 24.3). Although many of the hatcheries have facilities for maturation of brood stocks by eye stalk ablation, only a few actually use them to produce nauplii.

Coastal Aquaculture Authority

The development of Specific Pathogen Free brood stocks is recommended as an alternative for producing high quality seed (Sakthivel and Ramamurthy, 2003). Tables 24.4 and 24.5 list hatcheries in Tamil Nadu, Andhra Pradesh and Gujarat

TABLE 24.3 Shrimp Hatcheries in India

State	Hatcheries (No.)
Andhra Pradesh	178
Tamil Nadu / Puducherry	72
Kerala	25
Orissa	10
Karnataka & Goa	7
Maharashtra	6
West Bengal	2
Gujarat	1
Total	301

TABLE 24.4 Shrimp Hatcheries in Tamil Nadu and Andhra Pradesh Authorized by CAA to Import SPF Brood Stocks of *Penaeus vannamei* to March 2012

Hatchery	Location	Pairs to be imported
Oceanic Edibles International Ltd.	Alappakkam Village, Mandavai Post, Marakkanam via., Villupuram District, Tamil Nadu	1680
Oceanic Bio-Harvest Ltd	Poompuhar, Sirkali Taluk, Nagappattinam District, Tamil Nadu	
Vaisakhi Biomarine Pvt. Ltd.	Marakkanam, Tindivanam Taluk, Villupuram District, Tamil Nadu	1120
Babu Aquarists hatchery	Kothapalli Mandal, East Godavari District, Andhra Pradesh	
Best India Marine Harvests	Near Marakkanam, Villupuram District, Tamil Nadu	
BMR Industries Ltd.	ECR, Muttukadu village, Kancheepuram District, Tamil Nadu	980
Grobest Feeds Corporation Ltd.	Thenpattinam village, Cheyyur Taluk, Kancheepuram District, Tamil Nadu	840
Lotus Sea Farms	ECR, Krishnakaranai Village, Pattipulam Post, Kancheepuram District, Tamil Nadu	840
Sudith Shrimp hatchery	Koovathur, Cheyyur Taluk, Kancheepuram District, Tamil Nadu	420

CAA, Coastal Aquaculture Authority; SPF, Specific Pathogen Free.

TABLE 24.5 Shrimp Hatcheries in Andhra Pradesh and Gujarat Authorized by CAA to Import SPF Brood Stocks of *Penaeus vannamei* to March 2012

Hatchery	Location	Pairs to be imported
Sharat Industries Ltd.	TP Gudur Mandal, East Godavari District, Andhra Pradesh	1680
C.P. Aquaculture Pvt. Ltd.	Vakadu Mandal, Kotta, Nellore District, Andhra Pradesh	1400
Western Lotus Hatcheries	Kodinar Taluk, Junagadh District, Gujarat	
Wintoss Associates	Kothapatname Mandal, Prakasam District, Andhra Pradesh	1260
NSR Aquafarms P. Ltd.	Pyakaropet Mandal, Visakhapatnam District, Andhra Pradesh	980
Nellore hatcheries	Mypadu Village, Indukurpet, Nellore, Andhra Pradesh	
Samudra Hatcheries P. Ltd.	Thondagai Mandal, East Godavari District, Andhra Pradesh	980
Devi Seafoods Ltd.	Kothapatnam Mandal, Prakasam District, Andhra Pradesh	840
Avanthi Feeds Ltd	Somajiguda, Hyderabad, Andhra Pradesh	
DSF Aquatech P. Ltd.	Kothapalli Mandal, East Godavari District, Andhra Pradesh	840
Alpha hatchery	Indukurpet Mandel, Nellore Dist, Andhra Pradesh	560
SVR Hatcheries	Thondagai Mandal, East Godavari District, Andhra Pradesh	560
Mahitha Shrimp hatcheries	Mypadu Village, Indukurpet, Nellore, Andhra Pradesh	420
Sripa Aqua Marine P. Ltd	Kodur Village, Nellore Dist., Andhra Pradesh	420
Mas Aqua Techniks P. Ltd.	TP Gudur Mandal, East Godavari District, Andhra Pradesh	280

CAA, Coastal Aquaculture Authority; SPF, Specific Pathogen Free.

which have obtained permission from the Coastal Aquaculture Authority (CAA) to import brood stocks of Specific Pathogen Free (SPF) *Penaeus vannamei* to March 2012 (CAA, 2012a,b).

As permitted by the CAA, brood stocks of SPF *P. vannamei* can be imported from the agencies listed in Table 24.6.

TABLE 24.6 CAA-Authorized Suppliers of SPF *Penaeus vannamei* Brood Stocks

Supplier	Location	Reference
The Oceanic Institute	Hawaii, USA	http://www.oceanicinstitute.org
Kona Bay Marine Resources	Hawaii, USA	http://www.konabaymarine.com/Aboutus.html
Shrimp Improvement Systems	Florida, USA	http://www.shrimpimprovement.com
SyAqua	Bangkok, Thailand	http://www.syaqua.com/Forms/3.Products_and_Services/2.Broodstock.htm
Vannamei 101 Co. Ltd.	Phuket, Thailand	http://www.vannamei101.com
Charoen Pokphand Foods Public Co. Ltd.	Bangkok, Thailand	http://www.cpfworldwide.com/cpd/en/page/about/our_history.aspx
Shrimp Improvement Systems P. Ltd.	Singapore	http://www.shrimpimprovement.com
Shrimp Improvement Systems P. Ltd.	Hawaii, USA	http://www.shrimpimprovement.com
High Health Aquaculture Inc.	Hawaii, USA	http://www.spfgenetics.com

CAA, Coastal Aquaculture Authority; SPF, Specific Pathogen Free.

CONCLUSION

Trading in wild shrimp brood stocks is a lucrative business. Considering the major contribution by the shrimp industry to the country's economy, it is essential to have good management practices for the production of healthy and high quality seed. There are risks of wastage or damage to brood stocks during handling. They can also act as carriers for pathogens. Hence, use of Specific Pathogen Free (SPF) brood stocks is highly encouraged. This will not only promote sustainable aquaculture but will also reduce the catch of brooders from the wild, thus reducing the threat to existing wild stocks.

ACKNOWLEDGEMENTS

The contributions from Rajiv Gandhi Centre for Aquaculture, Neelankarai, Chennai and the Coastal Aquaculture Authority, Government of India are gratefully acknowledged.

REFERENCES

Arce, S.M, Moss, M.S., and Argue, B.J., 2000. Artificial insemination and spawning of pacific white Shrimp *Litopenaeus vannamei*: Implications for a Selective breeding program. UJNR Technical Report No. 28. US-Japan Cooperation Program in Natural Resources. 140.90.235.27.

Ayub, A., Ahmed, 2002. Maturation and spawning of four commercially important penaeid shrimps of Pakistan. Indian J. Marine Sci. 31 (2), 119–124.

CAA (Coastal Aquaculture Authority)., 2012a. List of approved suppliers for import of SPF broodstock of *P. vannamei*.(http://aquaculture.tn.nic.in).

CAA (Coastal Aquaculture Authority)., 2012b. List of Hatcheries that have been approved to import SPF *Penaeus vannamei*. The Shrimp list (a mailing list for shrimp farmers) (http://aquaculture.tn.nic.in).

Emadi, H., Matinfar, A., Negarestan, H., Mooraki, N., 2006. Growth comparison of progenies of cultured and wild spawners of Indian white prawn, Penaeus indicus in commercial farm ponds in north Persian Gulf, Bushehr Province. World Aquaculture Society, USA.(https://www.was.org/ Documents/../ AQUA2006/WA2006-647.pdf).

Esmaeili, A., Omar., I., 2003. Influence of Rainfall on optimal spawner catch for the shrimp fishery in Iran. North American J. Fisheries Manage. 23 (2), 385–391.

FAO., 2007. Improving *Penaeus monodon* hatchery practices. Manual based on experience in India. FAO Fisheries Technical Paper No. 446: 101. Food and Agriculture Organization, Rome.

Garcia, S., 1985. Reproduction, stock assessment models and population parameters in exploited penaeid shrimps. In: Rothlisberg, P.C., Hill, B.J., Staples, D.J. (Eds.), Second Australian National Prawn Seminar. NPS2, Cleveland, Australia, pp. 139–158.

Itami, I., Maeda, M., Suzuki, N., Tokushige, K., Nagagawa, H.O., Konto, M., Kosornchandra, J., Hirono, I., Aoki, T., Kusuda, R., Takahashi, T., 1998. Possible prevention of white spot syndrome (WSS) in Kuruma shrimp (*Penaeus japonicus*) in Japan. In: Flegel, T.W. (Ed.), Advances in Shrimp Biotechnology. National Centre for Genetic Engineering and Biotechnology, Bangkok, pp. 291–295.

Paul, M., Maheshwarudu, G., Varma, J.B., 2004. Trade of live brooders of the Black Tiger Shrimp *Penaeus monodon* Fabricius 1798 at Visakhapatnam, Andhra Pradesh. J. Indian Fisheries Assoc. 1, 37–46.

Sakthivel, M., Ramamurthy, B., 2003. What ails shrimp farming and its future development. Fishing chimes 22 (10&11), 33–36.

Sethi, S.N., Mahendran, V., Nivas, K., Krishnan, P., Dam Roy, S., Ram, Nagesh., Sethi, Shalini., 2011. Detection of white spot syndrome virus (WSSV) in brood stock of tiger shrimp (*Penaeus monodon*) and other crustaceans of Andaman waters. Indian J. Marine Sci. 40 (3), 403–1403.

Uma, A., Koteeswaran, A., Indrani, K., Iddya, K., 2005. Prevalence of white spot syndrome virus and monodon baculovirus in *Penaeus monodon* broodstock and postlarvae from hatcheries in southeast coast of India. Current Science Research Communications 89 (9), 1619–1622.

Valderrama, D., J.L. Anderson., 2011. Shrimp Production Review. Global Outlook for Aquaculture Leadership 2011. Global Aquaculture Alliance. November 6-9, Santiago, Chile.

Chapter 25

An Assessment of Faunal Diversity and its Conservation in Shipwrecks in Indian Seas

J.S. Yogesh Kumar,* S. Geetha,[†] C. Raghunathan,* and K. Venkataraman**

*Andaman and Nicobar Regional Centre, Zoological Survey of India, Port Blair, Andaman & Nicobar Islands, India; [†]Department of Zoology, Kamaraj College, Thoothukudi, Tamil Nadu, India; **Zoological Survey of India, Kolkata, West Bengal, India

INTRODUCTION

Coral reefs constitute a diverse and vulnerable ecosystem, characterized by a complex interdependence of plants and animals. These are massive limestone structures built up through the constructional cementing processes and depositional activities of animals of the class Anthozoa (Order: Scleractinia) as well as all other calcium carbonate secreting animals and calcifying algae (Venkataraman, 2000; Geetha and Kumar, 2012). Nowadays, worldwide coral reef ecosystems are declining continuously due to overfishing, pollution and other anthropogenic and natural disturbances (Venkataraman and Rajan, 1998; Venkataraman, 2000, 2006; Yogesh Kumar and Geetha, 2012 a,b,c; Kumar *et al.,* 2014). Artificial reefs and restoration are considered to be one of the major coral reef conservation techniques to mitigate the negative effects of anthropogenic activities on the reef ecosystem (Yogesh Kumar and Geetha, 2012c).

Sunken shipwrecks and man-made submerged structures may be called artificial reefs (Yan and Yan, 2003; Yogesh Kumar and Raghunathan, 2012). Shipwrecks offer substrata for settlement of fish and fouling organisms. This kind of artificial reef is common around the world, and artificial reefs have been used to increase fish production, to improve ecological and economic functions, to control beach erosion and trawl fishing and to conserve biodiversity (Seaman and Jensen, 2000; Baine, 2001; Perkol-Finkel and Benayahu, 2004, 2005). Although numerous studies have focused on artificial reefs, there are still many research gaps regarding their performance and potential applications (Svane and Peterson, 2001). The aim of the present study was to take a census of the faunal communities in five different shipwreck artificial reefs in Indian seas.

Marine Faunal Diversity in India. DOI: 10.1016/B978-0-12-801948-1.00025-2
441

Specifically, the authors compared the benthic communities of recent and old shipwrecks and examined possible correlations between data from vertical and horizontal portions in shipwrecks.

METHODS

The studies were conducted on the Vembar wreck (Gulf of Mannar), Zusis wreck (Goa), North Bay wreck (Port Blair, South Andaman), Peel wreck (Havelock, South Andaman) and Japan wreck (Car Nicobar). The coordinates, dimensions and depth of all the wrecks are presented in Table 25.1. The census of the benthic communities was made on the vertical and horizontal surfaces. The percentage of cover was recorded for hard corals, soft corals, sponges, tunicates, molluscs and algae in each shipwreck in 2 m long and 0.1 m wide flexible fibreglass measuring tape transects with the help of SCUBA, following line intercept transect and belt transect methods described in Perkol-Finkel and Benayahu (2004). To assess the fish diversity by visual census in each shipwreck, diversity and abundance of reef fish within 2.5 m on either side of the transect line were noted (English *et al.,* 1997). An underwater camera was used for video documentation for further studies.

Biodiversity indices such as species richness, evenness and Shannon-Wiener diversity index were compared among the shipwrecks. Margalef's species richness method was used for calculating species richness.

RESULTS

Transect surveys were conducted periodically from 2007 to 2014. Fouling invertebrates, including live corals and soft corals, sponges, ascidians, bryozoans, polychaetes, nudibranchia, barnacles, bivalves, molluscs, sea urchins, algae and other faunal communities (Figures 25.1 and 25.2) that intercepted the transects were recorded and the projected length of each on the transects was measured. Quantitative data on each benthic life form were recorded from each shipwreck and are presented in Table 25.2. Live coral cover on the vertical surfaces was calculated to be 6.5, 5.6, 3.7 and 3.7 percent at the North Bay, Zusis, Vembar and Japan wrecks, respectively. Live coral cover on the horizontal surfaces was 26.1, 24.9, 24.7, 21.5 and 9.5 percent at the Japan, Vembar, Zusis, North Bay and Peel wrecks, respectively. The Japan wreck exhibited the greatest live coral cover, while the Peel wreck exhibited the least live coral cover and soft coral cover. Generally, massive coral dominated compared with the *Acropora* branching corals.

Soft coral and barnacle were predominantly noted on the vertical surfaces compared with the horizontal. Maximum cover of sponges was reported for both orientations; other faunal communities produced a far smaller contribution to the fouling organisms (Table 25.2). The faunal community was identified underwater— mostly to species level, and when this was not possible they were assigned to genera and family.

The percentage of fish abundance (Figure 25.3) was assessed by family among the five shipwrecks (Table 25.3). The fish of the Lutjanidae, Pempheriidae,

TABLE 25.1 Features of the Studied Shipwrecks

	Vembar Wreck	Zusis Wreck	North Bay Wreck	Peel Wreck	Japan Wreck
Coordinates (latitude; longitude)	09°02'08.20"N; 78°24'03.02"E	15°21'00.43"N; 73°46'46.68"E	11°43'00.56"N; 92°45'60.60"E	12°03'84.20"N; 92°57'81.10"E	09°10'88.30"N; 92°50'12.30"E
Approx. age (year)	25–30	40	30–40	8–10	40–50
Place	Gulf of Mannar	Goa	Port Blair	Havelock	Car Nicobar
Dimensions (m): length × width × height	51 × 11 × 9	72 × 14 × 9	68 × 14 × 9	21 × 8 × 6	70 × 13 × 9
Depth (m)	9	9–11	10	9–12	28
No. of horizontal transects	25	20	20	10	15
No. of vertical transects	10	18	15	8	10
Present status	Fishing ground	Recreational diving	Fishing and recreational diving	Recreational diving	Fishing ground

FIGURE 25.1 Live corals on shipwreck. (A) *Favia favus*; (B) *Acropora hemprichii*; (C) *Pleurogyra sinulosa*; (D) *Platygyra lamellina*; (E) *Symphyllia recta*; (F) Different types of coral on horizontal surface. (Please see color plate at the back of the book.)

FIGURE 25.2   Other reef-associated fauna on shipwreck. (A) Sponges; (B) Nephtheidae (soft corals); (C) *Macrobynchia philippina* (hydrozoans); (D) *Dynamena*; (E) *Risbecia ghardaqana* (nudibranchs); (F) *Pinctada margarifera* (bivalve molluscs). (Please see color plate at the back of the book.)

TABLE 25.2 Percentage Benthic invertebrate diversity for Each Shipwreck

	Vembar wreck		Zusis wreck		North Bay wreck		Peel wreck		Japan wreck	
	V	H	V	H	V	H	V	H	V	H
Live coral	3.7	24.9	5.6	24.7	6.5	21.5	0.0	9.5	3.7	26.1
Soft coral	18.7	1.1	20.9	6.5	23.4	14.2	10.9	8.8	23.8	4.5
Sponges	14.9	20.3	14.4	10.9	12.0	9.5	5.4	12.4	15.8	10.5
Ascidians	2.4	3.8	5.3	3.7	2.5	3.1	1.9	2.6	9.5	4.5
Bryozoans	13.5	8.7	4.3	10.4	4.5	8.9	12.3	17.5	3.6	9.7
Polychaetes	1.2	2.6	2.0	4.4	0.0	4.0	1.7	0.0	2.7	4.1
Nudibranchia	1.3	1.1	2.4	1.8	2.5	1.6	0.0	2.6	1.9	1.7
Barnacles	24.5	11.7	24.5	8.6	27.0	9.9	40.7	11.6	14.9	12.3
Bivalves and molluscs	4.9	14.7	8.7	10.7	9.1	9.6	10.9	13.2	10.2	11.6
Sea urchins	1.3	0.0	2.4	2.7	2.5	2.4	0.0	2.9	0.8	2.6
Algae	1.2	1.3	2.0	5.9	2.0	4.9	3.6	4.7	3.7	3.9
Others	12.5	9.6	7.7	9.8	8.0	10.3	12.7	14.3	9.3	8.3

H, horizontal; V, vertical.

FIGURE 25.3   Reef-associated fish in shipwrecks. (A) *Plectorhinchus multivittatum*; (B) *Scarus psittacus*; (C) *Platax pinnatus*; (D) *Apogon* sp.; (E) Lethrinidae; (F) Scorpaenidae.

TABLE 25.3 Percentage Fish Diversity for Each Shipwreck

Sl. No.	Family	Common name	Vembar wreck	Zusis wreck	North Bay wreck	Peel wreck	Japan wreck
1	Apogoniidae	Cadinalfish	11.6	8.5	7.8	16.3	14.6
2	Acanthuridae	Surgeonfish	1.2	0.8	1.9	3.3	1.8
3	Carangidae	Jacksfish	7.0	4.6	4.7	0.0	7.3
4	Chaetodontidae	Butterflyfish	7.0	5.8	5.6	6.5	3.5
5	Cirrhitidae	Hawkfish	1.2	2.3	2.8	1.1	1.2
6	Ephippididae	Batfish	1.2	2.7	2.5	2.2	2.3
7	Balistidae	Triggerfish	2.9	5.4	3.8	8.7	0.9
8	Bleniidae	Blennies	0.0	0.8	1.6	1.1	2.0
9	Diodontidae	Porcupinefish	0.6	0.4	0.3	0.0	0.0
10	Gobiidae	Gobies	0.0	0.8	1.6	0.0	1.8
11	Labridae	Wrasses	12.2	5.8	5.6	5.4	4.4
12	Lutjanidae	Snappers	23.3	19.3	25.1	21.7	35.1
13	Muraenidae	Moray eel	1.2	1.5	0.3	0.0	0.3
14	Mullidae	Goatfish	2.3	3.1	1.9	5.4	2.9
15	Ostraciidae	Boxfish	1.2	1.9	1.3	1.1	1.2
16	Pempheridae	Sweepers	12.8	21.6	18.8	13.0	12.3
17	Pomacanthidae	Angelfish	1.2	1.5	2.5	2.2	2.3
18	Scaridae	Parrotfish	4.7	3.1	3.8	6.5	2.3
19	Scorpaenidae	Scorpionfish	4.1	4.6	3.1	2.2	0.9
20	Serranidae	Groupers	4.1	3.9	2.5	1.1	1.8
21	Siganidae	Rabbitfish	0.6	1.5	2.5	2.2	1.2

TABLE 25.4 Diversity Indices for Fish at Each Shipwreck

Parameter	Vembar wreck	Zusis wreck	North Bay wreck	Peel wreck	Japan wreck
No. of families	19	21	21	17	20
Total no of individuals	172	259	319	92	342
Diversity index, H'	2.45	2.56	2.53	2.4	2.27
Species richness, d	3.49	3.59	3.46	3.54	3.26
Species evenness, J	0.8	0.84	0.83	0.79	0.75

Apogoniidae and Labridae families dominated in all of the shipwrecks, followed by Carangidae, Chaetodontidae, Scaridae, Scorpaenidae, Serranidae, Balistidae, Mullidae, Pomacanthidae, Ostraciidae, Muraenidae, Ephippididae and Cirrhitidae.

The maximum number of individuals visually recorded was from the Japan wreck (342), followed by the North Bay wreck (319), Zusis wreck (259), Vembar wreck (172) and Peel wreck (92). The diversity indices Shannon-Wiener diversity index (H'), Pielou's evenness (J) and Margalef's species richness (d) were determined. H' was maximum for the Zusis wreck (2.56) and minimum for the Peel wreck (2.4); J ranged from 0.75 for the Japan wreck to 0.84 for the Zusis wreck; d was maximum (3.59) for the Zusis wreck and minimum (3.26) for the Japan wreck (Table 25.4).

DISCUSSION

Shipwrecks and other unplanned artificial reef structures offer a substratum for settlement of benthic invertebrates and fish. These structures mimic natural reef and attract a variety of fish, all kinds of marine life and recreational SCUBA divers (Black, 2001; Black and Mead, 2001; Sutton and Bushnell, 2007; Perkol-Finkel and Benayahu, 2004, 2009). The present surveys carried out at shipwrecks in Indian waters have revealed diversity and abundance of fauna including fish, live corals, soft corals, sponges, ascidians, bryozoans, polychaetes, nudibranchia, barnacles, bivalves, molluscs, sea urchins and algae. The live coral, soft coral and fish were more dominant in old shipwrecks (Japan, North Bay, Zusis and Vembar) than in the more recent shipwreck (Peel). More live corals were observed on horizontal surfaces than on vertical. But the percentage of soft coral was high for the vertical and low for the horizontal. This may be due to the environment, size, structure, geographical location, depth, surrounding substratum, type of materials and age (Baine, 2001; Rilov and Benayhu, 2000;

Sheng, 2000; Perkol-Finkel and Benayahu, 2005). Wrecks more than a hundred years old have been reported to have a hard coral cover similar to their adjacent natural reef (Riegl, 2001).

Artificial reefs attract reef fish. Therefore, consideration may be given to deploying such artificial reefs to increase fishery productivity (Einbinder et al., 2006). Maximum abundance and distribution of reef fish at artificial reefs have been reported to be higher than in the surrounding coral reef environment (Wilhelmsson et al., 1998). In the present study, we reported 21 families from the shipwrecks during the study period. Snapper (Lutjanidae) and grouper (Serranidae) were dominant among economically important and edible reef fish species at all the study station. This may be due to the availability of macro algae cover, fish young, nutrient richness and low disturbance (Williams and Polunin, 2001).

The present study locations are being actively used for recreational SCUBA diving and fishing. Therefore, fresh deployment of man-made artificial structures in many places on the ocean bottom could create new fertile habitats with higher biodiversity, and overexploitation caused by continuous fishing activities and trawler fishing could thus be reduced. Construction of artificial reefs in substantial numbers in near-shore areas, natural coral reefs would reduce the impact of anthropogenic activities on natural reefs. In this way, we can also protect and conserve these fragile ecosystems for future generations.

ACKNOWLEDGEMENTS

The authors are thankful to the Department of Science & Technology, Science & Engineering Research Board for providing financial support and also to the Ministry of Environment and Forests, Government of India for providing necessary permission for this study. The authors also wish to express their sincere gratitude to the executive director Mr Rajendraprasad, People's Action for Development, Vembar and PADI Instructor Mr Venkadash Charlo, Barracuda Dive Centre, Goa.

REFERENCES

Baine, M., 2001. Artificial reefs: a review of their design, application, management and performance. Ocean Coastal Manage 44, 241–259.

Black, K.P., 2001. Artificial surfing reefs for erosion control and amenity: theory and application. Special Issue 34 (ICS 2000). J. Coastal Res., 1–7.

Black, K.P., and Mead, S., 2001. Design of Gold Coast artificial surfing reef surfing aspects. In Black, K., (Ed) Natural and Artificial Reefs for Surfing and Coastal protection. Special Issue 29, J. Coastal Res., 115–130.

Einbinder, S., Perelberg, A., Ben-Shaprut, O., Foucart, M.H., Shashar, N., 2006. Effects of artificial reefs on fish grazing in their vicinity: evidence from algae presentation experiments. Marine Environ. Res. 61, 110–119.

English, S., Wilkinson, C., Basker, V., 1997. Survey Manual for Tropical Marine Resources. Australian Institute of Marine Science, Townsville, Australia, p. 390.

Geetha, S., Kumar, J.S.Y., 2012. Status of corals (Order: Sclerectinia) and associated fauna of Thoothukudi and Vembar group of Islands, Gulf of Mannar. India. Int. J. Sci. Nat. 3 (2), 340–349.

Hassan, M., Kotb, M.M.A., Al-Sofyani, A.A., 2002. Status of coral reefs in the Red Sea–Gulf of Aden. pp. 45-53. In: Wilkinson, C. (Ed.), Status of Coral Reefs of the World, 2002. Australian Institute of Marine Science, Townsville, Australia.

Kumar, J.S.Y., Marimuthu, N., Geetha, S., Satyanarayana, C.H., Venkataraman, K., Kamboj, R.D., 2014. Longitudinal variations of coral reef features in the Marine National Park, Gulf of Kachchh. J. Coast. Conserv. 18, 167–175.

Marimuthu, N., Dharani, G., Vinithkumar, N.V., Vijayakumaran, M., Kirubagaran, R., 2011. Recovery status of sea anemones from bleaching event of 2010 in the Andaman waters. Curr. Sci. 101 (6), 734–736.

Perkol-Finkel, S., Benayahu, Y., 2004. Community structure of stony and soft corals on vertical unplanned artificial reefs in Eilat (Red Sea): Comparison to natural reefs. Coral Reefs 23, 195–205.

Perkol-Finkel, S., Benayahu, Y., 2005. Recruitment of benthic organisms onto a planned artificial reef: Shifts in community structure one decade post deployment. Mar. Env. Res. 59, 79–99.

Perkol-Finkel, S., Benayahu, Y., 2009. The role of differential survival patterns in shaping coral communities on neighboring artificial and natural reefs. J. Experiment. Marine Biol. Ecol. 369 (1), 1–7.

Riegl, B., 2001. Degradation of reef structure, coral and fish communities in the Red Sea by ship groundings and dynamite fisheries. Bull. Mar. Sci. 69, 595–611.

Rilov, G., Benayahu, Y., 2000. Fish assemblage on natural versus vertical artificial reefs: the rehabilitation perspective. Marine Biology 136, 931–942.

Seaman, W., Jensen, A.C., 2000. In: Seaman, Jr., W. (Ed.), Purposes and practices of artificial reef evaluation. Artificial Reef Evaluation With Application to Natural Marine Habitat. CRC Press, Boca Raton, FL, pp. 2–19.

Sheng, Y.P., 2000. Physical characteristics and engineering at reef sites. In: Seaman, Jr., W. (Ed.), Artificial Reef Evaluation With Application to Natural Marine Habitat. CRC Press, Boca Raton, FL, pp. 51–94.

Sutton, S.G., Bushnell, S.L., 2007. Socio-economic aspects of artificial reefs: considerations for the great barrier reef marine park. Ocean Coastal Manage. 50, 829–846.

Svane, I.B., Petersen, J.K., 2001. On the problems of epibiosis, fouling and artificial reefs, a review. Mar. Ecol. 33, 169–188.

Tamelander, J., Rajasuriya, A., 2008. Status of coral reefs in South Asia: Bangladesh, Chagos, India, Maldives and Sri Lanka. In: Wilkinson, C. (Ed.), Status of Coral Reefs of the World, 2008. Global coral reef monitoring network and reef and rainforest research centre, Townsville, Australia, pp. 119–130.

Venkataraman, K., 2000. Status of coral reefs of Gulf of Mannar, India. In: Proc. 9th International Coral Reef Symposium, Bali, 23-27 October 2000. International Society for Reef Studies. 35.

Venkataraman, K., 2006. Coral Reefs in India. National Biodiversity Authority, Chennai, p. 18.

Venkataraman, K., Rajan, P.T., 1998. Coral reefs of Mahatma Gandhi Marine National Park and Crown of Thorn phenomenon. In: Gangwar, B., Chandra, K. (Eds.), Symposium Proceedings Island Ecosystem and Sustainable Development. Andaman Science Association and Department of Science and Technology. Anadman and Nicobar Administration, Port Blair, pp. 124–132.

Venkataraman, K., Wafar, M., 2005. Coastal and marine biodiversity of India. Ind. J. Mar. Sci. 34 (1), 57–75.

Wilhelmsson, D., Ohman, M.C., Stahl, H., Shlesinger, Y., 1998. Artificial reefs and dive tourism in Eilat, Israel. Ambio 27, 764–766.

Williams, I.D., Polunin, N.V.C., 2001. Large-scale associations between macro algal cover and grazer biomass on mid-depth reefs in the Caribbean. Coral Reefs 19, 358–366.

Yan, T., Yan, W.X., 2003. Fouling of offshore structures in China – a review. Biofouling 19, 133–138.

Yogesh Kumar, J.S., Geetha, S., 2012a. Seasonal changes of hydrographic properties in sea water of coral reef islands, Gulf of Mannar. India. Int. J. Plant. Anim. Environ. Sci. 2 (2), 135–159.

Yogesh Kumar, J.S., Geetha, S., 2012b. Seasonal variations of trace metal accumulation on coral reef in Gulf of Mannar. India. Int. J. Appl. Biol. Pharma. Tech. 3 (3), 61–88.

Yogesh Kumar, J.S., Geetha, S., 2012c. Fouling communities on ship wreck site in the Gulf of Mannar, India. Int. J. Appl. Biol. Pharma. Tech. 3 (3), 259–264.

Yogesh Kumar, J.S., Raghunathan, C., 2012. Recovery status of scleractinian corals and associated fauna in the Andaman and Nicobar islands. Phuket Marine Biol. Center Res. Bull. 71, 63–70.

Chapter 26

Saltwater Crocodiles in Andaman and Nicobar Islands, with Special Reference to Human–Crocodile Conflict

C. Sivaperuman

Andaman and Nicobar Regional Centre, Zoological Survey of India, Port Blair, Andaman and Nicobar Islands, India

INTRODUCTION

The Andaman and Nicobar Islands (ANI) comprise 572 islands and extend over 800 km. These islands were once a part of the Asian mainland, but were detached some 100 million years ago during the Upper Mesozoic period due to geological upheaval. The existing groups of islands constitute the physiographic continuation of the mountainous ranges of the Naga and Lushai Hills, and Arakan Yoma of Burma, through Cape Negrais to the ANI, and southeast Sumatra. The chains of these islands are the seven 'camel backs' of submerged mountain ranges projecting above sea level, running north to south between 6° 45' N and 13° 30' N latitudes, and 90° ° 20' E and 93° 56' E longitudes. These islands are tropical, with a warm, moist and equable climate. The proximity of the sea and the abundant rainfall prevent extremes of heat. The mountainous parts of the southern islands receive about 300 cm of rain annually, whereas the northern islands receive less. The period from December to February is comparatively cool due to the effect of the northeast monsoon. Warm weather extends from March to April, the driest months. In May, the southwest monsoon breaks over the area, and continues until October. The variation of temperature over the islands is relatively small: 23–31°C.

The crocodiles are among the only living remnants of reptiles which ruled during the Mesozoic era. Crocodiles are top predators and, as such, perform an important role in maintaining the structure and function of ecosystems (Glen *et al.*, 2007; Leslie and Spotila, 2001; Ross, 1998). In the Indian subcontinent, three species of crocodile occur: the Gharial (*Gavialis gangeticus*), the saltwater crocodile (*Crocodylus porosus*), and the mugger crocodile (*Crocodylus palustris*). The diet of crocodiles varies with age, with small crocodiles depending mainly on invertebrates and fish, and adults feeding on large animals including livestock and humans (Ross, 1998).

The saltwater crocodile (*Crocodylus porosus*) is the largest of all living reptiles. It is found in suitable habitat throughout southeast Asia and northern Australasia. Saltwater crocodiles are severely depleted in numbers throughout the vast majority of their range, with sightings in areas such as Australia, Bangladesh, Brunei, Cambodia (extinct?), China, India, Indonesia, Malaysia, Myanmar, Palau, Papua New Guinea, Philippines, Singapore (extinct?), Sri Lanka, Solomon Islands, Thailand (extinct?), Vanuatu, and Vietnam (extinct?) (Webb *et al.*, 2010). The saltwater crocodile occurs in the ANI, where it grows to over 6 m in length and can be encountered in open sea, near the shore, in mangrove creeks, freshwater rivers and swamps. The aim of this study was to assess saltwater crocodile populations and to examine issues of human–crocodile conflict and management of problem crocodiles.

METHODS

To assess the population of crocodile in the ANI, data were collected by various methods: extensive interviews were carried out to confirm the present population with forest department staff and local people, and the available published literature was also consulted. The information on human–crocodile conflicts was quantified by interviewing the victims and by visiting the site where the attack occurred.

RESULTS AND DISCUSSION

The details of the population of saltwater crocodiles in the ANI, based on personal observation, questionnaire and available literature, are presented in Table 26.1. The highest number of crocodiles reported was from North Andaman Islands, followed by Landfall Island.

Human–Crocodile Conflicts

In the ANI, crocodile attacks were reported from 1986 onwards; about 10 cases of crocodile attacks were reported between 1986 to 1993. Details of the crocodile attacks in various locations of the ANI are presented in Table 26.2. During the period of the present study, three crocodile attacks happened and two people died. No specific temporal pattern of attacks was observed. Attacks occurred near the shore and mangrove creeks. In some cases, there is a relationship with the dumping of waste food on the sea shore. Following a regular pattern of activity may have helped the crocodiles to locate humans for attack and encouraged them to wait for their arrival. All the attacks followed the known pattern

TABLE 26.1 Crocodiles in Andaman and Nicobar Islands

Islands	Number of individuals	Reference
North Andaman	15 breeding females & 100–200	Whitaker and Whitaker, 1978
North Andaman	50 breeding females	Choudhury, 1980; Choudhury and Bustard, 1979
North Andaman (excluding Landfall Island)	95	Andrews and Whitaker, 1994
Landfall Island	38 adults	Andrews and Whitaker, 1994
North Andaman Islands, North Reef, and Interview Island	31 adult & 10 nests	Andrews and Whitaker, 1994
Middle Andaman	17 adults, 9 sub-adults and 15 juveniles	Andrews, 1997
Little Andaman	27 adults, 11 sub-adults	Andrews, 1997
Rutland, Tarmugli, Hutbay	19 adults,35 sub-adults	Andrews, 1997
Baratang Island	2 adults	Sivaperuman, 2008 (pers. observation)
Baratang Island	2 adults	Sivaperuman, 2013 (pers. observation)
South Andaman (Whiberligunj, Tushnabad, Marina Park, Corbyn's Cove, Caddlegnj)	5 adults 1 sub adult	Senthil Kumar, 2011 (pers. observation)
Great Nicobar	6 adults, 3 sub-adults	Sivaperuman 2008 (pers. observation & questionnaire)
Ritchie's Archipelago	3 adults	Sivaperuman, 2009 (pers. observation)
Long Island, Middle Andaman	1	Sivaperuman, 2013 (pers. observation)

of hunting behavior reported in crocodiles (Daniel, 1983; Jayson *et al.*, 2006). As seen from the case studies, large crocodiles over 3 m length were involved in all the major and fatal attacks on humans. Problem crocodiles are defined broadly as those individuals that occur within areas of recreational use or human habitats.

TABLE 26.2 Number of Crocodile Attacks in Different Parts of Andaman and Nicobar Islands

Location	Number of attacks
North Andaman (Kalighat, Kishorinagar and Paschimsagar)	4
South Andaman (Tirur Creek and Shoal Bay Creek)	3
Middle Andaman (Kadamtala Creek, CFO Nallah, Rangat Nallah and Bakultala)	4
South Andaman (Havelock, Whiberligunj and Tushnabad)	3

Recent Crocodile Attacks

Human–Crocodile Attack in Havelock

A 25-year-old girl from the USA, who was reported missing, was found to have been attacked by a crocodile from Radha Nagar Beach, Havelock. Andaman and Nicobar police confirmed that US tourist Lauren Elizabeth Failla was missing from Havelock Island. A massive search operation was launched by the police to trace Elizabeth, and her crocodile-eaten and decomposed body was recovered. The Department of Environment and Forests (DEF) decided to capture the problem crocodile immediately, and a team was constituted for the capture and translocation of the animal. The identification of the crocodile was confirmed from a video recording by an underwater camera which was recovered from the place of attack, and the characteristic features of the animal were studied thoroughly. The animal was monitored by direct and indirect observation in the shallow water, mudflats, mangroves and creeks.

Human–Crocodile Attack in Bakultala

A saltwater crocodile attacked Shri Ajay Kulla, a 23-year-old youth of Bakultala village, middle Andaman, on 1st August 2012 at 0900 hours, when he was fishing in a nearby nallah along with his friend. His friend who was in the spot saw the attack and alerted nearby villagers. When they reached the spot, Ajay was found to be missing. Despite the best efforts of officials of the DEF and the police, the body of the victim could not be traced. This was the fifth fatal attack by a crocodile within a period of 28 months in the ANI.

After a month-long operation the problem crocodile was captured from Shyamkunj mangrove creek by a team of the DEF and moved successfully to the Mini Zoo, Port Blair, as described below.

Capturing Problem Saltwater Crocodiles

Both problem crocodiles were captured using indigenous techniques and taken to the Mini Zoo at Port Blair.

Capturing the Crocodile at Havelock
Equipment Used to Capture the Crocodile

The following locally-available and indigenously-developed materials were used to capture the crocodile: floating cage of cane and bamboo, harpoons, wire mesh, nylon rope, bamboo pole, jerry cans, bundles of discarded PET bottles, fishing buoys, dinghy, and chicken for bait. Three traps were placed in potential sites used by the crocodile (Kumar *et al.*, 2012).

Capture of the Problem Crocodile

After 2 months of attempts, the problem crocodile was captured. An adult male, it was 4.25 m long and weighed 480 kg. Three floating cages and two sets of net and noose traps were laid in the territory of the crocodile. On 6th June 2010 at 2245 hours, we started routine monitoring activity from Char Nariyal Camp by small fibre boat loaded with harpoon, search lights, reserve fuel, ropes, mosquito repellents and water. The search progressed, and around 0115 hours we reached the last cage and were alerted by the sound of splashing water. Our Forest Guard was first to spot the animal, which was struggling in the cage, and alerted the team. The animal sensed our presence and warned us by groaning and made a couple of dead rolls in a desperate attempt to cut the jaw rope. At that moment, we decided to harpoon the animal to prevent its escape. The harpoon was a small piece of metal with two sharp inward curving hooks, tied to a 20 m nylon rope and buoy. Our team made an attempt to place an additional top jaw rope from the boat using a pole. Unfortunately, the jaw rope already present on the crocodile was on the initial portion of the snout and prevented the insertion of another noose. Two of our Forest Guards decided to risk their lives by walking close to the animal from the mangrove side to insert the noose on the inner jaw of the crocodile using a stick. This worked, and two additional top jaw ropes were placed. Immediately, the tail was also secured by another team, and thus the animal's movement was controlled. Thereafter, the animal was secured to nearby trees at 0400 hours. The captured crocodile was secured in the stretcher, transported to Havelock jetty by dinghy, and subsequently taken to the Mini Zoo at Port Blair.

Capturing the Crocodile at Bakultala

The DEF advised locals to be vigilant for crocodiles around Bakultala village, and a warning sign board was placed at a nearby mangrove creek. Consequently, a squad was constituted on 1st August 2012 under the direction of the Principal Chief Conservator of Forests (Wildlife) to capture the problem saltwater crocodile. The squad members included staff of the DEF, local people and the police. The animal was monitored by direct and indirect observation (footprint, faecal, etc.) on the shore and in shallow water, mudflats and mangrove creeks (Viswakannan *et al.* In Press).

Capture of the Problem Crocodile

The crocodile, an adult male, was 4.10 m long and weighed 450 kg. The team members thoroughly combed the entire area, singled out the problem animal based on various characteristics, and set traps to capture it. Three methods were employed. First, a cage constructed of bamboo and cane was placed in the mangrove creek, but this attempt was unsuccessful, as the cage was too small. Second, a snare trap was deployed. This is one of the effective techniques to capture the crocodiles, used nationally and internationally. In this method, a stiff wooden dowel is placed inside a dead chicken, and a rope extends from the chicken body with buoys. The team waited for the crocodile to eat the bait, and it was eaten by the crocodile. Subsequently, the team followed the crocodile for more than 2 hours but could not capture it, because of accessibility issues within the mangrove creek area and also because the animal regurgitated the bait.

Thereafter a metal cage was devised by the team, which enabled successful completion of the operation. Initially the metal cage was placed on the shore with a chicken as bait, but the crocodile was not attracted. Then the cage was placed in the territory of the crocodile in open water, using floater logs, and baited with dead chicken. After 27 days the target crocodile was caught. It was the second-largest crocodile suspected to be a man-eater captured in the ANI. The captured individual was transported to Shyamkunj Jetty by dinghy. Thereafter, adaptation of the cage was improvised, using planks to avoid any injury to the animal during the transportation. Finally the animal was transported to Port Blair by road and relocated into the Mini Zoo at Port Blair.

This immediate action by the DEF restored safety and wellbeing to the local population in Bakultala village, at least in the short term. However, in such a situation, removal of the problem crocodile may only solve such a problem for the time being. Following battles for supremacy, another male will inevitably dominate the area, and may again pose a threat to people and livelihoods.

Possible Reasons for Attacks

Habitat destruction of crocodiles and sharing of the same habitat by humans and crocodiles are the major reasons for such human–crocodile conflict. The increasing human activities, such as fishing in the mangrove areas, and crossing of creeks without adequate protection, increase the risk of crocodile attacks on humans. One of the possible motives for attacks on people is territorial defence. The presence of livestock and other domestic animals on the sea shore may also attract crocodiles to inhabited areas. In addition, the dumping of waste food materials on the sea shore provides an added attraction for the crocodiles. It is possible to manipulate the size distribution of the crocodiles by removing some of the larger and more dangerous individuals to other locations in the Islands (Ross, 1998). The relocation of problem crocodiles has been suggested as a management strategy in Australia (Walsh and Whitehead, 1993). However, experiences in relocating large specimens of *C. porosus* in Darwin, Australia suggest that some of these animals return to their capture area after relocation by up to 100 km. The best solution is to change people's behavior so that they are unlikely to encounter crocodiles. In Sri Lanka, communities construct riverside barriers from local materials to provide safe areas for bathing and water collection. Education about the presence and danger of crocodiles, improved (larger) fishing craft and careful disposal of attractive waste such as fish and meat offal can all reduce the risk of crocodile attack. Presenting such information in local languages and in forms familiar to local communities (dance, music, theatre, religious discussion, folk belief and folk tales) and by local communicators, leaders and opinion leaders assists effective transfer of these messages.

Community Awareness and Participation

The Department of Environment and Forests in the Andaman and Nicobar Islands promotes crocodile awareness among residents and visitors by disseminating educational information via brochures, pamphlets and warning boards. A public awareness campaign is repeated regularly to minimize crocodile attacks, with sign boards placed at popular beaches. A research programme is recommended, to monitor the effectiveness of policies and human–crocodile relationships in the Andaman and Nicobar Islands, in order to minimize human–crocodile conflict in the future.

REFERENCES

Andrews, H.V., 1997. Population dynamics and ecology of the saltwater crocodile (*Crocodylus porosus* Schneider) in the Andaman and Nicobar Islands. Interim report. Phase III. Submitted to the Andaman and Nicobar Forest Department and the Centre for Herpetology (AN/C-3-97). p. 6.

Andrews, H.V., Whitaker, R., 1994. Population dynamics and ecology of the saltwater crocodile (*Crocodylus porosus* Schneider) in the Andaman and Nicobar Islands. Interim survey report. Phase II. Submitted to the Andaman and Nicobar Forest Department and the Centre for Herpetology (AN/C-2-94). p. 18.

Choudhury, B.C., 1980. The status, conservation and future of the saltwater crocodile (*Crocodylus porosus*, Schneider) in North Andaman Island. Union Territory of Andaman & Nicobar Islands. Indian Crocodiles – Conservation and Research. Occl. Publ. No. 1, 1–7.

Choudhury, B.C., Bustard, H.R., 1975. Restocking Mugger crocodile *Crocodylus palustris* (Lesson) in Andhra Pradesh: Evaluation of a pilot release. J. Bombay Nat. Hist. Soc. 79, 275–289.

Daniel, J.C., 1983. The Book of Indian Reptiles. Bombay Natural History Society, p. 11.

Glen, A.S., Dickman, C.R., Soule, M.E., Mackey, B.G., 2007. Evaluating the role of the dingo as a trophic regulator in Australian ecosystems. Austral Ecology 32, 492–501.

Jayson, E.A., Sivaperuman, C., Padmanabhan, P., 2006. Review of the reintroduction programme of the Mugger crocodile *Crocodylus palustris* in Neyyar Reservoir. India. Herpetological Journal 16, 69–76.

Kumar, S.S., Sivaperuman, C., Yadav, B.P., 2012. Management of problem saltwater crocodiles (*Crocodilus porosus* Schneider) – A case study in the Andamen and Nicobar Islands, India. Herpetological Bull 120, 9–15.

Leslie, A.J., Spotila, J.R., 2001. Alien plant threatens Nile crocodile (*Crocodylus niloticus*) breeding in Lake St. Lucia. South Africa. Biological Conservation 98, 347–355.

Ross, J.P., 1998. Crocodiles. Status Survey and Conservation Action Plan, 2nd edition. IUCN/SSC Crocodile Specialist Group. IUCN, Gland, Switzerland and Cambridge UK.

Viswakannan, P., C. Sivaperuman, S. Senthil Kumar and Shashikumar (In Press). Capturing problem Saltwater crocodiles (*Crocodylus porosus* Schneider) using indigenous techniques in the Andaman islands, India. J. Bombay Nat. Hist. Soc.

Walsh, B., Whitehead, P.J., 1993. Problem crocodiles, *Crocodylus porosus*, at Nhulunbuy, Northern Territory: An assessment of relocation as a management strategy. Wildl. Res. 20(1): 127–135.

Whitaker, R., Whitaker, Z., 1978. A preliminary survey of the saltwater crocodile (*Crocodylus porosus*) in the Andaman Islands. J. Bombay Nat. Hist. Soc. 76, 311–323.

Webb, G.J.W., Manolis, S.C., Brien, M.L. (2010). Saltwater Crocodile *Crocodylus porosus*. pp. 99–113. In:, Crocodiles, Status survey, conservation action, plan., 3rd edn, (Eds.) S.C. Manolis and C. Stevenson. Crocodile Specialist Group, Darwin.

Chapter 27

Conservation Status of Marine Faunal Diversity in India: An Analysis of the Indian Wildlife (Protection Act) and IUCN Threatened Species

P.T. Rajan
Andaman and Nicobar Regional Centre, Zoological Survey of India, Port Blair,
Andaman and Nicobar Islands, India

INTRODUCTION

The oceans are home to a large percentage of Earth's biodiversity, occupying 70 percent of its surface and, when volume is considered, an even larger percentage of habitable space. The oceans drive weather, shape planetary chemistry, generate 70 percent of atmospheric oxygen, absorb most of the planet's carbon dioxide, and are the ultimate reservoir for replenishment of fresh water to land through cloud formation. Trouble for the oceans means trouble for humankind. In recent years, there has been growing concern in the scientific community that a broad range of marine species could be under threat of extinction and that marine biodiversity is experiencing potentially irreversible loss due to overfishing, climate change, invasive species and coastal development (Dulvy *et al.,* 2003; Roberts and Hawkins 1999). Governmental and public interest in marine conservation is increasing, but the information needed to guide marine conservation planning and policy is seriously deficient (Edgar *et al.,* 2008). The International Union for Conservation of Nature (IUCN) Red List of Threatened Species is the most commonly used global dataset for identifying the types of threat, and the levels of extinction risk to marine species (Hoffmann *et al.,* 2008; Rodrigues *et al.,* 2006). It forms the foundation for determining and validating marine conservation priorities, for example through the planning and management of protected area systems designed to reduce extinction risk in the sea. However, the number of marine species assessed for their probability of extinction has lagged far behind that of the terrestrial realm; out of more than 41,500 plants and animals currently assessed under the IUCN Red List Criteria, only approximately

Marine Faunal Diversity in India. DOI: 10.1016/B978-0-12-801948-1.00027-6

1500 were marine species, including all of the world's known species of sharks and rays, groupers, and reef-building corals (Carpenter *et al.*, 2008). In many regions around the world, biodiversity conservation in the seas is currently taking place without the essential species-specific data needed to inform robust and comprehensive conservation actions. Protection of our rapidly declining ocean ecosystems and species is one of the greatest challenges we face as stewards of our planet.

India being a signatory of international environmental instruments has its obligation to fulfil the environmental mandate: i.e., the retention of traditional knowledge and experience gained over the years together with the law of conservation aims at sustainable development and intergenerational equity, so that the natural resources are preserved for future generation. In this regard, Article 48A, part of the Directive Principles of State Policy, in the Constitution mandates that the state should endeavor to protect and improve the environment and safeguard the forests and wildlife of the country. Furthermore, India being a signatory to the International Environmental instruments—the Convention of International Trade in Endangered Species (CITES) and the Convention on Biological Diversity (CBD)—is obliged to take steps to fulfil the environmental mandate (Ramakrishna and Dey, 2003). The main aim of all these conventions is to reduce the biotic pressure, with the assumption that any such pressure is detrimental to wildlife resources. The International Union of Conservation of Nature (IUCN) encourages and assists societies throughout the world to conserve the integrity and diversity of nature and to ensure that any use of natural resources is equitable and ecologically sustainable. The Wildlife (Protection) Act, 1972 and Wildlife Protection Amendment Act, 1991, 2001 are intended to provide a comprehensive national legal framework for wildlife protection with a two-pronged conservation strategy as a part of the environmental mandate of the country, with a major goal that specified endangered species are protected regardless of location and all species are protected in specified areas. In essence the Act prohibits the hunting of wildlife, protects their habitats and restrains trade in wild animals and trophies, etc. The Ministry of Environment and Forests, Government of India frames the guidelines, and the implementation authorities are the Wildlife Advisory Board, Wildlife Wardens and their staff of the state governments.

THREATENED AND ENDANGERED SPECIES

A total of 792 marine species on the IUCN Red List of Threatened Species are listed as either Critically Endangered (100 species/subpopulations), Endangered (155 species/subpopulations), or Vulnerable (537 species/subpopulations). Near Threatened, Least Concern and those listed as Data Deficient have been excluded, though they may end up on this list in the future. These are just the marine species known to be in trouble and are likely to be just the *tip of the iceberg*, so to speak, based on how little we currently know about life in the ocean. In India, 42 marine species are protected under the Wildlife (Protection) Act, 1972: 3 species of mammal, 6 species of reptile, 9 species of fish and 24 species of mollusc (Table 27.1).

TABLE 27.1 Scheduled[1] Marine Animals of India

Scientific name	Common name	Distribution	Status
Mammals			
Dugong dugon	Dugong, Sea Cow	Indo-Pacific. India: Gulf of Kutch, Mannar, Palk Bay, ANI	Schedule I
Oreaella brevezastris	Irrawaddy/Snubfin Dolphin	Near sea coasts and in estuaries and rivers in parts of the Bay of Bengal and southeast Asia.	Schedule I
Physeter macrocephalus	Sperm Whale	Cosmopolitan species	Schedule I
Reptiles			
Crocodylus porosus	Saltwater Crocodile	Indo-Pacific. India: Orissa, West Bengal, ANI	Schedule I
Dermochelys coriacea	Leatherback Sea Turtle	Atlantic, Pacific, and Indian Oceans. India: ANI	Schedule I
Caretta caretta	Loggerhead Sea Turtle	Atlantic, Pacific, and Indian Oceans. India: ANI	Schedule I
Lepidochelys olivacea	Olive Ridley Sea Turtle	Atlantic, Pacific, and Indian Oceans. India: ANI.	Schedule I
Eretmochelys imbricata	Hawksbill Sea Turtle	Atlantic, Pacific, and Indian Oceans. India: ANI, the coasts of Tamil Nadu, Orissa.	Schedule I
Chelonia mydas	Green Sea Turtle	Atlantic, Pacific, and Indian Oceans. India: ANI.	Schedule I
Fish			
Rhincodon typus	Whale Shark	All tropical and warm-temperate seas. India: Gulf of Kutch, Palk Bay, ANI	Schedule I; Part IIa Fishes
Anoxypristis cuspidate	Knifetooth Sawfish	Indo-West Pacific. India: ANI	Schedule I; Part IIa Fishes
Carcharhinus hemiodon	Pondicherry Shark	Indo-Pacific. India: Pondicherry	Schedule I; Part IIa Fishes
Glyphis glyphis	Speartooth Shark	Northern Australia and New Guinea. Not Reported from India	Schedule I; Part IIa Fishes
Himantura fluviatilis	Ganges Stingray	India: River Ganga	Schedule I; Part IIa Fishes
Pristis microdon	Largetooth Sawfish	Indo-Pacific	Schedule I; Part IIa Fishes
Pristis zijsron	Longcomb Sawfish	Indo-Pacific	Schedule I; Part IIa Fishes
Rhynchobatus djiddensis	Giant Guitarfish	Indo-Pacific	Schedule I; Part IIa Fishes

(Continued)

TABLE 27.1 Scheduled[1] Marine Animals of India *(cont.)*

Scientific name	Common name	Distribution	Status
Urogymus asperrimus	Porcupine Ray	Indo-Pacific	Schedule I; Part IIa Fishes
Molluscs			
Cassis cornuta	Horned Helmet	Indo-Pacific: Andaman Islands, Gulf of Mannar (Tamil Nadu)	Schedule I; Part IVb
Cypraecassis rufa	Bull mouth Helmet	Indo-Pacific. India: ANI	Schedule I; Part IVb
Charonia tritonis	Trumpet Triton	Indo-Pacific. India: ANI, Lakshadweep	Schedule I; Part IVb
Tudicla spirallus	Spiral Vase	India: South India (Pondicherry)	Schedule I; Part IVb
Conus milne-edwardsi	Glory of India	Indian Ocean. India: ANI.	Schedule I; Part IVb
Nautilus pompilius	Chambered Nautilus	Indian Ocean, Western & Central Pacific. India: Andaman Islands	Schedule I; Part IVb
Tridacna maxima	Elongate Giant Clam	Indo-Pacific. India: ANI	Schedule I; Part IVb
Tridacna squamosa	Fluted Giant Clam	Indo-Pacific. India: Andaman Islands	Schedule I; Part IVb
Hippopus hippopus	Bear Paw Clam	Indo-Pacific. India: Nicobar Islands	Schedule I; Part IVb
Trochus niloticus	Commercial Trochus	Indo-Pacific. India: ANI	Schedule IV; Part 19, XIV
Turbo marmoratus	Great Green Turban	Indo-Pacific. India: Andaman Islands	Schedule IV; Part 19, XIV
Strombus (Dolomena) plicatus sibbaldii	Sibbald's Conch	Gulf of Aden to Sri Lanka. India: Kerala Coast	Schedule IV; Part 19, XIII
Lambis (Harpago) chiragra chiragra	Chiragra Spider Conch	Indian Ocean to Eastern Polynesia. India: ANI	Schedule IV; Part 19, VI
Lambis (Harpago) chiragra arthritica	Arthritic Spider Conch	East Africa to Central Indian Ocean. India: Pondicherry	Schedule IV; Part 19, VII
Lambis (lambis) crocata	Orange Spider Conch	Indo-West Pacific. India: Andaman Islands.	Schedule IV; Part 19, VIII
Lambis (lambis) truncata truncata	Truncate Spider Conch	East African Coast to Bay of Bengal and Cocos Keeling Atoll. India: ANI	Schedule IV; Part 19, XI
Lambis (Millepes) millepeda	Millipede Spider Conch	South-Western Pacific. Indian Ocean	Schedule IV; Part 19, IX

TABLE 27.1 Scheduled[1] Marine Animals of India *(cont.)*

Scientific name	Common name	Distribution	Status
Lambis (Millepes) scorpius scorpius	Scorpio Conch	Indonesia and Ryukyu Islands to Samoa. India: This subspecies not recorded from Indian waters. *L. (M.) scorpio indomaris* Abott. Occurring in ANI.	Schedule IV; Part 19, X
Cypraea limacina	Limacina Cowrie	Indo-West Pacific. India: Tamil Nadu, Gulf of Mannar.	Schedule IV; Part 19(i)
Cypraea mappa	Map Cowrie	Indo-Pacific. India: ANI	Schedule IV; Part 19(ii)
Cypraea talpa	Mole Cowrie	Indo-Pacific. India: ANI	Schedule IV; Part 19(iii)
Pleuroploca trapezium	Trapezium Horse Conch	Indo-Pacific. India: Andaman Islands, West Bengal Coast.	Schedule IV; Part 19(iv)
Harpulina arausiaca	Vaxillate Volute	Sri Lanka & Southern India: Gulf of Mannar.	Schedule IV; Part 19(v)
Placuna placenta	Windowpane Oyster	Philippines; Southeast Asia. India: Andaman Island, Goa, Orissa, Tamil Nadu, West Bengal coasts	Schedule IV; Part 19(xii)
Echinodermata (all Holothurians)	Sea Cucumber	Indo-Pacific. India: Gulf of Kutch, Mannar, Palk Bay, Lakshadweep Islands, ANI.	Schedule I; Part IV-C
Coelenterates			
All Scleractinians	Reef Building Coral	Indo-Pacific. India: Gulf of Kutch, Mannar, Palk Bay, Lakshadweep Islands, ANI	Schedule I; Part IVA
All Antipatharians	Black Coral	Indo-Pacific. India: Gulf of Kutch, Mannar, Palk Bay, Lakshadweep Islands, ANI	Schedule I; Part IVA
Tubipora musica	Organ Pipe Coral	Indo-Pacific. India: Lakshadweep and Nicobar Islands	Schedule I; Part IVA
All *Millepora* species	Fire Coral	Indo-Pacific. India: Gulf of Kutch, Mannar, Palk Bay, Lakshadweep Islands, ANI	Schedule I; Part IVA
All Gorgonians	Sea Fan	Indo-Pacific. India: Gulf of Kutch, Mannar, Palk Bay, Lakshadweep Islands, ANI	Schedule I; Part IVA
All Calcareans	Sponges	Indo-Pacific. India: Gulf of Kutch, Mannar, Palk Bay, Lakshadweep Islands, ANI	Schedule III

ANI, Andaman and Nicobar Islands.
[1]*Wildlife (Protection) Act, 1972.*

Other groups such as sea cucumber, corals, calcareous sponges, sea fan and black corals are included, but not species-wise from India. This is an attempt to include such species which are threatened from India, the results of studies done on the coral reefs of the Andaman and Nicobar Islands for more than two and half decades. A total of 88 species of sea cucumber, 364 species of coral and 92 species of calcareous sponge are reported from India.

Members of this genus *Acropora* have a low resistance and low tolerance to bleaching and disease, and are slow to recover. These species is found typically in shallow water and are particularly vulnerable to bleaching and human impacts. *Acropora* species in the Andaman and Nicobar Islands were once very abundant, but in recent decades have remained at low levels of abundance, with no signs of recovery and, in some areas, continued decline. These species are believed to be most greatly threatened by disease, temperature-induced bleaching, and physical damage from hurricanes.

Six major groups of marine species—mammals, reptiles, fish (sharks and rays, groupers), shells, reef-building corals and sponges—are discussed here.

Marine Mammals

Marine mammals are a diverse group of species and include whales, dolphins and the dugong (order Sirenia). Most are cetaceans that are mainly encountered as individuals that have been stranded on beaches, or have been captured in fishing nets. One quarter of marine mammal species are in threatened categories. Major threats to these species include entanglement in fishing gear, directed harvesting, and the effects of boat strikes. In many regions, marine mammals are also threatened by water pollution, habitat loss from coastal development, loss of prey or other food sources due to poor fisheries management, and intensive hunting both historically and in place today. There are 79 species currently recognized in the world; 5 species of whale, 4 species of dolphin and one species of porpoise, *Dugong dugon* (sea cow), are recorded from India.

Marine Reptiles

Saltwater Crocodiles

The saltwater or estuarine crocodile, *Crocodylus porosus* (Figure 27.1), is the largest of all living reptiles. It is found in suitable habitats in the Indo-Pacific coast. Saltwater crocodiles generally spend the tropical wet season in freshwater swamps and rivers, moving downstream to estuaries in the dry season, and sometimes traveling far out to sea.

Marine Turtles

Six of the seven species of marine turtle (order Testudines) are listed in threatened categories. Threats to marine turtles occur at all stages of their life cycle. Marine turtles lay their eggs on beaches, which are subject to threats such

FIGURE 27.1   Saltwater crocodile (*Crocodylus porosus*).

as coastal development, sand mining, earthquake and tsunami. The eggs and hatchlings are threatened by pollution and predation by introduced predators such as pigs and dogs, and eggs are collected by humans for food in many parts of the Andaman and Nicobar Islands. Once at sea, marine turtles are faced with threats from targeted capture in small-scale subsistence fisheries, by catch largely by long-line and trawling activities, entanglement in marine debris, and boat strikes. Their life history characteristics of being long-lived, late to mature and with a long juvenile stage, combined with the many threats from human activities in the sea and on land that affect them at all stages of their life cycle, are among the reasons for their high risk of extinction. In addition, global climate change is now considered to be a serious, if not entirely understood, threat. Given their long generation times, global distributions, and the paucity of long-term data, assessing the risk of extinction for marine turtle species is challenging.

Threats to Turtles

Threats to turtles include poaching by feral dogs and humans, sand mining, incidental catch, habitat degradation, tourism and their status as an important food source for the original inhabitants of the Andaman and Nicobar Islands. The original inhabitants carried out subsistence hunting offshore or by capturing nesting turtles on beaches. They also collected turtle eggs, which formed a valuable protein source. In the Andaman Islands, all the species except the leatherback were hunted for meat. In Nicobar, cooked turtle meat is consumed regularly.

Fish

Sharks and Rays

Of the 1046 species of sharks and their relatives (class Chondrichthyes) of the world, 55 species are reported from the Andaman and Nicobar Islands, six of which, in the order Rajiformes (skates and rays), fall under the Schedule I of the Wildlife (Protection) Act, 1972: in family Pristidae (sawfish), *Anoxypristis cuspidata* (Figure 27.2), *Pristis zijsron* and *Pristis microdon*; in family

FIGURE 27.2   Knifetooth sawfish (*Anoxypristis cuspidata*).

Rhinobatidae (guitarfish), *Rhina ancylostoma*, *Rhinobatos granulatus* and *Rhynchobatus djiddensis* and the species number may increase in the near future if current threats are not reduced.

The Pondicherry shark (*Carcharhinus hemiodon*) is an extremely rare, possibly extinct species of requiem shark, family Carcharhinidae. A small and stocky grey shark, it grows not much longer than 1 m (3.3 ft) and has a fairly long, pointed snout. This species can be identified by the shape of its upper teeth, which are strongly serrated near the base and smooth-edged near the tip, and by its first dorsal fin, which is large with a long free rear tip. Furthermore, it has prominent black tips on its pectoral fins, second dorsal fin, and caudal fin lower lobe. Not seen since 1979, the Pondicherry shark was once found in Indo-Pacific coastal waters from the Gulf of Oman to New Guinea, and may have entered fresh water. The IUCN has listed the Pondicherry shark as Critically Endangered. If it still survives, it will be threatened by intense and escalating fishing pressure throughout its range.

The Ganges shark (*Glyphis gangeticus*) is a critically endangered species of requiem shark found in the Ganges river of India. The Ganges shark, as its name suggests, is largely restricted to the rivers of eastern and northeastern India, particularly the Hooghly River of West Bengal, and the Ganges, Brahmaputra, and Mahanadi in Bihar, Assam, and Odisha, respectively. It is typically found in the middle to lower reaches of a river. River sharks are thought to be particularly vulnerable to habitat changes. The Ganges shark is restricted to a very narrow band of habitat that is heavily impacted by human activity. Overfishing, habitat degradation from pollution, increasing river use and management, including construction of dams and barrages, are the principal threats.

The speartooth shark (*Glyphis glyphis*) is an extremely rare species of requiem shark, family Carcharhinidae. Only immature specimens, which inhabit the tidal reaches of large tropical rivers in northern Australia and New Guinea, are known. It is exclusively found in fast-moving, highly turbid waters over a wide range of salinities. The speartooth shark is threatened by incidental capture in commercial and recreational fisheries, as well as by habitat degradation. Given its small population, restricted range, and stringent habitat requirements, this species is highly susceptible to these pressures and has been listed as Endangered by IUCN.

The porcupine ray (*Urogymnus asperrimus*) is a rare species of stingray in the family Dasyatidae and the only member of its genus. This bottom-dweller is found throughout the tropical Indo-Pacific, as well as off west Africa. The porcupine ray has long been valued for its rough and durable skin, which was made into a shagreen leather once used for various utilitarian and ornamental purposes, such as to cover sword hilts and shields. It is caught incidentally by coastal fisheries. Unregulated fishing has led to this species declining in many parts of its range, and thus has been listed as Vulnerable by the IUCN.

Seahorses compose the fish genus *Hippocampus* within the family Syngnathidae, in order Syngnathiformes. They are mainly found in shallow tropical and temperate waters throughout the world. They prefer to live in sheltered areas such as sea grass beds, coral reefs, or mangroves. Seahorse populations are thought to have been endangered in recent years by overfishing and habitat destruction. Import and export of seahorses has been controlled under CITES.

Groupers

Groupers (family Serranidae) are found in rocky and coral reefs of the tropics and sub-tropics around the world (161 species), and are also subject to threats from over-exploitation from fishing, especially for the live fish trade, given their high commercial value. Among the commercially important marine fishes, the groupers emerged as a particularly vulnerable group of fishes. At least two of the Andaman Nicobar Islands' 50 grouper species are now in a vulnerable position. One species, the giant grouper (*Epinephelus lanceolatus*) (Figure 27.3) is protected in the Andaman and Nicobar Islands under Schedule I of the Wildlife (Protection) Act, 1972.

Molluscs

The phylum Mollusca is very large, consisting of thousands of species commonly known as shells (Figure 27.4). A total of 1763 species of mollusc are recorded from the Andaman and Nicobar Islands. They occur in terrestrial, freshwater and marine habitats, the last including mangroves, coral reefs, rocky coasts, sandy beaches, sea grass beds and also at greater depths in the sea.

FIGURE 27.3   Giant grouper (*Epinephelus lanceolatus*). (Please see color plate at the back of the book.)

(a)

(b)

(c)

(d)

(e)

FIGURE 27.4 Molluscs. (a) *Trochus niloticus*; (b) *Turbo marmoratus*; (c) *Nautilus pompilius*; (d) *Tridacna squamosa*; (e) *Charonia tritonis*. (Please see color plate at the back of the book.)

KEY MESSAGES

The preservation and protection of our ocean resources, not only for the marine species they contain, but also for the food, products and ecosystem services that they provide for billions of people around the globe, needs to become an urgent priority. Many of the threats listed for marine species are overlapping. The development of sustainable fisheries, including the elimination of harmful fishing or harvesting practices, the enforcement of current fishery regulations, and implementation of improved fishery technology, are essential for reduction of the extinction risks for marine species. Similarly, more attention needs to be aimed at reducing pollution and destructive development of coastal areas. The need to slow or reverse global climate change is becoming more important to protect our planet's resources and quality of life, not only for the survival of the plants and animals living in the ocean, but for those that live on land or in freshwater as well. The continued assessment of the status of marine species is essential for monitoring of the impact of threats to the ocean's health and survival. However, the conservation status of the vast majority of marine species has not yet been investigated. In many parts of India, there have been huge declines in population for commercially important marine resources. Plans to compile data on distribution, ecology, population numbers and trends, and threats to marine species are well underway. The groups include all marine vertebrates, as well as important habitat-forming primary producers such as the corals. The conservation status of species in several important invertebrate groups such as gastropod molluscs, bivalve molluscs and echinoderms (such as starfish, sea urchins and sea cucumbers) will also be assessed. It is the most commercially important marine species threat data ever attempted, and will provide essential information for the protection and conservation of India's vital marine resources.

REFERENCES

Carpenter, K.E., Abrar, M., Aeby, G., Aronson, R.B., Banks, S., Bruckner, A., Chiriboga, A., Cortés, J., Delbeek, J.C., DeVantier, L., Edgar, G.J., Edwards, A.J., Fenner, D., Guzmán, H.M., Hoeksema, B.W., Hodgson, G., Johan, O., Licuanan, W.Y., Livingstone, S.R., Lovell, E.R., Moore, J.A., Obura, D.O., Ochavillo, D., Polidoro, B.A., Precht, W.F., Quibilan, M.C., Reboton, C., Richards, Z.T., Rogers, A.D., Sanciangco, J., Sheppard, A., Sheppard, C., Smith, J., Stuart, S., Turak, E., Veron, J.E.N., Wallace, C., Weil, E., Wood, E., 2008. One-third of reef-building corals face elevated extinction risk from climate change and local impacts. Science 321, 560–563.

Dulvy, N.K., Sadovy, Y., Reynolds, J.D., 2003. Extinction and vulnerability in marine populations. Fish and Fisheries 4, 25–64.

Edgar, G.J., Banks, S., Bensted-Smith, R., Calvopiña, M., Chiriboga, A., Garske, L.E., Henderson, S., Miller, K.A., Salazar, S., 2008. Conservation of threatened species in the Galapagos Marine Reserve through identification and protection of marine Key Biodiversity Areas. Aquatic Conservation: Marine and Freshwater Ecosystems 18, 955–968.

Hoffmann, M., Brooks, T.M., da Fonseca, G.A.B., Gascon, C., Hawkins, A.F.A., James, R.E., Langhammer, P., Mittermeier, R.A., Pilgrim, J.D., Rodrigues, A.S.L., Silva, J.M.C., 2008. Conservation planning and the IUCN Red List. Endangered Species Research 6, 113–125.

Ramakrishna, Dey, A., 2003. Manual on the identification of schedule mollusks from India:. Director, Zoological Survey of India, Kolkata, p.40.

Roberts, C.M., Hawkins, J.P., 1999. Extinction risk in the sea. Trends in Ecology and Evolution 14, 241–246.

Rodrigues, A.S.L., Pilgrim, J.D., Lamoreux, J.F., Hoffmann, M., Brooks, T.M., 2006. The value of the IUCN Red List for conservation. Trends in Ecology and Evolution 21, 71–76.

Chapter 28

Macrofaunal Assemblages of Carbyn's Cove Mangroves, South Andaman

R. Mohanraju

Department of Ocean Studies and Marine Biology, Pondicherry University,
Port Blair, Andaman and Nicobar Islands, India

INTRODUCTION

Mangroves are woody plants that grow in tropical and subtropical latitudes between 30° S and 30° N along the land–sea interface in bays, estuaries, lagoons, backwaters and in rivers, reaching upstream as far as the water remains saline (Qasim, 1998). They are quite old in evolutionary terms, possibly arising just after the first angiosperms (Duke, 1992), and comprise 60 species worldwide (Singh *et al.*, 2012). Most mangroves are found in tropical regions between 5° N and 5° S (Giri *et al.*, 2011). These are often called "tidal forests," "coastal woodlands," or "oceanic rainforests." These plants and their associated organisms (microbes, fungi, plants and animals) constitute the "mangrove forest community."

The benefits obtained from these ecosystems are quite broad and encompass a variety of economic, environmental and social aspects, including carbon sequestration to combat global warming, and protection from erosion, flooding, cyclones, typhoons and tidal waves (Primavera, 2000). As they are situated at the border between land and open sea, mostly at river mouths or around lagoons, they protect the coastline against natural disasters, such as when the great M9 earthquake followed by tsunami struck the Andaman and Nicobar Islands on 26 December 2004 (Agarwal *et al.*, 2005).

The Andaman and Nicobar Islands (ANI) are a group of 572 islands and islets located geographically north to south between 6°45′ to 13°40′ N latitudes and 92°12′ to 93°55′ E longitudes. They extend over 800 km, with a coastline that covers over 1962 km (Sivaperuman and Raghunathan, 2012), constituting about 0.25 percent of total Indian land mass. The coast of ANI accommodates abundant mangrove vegetation over an extent of about 615 km, contributing 22.5 percent of the total for India, comprising about 33 species belonging to 15 genera and 12 families of true mangroves (Singh *et al.*, 2012).

Marine Faunal Diversity in India. DOI: 10.1016/B978-0-12-801948-1.00028-8

MANGROVE MACROFAUNA

The mangrove ecosystem contains a rich amount of flora, including bacteria, fungi, algae, lichens, sea grass, salt marsh and other vegetation. Fauna includes both terrestrial, semi-aquatic and aquatic organisms: zooplanktons, benthos, shrimps, crabs, molluscs, insects, finfish, birds, reptiles, amphibians and marine mammals (Kathiresan and Rajendran, 2005). The diversity of invertebrates in Indian mangroves is rich, with more than 500 species of insects and archnida, 229 species of crustacean, 212 species of mollusca, 50 species of nematodes, and 150 species of planktonic and benthic organisms (Gopal and Krishnamurthy, 1993).

The leaf litter detritus from mangroves is important to fisheries because it provides an essential source of nutrients for the trophic food web and juvenile fish. It is estimated that 90 percent of all marine organisms spend some portion of their life cycle within mangrove systems. Mangroves are known to be the nursery ground for several shellfish and finfish (Primavera, 1998; Acosta and Butler, 1997).

Studies of the faunal communities of the mangrove ecosystem have been made by several researchers. Fernando (1987) studied the intertidal faunal assemblage of the Vellar estuary and found that the most abundant faunal groups were polychaetes, followed by tanaids, bivalves and gastropods. The role of mangrove molluscans as transitional fauna between land and sea was studied by Plaziat (1984), who concluded that only a few species, such as *Terebralia palustris, Littorina scabra and Littorina angulifera*, are mangal exclusive. The faunal assemblage of the soft sediments of mangroves comprises mainly polychaetes, molluscs and crustaceans as dominant groups in order of abundance (Murugan and Ayyakkanu, 1991; Kumar, 1995; Pillai, 1997; Kari and Paul, 2004), followed by echiurids, echinoderms, nemertean worms, bryozoans, sipunculids and fishes (Ansari *et al*., 1994). Mohammed (1995) observed more than 80 species of benthic fauna in the soft bottom sediments on the western side of the Arabian Gulf.

The macrobenthic communities, their diversity and community ecology and behavior in the mangrove ecosystem have been studied worldwide (Cannicei *et al*., 2002; Maria *et al*., 2003; Gilberto *et al*., 2004). Pereira *et al*., (2004) studied the dominant fauna of mangroves at Seven Bungalows Beach, Versova and reported that crabs and molluscs were the dominant groups. The diversity and vertical zonation of benthic crustaceans in soft substrates of mangrove were studied by Silva and Almeida (2004). Macintosh *et al.* (2002) studied the role of mangrove faunal communities as indicator species to monitor ecological and environmental changes in the ecosystem. Terrigenous sedimentation from terrestrial sources into the mangrove ecosystem has a drastic impact on the physiology of mangroves and associated macrobenthic communities (Ellis *et al*., 2004). The present study is a case study of the macrofaunal assemblages of the mangrove ecosystem of Carbyn's Cove, Port Blair, South Andaman.

METHODS

Study Area

The study was carried out in the mangroves of Carbyn's Cove (11° 38.475' N latitude and 92° 44.600' E longitude), which is a small bay situated at the southern tip of Port Blair, Andaman. The important feature of this area is that a sandy beach adjoins the thick mangrove region and serves as a tourist place. The mangrove belt is about 220 m long and 22 m broad on average. The study site is frequently inundated by tidal flow from a channel. A rocky intertidal region separates the sandy beach from the mangrove region.

Data Collection

All the collections were made at low tide. The faunal communities were de-termined by the transect method suggested by English *et al.* (1997). The line intercept was laid for 20 m and then at regular intervals of 5 m, and four other intercepts were laid perpendicularly. Each quadrate measuring 1 m $\times$ 1 m was further divided into small squares measuring 20 cm each such that each quad-rate contained 25 squares. The survey was carried out throughout the mangrove region. The quadrate was placed at 0 m, 5 m and 10 m from the shoreline respectively, and samples were collected with the use of a 15 cm PVC corer. The corer had an area of 6 cm^2, and samples were taken to a depth of 10 cm. On every sampling occasion 5–10 sediment samples were taken randomly from the quadrate. The sediment samples were transferred into plastic covers and sieved at the nearby sandy beach. The samples were passed through a series of sieves measuring 4 mm, 2 mm, 1 mm and 62 μm. The organisms retained in the sieve were then transferred into plastic containers and preserved in 4 percent formalin and identified down to species level.

RESULTS AND DISCUSSION

Macrofauna

Epibenthic species were found to dominate the macrofauna. The *cerithids* dom-inated along with other gastropods; certain bivalves, crustaceans such as prawns and crabs, and fish, among them mudskippers and pipe fishes, were recorded in the sediment samples. Although many prawns and crabs burrows were observed in the mangrove sediment, it was difficult to capture them by means of the corer. The species observed are listed in Table 28.1, and examples are shown in Figures 28.1 and 28.2.

Sieving was a very tedious and cumbersome process as most of a typical sediment sample was filled with the roots and twigs of the mangroves. The mangrove canopy was thick, and little light reached the sediments, which were always wet and rich with mangrove detritus. Sediment was soft mud and ankle

TABLE 28.1 Macrofauna of Carbyn's Cove Mangroves

Family	Species name
Phylum Arthropoda	
Class Crustacea	
Penaeidae	*Penaeus* sp.
Geocarcinidae	*Cardisoma carnifex* (Herbst)
Diogenidae	*Clibanarius* sp.
Phylum Mollusca	
Class Gastropoda	
Potamididae	*Terebralia palustris* (Linnaeus)
	Cerithidea cingulate (Gmelin)
Ellobidae	*Cassidula nucleus* (Gmelin)
Nassaridae	*Nassarius lurida*
Littorinidae	*Littorina scabra scabra* (linne)
	Littorina undulate (Gray)
Neritidae	*Nerita grayana*
	Neritina violaceae (Gmelin)
Class Bivalvia	
Ostereidae	*Crassostrea* sp.
	Saccostrea cucullata (Born)
Tellinidae	*Tellina angulate* (Gmelin)
Phylum Pisces	
Class Osteichthyes	
Order Syngnathiformes	
Syngnathidae	*Halicampus dunkeri*
Order Perciformes	
Gobiidae	*Periopthalmus* sp. (Pallas)

deep at all times. Crabs and molluscs form an important link between the primary detritus and the base of the food web and consumers at higher trophic levels (Macintosh, 1984) and, because of their large abundance and biomass, the energy assimilated by macrofauna plays a significant role in nutrient cycling. Studies of mangrove macrofauna in southeast Asia (Ashton *et al.*, 1999) have described the abundance and distribution of benthic macrofauna in relation to environmental conditions. Molluscs and crabs are especially abundant and rich in species (Macnae, 1968; Sasekumar, 1974; Frith *et al.*, 1976).

Surface temperature was 27–28°C; salinity was 30–31‰; pH was 7.5–8. The sediment was anoxic, with the black mud being seen below 2–3 cm of the surface soil. The dissolved oxygen was 1.91–2.03 mg/L.

Crustaceans

Mangrove habitat is rich in shrimp and prawn resources. Commercial prawn fishery yields are greater on the coasts with luxuriant mangrove forests than where mangroves are absent. Mangrove leaf litter provides an important

(a)

(b)

(c)

(d)

(e)

(f)

FIGURE 28.1 Gastropods of Carbyn's Cove mangroves. (a) *Terebralia palustris*; (b) *Cerithidea cingulata*; (c) *Littorina scabra scabra*; (d) *Littorina undulata*; (e) *Cassidula nucleus*; (f) *Nassarius lurida*.

(Continued)

(g)   (h)

(i)

FIGURE 28.1 *(cont.)*   (g) *Nerita grayana*; (h) *Neritina violaceae*; (i) *Tellina angulata*.

nutrient base for food webs. Plenty of species make use of the mangrove eco-system at different levels: some are obligate, i.e., spend their entire life cycle in this ecosystem, or are a crucial part of it (e.g., *Penaeid* prawns), using it as shelter or as a source of food; others are facultative, i.e., are able to survive and reproduce even in the absence of mangroves but show preference for the habitats and nutrients provided therein. The mangal acts as a sink of settle-ment and early growth of shrimps and prawns, and it may also be a source of larvae that are transported to other habitats. During decomposition of man-grove litter, a large amount of nutrient is released, and detritus food is formed. Mangrove waters serve as an essential nursery ground for juveniles of many species of prawns and shrimps (Figure 28.3).

Mangrove crabs play an essential role in leaf litter degradation in the man-grove ecosystem (Robertson, 1986; Micheli, 1993). Surprisingly, crabs were very sparse in the Carbyn's Cove mangrove forest, contrary to the view of Robertson (1986) and Micheli (1993), who noted that the mangrove forest was dominated by sesarmid crabs. The burrowing activities of the crabs improve soil aeration (Smith *et al.*, 1991), allow seawater penetration and nutrient exchange (Paphavasit *et al.*, 2009) and alter the topography and textural properties of man-grove soils (Warren and Underwood, 1986).

(a)

(b)

(c)

(d)

(e)

(f)

FIGURE 28.2   Crustaceans, bivalves and fin fish of Carbyn's Cove mangroves. (a) *Cardisoma carnifex*; (b) *Clibanarius* sp.; (c) *Saccostrea cuccullata*; (d) *Crassostrea* sp.; (e) *Halicampus dunkeri*; (f) *Periopthalmus* sp.

Molluscs

Molluscs are among the dominant group in structuring the mangrove ecosystem and also constitute a component of fouling communities (Marquez and Jimenez, 2004; Vilardy and Polania, 2004). The molluscan species recorded at Carbyn's Cove were found to be similar to those recorded from peninsular Malaysia (Macnae, 1968; Sasekumar 1974), Thailand (Frith *et al.*, 1976). *Terebralia*

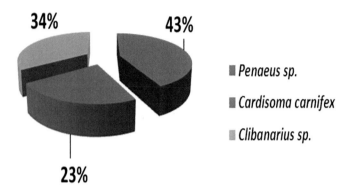

FIGURE 28.3 Species abundance of crustaceans in the study area.

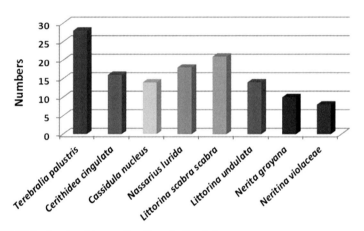

FIGURE 28.4 Species richness of molluscs in the study area.

palustris and *Littorina scabra scabra* were found to be the most dominant and co-dominant epifaunal communities (Figure 28.4). Mangrove molluscs are unusual in that, first, they are a transitional fauna between sea and land and, second, they are characteristically eurybiotic species belonging to both hard and soft substrates. There is no family that is specific to the mangal environment. Nevertheless some species, such as *Littorina scabra, Littorina angulifera, Terebralia palustris* and several melampids, are mangrove exclusive (Plaziat 1984). Berry (1964) has distinguished the following vertical zonation in mangroves: "Littorina" zone, or high zone; "Nerita" zone, or lower tree zone; and "Uca" zone, or ground zone. This phenomenon was clearly visible in the present study. *Saccostrea cucculata* or the "mangrove oysters" were observed to grow well on mangrove trees. Some soft bottom species such as cerithids and ellobids and also some melaniids were observed on the trunks in the Nerita zone. Brown has also shown that *Cerithidea cingulata* does not feed on the bark flora and moves down to feed on the mud, as do other species of this genus. The same is true for the climbing melampids, for instance for *Cassidula nucleus* of New Caledonia, and for the neritids.

Most of the suspension-feeding bivalves, together with the barnacles, do not extend beyond the mean water neap tide level. On the mud flats, filter feeders appear only exceptionally in the upper intertidal. Young specimens like *Terebralia palustris* are deposit feeders while the adults can feed by rasping fallen mangrove leaves or young root bark (Plaziat, 1984). It is well known that infralittoral *cerithids* are prey for many molluscs, sea stars and crabs. On the mangrove flats, crabs are obviously the major predators. Live gastropod shells frequently provide a substrate for sessile organisms. Oysters growing on potamids can seldom reach maturity because their increased feeding needs contrast with the fact that the specimens of *Terebralia* carrying them lie to higher intertidal levels with increasing age.

Fin Fish

Mangrove habitats usually contain a rich ichthyofauna, and in the present study two species of fishes were found: pipefish, *Halicampus dunkeri,* and mudskippers, *Periopthalamus* sp. Mangrove habitats provide an ideal niche for the juvenile fish, because of low water motion, soft substratum and an enormous amount of food, excellent shelter and protection from predatory organisms. Besides, the canopy of the mangroves provides a cool, stable and humid environment quite favorable to the young fish. The number of fin fish species in the mangroves exceeds that in all other habitats.

REFERENCES

Acosta, C.A., Butler, M.J., 1997. Role of mangrove habitat as a nursery for juvenile spiny lobster, *Panulirus argus* Belize. Mar. Freshwat. Res. 48, 721–727.

Agarwal, V., Agarwal, K.N., Kumar, R., 2005. Simulations of the 26th December 2004 Indian Ocean tsunami using a multi-purpose ocean disaster simulation and prediction model. Curr. Sci. 88, 439–444.

Ansari, Z.A., Shreepada, R.A., Kanti, A., Gracia, E.S., 1994. Macro benthic assemblage in the soft sediment of Marmagao harbor, Goa (central coast of India). Ind. J. Mar. Sci. 23, 225–231.

Ashton, E.C., Hogarth, P.J., Ormond, R., 1999. Breakdown of mangrove leaf litter in a managed mangrove forest in Peninsular Malaysia. Hydrobiologia 413, 77–88.

Berry, W.B.N., 1964. The Middle Ordovician of the Oslo region, Norway 16. Graptolites of the Ogygiocaris Series. Norsk Goel. Tidsskr. 44, 61–170.

Cannicei, S., Morino, L., Vannini, M., 2002. Behavioral studies for visual recognition of predators by mangrove climbing crab, *Sesarma leptosome*. Ani. Behav. 63, 77–83.

Duke, N.C., 1992. Coastal and Estuarine studies: Tropical Mangrove Ecosystems. American Geophysical Union, Washington DC, pp. 63-100.

Ellis, J., Nicholls, P., Craggs, R., Hofstra, D., Hewitt, J., 2004. Effect of terrigenous sedimentation on mangrove physiology and associated macro benthic communities. Mar. Ecol. Prog. Ser. 270, 71–82.

English, S., Wilkinson, C., Baker, V., 1997. Survey Manual for Tropical Marine Resources, 2nd edn. Australian Institute of Marine Science, Townsville, p. 390.

Fernando, O.J., 1987. Studies on the intertidal fauna of Vellar estuary. J. Mar. Biol. Ass. Ind. 29, 86–103.

Frith, D.W., Tantansiriwong, R., Bhatia, O., 1976. Zonation and abundance of macro fauna on a mangrove shore, Phuket Island. Phuket. Mar. Biol. Cent. Res. Bull. 10, 1–37.

Giberto, D.A., Bremec, C.S., Acha, E.M., Mianzan, H., 2004. Large scale spatial patterns of benthic assemblage in the south west Atlantic: the Rio de La Plata estuary and adjacent continental shelf waters. Estuar. Coast. Shelf Sci. 61, 1–13.

Giri, C., Ochieng, E., Tieszen, L.L., Zhu, Z., Singh, A., Loveland, T., Masek, J., Duke, N., 2011. Status and distribution of mangrove forests of the world using earth observation satellite data. Global Ecol. Biogeography 20 (1), 154–159.

Gopal, B., Krishnamurthy, K., 1993. Wetlands of south Asia. Whigham, D.F., Dykyjova, D., Hejny, S. (Eds.), Wetlands of the World: Inventory Ecology and Management, Vol 1, Springer, Dordrecht, pp. 345–414.

Kari, E.E., Paul, K.S.S., 2004. Spatial patterns of soft sediment benthic biodiversity in sub-tropical Hong Kong waters. Mar. Ecol. Prog. Ser. 276, 25–35.

Kathiresan, K., Rajendran, N., 2005. Mangrove ecosystem of the Indian Ocean region. Indian J. Mar. Sci. 34, 104–113.

Kumar, R.S., 1995. Macro benthos in the mangrove ecosystem of Cochin backwaters, Kerala (South west coast of India). Indian J. Mar. Sci. 24, 56–61.

Macintosh, D.J., 1984. Ecology and productivity of Malaysian mangrove crab populations (Decapoda: Brachyura). In: Soepadmo, E., Rao, A.N., Macintosh, D.J. (Eds.), In: Proceedings of the Asian Symposium on Mangrove Environment, Research and Management, Kuala Lumpur, 1984. University of Malaya and UNESCO, pp. 354–377.

Macintosh, D.J., Ashton, E.C., Havanon, S., 2002. Mangrove rehabilitation and intertidal biodiversity: a study in the Ranong mangrove ecosystem. Thailand. Estuar. Coast. Shelf Sci. 55, 331–345.

Macnae, W., 1968. A general account of flora and fauna of mangrove swamps and forests on the Indo-pacific region. Adv. Mar. Biol. 6, 73–270.

Maria, D.B.A., Niki, B.C., Carla, M., 2003. Analysis of macro benthic communities at different taxonomic levels, an example from an estuarine environment in the Ligurian sea (North west Mediterranean). Estuar. Coast. Shelf Sci. 58, 99–106.

Marquez, B., Jimenez, M., 2004. Mollusks associated to the submerged roots of the red mangrove, Rhizophora mangle, in the Gulf of Sante Fe, Sucre State, Venezuela. Rev. Biol. Trop. 50 (3–4), 1101–1112.

Micheli, F., 1993. Feeding ecology of mangrove crabs in north eastern Australia: mangrove litter consumption by *Sesarma messa* and *Sesarma smithii*. J. Exp. Mar. Biol. Ecol. 171, 165–186.

Mohammed, S.Z., 1995. Observations of benthic macro fauna of the soft sediment on western side of Arabian Gulf (ROPME sea area) with respect to 1991 Gulf war oil spill. Ind. J. Mar. Sci. 24, 147–152.

Murugan, A., Ayyakkanu, K., 1991. Ecology of benthic macro fauna in Cuddalore-Uppanar backwater, South east coast of India. Ind. J. Mar. Sci. 20, 200–203.

Paphavasit, N., Aksornkoae, S., Silva, J.D., 2009. Tsunami impact on mangrove ecosystem. Thailand Environment Institute, Nonthaburi, Thailand, p. 211.

Pereira, C., Rao, C.V., Krishnan, S., 2004. Study of the dominant fauna of mangroves at seven bungalows beach, Versova, Mumbai – a preliminary study. In: Proceedings of National Seminar on Creeks, Estuaries and Mangroves – Pollution and Conservation, 28-30 Nov 2002. Vidya Prasarak Mandal's B.N. Bandodkar College of Science, Thane, India, pp. 196–200.

Pillai, N.G.K., 1997. Distribution and seasonal abundance of Macro benthos of cochin back waters. Ind. J. Mar. Sci. 6, 1–5.

Plaziat, J.C., 1984. Mollusc distribution in the mangal. In: Por, F.D., Dor, I. (Eds.), Hydrobiology of the Mangal, The Ecosystem of the Mangrove Forests. Developments in Hydrobiology 20. Dr W. Junk Publishers, The Hague, pp. 111–143.

Primavera, J.H., 1998. Mangroves as nurseries: Shrimp populations in mangrove and non-mangrove habitats. Estuar. Coast. Shelf Sci. 46, 457–464.

Primavera, J.H., 2000. Philippines mangroves – Status, threats and sustainable development. Asia-Pacific Cooperation on Research for Conservation of Mangroves: Proceedings of an International Workshop, Okinawa, Japan, 26-30 March 2000. The United Nations University, Tokyo.

Qasim, S.Z., 1998. Mangroves. Glimpses of the Indian Ocean. University Press, Hyderabad, pp. 123–129.

Robertson, A.I., 1986. Leaf-burying crabs: their influence on energy flow and export from mixed mangrove forests (*Rhizophora* spp.) in North Eastern Australia. J. Exp. Mar. Biol. Ecol. 102, 237–248.

Sasekumar, A., 1974. Distribution of macro fauna on a Malaysian mangrove shore. J. Anim. Ecol. 43, 51–59.

Silva, J.R.R., Almeida, Z.S., 2004. Vertical zoning of benthic crustaceans in soft substratesx of mangroves from Quebra-Pote, Sao Luis, Maranhao. Brazil. Bolet. Tecnico. Cientifico Do. Cepene. 10, 65–83.

Singh, A.K., Ansari, A., Kumar, D., Sarkar, U.K., 2012. Status, biodiversity and distribution of mangroves in India: An overview. In: Proceedings of National Conference on Marine Biodiversity, Lucknow, 22 May 2012. Uttar Pradesh State Biodiversity Board, Lucknow, pp. 59–67.

Sivaperuman, C., Raghunathan, C., 2012. Fauna of Protected Areas in Andaman and Nicobar Islands. Zoological Survey of India, Kolkata, p. 26.

Smith, T.J., Boto, K.G., Frusher, S.D., Giddins, R.L., 1991. Keystone species and mangrove forest dynamics: the influence of burrowing by crabs on soil nutrient status and forest productivity. Estuar. Coast. Shelf Sci. 33 (4), 419–432.

Vilardy, S., Polania, J., 2004. Molluscan fauna of the mangrove root fouling community at the Colombian archipelago of San Andres & Old Providence. Wetland Ecol. Manag. 10, 273–282.

Warren, J.H., Underwood, A.J., 1986. Effects of burrowing crabs on the topography of mangrove swamps in New South Wales. J. Exp. Mar. Biol. Ecol. 102, 223–235.

Chapter 29

Status of Fauna in Mangrove Ecosystems of India

K. Kathiresan, N. Veerappan and R. Balasubramanian
Centre of Advanced Study in Marine Biology, Faculty of Marine Sciences, Annamalai University, Parangipettai, Tamil Nadu, India

INTRODUCTION

Mangrove forests are among the world's most productive ecosystems, lying at the boundary between land and sea in tropical regions. Mangrove forests are often called "tidal forests," "coastal woodlands" or "oceanic rain forests." The mangrove system supports genetically diverse groups of aquatic and terrestrial animals. Mangrove fauna are specially adapted to exist under wide ranges of salinities, tidal amplitudes, winds, temperatures and even in muddy and anaerobic soil conditions. The rich faunal diversity in mangrove systems arises from diversified habitats such as core forests, litter-forest floors, mudflats, adjacent coral reef and sea grass ecosystems, and contiguous water bodies: rivers, bays, intertidal creeks, channels and backwaters. The calm waters in the forests are ideal nursery and breeding grounds for fish and shellfish, while the aerial roots, lower trunks of trees and forest floor support a varied fauna of oysters, snails, barnacles, crabs and other invertebrates. The forest also supports terrestrial animals such as birds, reptiles, insects and mammals (Kathiresan and Bingham, 2001; Kathiresan and Qasim, 2005).

The mangroves are extraordinary ecosystems of great importance in protecting coral reefs, sea grass beds and seaweeds, providing nursery grounds, feeding and breeding sites for a variety of organisms, and in supporting coastal fisheries and livelihood (Kathiresan, 2011). A mangrove-rich area provides as much as seventy-fold greater catch of fish, with consequently higher income of fishermen, than does a mangrove-sparse area (Kathiresan and Rajendran, 2002). Monetary value of the mangroves is US$ 9990 per hectare per year, which is greater than that of coral reefs, continental shelves, or the open sea (Costanza *et al.,* 1998).

Mangrove habitats continue to disappear globally at a rate of 0.66 percent per year (FAO, 2007). About 90 percent of the mangrove forest cover is

Marine Faunal Diversity in India. DOI: 10.1016/B978-0-12-801948-1.00029-X

485

found in developing countries, but is nearing extinction in 26 countries. Long term survival of mangroves is at great risk because of fragmentation of the mangroves. It is possible that the ecosystem services offered by mangroves may be totally lost within 100 years (Duke *et al.*, 2007). Mangrove habitat loss has put 16 percent of mangrove plant species and 40 percent of mangrove-associated animal species at an elevated risk of extinction in the world (Polidoro *et al.*, 2010). The present chapter deals with the current status of research on mangrove-inhabiting fauna, threats and conservation in India.

MANGROVE-INHABITING FAUNA

Mangroves are a rare forest type in the world and have an estimated cover of 15.2 million hectares in 123 countries and territories (FAO, 2007). The mangroves have greater abundance and diversity of fauna along wetland coastlines and in deltas and estuarine areas.

Mangroves in India cover a total area of 4663 km^2, of which about 59 percent is on the east coast along the Bay of Bengal, 28 percent on the west coast bordering the Arabian Sea, and 13 percent on the Andaman and Nicobar Islands. Mangrove forest ecosystems in India support diverse groups of fauna comprising of 3091 species (Table 29.1). This is perhaps the largest

TABLE 29.1 Total Number of Faunal Species Reported in Mangrove Ecosystems of India

Faunal group	No. of faunal species			
	East coast	*A&N Islands*	*West coast*	*Total*
Prawns	36	17	29	55
Crabs	88	58	28	138
Insects	371	340	10	707
Molluscs	183	112	83	305
Other invertebrates	573	74	182	745
Fish parasites	7	0	0	7
Finfish	331	249	125	543
Amphibians	8	5	0	13
Reptiles	80	7	3	84
Birds	322	52	264	426
Mammals	62	8	3	68
Total	**2061**	**922**	**727**	**3091**

Source: Kathiresan (2000).

biodiversity record in world mangrove ecosystems (Kathiresan, 2000). Invertebrates are greater in number of species than vertebrates. The faunal species so far recorded were highest (2061) in the mangroves of the east coast, followed by 922 species in Andaman and Nicobar Islands and 727 species on the west coast.

The Sundarbans of India and Bangladesh is the largest single block, of about 10,000 km^2, of mangrove forest in the world. The Sundarbans is the only tiger-mangrove kingdom in the world and is home to globally threatened species such as the Bengal tiger, sea turtle, fishing cat, estuarine crocodile, Gangetic dolphin and river terrapin.

CONSERVATION STATUS OF MANGROVE FAUNA

The threatened faunal species that inhabit the mangroves in India are shown in Tables 29.2 to 29.7. Of 52 species of marine fish assessed, 9 are vulnerable and 2 are endangered (Table 29.2); of 41 invertebrates assessed, 4 species are endangered, 4 species are vulnerable and one species is critically endangered (Table 29.3). In the Sundarbans, 4 reptile, 3 bird and 5 mammal species are extinct, and 10 reptile, 3 bird and 2 mammal species are under threat (Tables 29.4 to 29.6). In Gujarat, 3 birds and 2 turtle species are threatened (Table 29.7).

TABLE 29.2 Threatened Species of Fish in Indian Mangroves

Sl. No.	Species	Family	IUCN status
1	*Dasyatis uarnak*	Dasyatidae	Vulnerable
2	*Arius subrostratus*	Ariidae	Vulnerable
3	*Boleophthalmus boddarti*	Gobiidae	Vulnerable
4	*B. dussumieri*	Gobiidae	Endangered
5	*Scartelaos viridis*	Gobiidae	Endangered
6	*Periophthalmus koelreuteri*	Gobiidae	Vulnerable
7	*Elops machnata*	Elopidae	Vulnerable
8	*Muraenichthys schultzei*	Ophichthidae	Vulnerable
9	*Psammoperca waigiensis*	Centropomidae	Vulnerable
10	*Leiognathus splendens*	Leiognathidae	Vulnerable
11	*Secutor ruconius*	Leiognathidae	Vulnerable

IUCN, International Union for Conservation of Nature.
Source: Rao *et al.* (1998).

TABLE 29.3 Threatened Species of Invertebrates in Indian Mangroves

Sl. No.	Species	Family	IUCN status
1	Gelonia erosa	Corbiculidae	Endangered
2	Meretrix casta	Veneridae	Vulnerable
3	Cardisoma carnifex	Gecarcinidae	Critically Endangered
4	Macrophthalmus convexus	Ocypodidae	Endangered
5	Pilodius nigrocrinitus	Xanthidae	Endangered
6	Sesarma taeniolatum	Grapsidae	Vulnerable
7	Uca tetragonon	Ocypodidae	Endangered
8	Penaeus canaliculatus	Penaeidae	Vulnerable
9	P. japonicus	Penaeidae	Vulnerable

IUCN, International Union for Conservation of Nature.
Source: Rao et al. (1998).

TABLE 29.4 Threatened and Extinct Reptile Species in Sundarbans

Sl. No.	Species	Family
1	Crocodylus porosus	Crocodilidae
2	Varanus bengalensis	Varanidae
3	V. salvator	Varanidae
4	V. flavescens	Varanidae
5	Chelonia mydas[1]	Chelonidae
6	Eretmochelys imbricata[1]	Chelonidae
7	Lepidochelys olivacea	Chelonidae
8	Caretta caretta[1]	Chelonidae
9	Dermochelys coriacea[1]	Chelonidae
10	Lissemys punctata	Trionychidae
11	Trionyx gangeticus	Trionychidae
12	T. hurum	Trionychidae
13	Batagur baska	Emydidae
14	Python molurus	Boidae

[1]Extinct species (Chaudhuri and Choudhury, 1994).

TABLE 29.5 Threatened and Extinct Bird Species in Sundarbans

Sl. No.	Species	Family
1	Pelecanus philippinensis	Pelecanidae
2	Theskiornis melanocephalus	Threskiornithidae
3	Leptoptilos javanicus[1]	Ardeidae
4	Ardea goliath	Ardeidae
5	Sarkiodornis melanotus[1]	Anatidae
6	Cairina scutulata[1]	Anatidae

[1]Extinct species (Chaudhuri and Choudhury, 1994).

TABLE 29.6 Threatened and Extinct Mammal Species in Sundarbans

Sl. No.	Species	Family
1	Panthera tigris	Felidae
2	Muntiacus muntjac[1]	Felidae
3	Bubalis bubalis[1]	Felidae
4	Rhinoceros sondaicus[1]	Felidae
5	Cervus deruchea[1]	Cervidae
6	Axis porcinus[1]	Cervidae
7	Platanista gangetica	Platinistidae

[1]Extinct species (Chaudhuri and Choudhury, 1994).

TABLE 29.7 Threatened Species in Mangroves of Gujarat

Sl. No.	Species	Family
Birds		
1	Platelia leucorodia	Threskiornithidae
2	Pelecanus philippensiscrispus	Pelecanidae
3	P. philippensis	Pelecanidae
Turtles		
4	Chelonia mydas	Cheloniidae
5	Lepidochelys olivacea	Cheloniidae

Source: Sunderraj and Serebiah (1998).

RECENT STATUS OF RESEARCH ON MANGROVE FAUNA

The faunal communities in the mangroves are rich in both resident and visiting or transient fauna. Most visiting terrestrial fauna are insects, birds, mammals and reptiles. The aquatic visiting fauna are mainly fish and crustaceans with some molluscs and echinoderms. The visitors invade mangroves from the adjacent habitats such as forests, coral reefs, estuaries, creeks and bays. Resident fauna of mangroves are mainly benthic fauna of intertidal habitats, which are grouped under two broad categories: infauna and epifauna. Infauna animals, which burrow and penetrate the substratum, predominantly comprise polychaetes, brachyuran crabs, wood-boring animals, mud burrowing bivalves and gobiid fish. Epifauna include commonly occurring gastropods and some sessile bivalves, such as oysters, *Modiolus* spp. and barnacle crustaceans. Resident terrestrial fauna include birds such as the black capped kingfisher (*Halcyon pileata*), brown-winged kingfisher (*Halcyon amauroptera*) and mangrove whistler (*Pachycephala grisola*) and insects such as *Polyura schreiber* (Lepidoptera: Nymphalidae) (Ramakrishna, 2008).

Faunal diversity in different sites of Indian mangroves continues to be studied. From the Puduvypu mangroves in Cochin of Kerala have been recorded 70 bird species, 10 mammals, 12 reptiles, 12 fish and three amphibians (Gopikumar *et al.*, 2008). From the mangroves of North Malabar have been recorded 109 bird species, of which 34 are migratory (Khaleel, 2008). In Ratnagiri and Sindhudurg districts of Maharashtra, 13 crustaceans, 12 gastropods, 11 bivalves, 36 estuarine fishes, 50 birds and three mammals have been registered (Bhosale, 2008).

Nocturnal insect diversity of Indian mangroves has been studied using solar powered light traps. The Coringa of Andhra Pradesh state is represented by 90 species of insects belonging to eight major orders. Hemipterans are dominant, followed by Coleoptera, Hymenoptera and Diptera. In Hemiptera, the family Notoectidae is the most dominant, with four different species, followed by Cydnidae, Fulgoridae and Pentatomidae. Family Staphylinidae belonging to the order Coleoptera is the dominant one, comprising 12 species, followed by Carabidae, Hydrophillidae and Trogossitidae. In the order Hymenoptera, the family Formicidae is the dominant one with respect to the number of species (Remadevi *et al.,* 2008b). A similar study on nocturnal entomo-fauna has been made in Karnataka state. In that study, Coleoptera was found to be the largest order, followed by Hemiptera, Diptera, Homoptera and Hymenoptera. However, on the east coast, Hemiptera was the major insect order, followed by Coleoptera, Diptera and Hymenoptera (mostly Formicidae) (Remadevi *et al.,* 2008c). Regarding the diversity of oribatid mites in mangroves of Calicut district of Kerala state, there are a total of 11 species belonging to 9 genera and 9 families at the mangroves of Beypore, while 13 species belonging to 13 genera and 10 families occur at the Kottakadavu. Some species such as *Javacarus kuehnelti foliates*, *Tegeocranellus* sp., and *Rostrozetes foveolatus* are abundant in both the mangrove ecosystems (Julie *et al.,* 2008).

Insect herbivory in the mangroves have been studied. In Karnataka state, a total of 8638 individual insects belonging to 13 orders and 305 species have been collected. Coleoptera provides the maximum diversity at species level, followed by Lepidoptera, Orthoptera and Diptera. The effect of herbivory on the mangrove plants varies with species and the effect of herbivores is significantly site-specific for *Avicennia officinalis* and *Sonneratia alba* but not significantly different for *Rhizophora mucronata* (Remadevi *et al.,* 2008a). Latheef *et al.,* (2008) have studied seed predation on *R. mucronata* propagules by a moth borer. About 51.6 percent of the propagules are attacked by the moth borer. Germination decreases with propagule damage. The propagules belonging to the damage classes of one hole, two holes, three holes and four holes show increasing loss of sprouting. Propagules with more than four holes exhibit a total loss of sprouting (Latheef *et al.,* 2008).

Mosquitoes are an important faunal component of mangroves. Mosquitoes breed in tree holes, crab holes and swamp pools of mangroves. In Indian mangroves, 62 species of mosquitoes belonging to 19 genera and 21 subgenera have so far been recorded. Tourism has increased the mosquito population in mangrove areas (Rajavel and Natarajan, 2008).

Marine woodborers constitute an important component of mangrove fauna and play a vital role in the biodegradation process. At least 27 species have so far been recorded from the mangroves of India, of which *Bactronophorus thoracites*, *Dicyathifer manni* and *Martesia nairi* are almost specific to mangrove habitats (Santhakumaran, 2008). Studies of marine wood borers at the mangroves of Kothakoduru and Bangarammapalem in Visakhapatnam district of Andhra Pradesh state have recorded the presence of 17 species of Teredinids belonging to six genera. The genus *Bankia*, represented by nine species, is the most dominant. Four species are new records to Indian waters and one new to the mainland of the peninsula. Despite this rich diversity in the study area, it is interesting to note the absence of *Teredo furcifera* and pholadids, which generally show universal distribution along Indian coasts. The frequency of another ubiquitous species, *Lyrodus pedicellatus*, is also meager in the localities explored (Rao *et al.,* 2008).

Crabs play an important ecological role in the productivity of mangroves. Their biodiversity has recently been studied in mangroves of Goa, Maharashtra and Kerala. There are a total of 35 species under 25 genera and 10 families in the study areas. The highest diversity of crab species is in Kerala (27 spp.), followed by Goa (17 spp.) and Maharashtra (12 spp.), and the highest diversity of species is encountered in the family Grapsidae (12 spp.) followed by Portunidae (8 spp.) and Ocypodidae (6 spp.) (Roy and Nandi, 2008). Crabs exhibit significant biomass in pre-monsoon and post-monsoon months in Nalallam and Kadalundi of North Kerala (Sasikumar 2008). The density of the crab species *Neosarmatium smithi* and *Parasesarma plicatum* increases with mangrove vegetation (Shanij *et al.,* 2008). The predatory effect of *N. smithi* has been experimentally studied on *A. officinalis* seedlings. After introducing the crabs into an enclosure set up in the natural habitat, the crabs caused more than 50 percent mortality of seedlings after predation. In the natural environment, the predation controls the density of the mangrove species (Praveen *et al.*, 2008).

In wetlands of southwest Bengal, there are 25 species of fish belonging to 18 genera and 7 families under the order siluriformes. Of these, 17 siluroid fish species are commercially important, with fishery potential, but many of these are threatened. The problems pertaining to the causes of decline over the decades are 10 in number, as derived from local people's perceptions: (1) habitat destruction occupied the first rank, followed by (2) use of unrestricted fishing gears, (3) fishing during breeding seasons, (4), water pollution by wastes from human settlements and industries, (5) floods, (6), excessive and injudicious use of water, (7) human population pressure, (8) use of ichthyotoxic materials, (9) siltation, and (10) catfish skin sensitivity because of the absence of scales (Chakraborty, 2005).

There are about 522 species of birds in coastal ecosystems of Karnataka state. Dakshina Kannada and Udupi districts harbor about 366 bird species. Uttara Kannada is home to 424 species. In a 1 hour study on the coast of Karnataka, 1611 individuals and 92 bird species were recorded. The bay owl was first reported from Uttara Kannada district. The white bellied sea eagle, which has now become rare, was breeding in Mangalore taluk of Dakshina Kannada district. However, vultures (king vulture, Eurasian griffon, white backed vulture, long billed vulture and scavenger vulture) that were commonly reported 20 years ago have drastically declined (Bhat, 2008). The Gangolli estuary complex (Kundapura taluk, Udupi district) has about 14 migratory birds and 21 resident birds, but in recent years some of the long distance migrants have altogether disappeared and the other migratory birds have either become rare or stopped visiting the mangroves. This is mainly due to intensive prawn culture, change in agriculture pattern, and habitat conversion (Madhyastha and Aravind, 2008).

THREATS TO MANGROVE FAUNA

General Threats

The most significant threat is human pressure on mangrove faunal resources. Major threats are over-exploitation of fishery resources, urban development and human settlement, tourism, pollution, discharge of industrial effluents, port and harbor development, mining, hypersalinity, siltation and sedimentation and conversion of land for agriculture or salt farming or aquaculture. A growing threat to mangroves is climate change, especially global warming and sea level rise. The mangrove fauna which are most vulnerable to sea-level change are located in areas with small islands, lack of rivers, tectonic movements, groundwater extraction, underground mining, micro-tidal and sediment-starved areas. The mangrove fauna which are least vulnerable to sea-level change are situated in riverine areas, macro-tidal and sediment rich areas and dense mangrove forest areas.

Specific Threats

Over-exploitation of juvenile fish is a serious problem, as it adversely affects the food chain and fishery resources. To cite an example, in Sundarbans, 540 million tiger prawn juveniles are collected every year by 40,000 fishers; during this operation, 10.26 billion other fish juveniles are killed. About 48 to 62 species of finfish juveniles are wasted per net per day. Annually, a single haul may destroy 17,947,050 kg of other fish juveniles. Undersized fish are harvested, and other fish at their reproductive stage are over-fished using nets of small mesh size (Mitra and Banerjee, 2005). A variety of molluscan species such as *Cerithidea cingulata* and *Telescopium telescopium* (Gastropods), *Anadara granosa* and *Meritrix* species (Bivalves) are illegally collected for shell, from which lime is manufactured in Andhra Pradesh and Tamil Nadu. These practices have serious impact on the ecological balance of the system (Kathiresan, 2000).

Another serious issue is water-related. Most river water is diverted by dam constructions, mainly for irrigation in the upstream areas. The freshwater supply that feeds coastal mangroves is reduced. Most of the river mouths are heavily silted, thereby reducing the influx of tidal water from the sea to the mangrove habitats. This water flow reduction may lead to high salinity which will reduce the growth of mangrove species and animal life. The changes in the water flux affect the fish in respect of their growth, migration and breeding, leading to a loss in fish resources. Few data in this regard are available, and this subject thus deserves much research. Monsoon failure frequently occurs in many places, leading to freshwater scarcity. In the future, the freshwater may not be let into the sea through estuaries or backwaters. In such a situation, many fish species may face extinction.

VALUES OF MANGROVE FAUNA

Biodiversity Values

Mangroves support biological diversity by providing habitats, spawning grounds, nurseries and nutrients for a number of animals. These include several endangered species ranging from reptiles (e.g., crocodiles, iguanas and snakes) and amphibians to mammals (the Royal Bengal tiger, deer, otters and dolphins) and birds (herons, egrets, pelicans and eagles). A wide range of commercial and non-commercial finfish and shellfish also depend on mangroves. The role of mangroves in the marine food chain is crucial.

Many rural communities have used mangroves to produce honey and traditional medicines. In particular, approximately 35,000 kg of honey is annually collected, mainly during April–May, by groups of local people from Indian Sundarbans. Mangrove leaves are largely used as fodder for camel and goats. Ecotourism activities are increasing in mangrove areas, providing further sources of sustainable income to local populations. In the absence of any alternative employment, the poor people depend largely on forest and fisheries resources and they tend to resort to illegal practices such as over-fishing, poaching and felling. The people are collecting the tiger shrimp seeds in large numbers, thereby causing damage to juvenile fish stocks. It is estimated that in the Sundarbans, each year, about five thousand fishermen and five hundred honey collectors and woodcutters enter the forests in search of livelihood, ignoring even the threat of attacks by tigers and crocodiles. Another example is camel herding which is one of the activities practised by the pastoral communities known as "Maldharis" in Gujarat. The Maldharis are in the habit of shifting along with their livestock to remote areas in search of fodder for their camels. The degradation and inaccessibility of mangroves has critically impacted their livelihoods (Kathiresan, 2009).

Bioprospecting Potential

Bioprospecting of mangrove ecosystems in search of valuable products and genes is of growing interest in India. In the mangrove wetlands of Sundarbans and Orissa, two species of horseshoe crab—*Carcinoscorpius rotundicauda* and *Tachypleus gigas*—are found during the pre-monsoon period of high salinity. These species exhibit excellent biomedical properties, as they form potential sources of bioactive substances—the Limulus Amoebocyte Lysate (LAL) and Tachypleus Amoebocyte Lysate (TAL). These substances are highly sensitive and useful for the rapid and accurate detection of Gram-negative bacteria, even if these are present in extremely minute quantities down to the level of 10^{-10} g. They can also be used for detecting endotoxin in several pharmaceutical products (Mitra and Bhattacharryya, 2001).

CONCLUSION

Mangrove ecosystems of India are blessed with rich faunal biodiversity. Site-specific information about the populations of mangrove-inhibiting fauna is highly warranted. It is necessary to understand basic biology, behavior and limiting factors of population reduction in the mangrove fauna which are at elevated level of extinction. This will help in implementation of the species, recovery. It is fair to conduct more research also on species which are less studied, such as the water monitor lizard, and to develop more breeding farms for economically important species such as crocodiles, reptiles and deer.

When mangrove forests are destroyed, reductions in local fish catches often result. Data on loss to fisheries due to mangrove loss are not adequately available. Establishment of strong geographic information systems (GIS) data is a need of the hour for fisheries resource management and for predicting effects of mangrove loss on fisheries. Effects of mangroves on fisheries of estuaries, shallow coastal waters and offshore have not been thoroughly studied. It is also necessary to develop sea-ranching programmes for the over-exploited fish.

It is a matter of urgency to protect the biodiversity of mangrove ecosystem, in particular faunal resources, with increasing man-made pressure and natural calamities. The local people should be trained and supported in adopting alternative and supplementary work for their livelihood, thereby reducing human pressure on mangrove resources. The local communities face hardship due to lack of infrastructure. An integrated management plan on strong scientific principles has to be drawn up for ecologically important areas of mangroves, with the participation of the local communities, to overcome the biodiversity loss and to mitigate the growing threat of climate change, especially in vulnerable sites.

The mangroves provide many goods and services for overall "biohappiness" of humans in terms of supporting fisheries and forestry products, removing carbon and pollutants, and ensuring the protection of coastlines from disasters. The December 2004 tsunami reaffirmed the importance of maintaining healthy mangrove ecosystems for disaster risk management. Bioprospecting of the mangroves for valuable genes and products deserves thorough study. There is also a dire need for economic valuation of the mangrove ecosystem.

ACKNOWLEDGEMENTS

The authors are thankful to the authorities of Annamalai University for providing facilities.

REFERENCES

Bhat, H.R., 2008. Bird diversity in coastal ecosystems of Karnataka. Institute of Wood Science and Technology, Bangalore, Abstracts, p. 28.

Bhosale, L.J., 2008. Biodiversity of mangroves of Ratnagiri and Sindhudurg districts of Maharashtra. Institute of Wood Science and Technology, Bangalore, Abstracts, p. 10.

Chakraborty, S.K., 2005. Studies on bioresource assessment and management of degraded mangrove ecosystem of Midnapore coast. Ministry of Environment & Forests Project, West Bengal, Final Report, p. 99.

Chaudhuri, A.B., Choudhury, A., 1994. Mangroves of the Sunderbans, Vol 1 India. CAB, Wallingford, Oxfordshire, p. 247.

Costanza, R.R., d'Arge, R., deGroot, R., Farber, S., Grasso, M., Hannon, B., Limburg, K., Naeem, S., O'Neill, R.V., Paruelo, J., Raskin, R.G., Sutton, P., van den Belt, M., 1998. The value of the world's ecosystem services and natural capital. Ecol. Econom. 25 (1), 3–15.

Duke, N.C., Meynecket, O.J., Dittmann, S., 2007. A world without mangroves? Letters. www.Sciencemag.org. p. 41.

FAO, 2007. The world's mangroves 1980–2005. Forestry Paper No. 153. Food and Agriculture Organization, Rome, p. 77.

Gopikumar, K., Sunil, P.K., Joseph, J.M., Hedge, H.T., 2008. Biodiversity of mangrove forests of Cochin coast of Kerala. Institute of Wood Science and Technology, Bangalore, Abstracts, p. 13.

Julie, E., Ramani, N., Sheeja, U.M., 2008. Diversity of oribatid mites in some ecosystems of Calicut District of Kerala. Institute of Wood Science and Technology, Bangalore, Abstracts, p. 36.

Kathiresan, K., 2000. Mangrove Atlas and Status of species in India. Report submitted to Ministry of Environment and Forest, Govt. of India, New Delhi, p. 235.

Kathiresan, K., 2009. Mangroves and Coral Reefs in India. Compilation of salient findings of research projects supported by Ministry of Environment & Forests (Govt. of India) during 10th Five Year Plan, p. 255.

Kathiresan, K., 2011. Eco-biology of Mangroves. In: James, N., Metras (Eds.), Mangroves: ecology, biology and taxonomy. Nova Science Publishers, New York, pp. 1–50.

Kathiresan, K., Bingham, B.L., 2001. Biology of mangroves and mangrove ecosystems. Adv. Mar. Biol. 40, 81–251.

Kathiresan, K., Qasim, S.Z., 2005. Biodiversity of Mangrove Ecosystems. Hindustan Publ Corp, New Delhi, p. 251.

Kathiresan, K., Rajendran, N., 2002. Fishery resources and economic gain in three mangrove areas on the south-east coast of India. Fisheries Manage. Ecol. 49 (5), 277–283.

Khaleel, K.M., 2008. Management strategies for the mangrove wetlands of North Malabar. Institute of Wood Science and Technology, Bangalore, Abstracts, p. 73.

Latheef, C.A., Remadevi, O.M., Chatterjee, D., 2008. Propagule predation on Rhizophora mucronata by a moth borer and its impact on the germination of the propagule. Institute of Wood Science and Technology, Bangalore, Abstracts, p. 30.

Madhyastha, N.A., Aravind, N.A., 2008. Birds of mangroves with particular reference to Gangolli Estuary complex (Kundapura Taluk, Udupi. Dt.). Institute of Wood Science and Technology, Bangalore, Abstracts, p. 33.

Mitra, A., Bhattacharryya, D.P., 2001. Biological diversity of mangrove wetlands in Sundarbans and scope for alternative source of income. Everymans Sci. 36 (2), 71–75.

Mitra, A., and Banerjee, K., 2005. Living resources of the sea: focus Indian Sundarbans. WWF, Parganas, West Bengal.

Polidoro, B.A., Carpenter, K.E., Collins, L., Duke, N.C., Ellison, A.M., Ellison, J.C., Farnsworth, E.J., Fernando, E.S., Kathiresan, K., Koedam, N.E., Livingstone, S.R., Miyagi, T., Moore, G.E., Vien Ngoc Nam, J.E., Ong, J.H., Primavera, S.G., Salmo, I.I.I., Sanciangco, J.C., Sukardjo, S., Wang, Y., Yong, J.W.H., 2010. The loss of species: Mangrove extinction risk and geographic areas of global concern. PLoS ONE 5 (4), 1–10.

Praveen, V.P., Shanij, K., Oommen, M.M., Nayar, .T.S., 2008. Predatory effect of the crab *Neosarmatium smithi* on *Avicennia officinalis* seedlings. Institute of Wood Science and Technology, Bangalore, Abstracts, p. 40.

Rajavel, A.R., Natarajan, R., 2008. Species diversity of mosquitoes in relation to the larval habitats in mangroves of India. Institute of Wood Science and Technology, Bangalore, Abstracts, p. 29.

Ramakrishna, 2008. Faunal resource and its distribution in mangrove ecosystem. Institute of Wood Science and Technology, Bangalore, Abstracts, 25.

Rao, A.T., Molur, S., Walker, S., 1998. Report of the workshop on "Conservation assessment and management plan for mangroves of India". Zoo Outreach Organization, Coimbatore, India, p. 106.

Rao, M.V., Balaji, M., Pachu, A.V., 2008. Occurrence and species diversity of marine wood borers at Kothakoduru and Bangarammapalem mangroves in Visakha patnam district. Institute of Wood Science and Technology, Bangalore, Abstracts, p. 31.

Remadevi, O.K., Latheef, C.A., Chatterjee, D., 2008a. Insect plant relationship with special reference to herbivory in the mangroves of Karnataka, India. Institute of Wood Science and Technology, Bangalore, Abstracts, p. 26.

Remadevi, O.K., Latheef, C.A., Chatterjee, D., Rajani, K., 2008b. Nocturnal insect diversity in Coringa mangroves–analysis of occurrence and distribution. Institute of Wood Science and Technology, Bangalore, Abstracts, p. 35.

Remadevi, O.K., Chatterjee, D., Latheef, C.A., 2008c. Solar light trap for the collection of nocturnal entomofauna in mangroves. Institute of Wood Science and Technology, Bangalore, Abstracts, p. 37.

Roy, M.K.D., Nandi, N.C., 2008. Diversity of crabs in the mangroves of the west coast. Institute of Wood Science and Technology, Bangalore, Abstracts, p. 41.

Santhakumaran, L.N., 2008. Role of marine wood borers in mangrove environment – Is it biodeterioration or degradation? Institute of Wood Science and Technology, Bangalore, Abstracts, p. 27.

Sasikumar, A., 2008. Effect of salinity on crab diversity of Nalallam and Kadalundi, North Kerala, India. Institute of Wood Science and Technology, Bangalore, Abstracts, p. 38.

Shanij, K., Praveen, V.P., Oommen, M.M., Nayar, T.S., 2008. A study on the density of two mangrove crabs: *Neosarmatium smithi* and *Parasesarma plicatum* (Crustacea; Decapoda; Grapsidae) in the mangrove forests at Kunhimangalam, Kerala. Institute of Wood Science and Technology, Bangalore, Abstracts, p. 39.

Sunderraj, S.F.W., Serebiah, J.S., 1998. Status of coastal wetlands in Gujarat – A briefing. Living on the edge, ENVIS publication Series. Annamalai University, India, pp. 6–15.

Index

A

Acabaria cinquemiglia, 24
Acabaria ouvea, 24
Acanthaster planci, 68
Acanthella cavernosa, 6
Acanthochitonidae, 41
Acanthogorgia breviflora, 21
Acanthogorgia ochracea, 21
Acanthogorgia spinosa, 21
Acanthuridae, 390
Acanthurus triostegus, 202
Acaudina molpadioides, 130
Accidental catch, as threat to marine
        mammals, 294
Aclididae, 45
Acmaeidae, 42
Acoustic surveys, as mammalian research
        technique, 292
Acropora cerealis, 425
Acropora cytherea, 125
Acropora formosa, 112
Acropora hemprichii, 442, 444
Acropora humilis, 315
Acropora palifera, 112
Acteonidae, 48
Actinaria, 317
Actinobacteria
    antibiotic production ability of, 412
    identification of, 412
    isolation of, 396
    overview, 395
    Streptomyces spp. as, 395
    from Tamil Nadu, 406
Actinocyclidae, 51
Aegialitis rotundifolia, 306
Aegiceras corniculatum, 196
Aeolidoidea, 53
    Euaeolidioidea, 53
    Eubranchidae, 53
    Facelinidae, 53
    Flabellinidae, 53
    Glaucidae, 53
    Pseudovermidae, 53
    Tergipedidae, 53
Aerial survey, as mammalian research
        technique, 292

Aeschynomene multicaulis, 315
Agarose gel electrophoresis, 378, 379
Agar-yielding seaweeds, 305
Aglajidae, 48
Albunea thurstoni, 315
Alcock, A. W., 172
Alcyonacea (Soft Corals and Telestids), 15
Aldisidae, 51
Allogastropoda, 47
Amathinidae, 47
Amphiboloidea, 54
Amphimedon chloros, 6
Amphipods, 325
Amphiprion ocellaris, 333
Amphiroa rigida, 116
Anadara granosa, 493
Anadoridoidea, 51
Ancilla ampla, 66
Andaman and Nicobar Islands (ANI), 303,
        304, 306, 352
    Exclusive Economic Zone (EEZ)
        region, 373
    fish fauna of, 352, 373
    fish trading practices in, 352
    mangrove vegetation in, 473
    marine mollusca, along, 64
    microbial biodiversity of, 395
    octocorals of, 21–28
    overview, 453
    population of crocodile, assessing, 454
    sponges of, 6
    taxonomic studies, 352
Anelassorhynchus, 321
ANI. *See* Andaman and Nicobar
        Islands (ANI)
Annella mollis, 27
Annella reticulate, 27
Anomalodesmata (bivalve), 59
    Clavagelloidea, 59
    Pandoroidea, 59
    Poromyoidea, 59
Anomura, 327
Anoxypristis cuspidata, 191, 467, 468.
        See also Knifetooth sawfish
Antarctic Minke whales, 288
Anthoplerua panikkarii, 317

Color Plates

Acanthella cavernosa (Dendy, 1922)

Amphimedon chloros (Ilan, Gugel & van-Soest, 2004)

Clathria (Thalysias) vulpina (Lamark, 1814)

Cliona varians (Duchassaing & Michelotti, 1864)

Liosina paradoxa (Thiele, 1899)

Monanchora unguiculata (Dendy, 1922)

Neopetrosia exigua (Kirkpatrick, 1900)

Oceanapia sagittaria (Sollas, 1902)

Plakortis simplex (Schulze, 1880)

Rhabderemia prolifera (Annandale, 1915)

Stylissa massa (Carter, 1887)

Xestospongia testudinaria (Lamarck, 1815)

FIGURE 1.1   Sponges of the Andaman and Nicobar Islands.

Color Plates

Favites abdita (Ehrenberg, 1834)   *Diploastrea heliopora* (Lamarck, 1816)

Favia speciosa (Dana, 1846)   *Goniastrea retiformis* (Lamarck, 1816)

*Stylophora pistillata (*Esper, 1797)   *Acropora cytherea* (Dana, 1846)

FIGURE 9.2   Scleractinian corals of Rutland Island.

Chicoreus ramosus
(Linnaeus,1758)

Conus nobilis
(Linnaeus 1758)

Conus straitus
(Hwass in Bruguiere, 1792)

Lambis (Lambis) lambis
(Linnaeus, 1758)

Nassa serta
(Bruguiere, 1789)

Rhinoclavis (Rhinoclavis)
sinensis (Gmelin,1791)

Strombus (Canarium) labiatus
(Roeding, 1798)

Strombus variabilis
(Swainson, 1820)

Terebralia palustrisa
(Linnaeus, 1758)

FIGURE 9.4   Molluscs of Rutland Island.

Color Plates

Cenometra bella
(Hartlaub, 1890)

Culcita noveguineae
(Muller and Troschel, 1842)

Echinometra mathaei
(de Blainville, 1825)

Fromia indica
(Perrier, 1869)

Stephanometra indica
(Smith, 1876)

Thelenota ananas
(Jaeger, 1833)

Holothuria (Halodeima) edulis
(Lesson, 1830)

Acaudina molpadioides
(Semper, 1868)

Linckia laevigata
(Linneaus, 1758),

Pearsonothuria graeffei
(Semper, 1868)

Ophiomastix annulosa
(Lamarck, 1816)

Stichopus vastus
(Sluiter, 1887)

FIGURE 9.6   Echinoderms of Rutland Island.

FIGURE 14.7 *Eviota sigillata*

FIGURE 14.8 *Koumansetta hectori*

FIGURE 14.9 *Valenciennea limicola*

FIGURE 19.1 **Marine macro-algae (seaweeds)** (a) *Gracillaria* sp.; (b) *Turbinaria* sp.

FIGURE 19.2 **Seagrass and dugong** (a) Seagrass meadow; (b) *Dugong dugon* grazing on seagrass.

FIGURE 19.6   **Sponges** (a) *Paratetilla bacca;* (b) *Xestospongia testudinaria.*

FIGURE 19.7   **Hydrozoa** (a) *Eudendrium* sp.; (b) *Macrorhynchia philippina.*

FIGURE 19.8   **Syphonaphora** *Physalia physalis* (Linnaeus, 1758).

Color Plates

FIGURE 19.9   **Corals** (a) *Acropora humilis* (Dana, 1846); (b) *Psammocora digitata* MED & H, 1851.

FIGURE 19.10   **Gorgonians** (a) *Nicella flabellata* (Whitelegge, 1897); (b) *Echinogorgia flora* (Nutting, 1910).

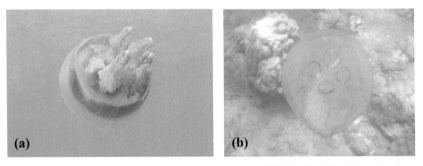

FIGURE 19.11   **Scyphomedusae** (a) *Aurelia aurita* (Linnaeus 1758); (b) *Crambionella mastigophora* (Maas 1903).

FIGURE 19.12 **Sea anemones** (a) *Entacmaea quadricolor* (Rüppell & Leuckart, 1828); (b) *Heteractis magnifica* (Quoy & Gaimard,1833).

FIGURE 19.13 **Ctenophore** *Ceonoplana* sp.

 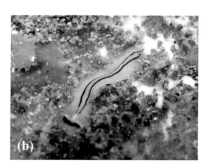

FIGURE 19.14 **Polyclads** (a) *Pseudobiceros hymanae*; (b) *Prosthiostomum trilineatum*.

FIGURE 19.15 **Polychaetes** (a) *Neries* sp.; (b) *Bispira brunnea*.

Color Plates

FIGURE 19.16   Peanut worm *Siponculus* sp.

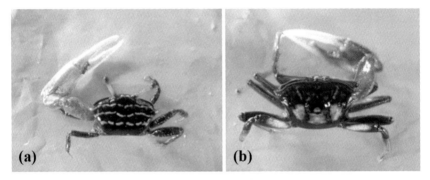

FIGURE 19.19   Brachyuran crabs (a) *Uca crassipes*; (b) *Uca coarctata*.

FIGURE 19.20   Molluscs (a) Chambered nautilus *Nautilus pompilius* (Linnaeus, 1758); lesser spider conch *Lambis scorpius indomaris* (Abbott, 1961).

FIGURE 19.21 **Sea slugs** (a) *Phyllidiella zeylanica*; (b) *Hypselodoris bullock*.

FIGURE 19.22 **Echinoderms** (a) *Mespilia globulus*; (b) *Comanthina nobilis*.

FIGURE 19.24 (a) Clownfish *Amphiprion ocellaris* (Cuvier, 1830); (b) Lionfish *Pterois antennata* (Bloch,1787).

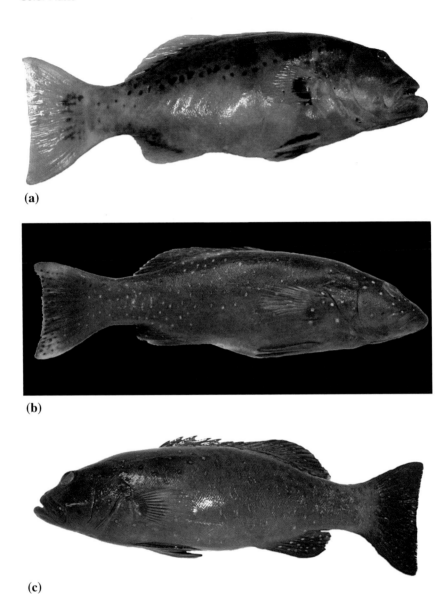

FIGURE 21.2 *Plectropomus* **species from South Andaman.** (a) *P. laevis* (total length 30 cm); (b) *P. leopardus* (total length 32 cm); *P. maculatus* (total length 30 cm).

FIGURE 23.3   Coral species commonly recorded on the North Bay reef. (a) *Acropora cerealis* (Dana, 1846); (b) *Favia matthaii* Vaughan, 1918; (c) *Favia truncatus* Veron, 2002; (d) *Favites spinosa* (Klunzinger, 1879); (e) *Fungia paumotensis* Stuchbury, 1833; (f) *Leptoseris striata* Fenner and Veron, 2002; (g) *Lobophyllia corymbosa* (Forskal, 1775); (h) *Pavona bipartita* Nemenzo, 1980; (i) *Pavona varians* Verrill, 1864; (j) *Physogyra lichtensteini* Milne Edwards and Haime, 1851; (k) *Pocillopora eydouxi* Milne Edwards and Haime, 1860; (l) *Pocillopora verrucosa* (Ellis and Solander, 1786).

Color Plates

FIGURE 25.1 Live corals on shipwreck (A) *Favia favus*; (B) *Acropora hemprichii*;
(C) *Pleurogyra sinulosa*; (D) *Platygyra lamellina*; (E) *Symphyllia recta*; (F) Different types of coral
on horizontal surface.

FIGURE 25.2 Other reef-associated fauna on shipwreck. (A) Sponges; (B) Nephtheidae (soft corals); (C) *Macrobynchia philippina* (hydrozoans); (D) *Dynamena*; (E) *Risbecia ghardaqana* (nudibranchs); (F) *Pinctada margarifera* (bivalve molluscs).

FIGURE 27.3   Giant grouper (*Epinephelus lanceolatus*).

(a)

(b)

(c)

(d)

(e)

FIGURE 27.4   **Molluscs.** (a) *Trochus niloticus*; (b) *Turbo marmoratus*; (c) *Nautilus pompilius*; (d) *Tridacna squamosa*; (e) *Charonia tritonis*.

Edwards Brothers Malloy
Thorofare, NJ USA
December 1, 2014